Comparative Aspects of Reproductive Failure

Comparative Aspects of Reproductive Failure

An International Conference at Dartmouth Medical School, Hanover, N.H.—July 25–29, 1966

Edited by

KURT BENIRSCHKE

Department of Pathology, Dartmouth Medical School, Hanover, N.H.

Springer Science+Business Media, LLC

1967

ISBN 978-3-642-48951-8   ISBN 978-3-642-48949-5 (eBook)
DOI 10.1007/978-3-642-48949-5

Library of Congress Catalog Card Number 67-12303

Title No. 1412
© 1967 by Springer Science+Business Media New York
Originally published by Springer-Verlag New York Inc. in 1967
Softcover reprint of the hardcover 1st edition 1967

ORGANIZERS, SESSION CHAIRMEN, AND SPONSORS

Organizers

K. BENIRSCHKE and T. C. JONES

Session Chairmen

K. BENIRSCHKE; V. H. FERM; M. GALTON; D. HOEFNAGEL, and G. MAR-
GOLIS, Department of Pathology, Dartmouth Medical School, Hanover, New Hampshire.

C. GILMORE, and T. C. JONES, Department of Pathology, Angell Memorial
Hospital and Harvard Medical School, Boston, Massachusetts.

Sponsors of Conference

CHARLES RIVER BREEDING LABORATORIES
ELI LILLY RESEARCH LABORATORIES
GEIGY PHARMACEUTICALS
LAKEVIEW HAMSTER COLONY
LEDERLE LABORATORIES
NATIONAL INSTITUTE OF CHILD HEALTH AND HUMAN
DEVELOPMENT (GRANT HD-02035).
POPULATION COUNCIL (GRANT M-66.031)
SCHERING CORPORATION
SMITH, KLINE AND FRENCH FOUNDATION
SYNTEX COMPANY
UPJOHN COMPANY

Speakers at Conference on Comparative Aspects of Reproductive Failure
(Hanover, N.H. USA, July 25–29, 1966)

Josimovich Kennedy Hoerlein Hutt   Frenkel   Ferm   Hancock   Blandau   Galton
            Carr   Silverstein   King   Metcalfe Hsia
            Ryan   Hafez   Benirschke Driscoll Böving   Bridges
            Gruenwald Jones   Hertig   Snyder   Warkany   Chang

LIST OF PARTICIPANTS

AINSWORTH, L., Western Reserve School of Medicine, Cleveland, Ohio

BENIRSCHKE, K., Dartmouth Medical School, Hanover, New Hampshire

BLANDAU, R. J., University of Washington School of Medicine, Seattle, Washington

BÖVING, B. G., Carnegie Institution of Washington, Baltimore, Maryland

BRIDGES, C. H., College of Veterinary Medicine, Texas A & M University, College Station, Texas

CARR, D. H., University of Western Ontario, Faculty of Medicine, London, Ontario, Canada

CHANG, M. C., Worcester Foundation for Experimental Biology, Shrewsbury, Massachusetts

DRISCOLL, SHIRLEY G., Harvard Medical School, Boston, Massachusetts

FERM, V. H., Dartmouth Medical School, Hanover, New Hampshire

FRENKEL, J. K., University of Kansas Medical Center, Kansas City, Kansas

GALTON, M., Dartmouth Medical School, Hanover, New Hampshire

GRUENWALD, P., The Johns Hopkins University, School of Medicine, Baltimore, Maryland

HAFEZ, E. S. E., Washington State University, Pullman, Washington

HANCOCK, J. L., Agricultural Research Council, Animal Breeding Research Organization, Edinburgh, Scotland

HERTIG, A. T., Harvard Medical School, Boston, Massachusetts

HOERLEIN, A. B., College of Veterinary Medicine, Colorado State University, Fort Collins, Colorado

HOLM, L. W., University of California School of Veterinary Medicine, Davis, California

HSIA, D. Y. Y., Northwestern Medical School, Chicago, Illinois

HUTT, F. B., New York State College of Agriculture, Cornell University, Ithaca, New York

JONES, T. C., Harvard Medical School, Boston, Massachusetts

JOSIMOVICH, J. B., University of Pittsburgh School of Medicine, Pittsburgh, Pennsylvania

KENNEDY, P. C., University of California School of Veterinary Medicine, Davis, California

KING, C. D., University of Wisconsin, Wisconsin Regional Primate Research Center, Madison, Wisconsin

KING, J. E., Kenya Game Department, Nairobi, Kenya

LIGGINS, G. C., University of California School of Veterinary Medicine, Davis, California

MARGOLIS, G., Dartmouth Medical School, Hanover, New Hampshire

MEDEARIS, D. N. Jr., University of Pittsburgh School of Medicine, Pittsburgh, Pennsylvania

METCALFE, J., University of Oregon Medical School, Portland, Oregon

NOVY, M. J., University of Oregon Medical School, Portland, Oregon

PETERSON, E. N., University of Oregon Medical School, Portland, Oregon

SILVERSTEIN, A. M., The Wilmer Institute, The Johns Hopkins University, School of Medicine, Baltimore, Maryland

SNYDER, R. L., Penrose Research Laboratories, Philadelphia Zoological Garden, and University of Pennsylvania, Philadelphia, Pennsylvania

WARKANY, J., University of Cincinnati, College of Medicine, Cincinnati, Ohio

PREFACE

To many, the contents of this conference may not seem appropriate at a time when the minds are preoccupied with a "population explosion." To the participants and guests of this conference, however, this was a week of fascinating discussions. While quantitative aspects of reproduction were touched upon, it was mostly a search for an understanding of the qualitative aspects of reproduction and its failure. Only when we understand these more completely will it be possible to render optimum care and have the foundations for meaningful population control.

The conference was conceived in discussions at the Committee on Pathology of the National Academy of Sciences, Washington, in 1965. It was felt that investigators in medicine and the veterinary fields would profit greatly from a closer liaison. All too frequently, we work relatively isolated in our respective fields and, with the burgeoning information filling our journals, we have not enough time and leisure to stand back and attempt a comparative look at the subject of study. Often we are not familiar with the techniques other disciplines use, and which we could well employ to great advantage. While this applies to many aspects of medicine, a comparative approach to the study of reproductive failure seemed most advantageous at this time.

This conference was therefore planned to provide, in the relaxing atmosphere of a small college community, an opportunity for presentation of important topics, to allow discussion, and to promote contact among participants and guests by one week of dormitory living. We were fortunate to obtain the financial support from the sources indicated and are most grateful for this endorsement of our purpose. In order not to delay publication we have chosen not to include the discussion of the manuscripts and we are grateful to the contributors for being so prompt in submitting their papers. Unfortunately, Dr. J. Warkany felt that his reviews of congenital anomalies in man were published so recently (*New England Journal of Medicine* **265**:993 & 1046; 1961) that his contribution should not be reprinted. The subject of comparative placental pathology, while deemed extremely important, could be covered only in potpourri fashion. The great dearth of knowledge was particularly apparent in this area, and the conference may stimulate someone to undertake a comprehensive study. Moreover, the excellent motion pictures shown by Dr. Blandau on ovulation, tubal egg transport, large animal immobilization with Cap-Chur gun, and employing anectine, can only be referred to in these pages. To the participants of the conference they proved to be a most stimulating medium of communication.

There was much enthusiasm expressed for such an interdisciplinary conference at this long meeting, and it is hoped that similar gatherings will be held in the future. Only when the topics began to be unfolded did we realize how superficial the treatment of details had to be and how many serious omissions had occurred. Most of all it was felt that an exposure in greater depth to new techniques would be a rewarding experience for future conferences.

K. BENIRSCHKE
T. C. JONES

Hanover, N.H.
Boston, Mass.
August 1966

CONTENTS

Introduction to Comparative Reproduction

E. S. E. HAFEZ

The Overall Problem in Man

A. T. HERTIG

Reproductive Failure in Domestic Mammals

E. S. E. HAFEZ

Contents

Cytogenetics of Abortions

D. H. CARR

Genetic and Biochemical Aspects of Reproductive Failure

D. Y. Y. HSIA

Chemo-Mechanics of Implantation

B. G. BÖVING

Comparative Aspects of Steroid Hormones in Reproduction

K. J. RYAN and L. AINSWORTH

Protein Hormones and Gestation

J. B. JOSIMOVICH

Prolonged Gestation

P. C. KENNEDY, G. C. LIGGINS and L. W. HOLM

Contents

Oögenesis—Ovulation and Egg Transport

R. J. BLANDAU

Experimental Hybridization

M. C. CHANG and J. L. HANCOCK

Sterility and Fertility of Interspecific Mammalian Hybrids

K. BENIRSCHKE

The Sterility of Two Rare Equine Hybrids

J. M. KING

Developmental Malformations as Manifestations of Reproductive Failure

V. H. FERM

Malformations and Defects of Genetic Origin in Domestic Animals

F. B. HUTT

Contents

Contents

Reproduction at High Altitudes

J. METCALFE, M. J. NOVY, and E. N. PETERSON

Fertility and Reproductive Performance of Grouped Male Mice

R. L. SNYDER

Closing Remarks

T. C. JONES

INTRODUCTION TO COMPARATIVE REPRODUCTION

E. S. E. HAFEZ

Professor, Reproduction Laboratory, Washington State University, Pullman, Washington

During the course of mammalian evolution, there have been marked anatomic, endocrine and physiologic changes, the main tendency of which were to insure the protection of young and the survival of the species. These changes involved economy in the production of gametes, reduction in the size of the egg, internal fertilization, the development of the corpus luteum as a temporary endocrine organ and the development of the placenta as a nutritive, excretory, endocrine and protective organ.

The mammalian species discussed in this conference have viviparous reproduction which is characterized by internal fertilization and gestation. There are several thousands of mammalian species, but the reproductive biology has been extensively studied in only 20 to 24 species. Some of these species are characterized by peculiar reproductive phenomena, such as restricted sexual season, absence of estrus, presence of menstruation, dissociation of ovulation and estrus, nonspontaneous ovulation, spontaneous multiple ovulation with limited implantation, delayed implantation or ovulation during pregnancy.

The efficiency of reproduction of a given species depends on the duration of the sexual season, the frequency of recurrence of the sexual cycle, ovulation rate, the duration of pregnancy, litter size, the suckling period, the age at puberty and the duration of the reproductive period in the animal's life. In general, the age at which puberty is attained is earlier in small-sized species than in large ones. Sexual maturity is usually reached later in the male than in the female. In the male, there is no definite age at which the reproductive functions cease. In the human female, the reproductive functions cease abruptly during life; this constitutes menopause or climacteric. In other mammals, the animals die before the arrest of the reproductive functions.

1

I. Seasonality of Reproduction

The nature and duration of the sexual season of any species is related to the extent of its domestication, its geographical origin, and the biology of the species. The varying capacity of domesticated species to adapt autonomous breeding rhythm to changed environment is revealed by the way breeding habits are adjusted to captivity.

One major effect of domestication is to spread the sexual season over the year, a result of artificial selection under improved shelter, disease control, and better nutrition. There is, however, a complete gradation from the typical monoestrous condition to the extreme of polyestrus. The wild dog exhibits estrus only in the winter. Domestic dogs may show estrus at any time of the year, but the frequency of estrus has a bimodal pattern; most dogs come into estrus during the winter and late spring.

Mammals have been classified into three major classes according to the photoperiodic requirement of the sexual season. Sheep and goats show estrus during the season of long days. Species which show estrus the year round include the other highly domesticated animals such as cattle, buffaloes, swine, rabbits, and those which originate in the tropics (little seasonal fluctuation in climate) such as camels. Sexual functions are maintained throughout the year in most races of man with some fluctuation in conception rate in certain localities.

In general, the shortest days of the year coincide with the severe climatic conditions (winter months). In short-day breeders, breeding occurs during the shortest days. The gestation period in these species is one half-year cycle, as in sheep (150 days). Consequently, the time of birth coincides with the most favorable conditions for the survival of young. In long-day breeders, however, mating only occurs during the longest days. The gestation period is two half-year cycles as with the horse and donkey. Consequently, the young are born during the longest days the next year. On the other hand, in species which have been highly domesticated, such as farm animals, the animals no longer have acquired any photoperiodic responses, because the conditions (feed, housing, and disease control) for the survival of young are usually favorable. Matings and parturitions occur throughout the year irrespective of the gestation length or the day length. The same trend applies to man and most monkeys.

II. The Egg

Although mammalian eggs do not show such marked structural peculiarities as the sperm, there are distinct differences in their size and membranes not only in the different families, but even within a given family.

The differences in size depend almost entirely on the amount of the deutoplasm. With the exception of the eggs of monotremes the diameter of the vitellus (intra-zonal) of the mammalian egg at the time of ovulation ranges between 60 to 185 μ. In Eutherian eggs, the diameter is usually less than 150 μ. With the completion of the second maturation division and the expulsion of the second polar body, the size of the vitellus diminishes, but the intra-zonal measurement is not appreciably affected.

Ovulation rate, the number of follicles which ovulate at each estrus, varies widely from species to species. Many species, including man, the larger ungulates, the elephant, and the whale ovulate as a rule only one follicle at each cycle (monotocous). On the other hand, many species, smaller rodents, insectivores, pigs and large carnivores ovulate a number of follicles simultaneously (polytocous). The number of follicles ovulated is slightly greater than the number of young born in any species, since some of the eggs are not fertilized or die during gestation.

III. IMPLANTATION AND PLACENTATION

Species differ in the way in which the blastocyst implants, the depth of implantation and the site where the blastocyst becomes implanted. In ungulates, carnivores, the leporid family of rodents, the Lemuroidea, Tarsoidea, platyrrhine and catarrhine monkeys, the implantation is central and the blastocyst remains in the uterine horn and eventually expands to fill the lumen. In the guinea pig, the implantation is interstitial and the blastocyst implants by passing through the uterine epithelium and becomes completely cut off from the uterine lumen. In eccentric implantation, the blastocyst remains small and comes to occupy a small diverticulum or cleft of the uterine lumen.

In man and the rhesus monkey, the part of the trophoblast which establishes the first attachment overlies the inner cell mass and is on either the dorsal or ventral uterine wall (positions corresponding to the antimesometrial side of a bicornuate uterus). In most rodents and most insectivores the antimesometrial side of the uterus is the place where implantation occurs.

Different families of mammals develop very diverse types of placentae, depending on litter size, the internal structure of the uterus, and the degree of fusion between maternal and fetal tissues. Grosser maintains that the "placental membrane" is the structure of greatest functional significance in any type of placenta (Table 1). Placental permeability is usually judged by the mode of transfer of four groups of materials: bacteria and viruses; colloidal and crystalloid dyes; unhydrolysed organic colloidal molecules, *e.g.*, proteins and fats; and gases and hydrolysates.

The mucosa of the pregnant uterus is referred to as the *decidua*, a term which was first used to designate the maternal tissue which is shed with

Table 1

ESTRUS, OVULATION AND SOME RELATED PHENOMENA IN SEVERAL MAMMALS (Compiled from the literature cited)

Phenomenon	Age at Puberty (mos.)	Cycle length (days)	Duration of Estrus	Time of Ovulation (hrs. after onset of estrus)	Diameter of Ovulatory Follicle (mm)	Diameter of Mature Egg (vitellus) (μ)	Fertilizable Life of Egg (hrs.)
Species							
Cat							
Felis catus	5 (7–12)	15–21	4 days	24–50 hrs. after coitus		120–130	
Cattle							
Bos taurus	6–14	21 (14–24)	17 (4–30) hrs.	12 (2–26) hrs. after end of estrus	12–19	138–143	12–20
Dog							
Canis familiaris	6–24	2 cycles a yr.	9 (6–14) days	1–3 days after receptivity	6	130–140	A
Donkey							
Equus asinus	12	21–28	2–7 days		10–30		
Goat							
Capra hircus	16	18–21	39 hrs.		8	145	
Guinea pig							
Cavia porcellus	2.3 (1–4)	16.5–17.5	12 hrs.	10 hrs.	0.8	65	
Hamster, golden							
Cricetus auratus	2	4	27 hrs.			72	5
Horse							
Equus caballus	15–24	19–23	6 (5–10) days	24–48 before end of estrus	25–65	105–141	12

Man (USA) *Homo sapiens*	15.45 yrs.*	28.3	None (5 days bleeding)	14 (11–18) days of cycle	2–4	130–140	6–24
Monkey, rhesus *Macaca mulatta*	24	28	None†	13 days of cycle		125–143	23
Mouse *Mus musculus*	1.1	4.5		2–3 hrs.	0.4	75–88	6–12
Pig *Sus domesticus*	6–7	21 (16–24)	55–70 hrs.	1st or 2nd day of estrus	8–12	120–140	12–24
Rat *Rattus norvegicus*	1.5–2.5	4–6	13–20 hrs.	8–11 hrs.	0.9	70–75	12
Rabbit *Oryctolagus cuniculus*	6–8	hypothetical (7)	continuous	10 hrs. after coitus	1.5	120–130	6–8
Sheep *Ovis aries*	7–10	16–18	1–2 days	12–24 before end of estrus	5–8	147	12–24

Figures in parentheses indicate the range of several breed averages.

* Menopause at 49 years.

† Maximum desire 2 days before ovulation.

A = First polar body is not extruded from the egg for some days after ovulation; egg not ready for fertilization for 3 days after ovulation.

Table 2

BREEDING, GESTATION AND PRENATAL DEVELOPMENT IN SOME MAMMALS (Compiled from the literature cited)

Phenomenon Species	Birth Weight	Puberty Weight	Adult Weight	Gestation Period (days)	Type of Placentation			Litter Size
					Gross Shape	Relation to Endometrium A	Relation to Endometrium B	
Cat								
Felis catus	80–120 g	400–600 g	1000–1500 g	58–71	zonary to discoid	deciduate	endotheliochorial	4
Cattle								
Bos taurus	25–46 kg	200–300 kg	500–700 kg	280	cotyledonary	transitional	syndesmochorial	1 (twins 1:50 births)
Dog								
Canis familiaris	.2–.4 kg	1.5–5 kg	6–25 kg	58–63	zonary to discoid	deciduate	endotheliochorial	7–10
Donkey								
Equus asinus	10–15 kg	100–200 kg	200–550 kg	365	diffuse	nondeciduate	epitheliochorial	1
Goat								
Capra hircus	3–4 kg	25–30 kg	50–80 kg	146–151	cotyledonary	transitional	syndesmochorial	2 (1–3)
Guinea pig								
Cavia porcellus	80–90 kg	400–500 g	700–1000 g	67–68	discoid	deciduate	hemoendothelial	3–4
Hamster, golden								
Cricetus auratus	2.7 g	80 g	110 g	16	diffuse			6 (1–12)
Horse								
Equus caballus	20–45 kg	250–350 kg	400–650 kg	329–345	diffuse	nondeciduate	epitheliochorial	1 (twins 1:900)

Man (USA) *Homo sapiens*	3.4 (2.8–3.9) kg	40–60 kg	57–72 kg (20–24 yrs.)	280 discoid	deciduate	hemochorial	1 (twins 1 in 95 births) *
Monkey, rhesus *Macaca mulatta*	0.5 (.3–.7) kg	3.5 (2.7–4.8) kg	11 (8.8–12.1) kg	158–175 discoid	deciduate	hemochorial	1 (twins 1: 100)
Mouse *Mus musculus*	.9–1.5 g	20–25 g	24–50 g	19			4.5–7.4
Pig *Sus domesticus*	1.5 kg	60–110 kg	150–210 kg	112–116 diffuse	nondeciduate	epitheliochorial	6–12
Rat *Rattus norvegicus*	5–6 g	150–250 g	250 g (300–400) g	21–22 discoid	deciduate	hemoendothelial	8 (7–9)
Rabbit *Oryctolagus cuniculus*	.06 kg	3–4 kg	5–6 kg	30–32 discoid	deciduate	hemoendothelial	4–10
Sheep *Ovis aries*	3–45 kg	25–42 kg	70–80 kg	144–152 cotyledonary	transitional	syndesmochorial	1 (1–3)

Figures in parentheses indicate the range of several breed averages.
A = Huxley's Classification.   B = Grosser's Classification.
* 25–29% of twins born to white parents are monozygotic.

the after-birth at parturition. Decidual cells vary a good deal in their size, structure, mode of origin, and rate and degree of cell formation. The physiological significance of the decidual cells still remains one of the most obscure questions of placentation. In species with *deciduate* placentae, the fetal and maternal tissues do not separate and are expelled together at parturition. In species with *non-deciduate* placentae, the fusion is incomplete, and the fetal placenta separates easily from the uterus leaving the endometrium behind.

IV. THE NEONATE

As a rule, the neonate in placental mammals is very immature, develops slowly and depends on maternal care. The stage of development at which the neonate is born varies greatly in different species and determines the extent to which maternal care is required. In the rat and rabbit, the neonate are born blind, naked and with poor thermoregulatory ability; thus, they require a warm maternal nest. In the guinea pig, the young are born in an advanced stage of development and can fend for themselves in a few days. The extent of mother-young social interactions in different species, therefore, varies widely. These interactions are necessary for the full development of the physical and mental characteristics of the species.

V. COMPARATIVE VALUES

The major reproductive characteristics discussed in this conference are summarized from the literature in Tables 1 and 2. These averages vary according to the breed, season of year, plane of nutrition, management of the animal and technique used in the investigation. For example, the gestation period is longer in saddle horses than in draft horses, and the gestation terminating in the winter is about 20 days shorter than one terminating in the spring. On the other hand, the pattern of reproduction may be quite similar in two species as in the European cattle (*Bos taurus*) and the Brahman cattle (*Bos indicus*). The duration of estrus, for example, may be determined by several techniques, namely vaginal smear in rodents, swelling of the sexual skin in monkeys, and the behavioral responses of the female to the male in ungulates. These factors should be kept in mind when different species are compared.

The voluminous literature on comparative reproduction has been reviewed in several volumes published in the last decade. The bibliography cited below is a selected list of references.

References

Armstrong, C. N., and A. J. Marshall (eds.) : Intersexuality in Vertebrates including Man. New York: Academic Press, 1964.

Asdell, S. A.: Patterns of Mammalian Reproduction. Ithaca: Comstock, 1964.

Austin, C. R.: The Mammalian Egg. A Study of a Specialized Cell. Oxford: Blackwell Sci. Publication, 1961.

Beatty, R. A.: Parthenogenesis and Polyploidy in Mammalian Development. Cambridge: Cambridge University Press, 1957.

Cole, H. H., and P. T. Cupps (eds.): Reproduction in Domestic Animals. Vols. I and II. New York: Academic Press, 1959.

Colloque de la société nationale pour l'étude de la stérilité et de le fécondité. "Les Fonctions de Nidation Utérine et leurs Troubles." Paris: Masson et Cie., Libraire de l'Academie de Medécine, 1960.

Ciba Foundation Symposium—Mammalian Germ Cells. Boston: Little, Brown, 1953.

Ellenberger, W., and H. Baum: Handbuch der vergleichenden Anatomie der Haustiere. O. Zietzchmann, E. Ackerknecht, and H. Grau (eds.) 18th ed. Berlin: Springer, 1943.

Hafez, E. S. E. (ed.): Reproduction in Farm Animals. 2nd edit. Philadelphia: Lea & Febiger, 1967.

Hafez, E. S. E. (ed.) : The Behaviour of Domestic Animals. London: Bailliere, Tindall & Cox, 1962.

Krolling, O., and H. Grau: Lehrbuch der Histologie und vergleichenden mikroskopischen Anatomie der Haustiere. 10th ed. Berlin: Parey, 1960.

Lloyd, C. W.: Recent Progress in the Endocrinology of Reproduction. New York: Academic Press, 1959.

————: Human Reproduction and Sexual Behavior. Philadelphia: Lea & Febiger, 1964.

Nalbandov, A.: Reproductive Physiology; Comparative Reproductive Physiology of Domestic Animals, Laboratory Animals and Man. San Francisco: Freeman, 1965.

Nishikawa, Y.: Studies on Reproduction in Horses. Singularity and Artificial Control of Reproductive Phenomena. Tokyo: Japan Racing Association, 1959.

Overzier, C. (ed.): Intersexuality. New York: Academic Press, 1963.

Parkes, A. S. (ed.): Marshall's Physiology of Reproduction. Vols. I & II. London: Longmans, 1952, 1960.

Raven, C. P.: Oogenesis: The storage of developmental information. New York: Pergamon Press, 1961.

Rheingold, Harriet L. (ed.) : Maternal Behavior in Mammals. New York: John Wiley, 1963.

Roberts, S. J.: Veterinary Obstetrics and Genital Diseases. Ithaca: S. J. Roberts, 1956.

Romanoff, A. L., and A. J. Romanoff: The Avian Egg. New York: John Wiley, 1949.

Rowlands, I. W. (ed.): Comparative Biology of Reproduction in Mammals. New York: Academic Press, 1966.

Salisbury, G. W., and N. L. Vandemark: Physiology of Reproduction and Artificial Insemination of Cattle. San Francisco: Freeman, 1961.

Sisson, S.: Anatomy of Domestic Animals. Revised by D. Grossman. 4th ed. Philadelphia: Saunders, 1953.

Transactions of the Annual Conference on "Gestation" starting in 1954. New York: Sponsored by the Josiah Macy, Jr. Foundation.

Velardo, J. T. (ed.) : The Endocrinology of Reproduction. New York: Oxford University Press, 1958.

Villee, C. A. (ed.) : Control of Ovulation. Oxford: Pergamon, 1961.

Young, W. C. (ed.): Allen's Sex and Internal Secretions. 4th ed. Baltimore: Williams & Wilkins, 1961.

Zuckerman, S. (ed.): The Ovary. Vols. I. and II. New York: Academic Press, 1962.

THE OVERALL PROBLEM IN MAN *

ARTHUR T. HERTIG

*Shattuck Professor of Pathological Anatomy, Department of Pathology,
Harvard Medical School, Boston, Massachusetts*

INTRODUCTION

Reproductive failure occurs in all forms of life, plant and animal. I
have been asked to make a few introductory or general remarks about the
problem in man. Whatever I have to say is based upon 35 years' experience in obstetric and gynecologic pathology, seasoned by interest in
the early embryology of the human and interspersed with a nine-year
interval as clinical obstetrician. These remarks are in no sense an exhaustive review of reproductive anatomy, physiology and pathology, even as
applied to the human, let alone to mammals in general. Comments will
be confined to my observations on the early or potentially abortive phases
of human development and a pathologic evaluation of a thousand consecutive human abortions, most of them spontaneous.

The findings on the 34 fertilized human ova, 1 to 17 days of age, which
Dr. John Rock and I discovered at the Free Hospital for Women have
been written up individually and collectively on many occasions. The best
single complete reference to this unique series of specimens is to be found
in the Festschrift Volume prepared by Dr. George Washington Corner's
professional colleagues to honor his 60th birthday. This appeared in the
American Journal of Anatomy (Hertig, et al., 1956). It contains all of the
necessary bibliography for the embryological material which might be of
interest to this audience.

The problem of human reproductive failure, using some of this material
for illustration, was discussed in the first Teratologic Conference, organized by Drs. Wilson and Warkany and held in Bethesda, Maryland, during
1957. It was published later in *Pediatrics* (Hertig, et al., 1959).

The abortion material has been published in *The Annals of Surgery*

* Aided by United States Public Health Service Research Grant HD-00137.

(Hertig and Sheldon, 1943). This was written as part of a symposium on trauma and hence was an attempt to show that bona fide traumatic abortion was indeed a rare bird; occurring only once in a thousand human spontaneous abortions. These three general references, together with specific additional references, will cover the material I shall present. A supplementary list of modern references germane to the general subject of human fetal wastage is appended for completeness and reader interest. Many of them will undoubtedly be duplicated elsewhere in these proceedings by other conference members.

I. Pathogenesis of Abortion

It has long been evident that in man and animals there is a spectrum of fertility ranging from sterility to the so-called individual "population explosion." The various parameters of comparative reproductive physiology have been admirably covered by the proceedings of a conference on this subject held at West Point, New York, in 1959. These proceedings were edited by the eminent physiologist Carl G. Hartman and published in 1963. It has a complete biliography, summary of what was known at that time and a superb list of questions which need to be answered before the process of mammalian reproduction is well understood or controlled. Such conferences have a way of attracting many of the same participants the world over, the present one here at Dartmouth being no exception.

It is not within my sphere of interest or competence to speak of the endocrine control of ovulation even though that phenomenon is the *sine qua non* of female fertility. (My colleague, Eleanor C. Adams, and I are, however, working on the histochemistry and ultrastructure of human ovaries and oocytes before, during and after ovulation.)

A simple question, however, might well be asked now. What is the optimum time for human ovulation in terms of fertilizability of the oocyte? We suspect in the human from good data in the guinea pig (Blandau and Young, 1936) that 12 hours is the optimum time for fertilization of the oocyte *after* it has ovulated. But what effect does delay in ovulation have upon the fertilizability of the oocyte which is champing at the bit to get out of its "follicle of the month"? Until three years ago I would have said that the time of ovulation did not make any appreciable difference as long as the oocyte came from a normal follicle which would form a good corpus luteum, the endometrium was normally proliferative, the first polar body had been extruded and the oocyte met a spermatozoon within a few hours.

Apparently this is not entirely true, judging from a statistical reanalysis of the quality of the 34 fertilized ova in terms of their apparent time of ovulation (Fig. 1). These data were first published by Hertig in 1965 as part of the proceedings of a conference on Human Ovulation held in

HUMAN OVULATION

Distribution of Presumed Day of Ovulation Based on Endometrial Histology and Ovular Development of 34 Fertilized Ova; 21 Normal and 13 Abnormal

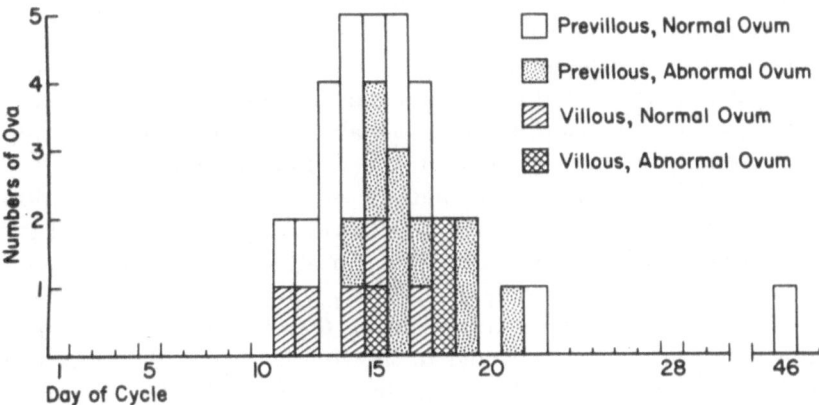

Fig. 1. A bar graph to show the distribution of types of fertilized human ova in terms of their apparent time of ovulation. (From Hertig, Human Ovulation, Boston: Little, Brown and Company, Inc., 1965.)

1963. I say "apparent" time of ovulation since no one has ever actually observed a human ovary during ovulation. The formula for ovulation used by Rock says that E.O.T. (estimated ovulation time) is 16 to 12 days prior to the next expected menstrual period. These data were available in the 210 patients that Rock and Hertig studied during the search for early human ova. Also the endometria were available and dated according to the criteria originally established by Rock and published *in extenso* by Noyes, *et al.,* in 1950. Moreover, all patients in whom fertilized ova were found, whether normal or abnormal, had had a coital date recorded within 24 hours of presumed ovulation. The developmental age of the conceptus was also used in helping to estimate the actual time of ovulation. The corpus luteum was of some, but not critical, help in estimating time of ovulation.

In any event, irrespective of whether one can pinpoint the time of ovulation, it appears from reanalysis of the Rock and Hertig data, that if an oocyte lingers longer than day 14 in the follicle it has an increasing chance of becoming a "bad egg" when fertilized. Thus, Figure 1 shows that the fertilized ova *apparently* ovulated on or before day 14 had only 1 chance in 13 of being defective, whereas if ovulated after day 14 they had 12 chances in 21 of being defective—a seven-fold difference in favor of ovulating "on time." Stated differently, an intrafollicular oocyte is very likely to stay fresh and healthy if it ovulates "on time" or even a bit early whereas it has only about a 50–50 chance of amounting to a useful citizen if it is delayed beyond the 14th day of the cycle. This is not really surprising when one studies histochemically and ultrastructurally the complexity of growth and organization of the various organelles within the developing oocyte. Such changes are better documented in the guinea pig (Adams and Hertig, 1964) than in the human but we know enough about the latter to realize that the human oocyte is equally complex as it matures.

Assuming, however, that a human female ovulates on time and that the secondary oocyte meets a properly capacitated spermatozoon in the proper place, presumably the ampulla of the fallopian tube, and that the pronuclei of the zygote fuse, what are some of the morphologically verifiable events that subsequently take place? The normal stages of early human development are well illustrated in all standard texts of Obstetrics, Embryology and Anatomy. Hence they will not be illustrated in the text although a few will be shown during the conference. The majority of this discussion will be devoted to showing examples of the fertilized ova that were so palpably bad that there is no doubt that they would have aborted spontaneously or caused congenital anomalies of development had they gone to term. It must be emphasized that all 34 of the ova described by Rock and Hertig were discovered in women of proven fertility. In Table 1 are given the pertinent clinical data from the 107 (of the 210) patients in whom the conditions for pregnancy were apparently optimal. The 210 patients were under the age of 40, menstruating normally, living with their husbands, had one or more coital dates during the period of estimated ovulation time and had some clinical condition indicating the necessity for hysterectomy. Of these 210, 107 were probably optimal owing to: (1) demonstrated ovulation; (2) recorded coital dates within 24 hours before or after ovulation; and (3) the absence of pathologic conditions in the tubes, ovaries or uterus that would interfere with conception.

Owing to the difficulty of finding an early human conceptus at certain

Table 1

CLINICAL DATA CONCERNING 107 PATIENTS WITH
OPTIMAL PROBABILITY OF CONCEPTION

	Normal Ova (24)	Abnormal Ova (10)	Failures (73)
Age—range (yr.)	26–42	29–39	25–43
Age—average (yr.)	33.8	33.1	35.6
Labors (No.)	5.4	5.3	4.5
Abortions (No.)	0.62	0.9	0.57
Labors/Abortions	8.7/1	5.9/1	7.9/1

times in its development because of the size of the specimen, its optical properties or the difficulty of technique it was decided to analyze 36 patients whose uteri yielded 21 specimens: 15 normal and 6 abnormal. The criteria for these perfect patients were the same as for the 107, except (1) no anatomic pathology of any sort was demonstrated and (2) the endometrium was 25 days or older. We were thus assured that all conditions were favorable for finding an implanted ovum if it were there. In this very select series of patients nothing could have precluded pregnancy except subtle physiologic, genetic or immunologic factors. Moreover, the ovum was big enough to be seen by anyone who looked with care and knew what he was looking for. The clinical data from this perfect series of 36 patients are to be found in Table 2.

Table 2

CLINICAL DATA CONCERNING 36 PATIENTS WITH OPTIMAL PROBABILITY OF CONCEPTION AND WITH ENDOMETRIUM 25 DAYS OR OLDER

	Normal (15)	Abnormal (6)	Failures (15)
Age—range (yr.)	26–38	29–37	25–41
Age—average (yr.)	32.7	33.7	34.3
Labors (No.)	5.4	5.7	3.9
Abortions (No.)	0.6	1	0.7
Labors/Abortions	9.0/1	5.7/1	5.6/1

Prior to discussing a subject upon which we have data, defective segmenting ova, it might be well to mention a subject upon which we have no data: the failure of fertilization of a human secondary oocyte exposed to a normal spermatozoon in a suitable environment. It is likely that this factor in sterility and infertility obtains in the human as it does in domestic animals. Corner, in 1923, showed clearly from corpus luteum counts that 9.6 per cent of domestic pig eggs failed to become fertilized even though exposed to spermatozoa. Another 14.4 per cent were abnormal and 23 per cent of the eggs were not found, whether they were fertilized, unfertilized, normal or abnormal. This raises the question of whether, at least in the human, the oocyte always gets into the fallopian tube following ovulation. No data are available but it seems unlikely that such a complicated process should always work perfectly every time. Certainly no other biologic process works perfectly all of the time! In any event, it would seem from the hypothetical curve of early fertility in Figure 9 that about 15 per cent of human eggs either fail to get into the tube or fail to become fertilized once they arrive there.

A. Defective Segmenting Ova

Of critical interest and importance are the six days of human development prior to implantation. It is generally agreed that a relatively high proportion of fertilized ova disappear during this stage of development. This has been demonstrated to be so for the domestic sow by Corner in 1923, and by Perry in 1954; for the wild rabbit by Brambell in 1948; for the rat by Frazer in 1955; and for the macaque monkey by Lewis and Hartman in 1941, by Heuser and Streeter in 1941 and for the Rhesus monkey by Corner and Bartelmetz in 1954. No doubt Dr. Hafez will comment upon this problem in the domestic animals. Hertig, *et al.,* in 1956 showed that four of the eight segmenting, pre-implantation human ova were so defective that they undoubtedly would have disappeared prior to or sometime after implantation.

All defective segmenting eggs seem to have one or more of the following features in common:

1. necrotic blastomeres
2. multinucleated blastomeres
3. distortion or malposition of embryonic or trophoblastic blastomeres
4. distortion of overall shape of the morula or blastula

In addition, their rate of growth is abnormal, usually slower than normal. Nevertheless, these ova are not dead, since mitoses may be observed in even the most defective-appearing ones. Thus, they are embryologically stunted in the sense used by the embryologists at the Carnegie Institution of Washington's Department of Embryology in Baltimore. Mall and Meyer described their classification of abortuses in their classic monograph of 1921. (It is essentially the same classification used by me and my trainees at the Boston Lying-In Hospital in classifying the abortuses to be described later in this presentation.)

Some features of an abnormal segmenting 8-cell morula are to be seen in Figure 2. The unanswered question is: "what proportion of obviously defective ova disappears prior to implantation and what proportion implants only to become the obviously defective, potential abortuses as seen in Figures 3 to 6?" The answer to this question will determine the ultimate shape of the peculiar looking curve seen in Figure 9.

It is evident that if only the good eggs implant (4 out of 8) the fertility curve will start at something over 100 per cent, a manifest impossibility. If all of segmenting eggs implant (8 out of 8) the curve will start at 58 per cent, the point the curve reaches at the time of the missed menstrual period, an unlikely probability. Probably the truth lies somewhere in between. The initial point of the curve is probably at about 85 per cent, that is, assuming that half of the bad eggs fail to implant and half of

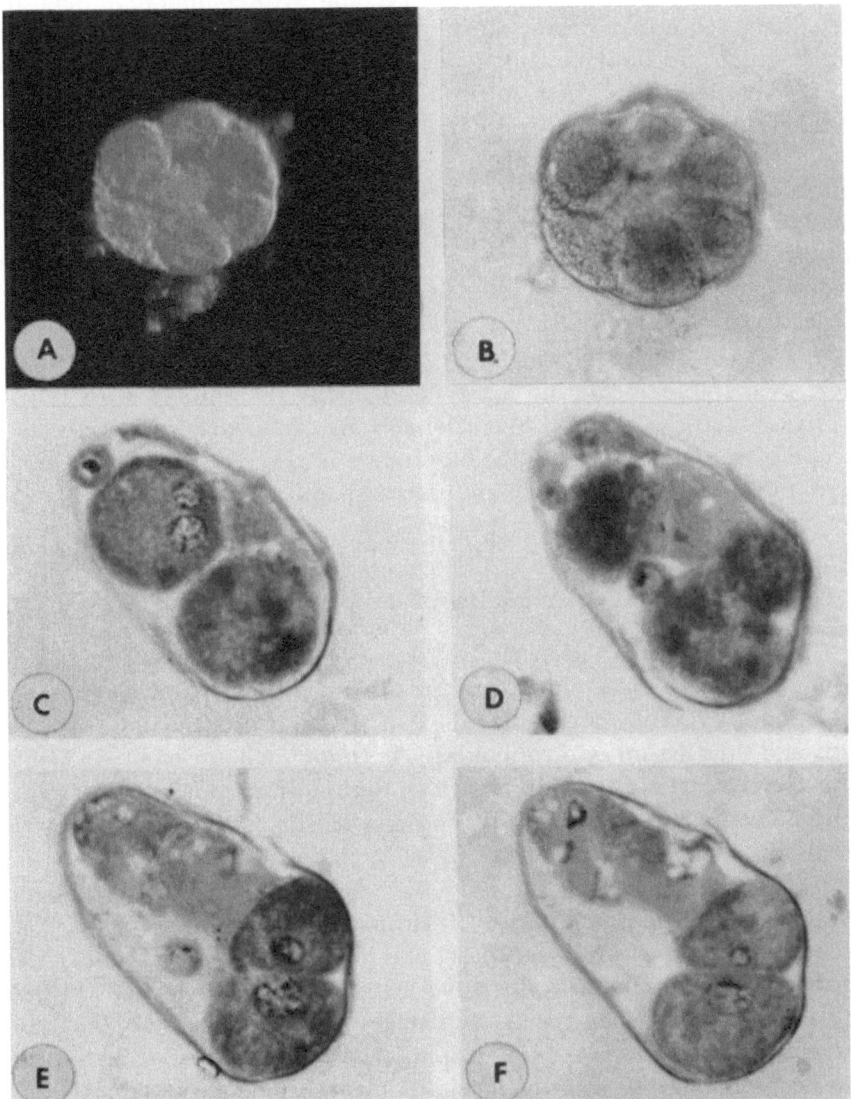

Fig. 2. The gross and microscopic features of an abnormal segmenting 8-cell human morula found within the uterine cavity. It is about 4-days' developmental age. Carnegie No. 8450. (A) Intact specimen illuminated by oblique light. Sequence 5 ×250. (From Hertig and Rock, 1950.) (B) The same intact specimen photographed by transillumination. Note absence of a blastomere at 8 o'clock. Sequence 8 ×300. (C–F) Four serial sections through the main mass of ovum to show multinucleation of blastomere, necrosis of blastomeres, multiple polar bodies or abortive micro-blastomeres and general discoid rather than rounded appearance of morula. The zona pellucida is normally intact, but shrunken slightly by fixation (Bouin's) and partial dehydration. Sections 5, 6, 7 and 8 from left to right and top to bottom. All ×500. (C, D, E from Hertig, *et al.*, 1954. F from Hertig and Rock, 1949.)

them do implant. This would be in keeping with comparative data and also make sense in terms of the 6 obviously defective ova out of a total of 21 which *did* implant and were found after day 25 of the cycle. The variety of defects in these 6 implanted ova is so varied that it seems unlikely that all of them originated *de novo* after implantation. Indeed the variety of defective segmenting ova is in very good keeping with the kinds of defects found in the series of implanted ova, small though it may be. Even though this series of 34 human ova is small, 24 good and 10 bad, it is the only such series available. (I am assured by competent biomathematicians that the series is large enough to be satistically valid, at least as far as interpreting data about ovulation.)

It is well known that during segmentation trophoblast differentiates first, then the segmentation cavity, and finally the embryo with respect to its ectoderm, endoderm and probably its primordial germ cells. It is proposed to describe briefly the 6 defective implanted ova and to discuss how and to what degree they are defective and what their probable pre-implantation stages might have looked like.

B. Ova Consisting of Trophoblast Only

Figure 3 represents a specimen consisting of trophoblast only. It is about 11 days of age, implanted in normal endometrium undergoing gestational hyperplasia (Hertig, 1964). It consists of syncytiotrophoblast only and thus has no cytotrophoblast, chorionic cavity or embryo. Trophoblast is the first tissue to differentiate, the cytotrophoblast giving rise to syncytiotrophoblast as shown by Tao and Hertig in 1964. The syncytiotrophoblast in this specimen has reached the stage of lacuna formation which takes place during the 9–12 day stages. The lacunae coalesce to form the intervillous space into which maternal blood flows. This specimen shows blood within these spaces and hemorrhage into the underlying eroded gland. Developmentally it lies between 11 and 13 days of age since it has some of the features of both stages of development. It obviously arose from one or more pure trophoblastic blastomeres. It had never reached the stage of morula formation with segmentation cavity. The embryonic blastomeres, if present, failed to develop. There is no evidence of an embryo in any of the perfect series of serial sections. Its grossly shrunken nature is in striking contrast to the normal 11-day stage shown for comparison. (Incidentally, other primates, the chimpanzee, for example, show this same form of defective development. One of the specimens in the Carnegie Embryological Collection (Number C-642) shows such a pure trophoblastic object in the act of early implantation.)

Obviously this human specimen is destined to abort. Whether the patient would have missed her expected menstrual period is unknown

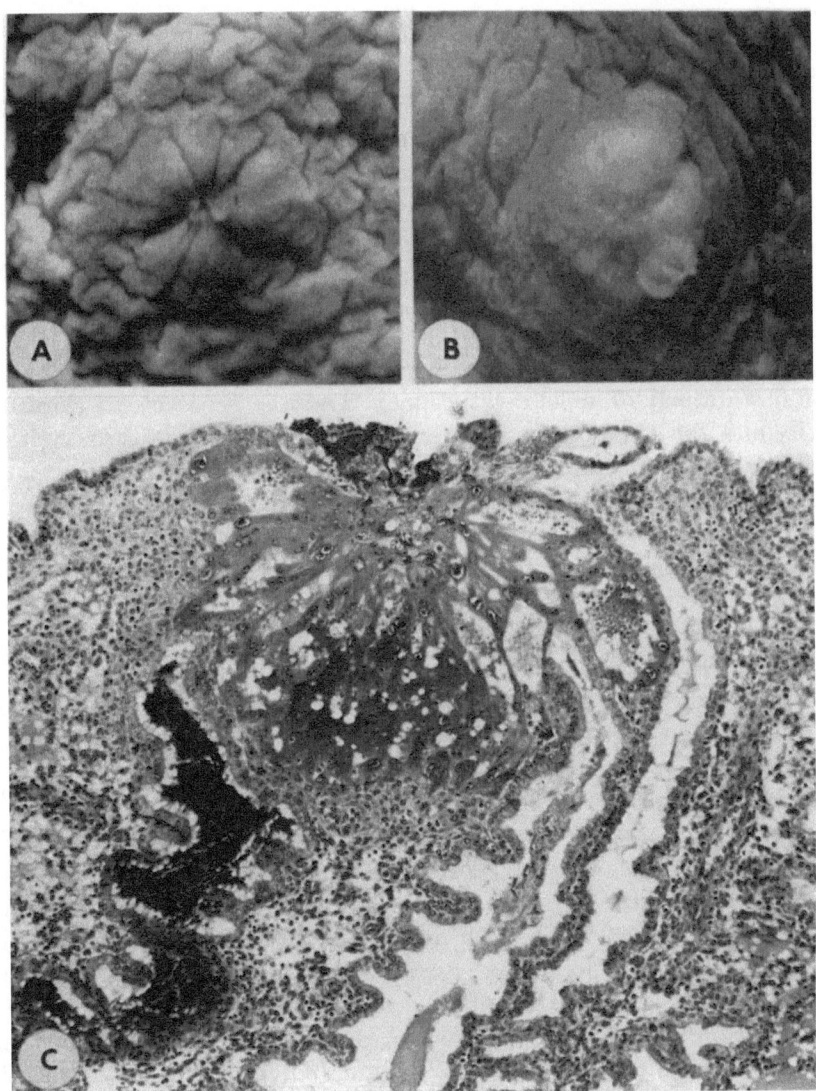

Fig. 3. An extremely defective human ovum of about 11 days showing only pure syncytiotrophoblast, but no cytotrophoblast, chorionic cavity or embryo. A is to be compared grossly to a normal implantation site B of the same developmental age. (A) Gross implantation site of defective 11-day ovum. Note wrinkled depressed elevation with multiple radiating furrows originating at the central area. The latter, covered by a scab, indicates the site at which the ovum gained entrance to the endometrium. Compare with B, a normal ovum, and with C for microscopic details. Carnegie No. 8329, sequence 2 ×22. (From Hertig and Rock, 1949.) (B) Gross implantation site of normal 11-day ovum to compare with A. Note bulging appearance owing to normal ovum beneath. The triangular defect, with teardrop of coagulum, is the site of epithelial repair. Carnegie No. 7699, sequence 8 ×22. (From Hertig and Rock, 1941.) (C) A microscopic detail of the defective ovum through its central portion. Note the presence of pure syncytiotrophoblast, but no other features of a normal 11-day specimen. The hemorrhage within the endometrial gland is normal for a 13-day ovum. Carnegie No. 8329, sect. 12-2-3 ×100. (From Hertig and Rock, 1949.)

but probable. I have seen such defective ova in decidual casts. They are hard to find and require patient search under the dissecting binocular microscope. It is likely that the patient would have had a slightly delayed period of profuse amount without the passage of decidual tissue. The corpus luteum was well formed but did not show the degree of secondary hyperplasia that normal pregnancies of this age exhibit.

C. Ova Containing No Chorionic Cavity

At first glance the specimen in Figure 4 appears to be a tangential section of a normal 10- to 11-day specimen. There is, however, no chorionic cavity and the trophoblast is laminated instead of being concentrically arranged. Normally the syncytiotrophoblast is peripheral to the inner-lying cytotrophoblast. Here, in this specimen, the syncytiotrophoblast is all at the implantation pole, a temporary advantage. In contrast, all of the cytotrophoblast is at the abembryonic pole and hence cannot participate in orderly chorionic villus formation. In spite of the relative advancement of ovular development, the embryo consists only of ectoderm. Owing to the lack of chorionic cavity (segmentation cavity of blastula), there is no formation of auxiliary structures such as chorionic mesoblast, blood vessels, amnion or primary yolk sac. These structures delaminate *in situ* from the cytotrophoblast lining a chorionic cavity as shown by Hertig in 1935. Although this is a growing and viable ovum at present, it is destined to abort. The embryo for some days lives in a large tissue culture-like chamber with no blood supply of its own. It is impossible for this embryo to develop properly. Although the ovum has reached the 12-day stage, the embryo is still in the 7-day stage. The embryo will die and be lost. The patient will probably miss her menstrual period and pass a bit of decidua, even a cast, a week or so later. This implantation site will be easier to find than that of Figure 3, but there will be no chorionic cavity or well organized chorion. It will consist only of a disorganized mass of trophoblast. One wonders if this is the sort of ovum that might give rise to a choriocarcinoma in the patient who is unaware that she was or had been pregnant. Such cases do occur, although rarely.

D. Ova Showing Hypoplasia of Trophoblast

Two examples of this relatively common abnormality are seen in Figures 5 and 6. The former shows a mild degree of hypoplasia whereas the latter shows a severe trophoblastic defect. As in all biologic phenomena, there is a spectrum of form and function. In terms of this paper, and in designing Figure 9, the first specimen referred to is classified as being

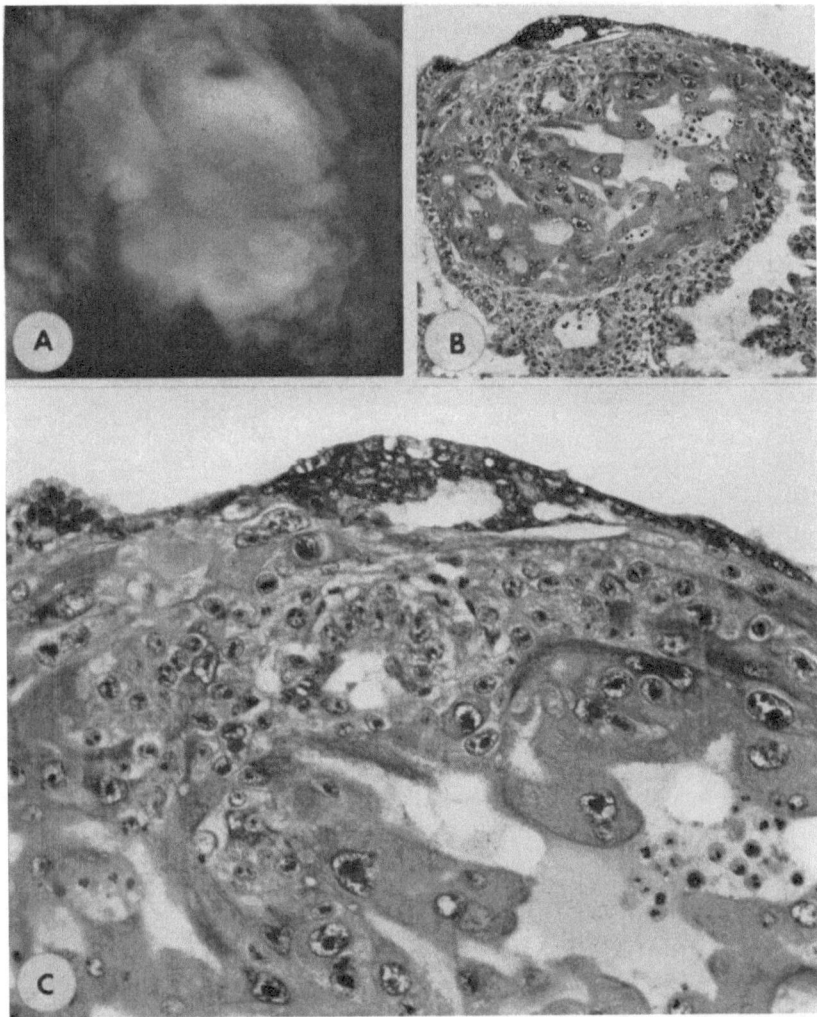

Fig. 4. An abnormal human ovum of about 11- to 12-days to show gross and microscopic features of absent chorionic cavity and laminated rather than circumferentially arranged trophoblast. Carnegie No. 8370. (A) Implantation site photographed under fluid, 70% alcohol. Note concavity or flattening of surface owing to lack of chorionic cavity. Sequence 6 ×46.5. (B) Lower power view of entire ovum and some of surrounding endometrium. The latter is normal but the ovum is solid with laminated trophoblast. The cytotrophoblast is above, containing the embryo, whereas the syncytiotrophoblast is below at the implantation pole. See Fig. C for details of ovum and embryonic disk. Section 6-3-5 ×100. (From Hertig and Rock, 1949.) (C) Detailed histology of same ovum to show clearly the ventrally curved germ disk, consisting entirely of ectoderm, closely surrounded by cytotrophoblast. There should be a chorionic cavity occupying about 60% of the entire chorionic mass. The embryo has no "Lebensraum." The lacunar network, the primitive intervillous space, is clearly seen within the syncytiotrophoblast below. Section 6-3-5 ×300. (From Hertig and Rock, 1949.)

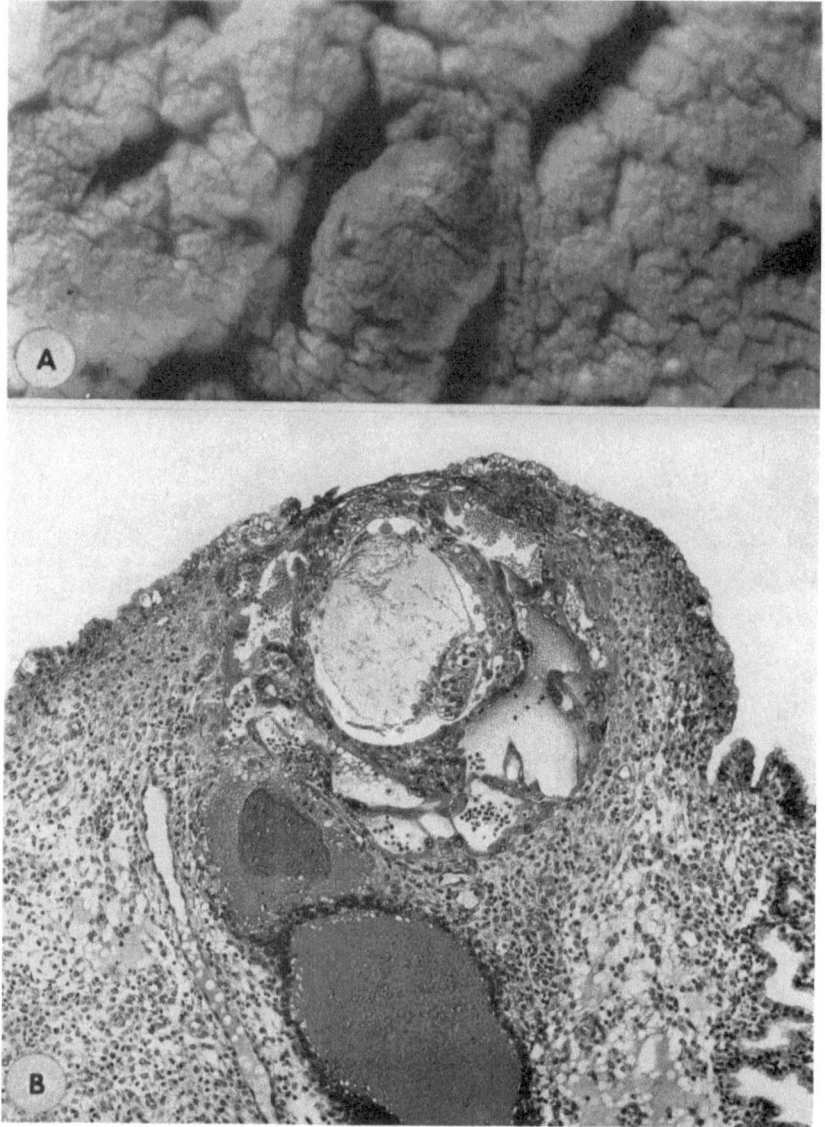

Fig. 5. The gross and microscopic appearance of a 12-day human ovum with minimal hypoplasia of trophoblast. In general, this ovum may be used as an essentially normal control for those in Figs. 2, 3, 4 and 6 even though the trophoblast is a bit hypoplastic. Carnegie No. 7770. (From Hertig and Rock, 1944.) (A) Gross implantation site photographed under 70% alcohol. The general rounded elevation with superficial ulceration is within normal limits. Sequence 5 ×22. (B) A mid-cross section of ovum to show: large cavity (missing in Fig. 4); primordial villi composed of inner cyto-trophoblast; and an early intervillous placental circulation within the outer syncytio-trophoblast. The hemorrhage within the gland is normal but perhaps too early for this stage of development. The endometrium is normal. Section 12-3-5 ×100.

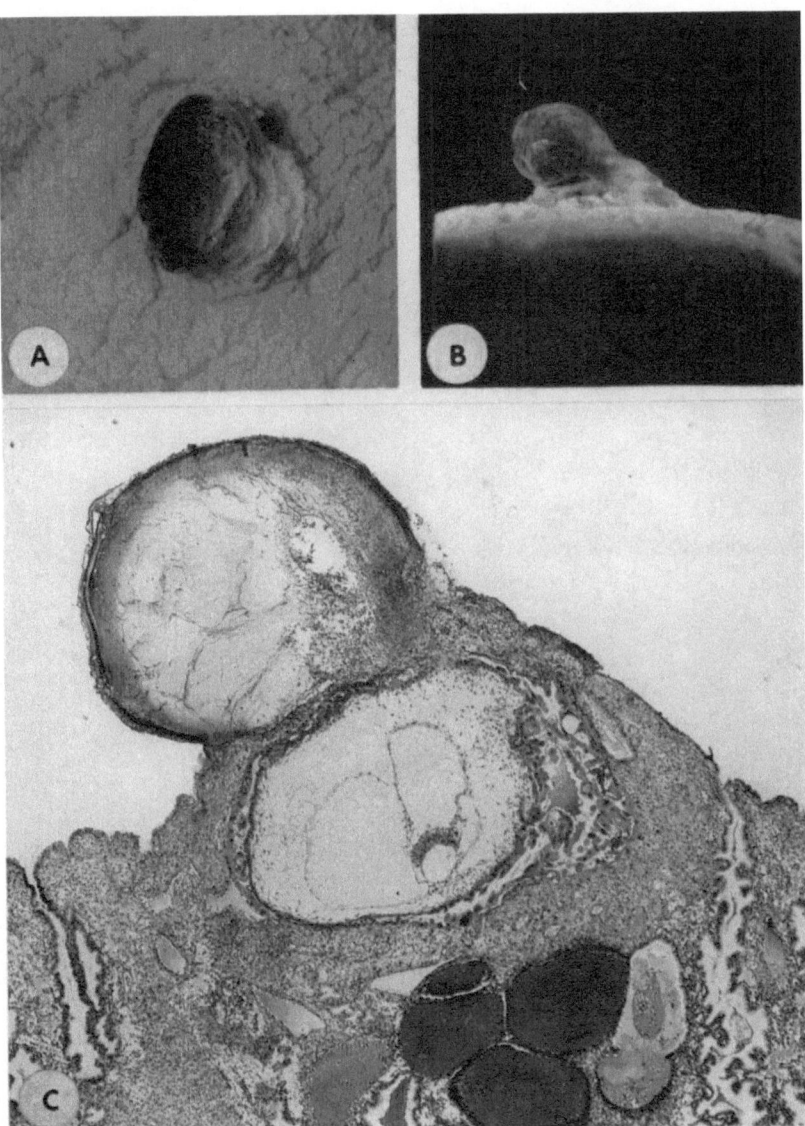

Fig. 6. A 13-day human ovum with severe hypoplasia of trophoblast which will undoubtedly result in embryonic death and ultimate abortion of an ovisac containing an amnion and yolk sac, but no embryo. The chorionic villi will be few in number, a "bald" chorion. Carnegie No. 7800. (From Hertig and Rock, 1944.) (A) Surface view of implantation site to show hemorrhage, a feature within normal limits for this stage of pregnancy. Sequence 2 ×8. (B) Lateral view of same implantation site to show normal elevation made by interstitially implanted ovum surrounded by massive hemorrhage. Sequence 3 ×8. (C) A mid-cross section of ovum and surrounding endometrium. The latter is normal, but the chorionic trophoblast is markedly deficient. There are no simple chorionic villi of which there should be about 12–15 in number in this plane of section. (Please refer to Fig. 8 for details of a villous ovum slightly older than this specimen.) The embryo with amnion, germ disk and yolk sac, is normal. Section 15-1-3 ×35.

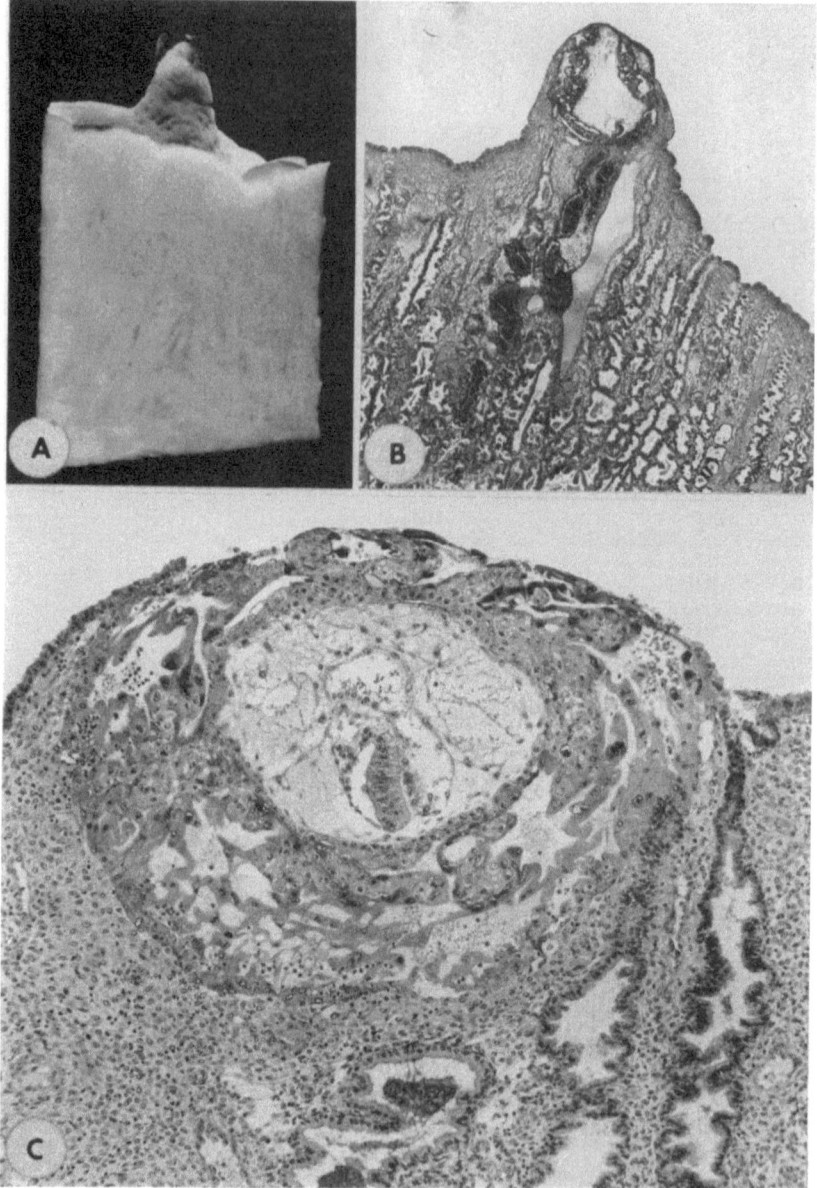

Fig. 7. (A) Gross lateral view of superficially implanted 12-day ovum. Note hemorrhage from top of implantation site. See microscopic features in Fig. 7B. Carnegie No. 8290, sequence 7 ×5. (From Hertig *et al.*, 1956.) (B) A superficially implanted ovum of about 12 days of age. In general the trophoblast, comparable to that in Fig. 5, is normal except at the implantation pole. At this stage of development, the ovum should be well embedded as are those in Figs. 3, 4, 5, 6 and 7C. Whether this ovum would have aborted is problematical. Had it gone toward term it certainly would have shown some degree of placenta circumvallata. As such it might

within normal limits although it is not as perfect as some of the other 11- and 12-day specimens reported by Hertig, *et al.,* in 1956. Indeed, it may well abort or form a bizarre shaped placenta with alternate thin and thick areas should it approach term. The other specimen, Figure 6, is certainly destined to abort because there is insufficient trophoblast present to form an adequate chorion or placenta. The embryo, now about 13 days old, will ultimately die and the chorion will be "bald" or devoid of chorionic villi.

Such chorionically defective abortuses are found in routine obstetric pathology. The mechanism of abortion in this case would be the failure of secondary luteal hyperplasia owing to the lack of H.C.G. (human chorionic gonadotropin) acting directly, or indirectly through the pituitary, upon the corpus luteum. It would appear that this defective trophoblast arose *after* implantation rather than before. Otherwise there would be no such perfect chorionic cavity nor such normal implantation consistent with the history and degree of development, namely, 13 days. This specimen may be compared to an 18- to 19-day stage, Figure 8.

E. Superficial Implantation of an Otherwise Reasonably Normal Ovum

Such a specimen is seen in Figures 7A and B. This was grossly a polypoid implantation site with considerable hemorrhage through the *decidua capsularis* from the underlying chorion. The trophoblast at the implantation pole is deficient. Whether this caused the shallow implantation is problematical because another similar specimen (not illustrated) had normal appearing trophoblast at the implantation pole. The significance of such shallow implantation lies in the probability that the abnormality precedes the common circumvallate placenta seen in Figures 11 and 12. This important anomaly of placentation, so serious throughout pregnancy for both baby and mother, will be discussed and illustrated in the following section on spontaneous abortion.

An 18- to 19-day villous ovum in Figure 8 (not from the Hertig-Rock

---

have aborted at midterm (Fig. 11) or separated prematurely (Fig. 12). Carnegie No. 8290, section 25-2-4 ×12. (From Hertig and Rock, 1949.) (C) This essentially normal 12-day previllous ovum shows a 90 degree tipping or rotation of the germ disk as compared to its normal position. This specimen is used to illustrate a normal chorion, which may be compared to Figs. 7B, 5B, 4B and 4C. Whether this otherwise normal, but malpositioned embryonic disk would have resulted in a congenital anomaly of the body stalk or yolk sac is unknown. Note normal outer syncytiotrophoblast showing intervillous space formation and focal proliferation of inner cytotrophoblast to form primordial chorionic villi. Carnegie No. 8299, Section 11-5-4 ×100. (From Hertig et al., 1956.)

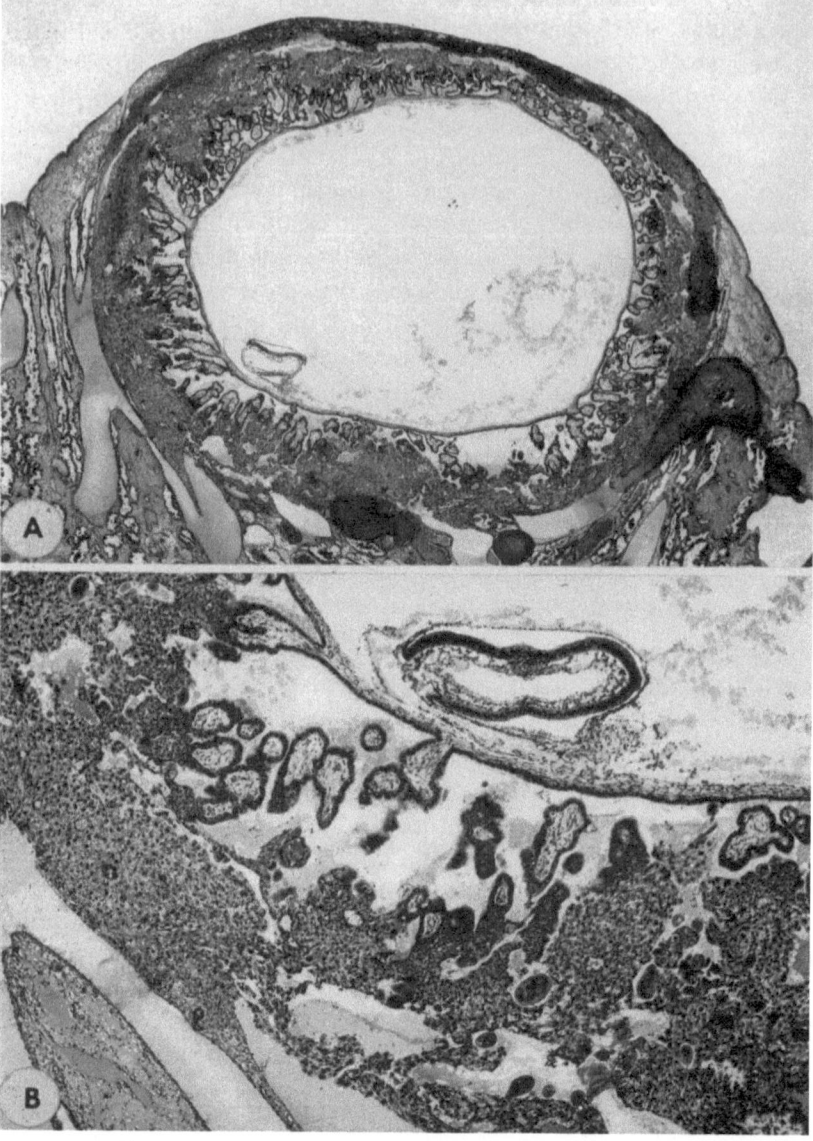

Fig. 8. An 18- to 19-day villous ovum, developmental age, which shows two abnormalities. The chorion is normal and should be compared to Fig. 6C, a hypoplastic chorion whose developmental age is 5- to 6-days younger. The abnormality lies in the embryo; its position is eccentric, and it consists of a conjoined germ disk with a single amnion. Carnegie No. 8727. (A) A low power view of the entire chorion and surrounding implantation site. Note development of normal simply-branched villi around the entire circumference, a normal phenomenon at this time of gestation. The embryo is located about 45 degrees away from the implantation pole and would have resulted in a marginal insertion of the cord, an anomaly of clinical significance. Details of the conjoined monoamniotic twins are to be seen in Fig. B. Section 82-2-2

series) contains a conjoined germ disk, actually one which never separated into its two potential embryos. It lies within a single amniotic membrane and is to be seen in Figure 8B. The importance of this specimen, the earliest of its kind known to the author, is that it emphasizes mechanisms of single ovum twin formation. It also highlights the fact that although conjoined twins are always within a single amnion, identical twins may also lie within one amniotic sac. The mechanism is merely that a twin germ disk continues its longitudinal separation *after* the amnion forms. Ordinarily, monochorionic (identical) twins form prior to 7 days, at which time the amnion begins to delaminate from cytotrophoblast in juxtaposition to the germ disk. Thus each embryonic disk induces its own amnion formation. It is common knowledge that identical twins usually have separate amniotic sacs and only rarely a single one. The obstetrical importance of monoamniotic twins lies in the fact that the twins become locked and/or intertwined in each other's cords. Obviously this contributes to fetal mortality.

In summarizing the data just presented, it would appear that normally fertile women (couples, actually, since we have no way of delineating male or female zygotic defects in this material) have a spectrum of fertility or infertility. In any one month, when conditions are optimal about 15 per cent of oocytes fail to become fertilized, about 10 to 15 per cent segment but fail to implant, about 70 to 75 per cent (at least 58 per cent) implant but only 42 per cent are of such viability as to cause the patient to miss her expected menstrual period. Stated somewhat differently: *during any one menstrual cycle* there is a 42 per cent sterility, a 16 per cent infertility and a 42 per cent fertility. It will be interesting to hear Dr. Hafez's comments along similar lines for domestic animals that customarily produce only one secondary oocyte as does the human female.

In terms of a patient who actually misses or delays her period owing to a pregnancy, whether good, bad or indifferent, the probability of an abortion is 27.6 per cent and that of going to term is 72.4 per cent. As indicated before, some abortions are not recognized as such by the patient or her physician unless tissue is actually passed and examined. What the true spontaneous abortion rate is no one knows precisely but it must lie between about 25 per cent of histologically proven early pregnancies and 10 per cent of recognized clinical pregnancies.

×9. (B) A medium power detail of the conjoined germ disk in the early stage of primitive streak formation. The primitive streaks are cut transversely. Note single amnion. This anomaly of the amnion, with or without separation of the twins, is of great clinical importance and frequently results in obstetrical difficulties and fetal death. This is the earliest stage of either of these anomalies known to the author. Section 84-1-3 ×35.

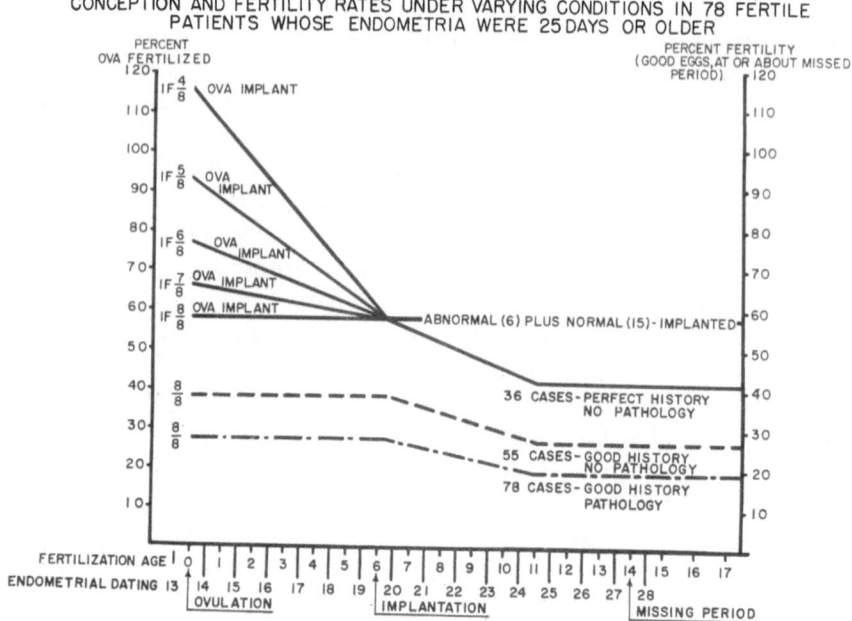

Fig. 9. A schematic, composite graph to indicate human fertility, infertility and sterility in any *one* menstrual cycle. This is based upon all of the clinical, pathological and embryological data from the 34 human ova obtained from normally fertile women described by Hertig *et al.* in 1956. The data from the 25th day on are accurate, that is, 42% fertility, 16% infertility and 42% sterility at the time of the missed period. The probable curve of fertility prior to day 25 is conjectural, based on the 8 segmenting ova (4 normal and 4 abnormal) from the tube and uterus and the 5 early implanted ova of 7 and 9 days of age. It is supposed that the ultimate curve would resemble that of any domestic animal which ovulates but one oocyte at a time. (From Hertig *et al.,* 1959.)

II. Spontaneous Abortion

Spontaneous abortion or miscarriage is the most common cause of reproductive failure. Its incidence depends upon the social stratum of the patient; it is about 5 per cent in ward patients and 10 per cent in private patients. The discrepancy lies in the time at which the patient first presents herself to the physician for her initial physical examination. Ward patients tend to come later and private patients earlier. Therefore there is more objective evidence of pregnancy in the latter group. The ward patients are wiser or busier and hence tend to come after the third missed period when most of the defective ova and other misconceptions have miscarried.

The only accurate statistic known to me is that from my own personal examination of every abortus from the private practice of the late Dr.

Paul Gustafson. Over a 5-year period, the incidence of morphologically proven miscarriage was 11.6 per cent of this middle-class private practice, consisting mainly of white patients. The discrepancy between these data and those from the Hertig-Rock series of embryos (27.6 per cent) is due to the clinical criteria of pregnancy. If a patient misses her period and shortly thereafter bleeds profusely, it is hard to prove pregnancy. The early decidua sloughs off and is lost; there is no tissue for examination by the pathologist.

Incidentally, the legal definition of an abortion is a gestation delivered on or before the twentieth week. Medical viability is at 28 weeks for all practical purposes. The fetus then weighs about 1000 grams and is 24.7 centimeters, crown rump length. Perhaps this medicolegal discrepancy is assuming less importance because the fetus of any age now has the legal right to sue through either parent for damage sustained in utero owing to damage alleged to have resulted from outside forces. Actually I was an expert witness for the defendant in Atlanta, Georgia, in 1958 or 1959 when such a case was admitted for trial. The plaintiff lost the case owing to the fact that I could show embryologically that the limb bud damage could not have taken place at the time of an automobile accident since there *were* no limb buds at that time. Nevertheless the case made legal history since it was admitted to trial.

Speaking of traumatic abortion, Hertig and Sheldon, 1943, showed that only one abortion in 1000 was due to external trauma even though 13 of these patients had a history of trauma; the other 12 having sustained the trauma *after* the death of the pregnancy. Thus one "traumatic" abortion in 13 is real; a fact that both doctors and lawyers should remember before undertaking court action.

It is not within the scope of this presentation to discuss induced abortion, either therapeutic or criminal. The latter is certainly common and has been estimated to be as high as 20 per cent of all pregnancies. There are about 5000 maternal deaths alone from this cause so that there may be a million or more self induced or criminal abortions per year in this country. At any rate, the problem is a large one legally, medically and sociologically. The interested reader is referred to a recent résumé of this important problem by Kummer and Leavy (1966).

Examination of material from an abortion requires:

1. an interest in the problem
2. fine dissecting instruments, mostly ophthalmological
3. a good binocular dissecting microscope
4. a knowledge of embryologic and chorionic pathology during the first trimester when most spontaneous abortions occur
5. the willingness to spend time and "tender loving care" in evaluating the specimen and selecting proper blocks for microscopic section.

It is better to examine the specimen fresh rather than after fixation. If a physician fixes the specimen first it should be done in plenty of 10 per cent formalin within a widemouthed container. It is difficult to remove and interpret a specimen which has been stuffed, while fresh, into a container which is too small or contains too little fixative.

Speaking of fresh specimens, tissues for tissue cultures for chromosome studies should include the embryo if possible. Often there is no embryo or only one which is macerated and stunted. Since all the auxiliary structures, amnion, primary yolk sac, chorionic mesoderm, chorionic blood vessels and the two types of trophoblast, are derived from trophoblastic blastomeres and *not* embryonic, care should be used in interpreting results of chromosomal studies. Whether the trophoblastic and embryonic blastomeres are genetically different is as yet unproven. They probably are.

It should also be pointed out that in selecting fresh sterile material from abortuses for tissue culture and chromosome studies, maternal decidua may be present as part of the placental "floor" or basal plate. To avoid decidua, the prosector must be sure that the tissue selected from the specimen under sterile saline shows the typical form of the chorionic membrane or chorionic villus.

The simple morphologic results of an analysis of 1000 consecutive spontaneous abortions examined in my laboratory from February, 1936 to December, 1941, are herewith recorded. These were published in 1943 by Hertig and Sheldon. During this 6-year period there was a total of 1416 abortuses examined by my residents and me. The 1000 were selected merely on the basis of adequate material and an adequate history. The etiology is a combination of the morphologic and clinical findings. The main findings will be listed and interspersed with running comment.

A. OVULAR FACTORS:
1. Pathologic ova, with absent or defective embryos   489
2. Embryos with localized anomalies   32
3. Placental abnormalities   96

         617

B. MATERNAL FACTORS:
1. Criminal abortions   21
2. Uterine abnormalities   64
3. Febrile and inflammatory diseases   20
4. Miscellaneous   12
5. Anatomically normal ova (classified)   265
6. Trauma (automobile accident)   1

         383

         1000

A. Ovular Factors

1. Pathologic Ova

The 48.9 per cent pathologic or "blighted" ova with absent or defective embryo are further classified as follows:

Group I. *Chorionic Villi Only.* This group is a matter of convenience only. The mere presence of villi alone gives little clue to the etiology or pathogenesis of the abortion unless they are hydropic or hydatidiform. In such a case the embryo was dead or missing as shown by Hertig and Edmonds in 1940.

Group II. *Empty Chorionic Vesicle.* This is common and the most pathologic of all "blighted" ova. It is often intact so that the pathologist may prove to himself that there is no derivative of the embryonic mass or germ disk.

Group III. *Chorion Containing Empty Amnion.* This is only slightly less pathologic than Group II. It is valid, common and often intact. Such a specimen is to be seen in Figure 10.

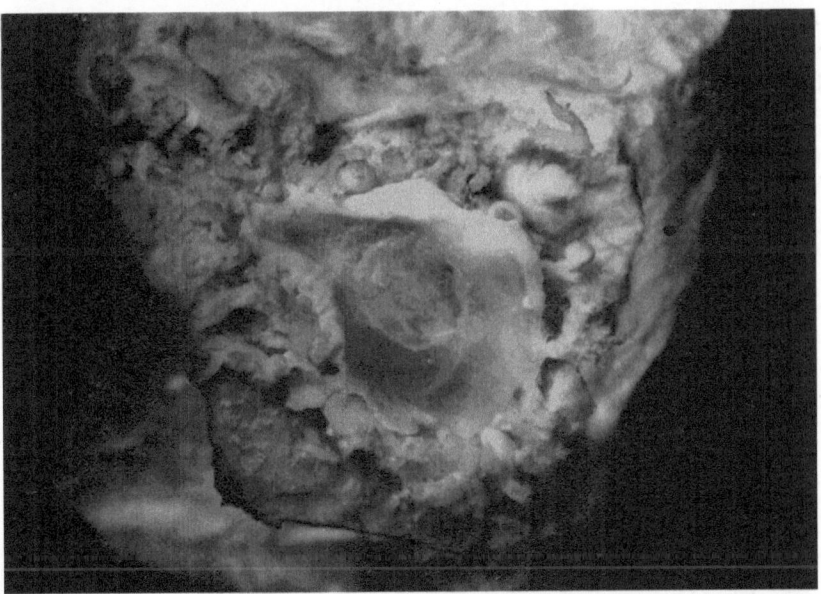

Fig. 10. A typical "blighted" or pathologic ovum of Carnegie Group III, a chorion containing an empty amnion. This chorion, still attached to the uterus, was dissected by the author and has had the decidua capsularis over the ovum removed and the chorionic cavity opened. Note empty amniotic sac within the chorion and the peripheral chorionic villi, some of which are swollen or hydropic. The myometrial cut surface of the fundus is at the top of the picture. Carnegie No. 8664, Sequence 1 ×6. F. H. W. S-49-145.

Group IV. *Chorion and Amnion Containing a Nodular Embryo.* This is to be separated from a similar specimen containing only the stump of an umbilical cord; the embryo having undergone autolysis. It is uncommon.

Group V. *Chorion and Amnion Containing a Cylindrical Embryo.* It is necessary to distinguish the head from the caudal end of the embryo. This is a rare group in my experience.

Group VI. *Chorion and Amnion Containing a Stunted Embryo.* It is uncommon. The embryo is stunted in both growth and development.

In general, the large and important group of pathologic ova are predestined to abort. Most of them are in Groups II, III and IV. About two-thirds of them show hydropic or hydatidiform swelling of the stroma of their chorionic villi. These constitute the early, abortive stage of hydatidiform mole. The latter is a true missed abortion and has a gestational age of about 20 weeks. The blighted or pathologic ova abort rather uniformly at between 9 and 10 weeks.

2. *Embryos with Localized Anomalies*

This important but small group, 3.2 per cent, indicates that abortuses in general have a higher proportion of anomalous fetuses than that of congenital anomalies among babies delivered at term. Why they should abort is not clear but probably due to endocrinologic malfunction of the feto-placental unit.

3. *Placental Abnormalities*

This is an important group of 9.6 per cent and may be broken down as follows:

a. Circumvallate placenta	45
b. Hypoplasia of the placenta	20
c. Placenta membranacea, partial	2
d. Velamentous insertion of umbilical cord	1
e. Hypoplasia of amnion	1
f. Rupture of marginal sinus	3
g. Premature senility of placenta	4
h. Breus' mole (intraplacental hematomata)	19
i. Succenturiate lobe with total infarction	1
Total	96

Figures 11, A and B, show the pathogenesis of the circumvallate placenta which accounts for 4.5 per cent of all spontaneous abortions and most of those occurring during the midtrimester. The mature counterpart of the circumvallate placenta is seen in Figure 12.

B. Maternal Factors

1. Criminal Abortion

2.1 per cent is obviously far too low. They were included since the material and history were complete. The hallmark of an induced abortion is a normal conceptus with an acute chorionitis. This finding rarely occurs spontaneously.

2. Uterine Abnormalities

a. Low implantation of placenta	56
b. Placenta accreta	2
c. Bicornuate uterus	2
d. Multiple leiomyomata of uterus	1
e. Retroversion of uterus, fixed	3
Total	64

The uterine abnormalities, 6.4 per cent, are important and fairly self-explanatory. The commonest one is low implantation of the ovum; a miniature placenta previa.

3. Febrile and Inflammatory Diseases

a. Bacterial inflammation of decidua, acute	12
b. Small pox	1
c. Pyelitis with horseshoe kidney	1
d. Fever of unknown etiology	5
e. Chronic endometritis	1
Total	20

Febrile and inflammatory diseases, 2 per cent, although few in number are real. It is well known that early pregnancy may be susceptible to abortion during febrile diseases as witness the number of abortions by patients during the large 1918 "Flu" epidemic. It is striking that uterine infection *per se* plays little role in abortion. Mall and Meyer in 1921 ascribed most of their abortions to infection. They misinterpreted the leucocytic response to sterile ischemic necrosis as infection. It is not. All abortions of spontaneous origin have thrombosis, necrosis and focal hemorrhage of decidua as a morphologic accompaniment of abortion. It is analogous to a delayed menstrual period and probably for the same reason, corpus luteum and/or H.C.G. failure. All abortuses show this irrespective of the underlying cause of the abortion. After all, the final expulsion of the ovum is the last of a series of sequential events that usually began some weeks or even months before.

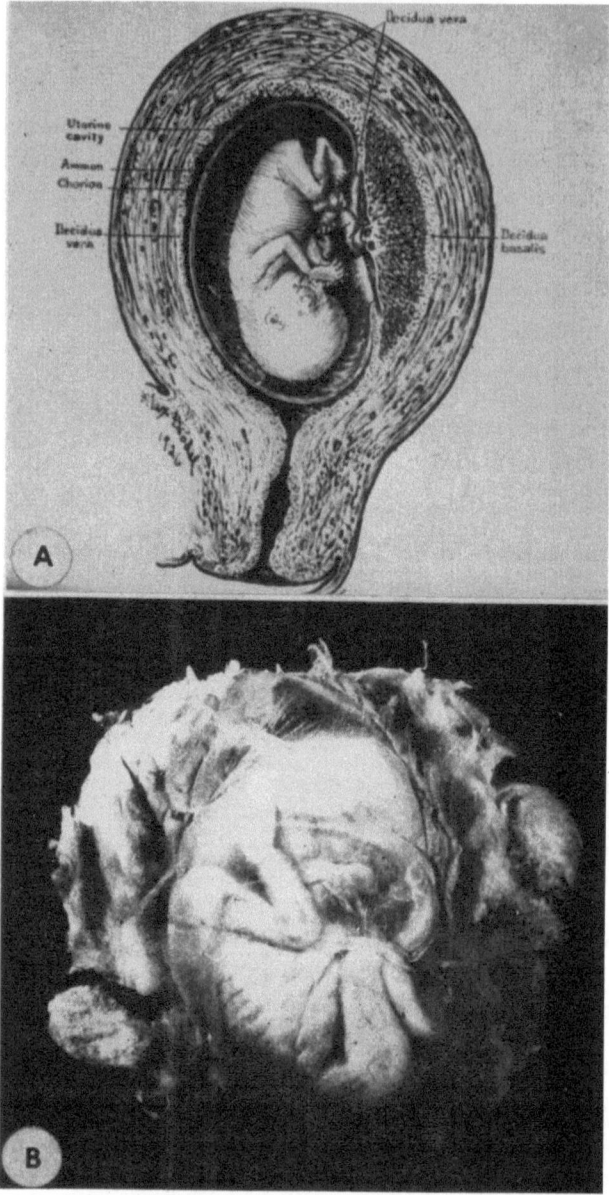

Fig. 11. Circumvallate placenta, immature, to show the abnormal placenta in situ and after a spontaneous miscarriage. (A) A midsagittal section of a pregnant uterus of about 3½ months menstrual age as drawn by the late great medical artist Max Brödel. Note relatively thick button-like mass of placental tissue whose membranes are attached toward the center of the fetal surface rather than at the edge of the placenta is undermining or invading the decidua. Note also that the membranes appear as though they had "herniated" or "prolapsed" out into the uterine cavity, covered only by thin decidua capsularis. Circumvallate placentas appear to arise from

4. Miscellaneous

a. Radiation effect on ovaries	2
b. Erythroblastosis fetalis	3
c. Surgical removal of corpus luteum	4
d. Blood dyscrasia	1
e. Interference with circulation of cord	2
Total	12

Miscellaneous are for convenience only and bear no relationship to each other. Any classification has a miscellaneous section.

5. Anatomically Normal Ova

a. Anatomically normal ova without disease   227

Fetus, macerated	146
Fetus, nonmacerated	74
Fetus, by history only	7
	227

b. Acute chorionitis, consistent with spontaneous premature rupture of membranes
    (Fetus macerated in 9 and normal in 5)   14

c. Positive Hinton and Wassermann tests   3
    (The syphilis was probably not responsible for the abortion)

d. Infarction of placenta, extensive   13
    (All fetuses were macerated)

e. Toxemia of pregnancy   5
    (Fetus macerated in 3 and normal in 2 cases)

f. Trauma (internal)   1
    (Two successive biopsies on sterility patient not
      known to be pregnant)
    Exploratory celiotomy   1
    Intrauterine lipiodol injection, 7 weeks prior to
      last menstrual period   1

      Total   265

This list of anatomically normal ova, 26.5 per cent, is a most important group. Several obviously might have been avoided, such as the lipiodol

small superficially implanted ova comparable to that seen in Fig. 7B. (From Williams, 1927.) (B) A spontaneous abortion associated with a complete circumvallate placenta. This early midtrimester abortion is typical. Most abortions at this time of pregnancy show this anomaly. The edge of the placenta at the bottom of the picture has separated, the ovisac is relatively small and the fetus is normal, albeit macerated. (From the museum of the Boston Lying-In Hospital.)

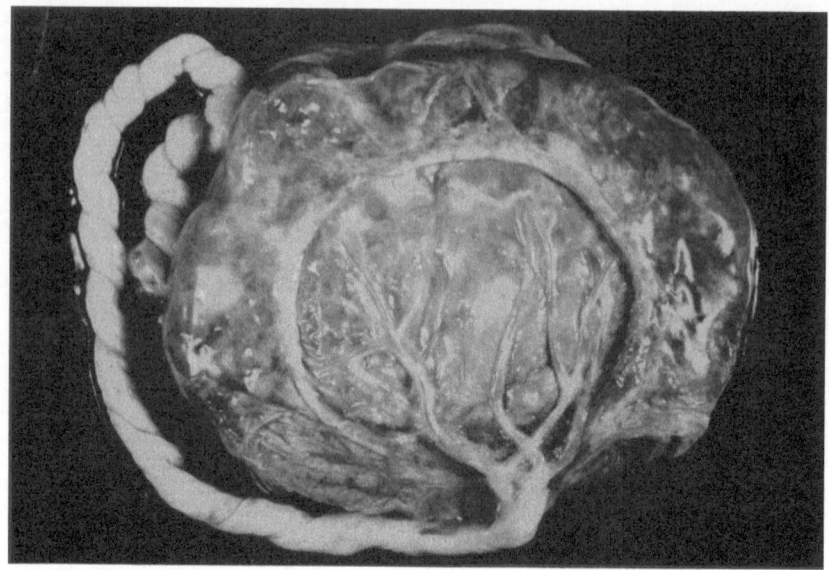

Fig. 12. A mature circumvallate placenta to show the characteristic rolled, opaque margin representing the origin of the membranes from the fetal surface of the placenta. The placentogenic area of the placenta is much smaller (and thicker) than normal owing to the initial small size of the ovum and its superficial implantation. The placenta, although misshapen, is about normal in mass for the stage of pregnancy. This type of placentation tends to result in the midtrimester abortion (Fig. 11), premature rupture of the membranes or premature separation of the placental margin. It is the severe or more extreme manifestation of the circummarginate placenta which is of little, if any, clinical significance. (From the Boston Lying-In Hospital.) (From Hertig, 1953.)

injection and biopsies during early pregnancy. It is a truism that all sterility patients may be pregnant and sometimes are. Biopsies should be done at the onset of the period. Lipiodol, if ever indicated, should be done 6 to 8 days after flow so that implantation and endometriosis will be avoided. Menstrual blood has been shown to contain viable fragments of endometrium which will grow in tissue culture.

The significance of this group lies in the fact that 7.4 per cent of the total were normal with living fetuses. Theoretically, these could be salvaged. Some may have been due to an incompetent cervix. That entity was unknown during the period when these abortuses were collected and examined. Shirodkar has called attention to this mechanism of midtrimester abortion of a normal conceptus and has devised an operation to prevent it.

In summary, there are many causes of human reproductive failure ranging from anovulation, failure of fertilization and failure of development after fertilization, and failure of development after implantation.

The several illustrations and tables indicate the variety of morphologic lesions which may be found in the fertilized ovum during the first half of pregnancy. It is hoped that the geneticists, virologists, mycologists, endocrinologists, teratologists, immunologists, biochemists and comparative pathologists here assembled will build upon these definite but still somewhat simple morphologic abnormalities of the conceptus in man and other vertebrates. From their researches will come a rational explanation of reproductive failure in man and animals.

Acknowledgments

Little, Brown and Company (Inc.) and Dr. Chester Keefer: Figure 1.
Carnegie Institution of Washington: Figures 2, A–F, 3, A–C, 4, A–C, 5, A, B, 6, A–C, 7, A–C, 8, A, B, 10.
The C. V. Mosby Company: Figures 2F, 3A, 3C, 4B, 4C, 5A, 5B, 6A–C, 7B, 11A.
American Association for the Aid of Crippled Children: Figure 12.
The Wistar Institute of Anatomy and Biology: Figures 7A, 7C.
J. B. Lippincott Company: Data on abortions.
American Academy of Pediatrics: Figure 9, Tables 1, 2.
Charles C Thomas: Figure 2A.

References

Adams, E. C., and A. T. Hertig: Studies on Guinea Pig Oocytes. I. Electron microscopic observations on the development of cytoplasmic organelles in oocytes of primordial and primary follicles. J. Cell. Biol. *21*:397, 1964.

Blandau, R. J., and W. C. Young: Ovum age and pregnancy in the guinea pig. Data from 462 artificially inseminated females. Anat. Rec. *67* (suppl. no. 1) :33, 1936.

Brambell, F. W. R.: Prenatal mortality in mammals. Biol. Rev. *23*:370, 1948.

Corner, G. W.: The problem of embryonic pathology in mammals, with observations upon intrauterine mortality in the pig. Am. J. Anat. *31*:523, 1923.

Corner, G. W., and G. W. Bartelmez: Early abnormal embryos of the Rhesus monkey. Contrib. Embryol. *35*:1, 1954.

Frazer, J. F. D.: Foetal death in the rat. J. Embryol. Exper. Morphol. *3*:13, 1955.

Hartman, C. G. (ed.) : Mechanisms concerned with conception. New York: The Macmillan Company, 1963.

Hertig, A. T.: Angiogenesis in the early human chorion and in the primary placenta of the macaque monkey. Contrib. Embryol. *25*:38, 1935.

————: The Pathology of Late Pregnancy Hemorrhage, in Prematurity, Congenital Malformation and Birth Injury. New York: Association for the Aid of Crippled Children, 1953.

————: Gestational hyperplasia of endometrium. A morphologic correlation of ova, endometrium, and corpora lutea during early pregnancy. Lab. Invest. *13*:1153, 1964.

Hertig, A. T.: Morphologic criteria of the time of ovulation in the human being. In: Keefer, C. S. (ed.): Human Ovulation. Boston: Little, Brown and Company, Inc., 1965.

Hertig, A. T., and H. W. Edmonds: Genesis of hydatidiform mole. Arch. Path. *30:*260, 1940.

Hertig, A. T., and J. Rock: Two human ova of the pre-villous stage, having an ovulation age of about eleven and twelve days respectively. Contrib. Embryol. *29:*127, 1941.

———, ———: On the development of the early human ovum, with special reference to the trophoblast of the previllous stage: a description of 7 normal and 5 pathologic human ova. Am. J. Obstet. Gynec. *47:*149, 1944.

———, ———: A series of potentially abortive ova recovered from fertile women prior to the first missed menstrual period. Am. J. Obstet. Gynec. *58:*968, 1949.

———, ———: Abortive human ova and associated endometria. Menstruation and its disorders (E. T. Engle, ed.). Springfield: Charles C Thomas, 1950.

Hertig, A. T., J. Rock, and E. C. Adams: A description of 34 human ova within the first 17 days of development. Am. J. Anat. *98:*435, 1956.

———, ———, ———, and M. Menkin: Thirty-four fertilized human ova, good, bad and indifferent, recovered from 210 women of known fertility. A study of biologic wastage in early human pregnancy. Pediatrics *23:*202, 1959.

———, ———, ———, and W. J. Mulligan: On the preimplantation stages of the human ovum: a description of four normal and four abnormal specimens ranging from the second to the fifth day of development. Contrib. Embryol. *35:*199, 1954.

Hertig, A. T., and W. H. Sheldon: Minimal criteria required to prove prima facie case of traumatic abortion or miscarriage. Ann. Surg. *117:*596, 1943.

Heuser, C. H., and G. L. Streeter: Development of the macaque embryo. Contrib. Embryol. *35:*1, 1954.

Kummer, J. M., and Z. Leavy: Therapeutic abortion law confusion. J.A.M.A. *195:*140, 1966.

Lewis, W. N., and C. G. Hartman: Tubal ova of the Rhesus monkey. Contrib. Embryol. *29:*15, 1941.

Mall, F. P., and A. W. Meyer: Study on abortuses: a survey of pathologic ova in the Carnegie embryological collection. Contrib. Embryol. *12:*3, 1921.

Noyes, R. W., A. T. Hertig, and J. Rock: Dating the endometrial biopsy. Fertil. and Steril. *1:*3, 1950.

Perry, J. S.: Fecundity and embryonic mortality in pigs. J. Embryol. Exper. Morphol. *2:*308, 1954.

Tao, T. W., and A. T. Hertig: Viability and differentiation of human trophoblast in organ culture. Am. J. Anat. *116:*1, 1965.

Williams, J. W.: Placenta circumvallata. Am. J. Obstet. Gynec. *13:*1, 1927.

Supplementary References

Ayre, J. E.: The impact of cytology and cytogenetics upon gynecology and obstetrics. Obstet. Gynec. Survey *19:*799, 1964. (233 references).

Becker, V.: Placental causes of premature delivery and fetal death. Deutsch Med. Wschr. *90*:1060, 1965. (20 references).

Benirschke, K.: Chromosomal studies on abortuses. Trans. New Engl. Obstet. Gynec. Soc. *17*:171, 1963.

Bonilla, E.: The anatomical factors in the spontaneous and habitual interruption of pregnancy. Rev. Clin. Inst. Matern. (Lisboa) *15*:77, 1964. (Sp.).

Bonham, D. G.: Perinatal mortality. Aust. New Zeal. J. Obstet. Gynaec. *5*:183, 1965.

Brill, A. G., and E. H. Forgotson. Radiation and congenital malformations. Amer. J. Obstet. Gynec. *90* (suppl. no. 1149) : 1964. (98 references).

Caldas, A. de C.: Toxi-infectious factors in the early and habitual interruption of pregnancy. Rev. Clin. Inst. Matern. (Lisboa) *15*:39, 1964. (Por.).

Carr, D. H.: Chromosome studies in spontaneous abortions. Obstet. Gynec. *26*:308, 1965.

Carvalho, W. D. de: Organic causes of female sterility. Matern. Infanc. (S. Paulo) *24*:239, 1965. (Por.).

Cordeiro, J. C.: Psychosomatic perspective of conjugal sterility. Acta Gynaec. Obstet. Hisp. lusit. *14*:165, 1965. (50 references) (Por.).

Cunha, D. P. da: Early and habitual interruption of pregnancy. Rev. Clin. Inst. Matern. (Lisboa) *14*:5, 1963. (187 references) (Por.).

Finley, W. H., and S. C. Finley: Cytogenetics in congenital malformations: a review. Alabama J. Med. Sci. *1*:5, 1964. (269 references).

Freire-Maia, A., and N. Freire-Maia: Genetic load and undetected abortions. Am. J. Hum. Genet. *17*:93, 1965.

Gerfeldt, E.: Incidence, etiology and prevention of congenital maldevelopment. Med. Klin. *59*:1287, 1965. (Ger.) .

Gibbs, C. E., and H. R. Misenhimer: Perinatal mortality review. Texas J. Med. *61*:39, 1964.

Hammond, E. I.: Studies in fetal and infant mortality. I. A methodological approach to the definition of perinatal mortality. Amer. J. Public Health *55*: 1012, 1965.

——: Studies in fetal and infant mortality. II. Differentials in mortality by sex and race. Amer. J. Public Health *55*:1152, 1965.

Harboe, M.: Infertility causes by immunological factors. T. Norsk Laegeforen *85*:308, 1965. (Nor.).

Hornstein, O.: Problem of so-called sperm immunity. Urologie *3*:211, 1964. (Ger.).

Kamal, I., D. Kandil, and M. Fathalla: On the pathogenesis of spontaneous abortion evaluation of chorio-decidual patterns. Preliminary report of 60 cases. J. Egypt. Med. Assn. *47*:631, 1964.

Kjessler, B.: Karyotypes of 130 childless men. Lancet *2*:493, 1965.

Krieg, H., and A. Eyquem. Immunology and sterility. Gynaek. Rundsch. *1*:243, 1964. (185 references) (Ger.) .

Kubo, H., M. Yano, K. Hanawa, and J. Kataoka: Statistical observations on perinatal death. Iryo *19*:123, 1965. (Jap.).

Lemtis, H.: Disorders of the "unity of fetus and placenta" and its effects on the child. Gerburtsh. Frauenheilk. *25*:428, 1965. (Ger.).

Lenz, W.: Epidemiology of congenital malformations. Ann. N.Y. Acad. Sci. *123*:228, 1965.

Lozzio, B. C. de: Congenital malformation and autosome abnormalities. Medicina (B. Air.) *24*:344, 1964 (Sp.).

Ludwig, K. S., and R. Schenk: Teratogenesis symposium of the Schweizerische Akademie der Medizinischen Wissenschaft. Schweiz. Med. Wschr. *94*:490, 1964. (Ger.).

MacPherson, C. R., and E. R. Zartman: Anti-M antibody as a cause of intrauterine death. Amer. J. Clin. Path. *43*:544, 1965.

Maier, W.: Our current knowledge on external injurious influences on the embryo and congenital defects. Deutsch. Z. Ges. Gerichtl. Med. *55*:156, 1964. (Ger.).

Malhotra, K. K.: The problem of male infertility. J. Indian Med. Assn. *45*:367, 1965.

Manclark, C. R., and M. J. Pickett: A proposed role of *in vivo* uterine anaphylaxis in the etiology of abortion and sterility. Int. Arch. Allerg. *27*:54, 1965.

Medearis, D. N., Jr.: Viral infections during pregnancy and abnormal human development. Am. J. Obstet. Gynec. *90* (suppl. no. 1140): 1964. (58 references).

Milham, S., Jr., and A. M. Qittelsohn: Parental age and malformations. Hum. Biol. *37*:13, 1965.

Niendorf, F.: Polyspermy and spontaneous abortion. Gynaecologia *158*:35, 1964. (Ger.).

Pigeaud, H.: The hormonal factor in repeated abortion. Rev. Clin. Inst. Matern. (Lisboa) *15*:65, 1964. (Fr.).

Plotz, E. J.: Virus disease. New York J. Med. *65*:1239, 1965. (94 references).

Raboch, J.: Spermiological findings in repeated spontaneous abortion. Zbl. Gynaek. *87*:194, 1965. (Ger.).

Rossberg, J.: Diagnosis and therapy of male fertility disorders. (Standard methods and review of literature from 1952–1962.) Derm. Wschr. *149*:107, 1964. (88 references) (Ger.).

Rozenszajn, L., M. Lancet, and D. Roch: The presence of anti-M in a woman with successive miscarriages. Israel Med. J. *23*:198, 1964.

Rümke, P.: Autospermagglutinins: a cause of infertility in man. Ann. N.Y. Acad. Sci. *124*:696, 1965.

Saldanha, P. H.: Genetic aspects of congenital malformations. Rev. Paul. Med. *65*:270, 1964. (Por.).

Sarma, V.: Congenital abnormalities of the foetus and their association with genital tract malformations in the mother. Brit. J. Clin. Pract. *19*:375, 1965.

Sato, H.: Chromosome studies in abortuses. Lancet *1*:1280, 1965.

Shapiro, S., L. J. Ross, and H. S. Levin: Infant and perinatal mortality in the U.S. Vital Health Statist. *3*:1, 1965.

Sroka, L.: Morphology and histochemistry of normal and abnormal sperm. Fertil. and Steril. *16*:613, 1965.

Stickle, G.: What priority, human life? Amer. J. Public Health *55*:1692, 1965.

Tartakow, I. J.: The teratogenicity of maternal rubella. J. Pediat. *66*:380, 1965.

Thiede, H. A., and S. B. Salm: Chromosome studies of human spontaneous abortions. Amer. J. Obstet. Gynec. *90*:205, 1964.

Varangot, J., and R. Pauphilet: Sterility caused by absence of ovulation. Rev. Prat. *14*:2891, 1964.

Weller, T. H., C. A. Alford, Jr., and F. A. Neva: Changing epidemiologic concepts of rubella, with particular reference to unique characteristics of the congenital infection. Yale J. Biol. Med. *37*:455, 1965.

Wesolowski, J.: Influenza and congenital malformations. Wiad. Lek. *18*:381, 1965.

Widok, K., and H. Widok: Causes of congenital malformations. Hippokrates *35*:417, 1964. (80 references) (Ger.).

Wingate, L.: Chromosome abnormality as a possible cause of habitual abortion. Bristol Mediochir. J. *80*:5, 1965.

REPRODUCTIVE FAILURE IN DOMESTIC MAMMALS[1]

E. S. E. HAFEZ[2]

Professor, Department of Animal Sciences, Washington State University, Pullman, Washington

The reproductive cycle consists of a chain of well-synchronized re-actions—follicle maturation, estrus, ovulation, sperm transport, sperm capacitation, fertilization, egg transport, endometrial progestation, blasto-cyst explansion, denudation of zona pellucida, implantation, placental formation, embryonic differentiation, fetal development, parturition, and lactation. The intricate interactions among the releasing factors from the hypothalamus, the gonadotropic hormones and the gonadal hormones control these reproductive processes. Lack of integrations of any phase of the hormonal sequences will result in reproductive failure. Genetic, anatomical, environmental, nutritional, metabolic, im-munological and pathological factors also contribute to such failure. The different syndromes of infertility may occur at five major stages of the reproductive cycle: (a) structural and functional development of the sexual organs, (b) prefertilization, (c) preimplantation, (d) postimplanta-tion, and (e) perinatal and neonatal (Fig. 1).

The purpose of this paper is to discuss the comparative aspects and the physiological mechanisms involved in reproductive failure in the female with special reference to domestic mammals. The gaps in our knowledge and the need for future research in the genetics, physiology, biochemistry, immunology and pathology of reproduction are emphasized.

[1] Invitational paper presented to the Symposium on "Comparative Aspects of Repro-ductive Failure" at Dartmouth Medical School, Hanover, New Hampshire, July 1966.

Unpublished data are part of an investigation supported in part by USPHS grants GM 10282-03, HD 00585-03, and HD 01013-03 from the National Institutes of Health.

Scientific Paper No. 2220, College of Agriculture Research Center, Projects 1695 and 1698.

[2] The manuscript was written while on sabbatical leave at the Department of Veteri-nary Physiology, the University of Sydney, Australia, as a NIH Special Fellow (IF3 HD-24, 411-01).

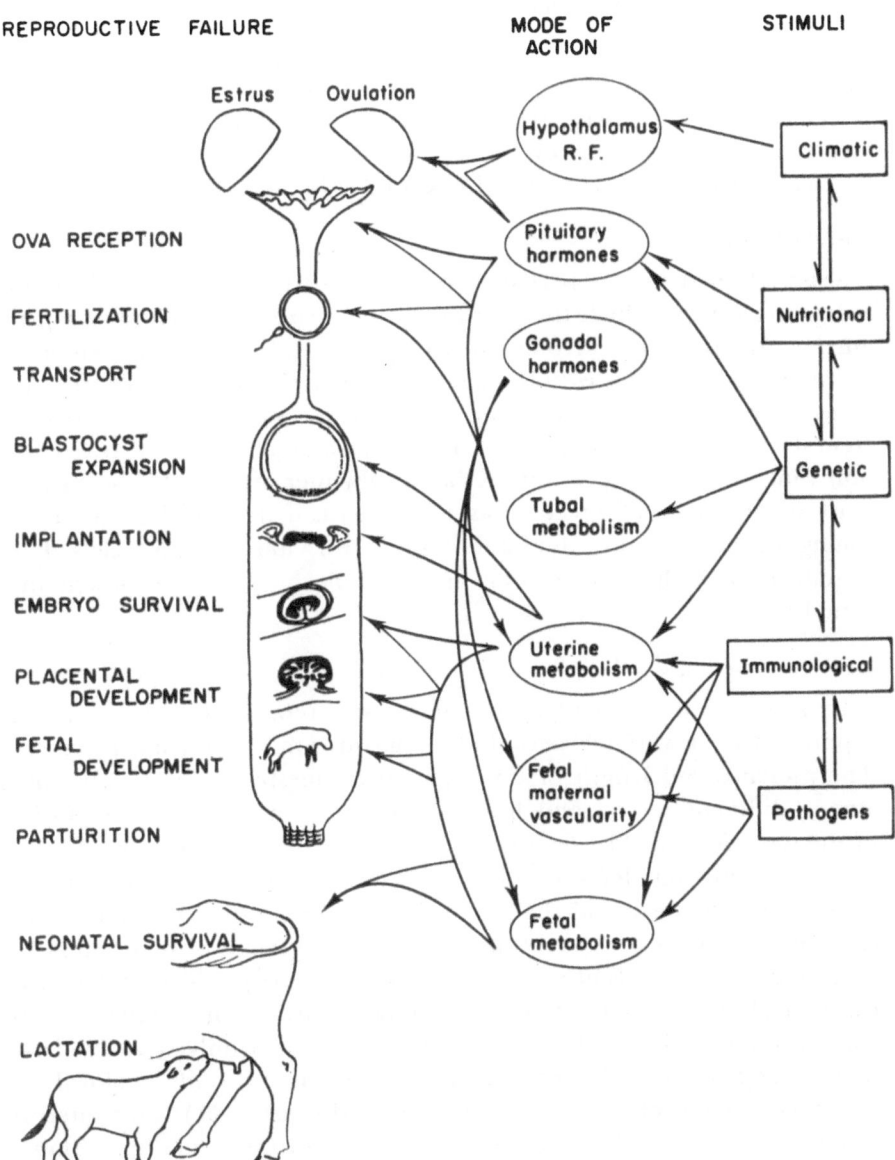

Fig. 1. Diagrammatic illustration of the different patterns of reproductive failure and the major physiological mechanisms involved. Reproductive failure in the female may involve estrus, ovulation, ova reception, fertilization, transport, blastocyst expansion, implantation, embryo survival, placental development, fetal development, parturition, neonatal survival or lactation. (Hafez, 1967.)

I. Failure in Ovarian Functions

A. Anestrus

Temporary or permanent absence of ovarian functions and in the recurrence of the estrous cycles are caused by seasonal changes in the physical environment, lactational stress or aging of the animal.

Seasonal Anestrus

Seasonal anestrus is a period of relative rather than absolute ovarian quiescence during which the changes in the ovaries and accessory reproductive organs are not accompanied by behavioral estrus. The extent of seasonal anestrus varies with the species, breed and environment. Anestrus is more pronounced in sheep and horses than in cattle, pigs, and most laboratory mammals. Sheep have an intrinsic annual rhythm of reproduction which reaches its peak in the fall and winter months. In the British mutton and long-wool breeds, seasonal anestrus is intermediate in length among the mountain breeds and the Dorset Horn and Merino breeds. The introduction of the ram into an anestrous flock appears to hasten the sexual season.

Anestrus During Lactation

In several species, ovulation and related reproductive activity are suppressed for a variable period after parturition and during lactation. The presence and length of this interval of anestrus vary greatly both among different species and breeds and in the succeeding pregnancies of the same female. The length of anestrus is also influenced by the season of parturition, level of milk production, number of litter being nursed, and the degree of the involution of the post-partum uterus. For example, Brahman cows suckling calves during periods of high temperatures and on low levels of energy intake are especially subject to anestrus during lactation. A tendency for cows nursing calves to have longer intervals of anestrus during lactation than similar cows milked twice daily may suggest an influence of nursing or frequency of milk removal on pituitary gonadotropic activity (Wiltbank and Cook, 1958). This interval of anestrus could be decreased by injections of progesterone in combination with estrogen (Ulberg and Lindley, 1960; Foote and Hunter, 1964).

In animals such as rabbits, ovulation may occur if the animals are bred following parturition, but implantation is suppressed if a large litter is nursed. In ruminants, there is a delay in the establishment of normal estrous cycles after parturition. This may arise from complete quiescence of the ovaries, or from quiet ovulations. The average length of interval from parturition to first estrus in cattle and sheep ranges from four to ten weeks. In the sow, ovulation and estrus normally do not occur during lactation, but estrus appears 1 to 2 weeks after weaning. Also, the post-partum anovulatory estrus is common in sows (Warnick, *et al.,* 1950).

The physiological interactions between lactation and reproductive failure have never been fully established. It is usually assumed that the depression of ovarian function and the failure to maintain the corpus luteum of pregnancy may be connected with pituitary dysfunction as a result of the requirements of both reproduction and lactation. Ovarian cysts may be caused by aberrations of gonadotropin secretion, probably as a result of disturbed function caused by abnormally high lactation. The physiological integrity of the uterus during lactational anestrus is influenced by the parity of the animal and the rate of uterine involution. Very little is known on the physiology and pathology of uterine involution as affected by age, season of parturition, nutrition, previous litter size and placental development. A minimum period, varying with the species and breed, is necessary for satisfactory reproduction following parturition.

Anestrus Due to Aging

Animals under wild conditions presumably die or are killed before reaching the reproductive processes decline due to natural aging. Little is known about the effect of aging on reproduction in domestic animals. In rats the incidence of irregular or anovulatory cycles rises steadily with age and even when they are no longer fertile, mating frequently results in pseudopregnancy (Asdell and Crowell, 1935). In some instances, the menopause in humans is associated with a transition from a polyestrous condition to one approaching the monoestrous state. The cessation of reproductive activity well before the end of life itself appears to be peculiar to man. Also, the cessation of reproductive activity is relatively more gradual in domestic animals than in man.

Ovarian Hypoplasia

Ovarian and testicular hypoplasia occur in Swedish mountain cattle. Cattle with bilateral ovarian hypoplasia have infantile reproductive tracts and never exhibit estrus. There is a marked tendency for the ovarian hypoplasia to be associated with white coat color. The syndrome is inherited as an autosomal recessive.

B. Abnormal Estrus

Anovulatory estrus, short estrus, prolonged estrus, "split estrus," and "quiet ovulations" are not uncommon in domestic mammals. Anovulatory estrus is more common in mares than in other animals and it has been attributed to a low LH output. In young animals, estrus may be of short duration and without well-marked signs, so it is undetected without the presence of the male. It may be so short as to be unobserved if it takes place during the night.

Split estrus commonly observed in healthy mares and during the early spring occurs at all stages of maturation of the ovarian follicles. Prolonged estrus, lasting 10 to 40 days, may also occur in early spring, and in mares used for heavy draft. These abnormalities are due to endocrinological or neuro-endocrinological factors. For example, small lesions induced in the preoptic area of the hypothalamus cause a temporary state of hypersexual behavior in the female rat (Hillarp, *et al.,* 1954).

"Quiet Ovulation"

This syndrome occurs frequently in all farm mammals. The first ovulation of the normal breeding season of the ewe, as with gonadotropin-induced or spontaneous ovulation in anestrus, is not normally accompanied by behavioral estrus. This type invariably ensues when ovulation occurs in the absence of a corpus luteum from a previous cycle. Quiet ovulation may also occur toward the end of the breeding season, probably as a result of estrogen insufficiency. Quiet ovulations occur intermittently between regular estrous periods. Subnormal behavioral estrus with apparently normal ovarian changes characteristic of estrus is a frequent occurrence in the mare.

There is a high incidence of quiet ovulations (68 per cent) in cows milked three times daily (Casida and Wisnicky, 1950). Quiet ovulations occur in 18 per cent of ovulations in cows in the period from parturition to first estrus (Trimberger, 1956). Animals bred at quiet ovulation have a normal conception rate. Quiet ovulation also occurs in your heifers that are at the bottom of the "peck order." The heifer may be frightened by a strange male or a very timid cow will be frightened by all other animals. The neurophysiological mechanisms of quiet ovulations are not known.

Cystic Ovaries

Cystic ovaries may contain numerous small cysts or one or more cysts that are many times larger (up to 4 or 5 cm.) than the normal pre-ovulatory follicles. The condition may be bilateral or unilateral; some follicles may ovulate and others become cystic on the same ovary, or all follicles may be cystic. The cysts go through cycles of development, *i.e.,* they grow to a point then regress. Cystic follicles may be associated with anestrus, prolonged estrous behavior, or cyclic estrous periods which appear normal. The occurrence of cystic follicles with the complete absence of estrus is less frequent and much more difficult to detect from external appearance.

Endocrine dysfunction, *e.g.,* LH deficiency, may cause cystic ovaries or cystic degeneration of the corpus luteum. Cupps, *et al.* (1956), studied the histology of the anterior pituitary and the adrenal cortex in infertile cows exhibiting irregular cycles or absence of estrus. They found a high incidence of atretic follicles. In advanced cases, there was a decreased percentage of small beta and alpha cells in the anterior pituitary, a narrowing of the fascicular zone of the adrenal cortex, together with a high incidence of extramedullary myelopoiesis in the reticular zone. In a study of cystic ovaries in Swedish Red-and-White cattle, there was evidence for a hereditary basis of this syndrome, since the daughters of the affected dams showed a significantly higher frequency of the syndrome than did daughters of normal dams (Henricson, 1957).

Nymphomania

Nymphomania is a common cause of sterility in dairy cattle. It occurs less frequently in beef cattle, horses, and swine. Nymphomaniacs show intense estrous-like behavior persistently or at frequent but irregular intervals. They are frequently as sexually aggressive as bulls in seeking out and in attempting to mount estrous cows, but rarely do they allow others to mount them. In this respect, their behavior differs from cows in true estrus. The nymphomaniac paws, bellows and otherwise exhibits male-like behavior. The voice and general body conformation become male-like. Characteristic symptoms of nymphomania are the relaxation of the pelvic ligaments and the elevation of the tail head. Nymphomania seems to be inherited although it is not associated with inherited milk-producing capacity in dairy cattle.

Several hypotheses have been offered to explain nymphomania (reviewed by Roberts, 1956). A primary dysfunction of the anterior pituitary may cause excess secretion of FSH and a deficiency in secretion of LH, causing excessive growth of the ovarian follicles, high estrogen secretion, but no corpus luteum formation (Garm, 1949). Cupps, *et al.* (1956), reported that nymphomania may be associated with two distinct conditions in the endocrine glands. In nymphomaniacs whose ovaries had no corpora lutea, there was an increase in small beta cells of the anterior pituitary and a hypertrophy of the fascicular zone of the adrenal cortex, together with a large number of follicles in the ovaries. Animals with ovaries containing corpus luteum lacked a large number of follicles, showed hyaline degeneration of many small beta cells of the anterior pituitary, and hypertrophy of the reticular zone of the adrenal cortex.

II. Fertilization Failure and Atypical Fertilization

Fertilization Failure

Fertilization failure may be the result of death of the ovum before sperm entry, structural or functional abnormality in the ovum or sperm, or of immunologic incompatibility between the ovum and sperm.

Aged and Abnormal Ova

There are considerable species and individual variations in the duration of the fertile life of the ova. As the unfertilized ova age, they retain their ability to be fertilized for a longer time than they retain their ability to produce viable embryos. The aging process is a gradual one, in which various functions are successively inhibited. An early effect of aging of the ovum is that the resulting embryo is not fully viable and is resorbed before birth (Young, 1953; Blandau, 1954). Further aging leads to abnormalities in fertilization involving particularly the pronuclei (Blandau, 1952). The rate of biophysical and biochemical reactions associated with sperm entry into the ovum becomes slow; this condition leads to increased polyspermy (Austin and Braden, 1953).

In horses and swine, the longer duration of estrus (2–3 days) makes the time of copulation more critical for fertilization. As the interval from onset of estrus to copulation in the pig increased, there was an increase in the number of ova with more than two pronuclei (Hancock, 1959, 1961). Copulation 44 hours after the onset of estrus was associated with 4 per cent heteroploid embryos (Thibault, 1959; Bomsel-Helmreich, 1962). In such instances the developing embryo degenerates at a later stage of development. The frequency of polyandry seems to be controlled by genetic factors. In rats hyperthermia induced 8 to 10 per cent polyspermic (dispermic) ova in the Sherman and Long-Evans strains, but only 3.5 per cent in the Wistar CF strain (Piko and Bomsel-Helmreich, 1960).

Both ovarian oocytes and tubal ova may undergo fragmentation (cytoplasmic division) apparently spontaneously and often in a manner that superficially resembles normal cleavage (*cf*. Austin, 1961). Fragmentation is more frequent in immature animals and in animals in which superovulation is induced by exogenous gonadotropins. An increase in the frequency of fragmentation occurs as a result of delayed insemination after the time of ovulation (Odor and Blandau, 1956), and as a result of the application of irradiations or radiomimetic agents to sperm before fertilization.

Failure to undergo fertilization and normal embryonic development may be due to inherent abnormalities of the ovum (Chang, 1962; Hartman, 1953) or to environmental factors. For example, fertilization rate is lowered in sheep exposed to elevated ambient temperature (90°F) prior to breeding (Dutt, *et al.*, 1959). Several types of morphologic and functional abnormalities have been observed in the unfertilized ovum, *e.g.*, giant ovum, oval-shaped, lentil-shaped, and ruptured zona pellucida (Fig. 2). In the sheep, morphologically abnormal ova are more frequent early in the sexual season than at later stages (Dutt, 1954; Hulet, *et al.*, 1956).

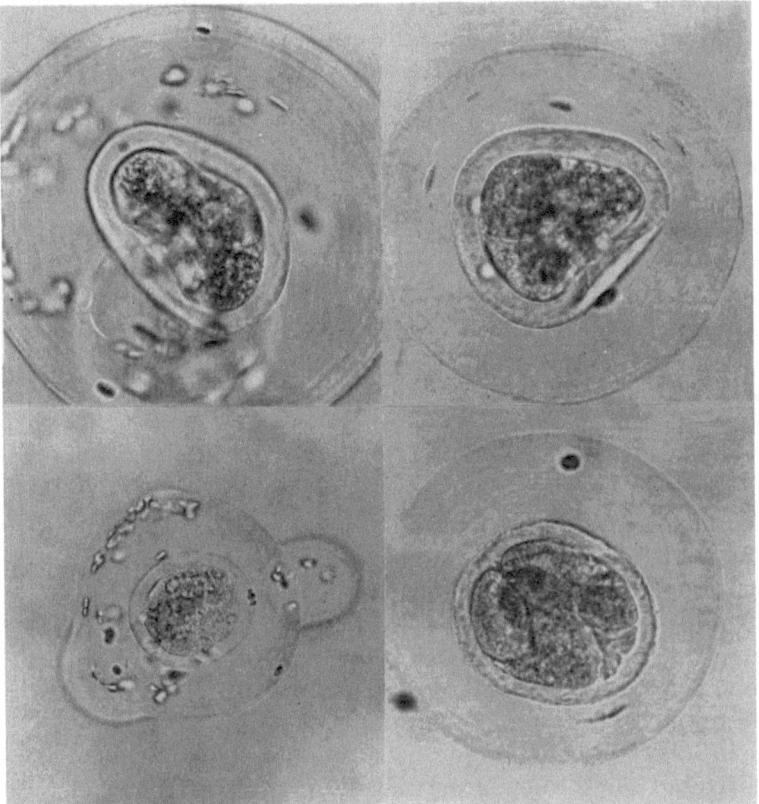

Fig. 2. Different types of degeneration in the fertilized egg collected from super-ovulated rabbits 64 hours after breeding (×200; ×120).

Aged and Abnormal Sperm

With the spread of artificial insemination, fertilization failure could result from qualitative or quantitative deficiencies in the semen. Pig semen stored for three days before insemination produced zygotes much more susceptible to early mortality (Dziuk and Henshaw, 1958). The number of sperm in the zona pellucida of cleaved ova was smaller when old semen was used as compared to fresh semen (First, *et al.*, 1963). It is possible that the reduced DNA in aged sperm is associated with the loss of embryo survivability.

The semen usually contains a varying percentage of morphologically abnormal sperm. Little research has been conducted on the physiological significance of abnormal sperm in relation to fertilization failure and atypical fertilization. With the progress in techniques in fertilization *in vitro,* such studies may be possible. The development of quantitative microscopy has led to the discovery of a characteristic type of abnormal sperm associated with low fertility in man and cattle. Morphologically these sperm appear to have been normally formed, but they contain abnormal and highly variable amounts of DNA and arginine in their nuclei (Leuchtenberger, *et al.*, 1956).

Since there is wide antigenic variability among individual animals, sperm may be regarded as foreign bodies within the reproductive tract. It is possible that certain foreign materials are rejected. This phenomenon may lead to infertility between interspecific matings as a result of antigenic incompatibility between the sperm and recipient female (Bishop and Walton, 1960; Ryle, 1957). The interaction between the sperm and the ovum prior to fertilization is also influenced by an antigen antibody mechanism, a reaction which is highly specific for rejection or acceptance.

Atypical Fertilization

The complex process of fertilization is subject to several possible aberrations: namely, polyspermy, monospermic fertilization (an ovum containing two female pronuclei), failure of pronucleus formation, gynogenesis or androgenesis. Atypical fertilization may occur spontaneously as a result of aging of the gametes or elevation of environmental temperature. It has also been induced experimentally by X-irradiation or the administration of certain toxic substances.

The incidence of polyspermy has been established in several laboratory mammals, but there a few reports in farm animals. Dispermic and trispermic zygotes are recognized by the presence of two or three sperm tails in the cytoplasm. Dispermic zygotes (trinucleated) have been recorded in cattle (Pitkjanen and Ivankov, 1956); pigs (Pitkjanen, 1955); sheep (Pitkjanen, 1958); and rabbits (Amoroso and Parkes, 1947). Such zygotes are probably triploid which perish before birth (Beatty, 1957).

Several types of abnormal fertilization allow some degree of embryonic development: incomplete maturation of the ovum, polyspermy, and failure of the nucleus of the ovum or the sperm to develop. Incomplete maturation of the ovum implies that obstruction of either or both of the polar bodies fails. In such cases, meiosis proceeds to telophase or anaphase, but both chromosome groups remain within the vitellus. The zygotes resulting from such ova are triploid or pentaploid which invariably degenerate during early pregnancy.

III. Prenatal Mortality

Prenatal mortality occurs at all stages of gestation. However, the most common is early embryonic mortality which occurs before or immediately following implantation and results in complete resorption of the conceptus. Early embryonic mortality is of two forms. In the first form, the fertilized ovum develops to the morula or early blastocyst stage but degenerates before the middle of the estrous cycle. The corpus luteum regresses as in normal cycle, when fertilization has not taken place, and the animal returns to estrus after an estrous cycle of a normal length. In the second form, the blastocysts degenerate beyond midcycle, prior to, or immediately following, implantation. In this form, the regression of the corpus luteum is thus delayed for a period which is longer than a length of one estrous cycle.

Estimates of prenatal loss are based on comparisons between the number of ova fertilized and the number of embryos surviving to a later stage, by obtaining data at two or more intervals during early pregnancy. Fertilization rate is calculated as the number of cleaved ova expressed as a percentage of the total number of ova shed. The rate of embryonic mortality is calculated as the number of non-viable embryos expressed as a percentage of the total number of fertilized ova. In domestic animals most of prenatal mortality occurs in the first four weeks of gestation (Table 1). In cattle, the majority of embryonic deaths occur between the 16th and 25th day of single pregnancies (Hawk, *et al.*, 1955). This period coincides with the rapid growth and differentiation of the embryo and the extra-embryonic membranes. The embryo is dependent for its survival on the absorption of nutrients contained in the uterine milk from about the 16th day of gestation (Greenstein and Foley, 1958), until the fetal-

Table 1

ESTIMATES OF PRENATAL MORTALITY IN A MONOTOCOUS AND A POLYTOCOUS SPECIES (From the literature and Hanly, 1961)

Species	Stage of Gestation (weeks)	Estimated Prenatal Mortality (%)
Cattle, nulliparous, with history of infertility	4–5	39–60
Cattle, multiparous, with history of infertility	4–13	15–21
Cattle, nulliparous, without history of infertility	4–5	44–54
Cattle, multiparous, without history of infertility	14–21	20–21
Cattle, multiple pregnancy	4–5	60–70
Swine, nulliparous	4–10	23–48
Swine, multiparous	4	39
Swine, multiparous	birth	41–44
Sheep, single pregnancy	2–3	20
Sheep, single pregnancy	5–6	30
Sheep, multiple pregnancy	5–6	40
Horses	5–6	11

maternal union is established at about 36 days of pregnancy (Foley and Reece, 1953). The death of the embryo during this period may be due to an unfavorable uterine environment. In polytocous species, death of all or part of the litter may occur at any time during gestation; the dead conceptus is frequently resorbed (Fig. 3). In the rabbit, some 10 per cent of the ova are lost before implantation and not less than 35 per cent of the implantations are lost *in toto* (Brambell, 1942). There is comparable prenatal mortality in the large litters of the pig (Lasley, 1957); sheep (Dutt, 1954); and cattle (Hafez, 1964). It seems that domestication has caused a reduction in the rate of embryonic mortality.

There are certain periods when the peak losses occur; these periods vary with the species. In the domestic rabbit, the embryonic loss amounted to 10 per cent of the ova before implantation and 18 per cent of the embryos after implantation; the peak of post-implantation mortality occurs following implantation (Adams, 1960). The significant loss which occurs shortly before implantation is probably due to the failure of synchronization of the transport of the morula through the uterotubal junction, the development of the blastocysts, and the progestational development of the endometrium (Noyes and Dickman, 1960). The marked prenatal mortality which occurs following implantation is

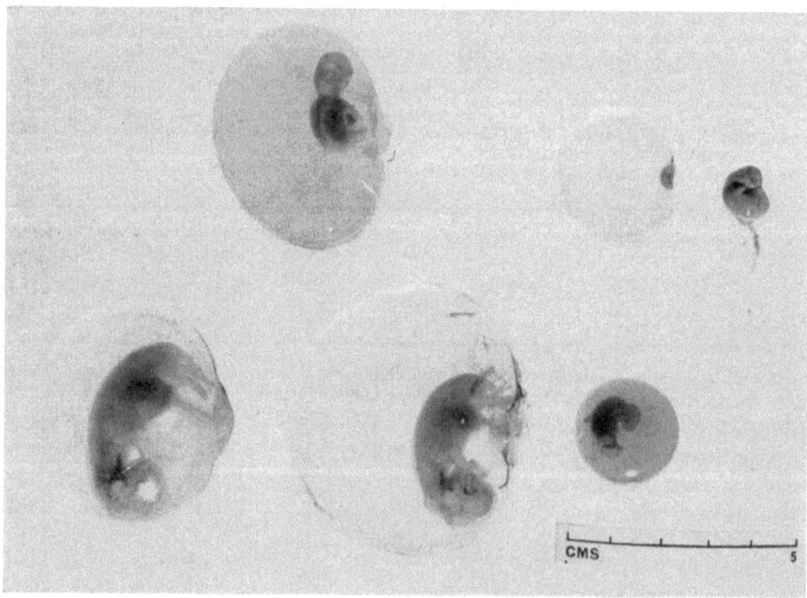

Fig. 3. Different types and degrees of early fetal mortality due to over-crowding *in utero* in sheep. The sheep were injected with pregnant mare serum to induce multiple pregnancy; animals were autopsied after five weeks of gestation.

probably due to deficiency in placental development or inadequacy of endometrial vascular supply (*cf.* Hafez and Tsutsumi, 1966).

A. Physiological Mechanisms and Implantation

Implantation failure occurs in a certain proportion of any population of normal, healthy females. Failure of implantation in a normal female does not necessarily indicate the presence of any permanent factors affecting fertility. A similar proportion of infertility is often encountered following natural breeding; the same animals will conceive later either to normal copulation or to embryo transfer. Adams (1962) transferred different numbers of rabbit embryos and found that varying the number of embryos had no marked effect on the pregnancy rate and that transfer to both uterine horns, as opposed to unilateral transfer, only slightly improved it. This would suggest that failure of implantation, in cases of transient infertility, is connected with systemic rather than with local factors.

Endocrine Factors

The transport of the fertilized ovum through the tube to the uterus is governed by a gradual fall of estrogen level after estrus and an increased production of progesterone as the corpus luteum replaces the ruptured follicle. For about three days after estrus in the rabbit and sheep, a "tube block" exists at the uterotubal junction preventing the embryo and the tubal fluids from entering the uterus (Black and Asdell, 1959; Edgar and Asdell, 1960). Accelerated or delayed transport of the ovum, as a result of estrogen-progesterone imbalance, causes preimplantation death.

The regression of the cyclic corpus luteum in the nonpregnant animal is explained as due to continuous luteolytic effect from the uterus. The presence of a normal embryo in the uterus prevents this action, and it is for this reason that the corpus luteum does not regress during pregnancy (Moor and Rowson, 1964). An abnormally undersized conceptus might not be able to counteract this luteolytic effect with consequent regression of the corpus luteum and termination of pregnancy. Retarded growth of the implanted embryo may be associated with lack of development of binucleate cells which appear in the ruminant trophoblast at about the normal time of corpus luteum regression and which may have either a luteotropic or an adhesive function.

The most critical period for the embryo survival is that of the late blastocyst. Normally, the developing corpus luteum, under the influence of the luteotropic hormone from the pituitary, secretes progesterone, the action of which on the female tract is closely correlated with the development of the embryos. The failure of the blastocyst to implant may be due to the progestational changes in the endometrium at the appropriate time. Thus the pregnancy rate was increased in normal cows by the injection of 100 mg. of progesterone on each of days 2, 3, 4, 6, 9 after breeding (Johnson, *et al.*, 1958).

The neural mechanisms of implantation are of academic significance. Implantation failure results from resection of the pituitary stalk, or from total section of the nerves supplying the uterus. These procedures interrupt the neural pathways from the hypothalamus to the pituitary gland and from the uterus to the hypothalamus respectively, thus preventing the signal, for the release of luteotropic hormone, from reaching the pituitary gland and hence interfering with the production of progesterone.

Implantation failure may result from inadequacies in the maternal environment which cause the death of the blastocyst. For example, the exposure to the odor of males belonging to a different strain from the fertilizing male causes failure of pregnancy in 70 to 80 per cent of recently mated mice (Bruce, 1960). The embryos develop up to blastocyst stage and then perish. Presumably the olfactory stimulus from the strange male causes some endocrine disturbance in the female, which produces an unsuitable uterine environment.

B. Causes of Prenatal Mortality

Prenatal mortality can be due to maternal factors, embryonic factors, or to fetal-maternal interactions. Maternal failure tends to fall on all the embryos as a whole, resulting in a complete loss of pregnancy. Embryonic failure tends to affect individual embryos of a given gestation resulting in partial loss of pregnancy. In other cases the maternal environment is unfavorable and not all implanted embryos can be supported. The heredity and nutrition of the dam, overcrowding *in utero,* hormonal imbalance and thermal stress all contribute to prenatal mortality.

Age of Dam

Prenatal mortality is influenced by the age (Finn, 1962) and parity of the dam (Shikata, 1960). There are species differences in the effect of parity on the incidence of embryonic loss. In cattle (Erb and Holts, 1958) and sheep (Edgar, 1962), embryonic loss is higher in young animals than in old ones. In swine, it is higher in sows than in gilts (Perry, 1956).

Nutrition of the Dam

Caloric intake and specific nutritional deficiencies affect ovulation rate, fertilization rate, as well as prenatal death. In swine, high caloric intake or continuous unlimited feeding increased ovulation rate but also increased the embryonic mortality up to implantation time (Haines, et al., 1955, 1959; Self, et al., 1955; Goode, et al., 1960). Following implantation, fetal death seems to be increased by limited feeding (Waldorf, et al., 1957). In sheep, full feeding before breeding increases ovulation rate and embryonic death (El Sheikh, et al., 1955; Foote, et al., 1959). The effect of caloric intake on prenatal death in cattle is controversial.

Nutrition $\times$ season interactions in prenatal mortality are evident. Rabbits of an inbred strain with a high proportion of fetal atrophy were, when adults, put on a high and low plane of nutrition (Hammond, 1965). The conception rate was lower in the fall and summer than in the winter and spring. Low plane of nutrition depressed the conception rate at all times of the year but particularly during the fall and summer months, while the high plane of nutrition raised the conception rate and reduced seasonal variation.

Embryonic mortality may be induced in mice by fasting (McClure, 1959, 1961). It is possible that the stress of fasting stimulates hypersecretion of the corticosterone in lethal amounts since the injection of corticosterone and ACTH can cause embryonic mortality in rats (Robson and Sharaf, 1952; MacFarlane, et al., 1957).

Heredity

From the biological standpoint, prenatal loss may be desirable in eliminating chromosomally defective individuals at a very early stage of their existence (Baier, 1955; Bishop, 1964). The frequency and repeatability of embryonic loss is partly determined by the genetic make-up of the sire (Perry, 1960) and the dam (Baker, et al., 1958), as well as the breeding system. In cattle, prenatal death in the first half of pregnancy is 27 per cent in inbred dams and some 11 per cent in outbred ones (Hawk, et al., 1955; Mares, et al., 1958).

In pigs the inbreeding of the dam seems to have a greater effect on litter size than the inbreeding of the embryo (Squiers, et al., 1952; King and Young, 1957). There is significant variation in the rate of prenatal loss of the daughters of different sires (Rathnasabapathy, et al., 1956; Reddy, et al., 1958). It may be possible to select sires on the basis of the uterine capacity and efficiency of their daughters. Inbreeding increases the embryonic mortality rate and crossbreeding reduces prenatal death losses (Squiers, et al., 1952). This depression is likely due to the combined losses resulting from maladjustment in the uterine environment and, to a lesser extent, an increase in the number of genetically defective embryos.

Breed differences in the pattern of prenatal mortality have been reported in sheep. There was less loss of embryos which were carried by Columbia ewes than those carried by Hampshires, irrespective of the breed of the embryo (Foot, et al., 1959). There was also an interaction between breed and feeding effects on embryonic mortality. When Columbia and Hampshire ewes carrying Columbia-sired embryos were fed grain, the effect was detrimental. Both breeds of ewes with Hampshire-sired embryos were adversely affected by grain feeding and lost a larger proportion of embryos (Bellows, *et al.*, 1963).

Different breeds of swine vary in their responses to restricted caloric intake. Feeding half the normal amount of protein to animals with high protein requirements resulted in increased embryonic mortality. Under the influence of exogenous progesterone, Chester White gilts have a greater response to PMS than did Poland China gilts (Rigor, *et al.*, 1963; First, *et al.*, 1963). This suggests that breed differences may depend on variation in the gonadal response to a given hormone.

Overcrowding in Utero

Embryonic mortality is caused by local and systemic factors. In the rabbit, there seems to be no uniform pattern for the distribution of early embryonic death along the uterine horn (Fig. 4). However, it was suggested that the end positions in the uterine horns are most favorable for embryonic growth in swine. Transuterine migration of embryo is of special physiological significance in certain polytocous species for equal distribution of conceptuses between the two uterine horns. In pigs there is a high incidence of embryonic mortality where migration does not occur (Smidt, 1962).

With advancing pregnancy, the embryo becomes increasingly dependent upon the placenta for its survival. The degree of placental development is primarily influenced by the availability of space and vascular supply within the uterus. As the number of implantations rise, the vascular supply to each site is reduced; this restricts placental development and causes high embryonic and fetal death (Eckstein, *et al.*, 1955; McLaren and Michie, 1960; Adams, 1962; Hafez, 1964) (Fig. 5).

A high percentage of fused placentae occurs in overcrowded uteri in polytocous species (Hafez, 1964). Different degrees of fused placentae (Fig. 6) were observed in 4 per cent of pregnant mice, more frequently when seven or more feti occupied one uterine horn (Sugiyama, 1961). Fused placentae invariably leads to embryonic or fetal loss in rodents. On the other hand, when the litter size is so small in polytocous species, the fetus becomes oversized and may cause fetal dystocia (Fig. 7).

Thermal Stress

Increased embryonic mortality occurs in a number of species following exposure of the female to elevated ambient temperature (Ulberg, 1958; Dutt, 1960). The rabbit embryo, in early stages of development, is directly affected by the increased maternal body temperature that oc-

Low Crowding   High Crowding

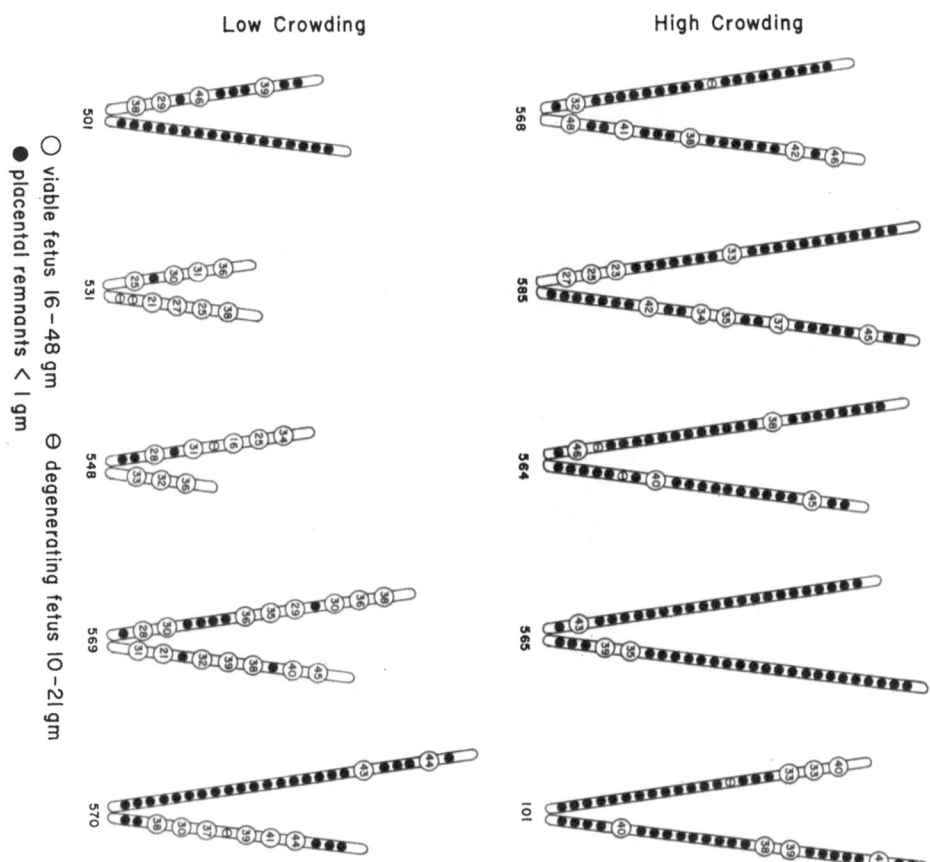

○ viable fetus 16–48 gm ⊖ degenerating fetus 10–21 gm

● placental remnants < 1 gm

Fig. 4. Diagrammatic illustration to show the relationship between over-crowding of implantations *in utero* at nine days *post-coitum* (*p.c.*) and fetal survival at 29 days *post-coitum* (2 days prior to delivery) in which large numbers of embryos were transferred to both uterine horns of recipients. Note that the number of viable fetuses was less in the case of low crowding than in high crowding. Also note that the location of the fetus along the uterine horn had no effect on the survival or weight of 29-day-old fetuses. The figures within circles denote fetal weight at 29 days *p.c.* (Hafez, 1964.)

companies thermal stress of the female (Alliston, *et al.*, 1965). Such effects may not become apparent until later stages of development. Other causes of prenatal mortality, whose mode of physiological action may or may not be the same, may exert a similar delayed effect. For example, ionizing radiation of gametes and zygotes causes increased post-implantation mortality (Chang and Harvey, 1964).

In monotocous species, high proportion of embryonic mortality occurs mainly in tropical areas. It has been estimated at 25 to 35 per cent for sheep under thermal stress. Yeates (1958) exposed ewes to 42°C for

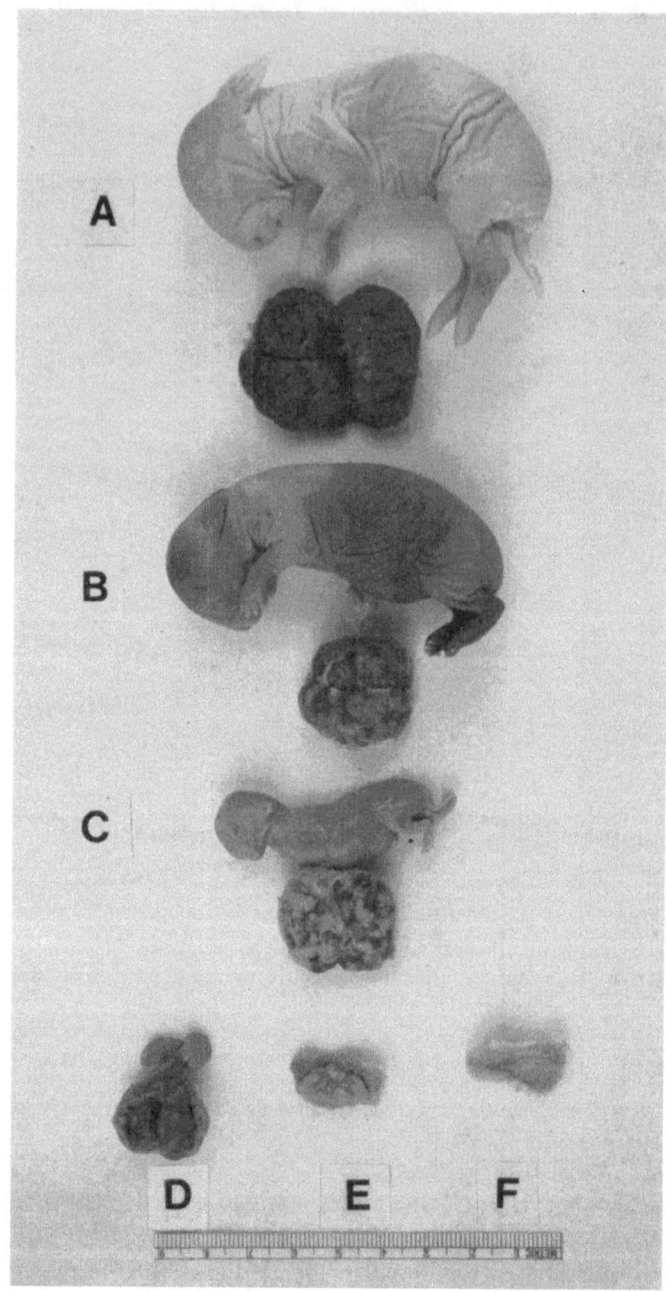

Fig. 5. Viable and degenerating fetuses recovered 29 days *post-coitum* from over-crowded uteri. (A) Viable fetus with a healthy placenta. (B) Fetus and placenta degenerated during late fetal stage. (C) Fetus and placenta degenerated during mid-fetal stage. (D) Fetus and placenta degenerated during early stage. (E) Fetus completely degenerated and the remnants including fetal and maternal placentae. (F) Maternal placenta; fetus and fetal placenta completely degenerated. (Hafez, 1964.)

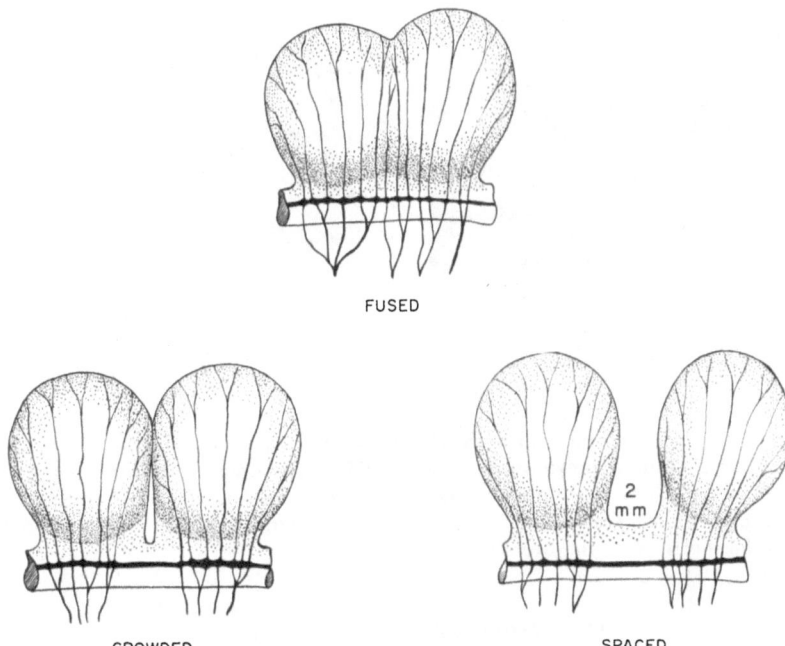

Fig. 6. Diagrammatic illustration of fused, crowded and spaced implantations in experimental recipients to which large numbers of embryos were transferred. (Hafez 1964.)

7 hours each day, starting the day following mating. Heat damage to the embryos was greater when the ewes were under continuous high temperature throughout pregnancy. Cows 4 to 6 months pregnant aborted two days after a 27-hour exposure to 38°C (Ragsdale, *et al.,* 1948). In swine, high temperatures for a few days during mid-pregnancy did not cause abortion or fetal resorption (Heitman, *et al.,* 1951). However, it was reported that larger litters were farrowed from sows following cooling by sprinkling during the summer months.

Semen Quality

The effect of semen characteristics on embryonic mortality is pronounced, especially where artificial insemination is practiced.

In the pig, increased embryonic death has been found with 3-day-old semen as compared to fresh semen (Dziuk and Henshaw, 1958). Within certain limits the volume of semen seems to be of greater importance than sperm number in relation to embryonic death in artificially inseminated sows (Stratman and Self, 1958). In sheep, there was a negative correlation between sperm motility of semen for artificial insemination and embryonic death (Dutt and Simpson, 1957).

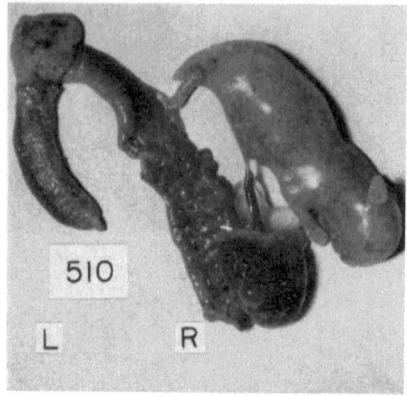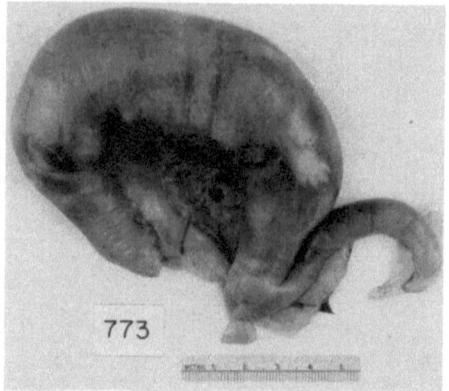

Fig. 7. Oversized rabbit fetuses developing *in utero* and causing fetal dystocia. (A small litter size, in a polytocous species, could be induced by transferring a small number of eggs into a recipient.) *Right:* One fetus in one uterine horn. Note the difference in size of gravid and nongravid uterine horns. *Left:* One fetus freed from the uterine horn. Note the oversized fetus and placenta.

IV. Other Pregnancy Disorders

1. Spontaneous Abortion

Spontaneous abortion is a common cause of reproductive loss in domestic animals; it is highest in cattle, particularly dairy cattle, but occurs also in sheep and horses.

Abortion may be caused by a variety of reproductive diseases as well as physiological disturbances which tend to occur if the animal is bred immediately after puberty or immediately after parturition. Foaling mares bred on the 9th day after foaling had a conception rate of 44 per cent and an abortion rate of 13 per cent; whereas mares bred after a longer interval of time had a 67 per cent conception rate with an abortion rate only one-fourth as great as in the former group (Jennings, 1941). During the fifth and tenth months of pregnancy, the mares are endocrinologically susceptible to abortion owing to hormonal deficiencies. It is well-known that abortions can be produced experimentally by alteration of the endocrine system, *e.g.*, removal of the corpus luteum or injection of massive doses of estrogen.

Abortion may be caused by hereditary factors which cause abnormal formation of some vital organ or general lowered viability of the fetus. These factors are not understood at the present time. Nutritional deficiencies in the pregnant female have been used to produce abortions experimentally. These deficiencies are usually much greater than that found in rations normally fed to animals. However, severe vitamin A deficiency causes abortion in swine.

2. Metabolic Disorders of Pregnancy

Metabolic disorders associated with reproductive failure in farm animals fall into two main groups. In the first group neuromuscular disorders are associated with disturbances in the metabolism of calcium, magnesium, and phosphorus. These syndromes are observed chiefly prior to and at the time of parturition; *e.g.*, lactation tetany of cattle; lambing sickness of sheep; milk fever of cattle, horses and pigs; and eclampsia in dogs. The second group is comprised of pregnancy toxaemia which is most common in sheep.

Acetonaemia in high-producing cattle is always associated with parturition and is sometimes accompanied by nervous symptoms. The syndrome is characterized by marked hypoglycaemia and acetonaemia and is thought to be due to a deranged carbohydrate metabolism. Pregnancy toxaemia (twin-lamb disease, ketosis of pregnancy, prepartum paralysis) occurs in sheep in the latter two-fifths of pregnancy as a result of multiple pregnancy, sudden change in caloric intake, or sudden restriction of exercise such as occurs with heavy snowfall. Very little is known about placental pathology associated with pregnancy toxaemia. In the human placenta, the essential lesion in pregnancy toxaemia is the premature degeneration of the synctial trophoblast, with nuclear disappearance, leaving a thin layer of cytoplasm covering the villus (Hertig, 1960).

Other metabolic diseases may also cause infertility. For example, diabetes in women causes abortion, fetal death, and premature delivery. Similar disorders occur in many mammalian species under experimental diabetes (Miller, 1947; Davis, *et al.*, 1947; Lindan and Morgan, 1950). The disease results in disturbance in gonadal function associated with ovarian structural changes (Shipley and Danley, 1947; Lawrence and Contopoulos, 1960).

3. Immunological Incompatibility

Immunological incompatibilities may block fertilization (prezygotic selection) or cause embryonic, fetal, or neonatal death. Homozygosity for blood groups is associated with an increase in embryonic death in cattle (Conneally, *et al.*, 1963).

Electrophoretic differences in the transferrin (β-globulin) in sera are genetically controlled; this has been expounded in terms of biochemical polymorphism (Ogden, 1959, 1961). Individuals can thus be classified according to their transferrin type. Transferrin locus influences reproductive performance in domestic animals. In cattle, this influence is exerted by reducing the fertilization rate and by increasing embryonic mortality in certain crosses of β-globulin genotypes (Ashton, 1961; Ashton and Fallon, 1962). From artificial insemination records of cattle, the figure for prenatal loss in the first third of pregnancy agrees with that which can be estimated as being due to fetal-maternal β-globulin incompatibility. In swine, reproductive failure in the BB $\times$ AB transferrin matings is due to higher embryonic mortality rather than to fertilization failures (Kristjansson, 1964).

Cattle sera can be classified quantitatively into J-antigen positive and J-antigen negative types; this phenomenon is associated with reproductive failure (reviewed by Schmid, 1963). As J-antigen negative cows could produce J-antibody, Jamieson (1960) postulated that cows capable of producing J-antibody might show a reproductive failure when inseminated with semen from a J-antigen bull or caused to conceive J-antigen embryos. Blood group systems closely related serologically to J are known in a variety of domestic animals (Stone, 1962), and these substances are closely related to human (Neimann-Sørensen, *et al.*, 1954).

Fetal-maternal incompatibility due to isoimmune antibodies has a significant effect on fetal mortality in cattle (Conneally, *et al.*, 1963).

4. Fetal Mummification

This syndrome occurs during early or late fetal life. The fetus becomes dry and compressed, the fetal membranes become tightly pressed around the fetus, and the placental fluids are resorbed. Some cases are due to infection in the gravid uterus, while other cases are not.

5. Retained Placenta

Retention of the fetal placenta following normal delivery of young is a common occurrence in ruminants with cotyledonary placenta. This occurs when the chorionic villi fail to become detached from the maternal caruncles. Normally the shrinkage of the villi after parturition occurs due to the cessation of fetal blood flow to that area.

Infections of certain organisms such as *Brucella abortus* or *Vibrio fetus* cause swelling of the placental tissues with subsequent retention of the placenta. Retained placentae may also occur independent of any specific infectious agent. In range cattle under poor nutritive conditions, calves may be born at term and normal with a high percentage of retained placentae in the dams.

Retention of the placenta causes pyometra and a delay in involution of the uterus and enclosure of the cervix.

6. Perinatal Mortality

Asphyxia of the offspring at the time of parturition, nutrition and age of the dam, hereditary factors and disease appear to be major factors contributing to perinatal mortality. In cattle, the incidence of perinatal mortality is greater in primiparous animals than in multiparae and in male fetuses than in female ones. In swine, the incidence of stillbirths increases in sows with prolonged farrowings; the frequency of stillbirths tends to decrease in sows injected with oxytocin after the spontaneous onset of uterine contractions to assist in, or to hasten, the completion of the farrowing process.

Subterranean clover of Western Australia is rich in plant estrogens and this feed has been implicated in large outbreaks of dystocia and perinatal mortality in sheep.

V. Neonatal Mortality

Failure in thermoregulatory functions and hemolytic diseases cause neonatal mortality in the first week of life. The neonate must make thermoregulatory adjustments in an environment where it is exposed to fluctuating conditions, in contrast to the relatively constant temperature and food supply *in utero* during pregnancy. The efficiency of such adjustments depends primarily on the degree of physiological immaturity of the species at birth. Swine and sheep at birth are particularly susceptible to low ambient temperature. Under hypothermia, the rectal temperature of lambs falls 2° to 3°C and that of piglets 2° to 5° in the first hour after birth.

Since a newborn is so much smaller than its parent, the absolute thickness of its body coat is less than that of the adult. The complement of subcutaneous fat, liver glycogen and coat cover of the neonate are greatly influenced by the degree of physiological maturity at birth. In pigs and many rodents, the neonate is naked and subcutaneous fat is often lacking. Such characteristics are of great importance for survival of the neonate under cold stress. Newborn sheep (*cf.* Alexander and Peterson, 1961) and cattle are not well adapted to withstand high temperature early in life; lambs between 2 and 7 days of age cannot survive longer than about 2 hours at 38°C dry bulb or 3 hours of exposure to direct solar radiation. The neonate is probably more susceptible to cold and excessive solar radiation than to heat.

Poor maternal behavior, when associated with a complicating factor, such as long labor, causes high neonatal mortality. Malpresentation contributes to long labor and stillbirth. Neonatal mortality in lambs results from starvation as indicated by empty stomachs, by depletion of perirenal adipose tissues, by low blood sugar and high blood urea levels immediately after death, and by conclusive symptoms (Alexander, 1961). Neonatal mortality may be a result of weakness of the mother or the young, or bacterial infection through the umbilical cord of young.

The hemolytic disease of the neonate in man is a result of isoimmunization during pregnancy (Race and Sanger, 1958). A similar syndrome is found in foals and mule foals; the antibodies formed by the mare against the cells of the fetus are transmitted to the newborn only through the milk, during the first 2 days of life (Caroli and Bessis, 1947). The use of foster mothers has become an effective control measure (Cronin, 1955). Hemolytic disease of the neonate has been induced in other species (reviewed by Stone and Irwin, 1963), *e.g.*, dogs and swine, in which the placenta is either slightly permeable (Myers and Segre, 1963) or not permeable to antibodies; the antibodies reach the neonate through the mother's milk. Thus, the antibodies are formed from immunization of the female with blood of the sire or similar blood. In other species, such as cattle and sheep, the neonate are protected by the neutralization of antibodies during their passage through the alimentary canal after ingestion of the mother's milk, preventing them from reaching the erythrocytes of the neonate causing their destruction. The physiological explanation of these species differences are not yet established.

VI. ENVIRONMENTAL STRESS AND REPRODUCTIVE FAILURE

Reproductive failure could be attributed to stress of the physical environment or to nutritional deficiencies.

1. Climatic Stress

Seasonal and diurnal fluctuations in the physical environment, particularly environmental temperature, exert pronounced effects on the breeding season, fertilization rate, embryonic mortality and prenatal development. Climatic effects on reproductive failure are not marked in moderate and cold stress. The extent of the effects of thermal stress on reproductive failure varies with the species, breed of animals, temperature fluctuations, and period of exposure of thermal stress.

High environmental temperatures cause structural abnormalities of ova, *e.g.*, shrunken cytoplasm, enlarged cytoplasmic globules, ruptured vitelline membranes or ruptured zona pellucida. Fertilization rate is reduced when sheep are exposed to 32°C on the 12th day of the cycle before breeding (Dutt, *et al.*, 1959; Dutt, 1960). Whether the harmful effect on fertilization results directly from hyperthermia or from an endocrine imbalance resulting in an unfavorable environment in the uterine tubes is not as yet established. Hyperthermia induced in rats before and during ovulation reduced the proportion of penetrated ova, whereas hyperthermia after ovulation increased polyspermy (Austin, 1956). In certain mammals, high temperature lowers pregnancy rate, decreases litter size, increases abortion and fetal resorption (Macfarlane, *et al.*, 1957).

Sheep, which are seasonal breeders, are most sensitive to climatic components causing reproductive failure. The failure may be due to disturbed endocrine mechanism causing regular estrus or faulty sperm transport or due to damage to the gametes. The initial cleavage stages are particularly sensitive to thermal effects (Ulberg, 1962; Woody and Ulberg, 1964).

Thermal stress is also detrimental to fetal survival. In pregnant animals kept in hot rooms the allantochorion was disproportionately large in relation to the size of the embryo, and it was suggested that this may have put an excessive stress on the embryonic heart. The adverse effects of high temperatures on embryonic survival could be reduced in sheep by the administration of thyroxine (Ryle, 1961, 1962, 1963) by preventing the retarded growth of the fetus.

The physiological integrity of the uterus is markedly affected by thermal stress. Embryos from donor rabbits heated for 6 days after conceptions developed normally when transplanted into unheated recipients; but when embryos from unheated donors were transplanted into heated pseudopregnant females, most of the young were resorbed (Shah, 1956).

Exposure of ewes to heat stress during pregnancy results in the birth of underweight lambs, the degree of reduction in weight being proportional to the length of the exposure period. That this dwarfing of lambs is indeed a specific effect of temperature and not a concomitant effect of reduced feed intake in the ewe has been demonstrated by similarity of body weights of the two groups of ewes in psychometric room studies (Yeates, 1958). Prostration of the neonate due to heat stress is of special economic significance in several animal breeding centers in the world.

2. Nutritional Stress

Reproductive failure may occur as a result of severe restriction in caloric intake, or of deficiencies in protein, minerals or vitamins. In Australia and South Africa, livestock being dependent mainly upon pasture for food, suffer from the limitations imposed on pasture growth by widely varying rainfall and temperatures. The resulting nutritional factors act directly, either alone or in combination with the high ambient temperatures.

In prepubertal animals, restricted caloric intake may cause hypoplasia of the reproductive organs and delayed puberty (Joubert, 1954; Crichton, *et al.*, 1959; Sorenson, *et al.*, 1959). In rats, restricting the caloric intake to 70 per cent of the *ad libitum* levels between weaning and the day of parturition reduces mammary gland growth more than total body weight (Sykes, *et al.*, 1948). In mature animals such restriction causes anestrus, irregular estrus, reduction in ovulation rate and conception rate, retardation in mammary gland development, and/or increased prenatal and perinatal death. In the adult sheep undernutrition may cause "quiet ovulation" especially near the onset of the breeding season. Fasting mice for as little as 48 hours terminates pregnancy within a few hours after the end of the fasting period. The fasting causes haemorrhage and necrosis of the deciduomata, leucocytic invasion, and liquefaction and absorption of the debris (McClure, 1961).

Protein deficiency has similar effects on low-caloric intake in cattle and swine. Heifers raised on protein-deficient diets show no symptoms of estrus, and the ovaries and uterus become infantile. The impairment of reproductive performance due to low-caloric or protein intake is reversible, since adequate feeding usually restores normal function.

Ruminants which consume roughage are more likely to suffer from phosphorus deficiency than from calcium deficiency, whereas swine, which feed chiefly on concentrates are more exposed to calcium deficiency. In range cattle, phosphorus deficiency causes ovarian dysfunction, which in turn results in delayed puberty and an increase in quiet ovulations, irregular estrus, and eventually complete cessation of estrus. A high-calcium intake may bring about such a reduction in the phosphate utilization. In swine and laboratory mammals, manganese deficiency causes disturbances in the ovarian function. Copper and cobalt deficiencies result in reduced viability of the offspring, the latter deficiency being normally restricted to ruminants. In certain areas of New Zealand, where sheep are grazing on pumice soils, selenium deficiencies cause embryonic mortality at 3 to 4 weeks of gestation.

Vitamin inadequacies may reduce ovulation rate in polytocous species. For example, ovulation rate in swine is reduced as a result of deficiencies in vitamin B_{12} (Johnsen, *et al.*, 1952) or other unidentified factors (Teague, 1955). Vitamin A and E deficiencies cause irregular estrus, anestrus, retardation in prenatal development and lowered viability of the neonate in swine. Cows deficient in vitamin A may produce calves with low viability while severe fetal resorption and fetal abnormalities are not uncommon. Keratinization of the vaginal epithelium may occur, predisposing to infection.

3. Plant Estrogens

Reproductive failure occurs in sheep and cattle grazing pastures and forages with estrogenic activity. In sheep, the syndrome is characterized by dystocia, uterine prolapse and neonatal mortality of lambs; and lactation in virgin and nonpregnant ewes. The endometrium is characterized by cystic glandular hyperplasia of variable degrees (Bennetts, *et al.*, 1946). The lesions in the uteri from affected ewes are characterized by the juxtaposition of markedly cystic and/or normal glands (reviewed by Moule, *et al.*, 1963). The epithelial lining of the cystic glands is flattened and devoid of mitotic activity. A similar hyper-estrogenic syndrome occurs in cattle fed alfalfa in Israel. The syndrome is characterized by infertility, nymphomania associated with both normal and abnormal estrous cycles, swollen vulva with hyperemic mucous membranes, cystic ovaries, hydrosalpinx, and hyperplasia of the epithelium of some endometrial glands (Adler and Trainin, 1960).

The physiological mechanism of the reproductive failure has not been fully elucidated. It may be due to aberrations in the estrous cycle, abnormal transport of ova or sperm, or to implantation failure. The cystic endometrial hyperplasia induced by clover feeding or by estrogen injections is more severe in ovariectomized than in intact guinea-pigs (Braden and Peterson, 1953). This led to the suggestion that ewes may be more susceptible to the estrogens obtained from clover pastures during the anestrous period than during the breeding season. Cystic glandular hyperplasia of the endometrium in guinea-pigs has been induced experimentally in guinea-pigs fed on dried subterranean clover for 1 to 2 months (Curnow and Robinson, 1948).

The estrogenic activity is usually due to plant estrogens—isoflavones and related substances with hydroxy groups. Most of these plant estrogens have relatively low activity, but are present in the plants in considerable amounts. Substances isolated from forages and shown to have estrogenic activity when fed to mice are genistein, biochanin A, daidzein, and coumestrol. The first three are isoflavones; coumestrol is a benzo-furano-coumarin. Genistein (5:7:4'-trihydroxy isoflavone) isolated from subterranean clover is estrogenic (Bradbury and White, 1954; Curnow, 1954). The possible biosynthetic routes of these compounds has been reviewed (Grisebach and Ollis, 1961).

Estrogenic activity or estrogenic substances are found in several plant materials used for human consumption, *e.g.*, barley grain (*Hordeum vulgare*), oat grain (*Avena sativa*), the fruits of the apple (*Pyrus malus*) and cherry (*Prunus avium*), the tuber of the potato (*Solanum tuberosum*) (Bradbury and White, 1954), and Bengal grain (*Cicer arietinum*) (Siddiqui, 1945). There are probably a number of plant compounds with estrogenic activity still awaiting investigation.

VII. ANATOMICAL AND HEREDITARY DEFECTS

1. Congenital and Acquired Defects

Several types of anomalies of development affect all the reproductive organs of all species. These syndromes are expressed in different degrees and some of them impair fertility in varying degrees. Among the more common acquired anatomical abnormalities are: tubal blocks, adhesions of the infundibulum to the ovary or uterine horns which interfere with the pick up of the ovum; or some mechanical destruction of part of the

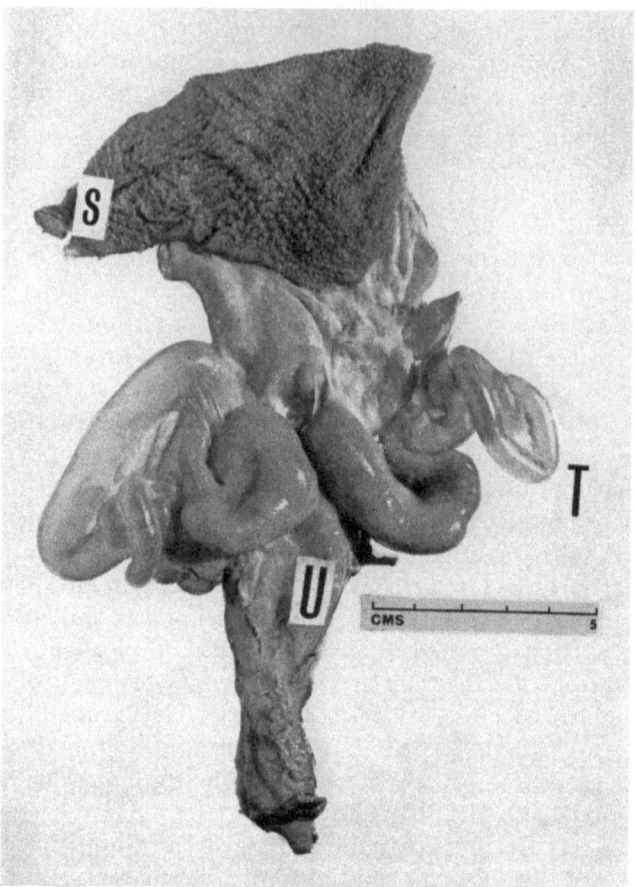

Fig. 8. Anatomical infertility in sheep due to hydrosalpinx, accumulation of luminal fluid within the oviduct (T). Note adhesions of oviduct to stomach (S). (U = uterus.)

reproductive duct system due to a previous pregnancy. *Hydrosalpinx* (the accumulation of fluid in the tubes) (Fig. 8) and *pyosalpinx* (accumulation of fluid with leucocytes and cellular debris) are the most common cause of anatomical sterility. It is not established whether these disorders are due to a congenital lesion or to other factors. Bilateral missing segments and occlusion of tubular parts of the reproductive tract are also causes of anatomical sterility.

Abnormal shape or position of the cervix, narrow cervical canal preventing the transport of sperm to the fallopian tubes, or torn cervix cause reproductive failure. Other congenital abnormalities, such as double os cervix, interfere with reproduction but do not always cause infertility. In animals with infantile ovaries, the ovarian follicles fail to reach ovulatory size and thus the duct system remains immature due to the absence of estrogen stimulation.

Small unilateral or bilateral cysts found in the mesosalpinx or on the fallopian tubes do not interfere with fertility. Large cysts may occlude the lumen of the tube and prevent egg transport. The origin of these cysts is obscure.

A classical congenital anomaly is the "white heifer disease" in cattle, in which the prenatal development of the Müllerian ducts is arrested. The vaginal canal is obstructed from the presence of an abnormally developed hymen. The syndrome is expressed in different degrees according to the time when prenatal development is arrested. The degree and area of hypoplasia differs so that varying anomalies of the fallopian tubes, uterus, cervix, and vagina form. The condition is differentiated from that of the freemartin by the presence of normal ovaries and the normal structure of the vulva and labia. This syndrome is associated with the possession of the gene for white coat color in the females.

2. Hereditary Defects

The phenotypic expression of genetic influence on reproductive failure includes the absence of gonadal tissue, the production of abnormal gametes, or the production of ova which can be fertilized and then give rise to inherently defective embryos (Fig. 9). The two latter types may

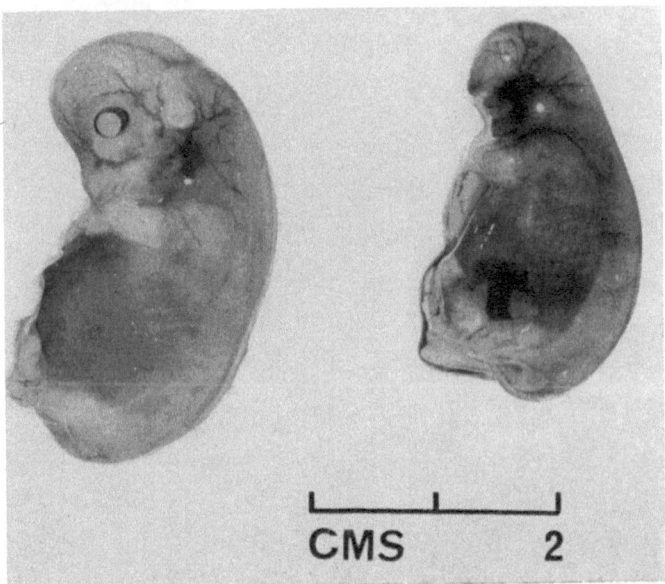

Fig. 9. Congenital abnormality (facelessness) expressed in the prenatal life of a five-week-old sheep fetus (*right*); and normal fetus (*left*).

degenerate at any stage of pregnancy or may be viable at birth but do not necessarily develop to maturity (Fig. 10). Lethal factors are Mendelian units which cause death of an organism prior to the reproductive stage (Hadorn, 1961). Traits of a recessive nature occur in many species as causing resorption and probably are the cause of reproductive failure usually associated with inbreeding.

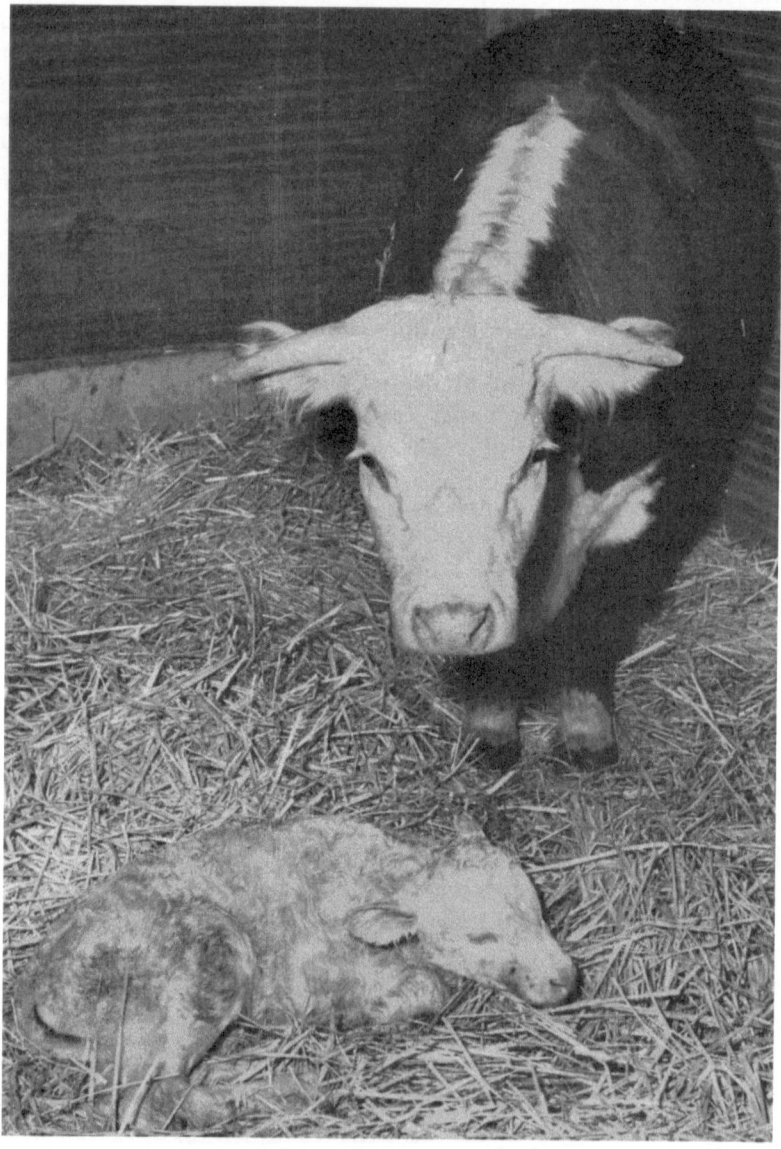

Fig. 10. Albino-dwarf neonate calf with its dam which was a purebred Hereford. This trait causes a high percentage of postnatal mortality in some herds.

Inherited lethal traits may be expressed before or at the time of birth. Some examples of inherited lethals or semi-lethals in the ewe include muscle contracture, earless, cleft palate, paralysis, rigid fetlocks, amputated, "gray lethal," dwarfism, nervous incoordination, congenital photosensitivity, and blindness (Rae, 1956). Certain lethals are breed characteristics. The bulldog calf (shortlegs) occurs in the Dexter dairy cattle in the heterozygous form. The mummification of the fetus during mid-pregnancy occurs in beef cattle breeds. The incidence of these, however, is low except where inbreeding is high. The gene for muscular hypertrophy (in Germany, Doppellendigkeit) has the great advantage of producing slaughter cattle with more lean meat, less fat and bone, and a higher proportion of high-priced cuts than normal animals (Mason, 1963). Meanwhile this syndrome results in poor fertility, difficult births, low milk yield and delicate calves (Koch, *et al.*, 1957).

Hereditary ovarian hypoplasia occurs in certain strains of Swedish mountain cattle, the left ovary being most frequently affected. The gene of white color in Shorthorns (Rendel, 1952) and in the Belgian Blue breed (Hanset, 1959–60) is closely associated with the "white heifer disease." The exact genetic situation is not clear but it is certain that the defect could be eliminated by discarding the white and roan animals and basing the breed on the whole color (red or black) alone.

VIII. Intersexuality

Intersexuality refers to congenital malformation of sexual development with some form of bisexual manifestation. Spontaneous intersexuality occurs at different stages of the prenatal development in all mammals. The experimental induction of intersexuality in domestic mammals by the injection of gonadal hormones into pregnant females or by parabiosis has been unsuccessful. However gonadal inversions by gonadal hormones were reported in amphibia, birds, and the opossum. Experimental induction of intersexuality in the mouse was also successful when eight-cell embryos of different strains became fused in culture, then transplanted at the blastocyst stage to recipient females (Tarkowski, 1961).

The classification of intersexuality is based on the nuclear sex, type of gonadal tissue, and extra genital structures and sexual behavior of the animal (Table 2). Freemartinism, true hermaphrodites and pseudo-hermaphrodites are the most common syndromes of intersexuality in domestic animals (Overzier, 1963). In true hermaphrodites both ovarian and testicular tissues are present. In bilateral hermaphrodites testicular and ovarian tissues are present on both sides. In unilateral hermaphrodites the testicular and ovarian tissue occur on one side; testicular or ovarian tissue on the other side. In lateral hermaphrodites the testicular tissue is found on one side, and the ovarian tissue on the other side.

Table 2

CLINICAL CLASSIFICATION OF INTERSEXUALITY IN DOMESTIC MAMMALS (Cf. Hafez and Jainudeen, 1966)

Syndrome	Nuclear Sex	Gonadal Tissue	Genital Tract	External Genitalia	Sexual Behavior
True hermaphrodite	female (common) male (rare)	bilateral* unilateral lateral	intersexual	female structures predominate	male, female or intersexual
Pseudohermaphrodite					
male	female (pigs)	testes	intersexual	female	male or female
female	female	ovary	intersexual	male	female
Freemartin	female	ovary or ovotestes	intersexual	female	male

* bilateral = testicular and ovarian tissue on both sides.
unilateral = testicular and ovarian tissue on one side; testicular or ovarian tissue on the other side.
lateral = testicular tissue on one side and ovarian tissue on the other side.

The histology of the hermaphrodite gonad ranges widely from testicular and ovarian components that are distinctly developed to ones that are completely undefinable. Pseudohermaphrodites possess gonads of one sex, but the external genitalia and secondary sexual characters resemble those of the opposite sex.

1. Freemartinism

Freemartinism is unique to the bovine, although a few cases have been reported in sheep, goats, pigs and horses. The freemartin is a female twin in which the development of the gonad has been controlled by the intercirculatory system of male and female twin fetuses. The syndrome is characterized by: (1) internal reproductive organs of both sexes; (2) ovotestes with varying degrees of transformation toward the male; and (3) external genitalia essentially like those of a normal female (Figs. 11, 12). Ninety to 95 per cent of the bovine twins are monochorial; i.e., the circulation of the two allantoic membranes join together resulting in a constant interchange of blood between the two conceptuses. The stage in prenatal development and the degree of vascular anastomosis vary with different freemartins. Thus the syndrome is expressed in varying degrees from females that are perfectly fertile to infertile types which possess external female genitalia and typically male sex cords but without germ cells.

The gonads of the freemartin are intra-abdominal and they never

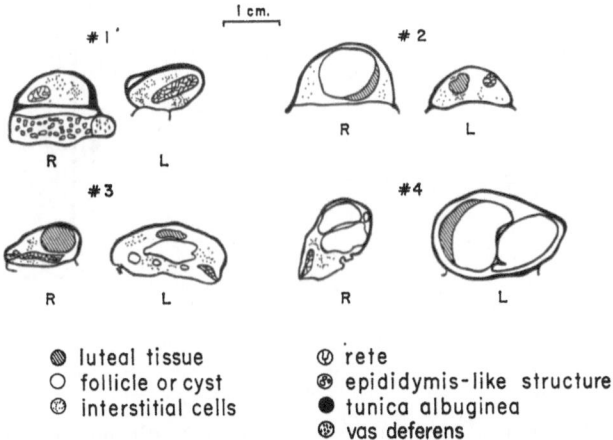

◐ luteal tissue	⊕ rete
○ follicle or cyst	⊛ epididymis-like structure
⊛ interstitial cells	● tunica albuginea
	⊕ vas deferens

Fig. 11. Cross section gonads of freemartin quintuplets (four female and one male) with low degree of transformation from female to male. Note the different degrees of transformation among the different animals. The gonads were taken from one-year-old freemartins. (Rajakoski and Hafez, 1963.)

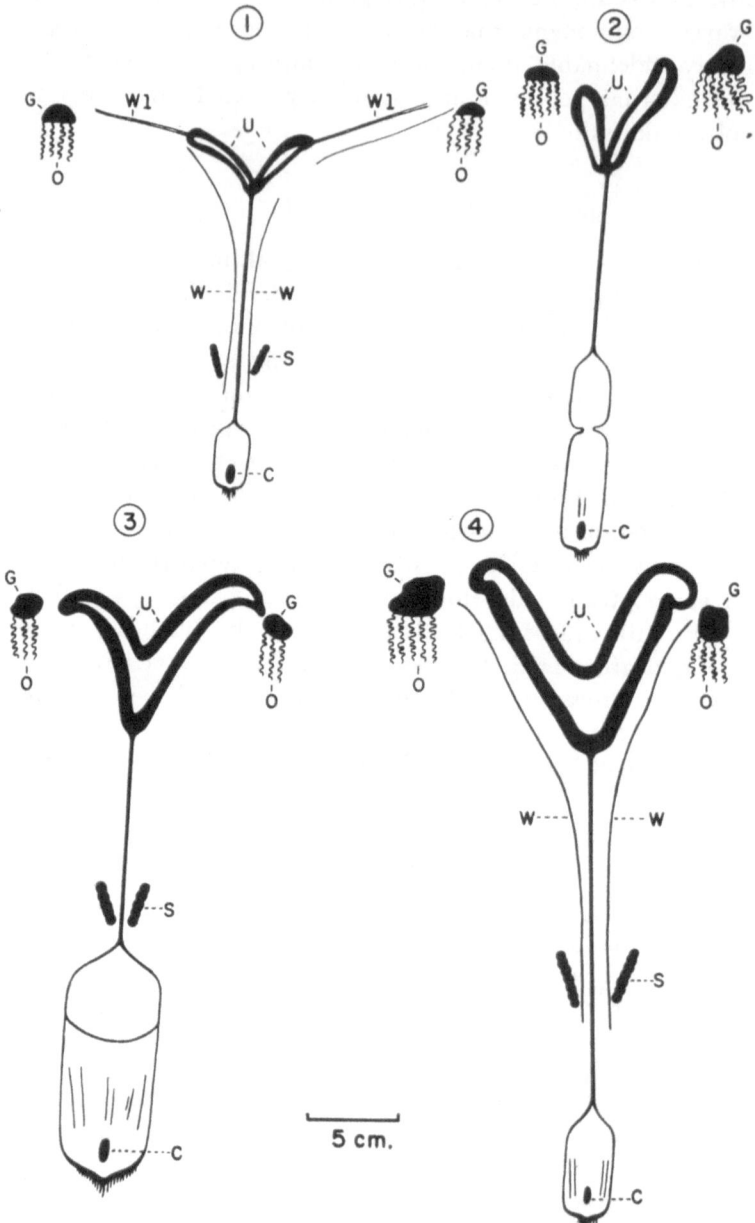

Fig. 12. Cross section of reproductive tracts of freemartin quintuplets (four female and one male) with low degree of transformation from female to male. Note the different degrees of transformation among the different animals. The reproductive tracts were taken from one-year-old freemartins. (Rajakoski and Hafez, 1963.)

descend within the pelvic cavity. The gonads resemble the testes in structure with an increase in the amount of interstitial cells. In extreme cases of freemartinism the gonads lack the germinal epithelium, secondary tunica albuginea, ovarian cortex, and follicles. The gonads may contain mesothelium, primary tunica, tunica vasculosa, interstitial cells, primary sex cords, centrally located rete, epididymis, and ductus deferens. Mild degrees of freemartinism are characterized by the partial presence of ovarian cortex with germinal epithelium, secondary tunica, and primordial ovarian follicles with oocytes and nuclei in prophase of heterotypic division and a more peripherally located rete. Histological transformation of gonads parallels the anatomical transformation of reproductive organs. One male embryo masculinized four female quintuplets to a lesser degree than that generally reported in the literature for twins and triplets (Rajakoski and Hafez, 1964; and Figs. 13, 14).

The cytological characteristics of the freemartin gonads are of special academic interest. The mitotic activity seems slower in freemartin fetal ovaries than in normal ovaries. The chromosomal stages, prophase, metaphase, anaphase, and telophase of the oocytes are less numerous in the freemartin fetuses than in normal (Rajakoski and Hafez, 1964). The testes of newborn male heterosexual twins exhibit chimerism: that is, they contain both XX and YY cells. However, Ohno, *et al.* (1962), found no XY metaphase figures in the gonads of three freemartin calves, even though these calves showed clear chimerism of their bone marrow cells. The gonad of adult freemartins apparently lacks germ cells of either XX or XY composition.

The Wolffian and Müllerian ducts in the freemartin are poorly developed. The abnormality is apparent very early in embryonic life. The external genitalia, with few exceptions, resemble that of the normal female. Frequently the clitoris is enlarged and the tuft of hair at the ventral commissure of the vulva is enlarged. The external genitalia leads into a short vagina. A rudimentary penis may take the form of a skinfold from the navel to a point above the rear attachment of the udder. The mammary glands are underdeveloped and could be felt by palpation.

Freemartins do not experience behavioral estrus; and although they mount estrous animals they are not mounted by other animals.

Physiological Mechanisms of the Syndrome

The sex chromatin pattern (Moore, *et al.*, 1957) and the sex chromosome pattern of the freemartin are those of a normal female. The physiological mechanisms responsible for freemartinism are not precisely identified. Two theories have been advanced to explain the freemartin condition: the "hormonal theory," and the "sex-chromosome mosaicism."

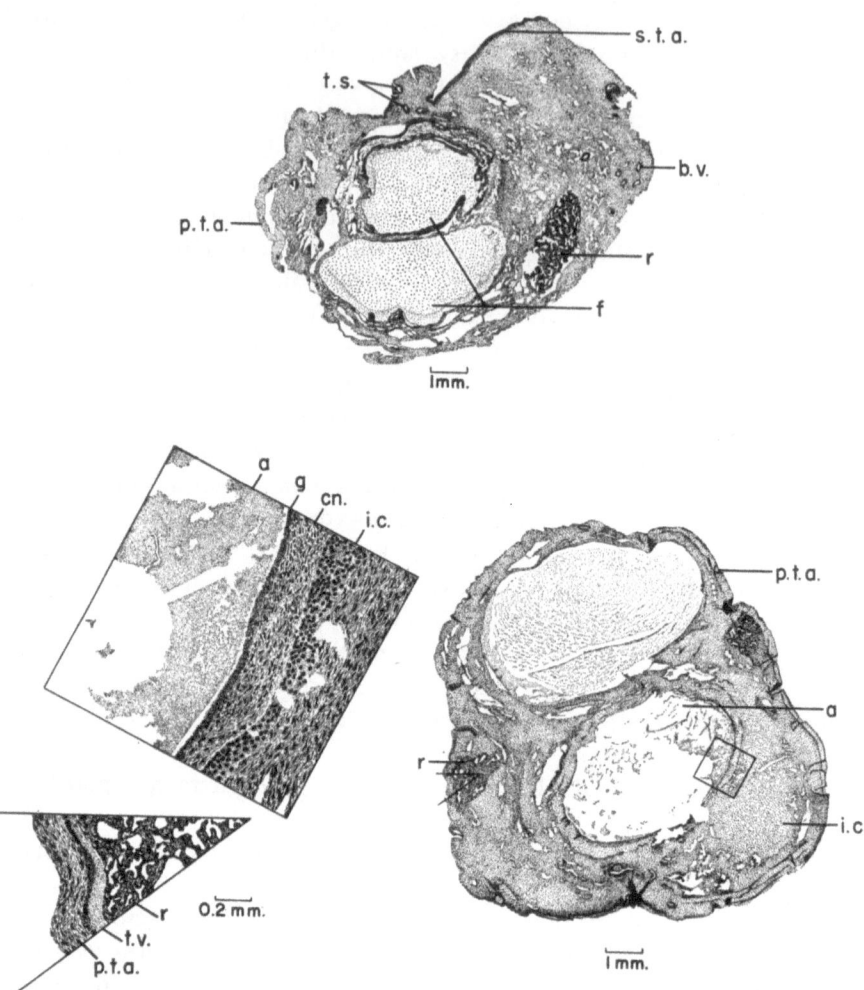

Fig. 13. Camera lucida drawing of two gonads of freemartin quintuplet cattle.
(Rajakoski and Hafez, 1963.)

a. = antrum	p.t.a. = primary tunica albuginea
b.v. = blood vessels	r. = rete
cn. = connective tissue	s.t.a. = secondary tunica albuginea
f. = follicle	t.s. = tubal system
g. = granulosa cells	t.v. = tunica vasculosa
i.c. = "interstitial cells"	

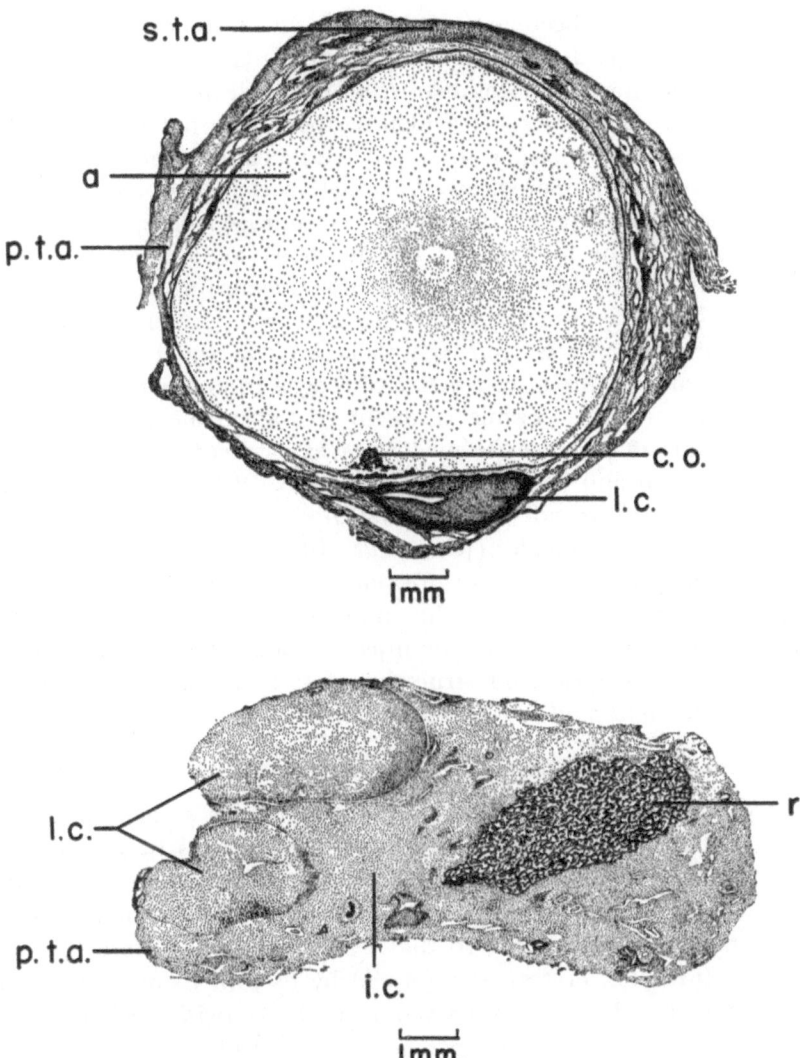

Fig. 14. Camera lucida drawing of two gonads of freemartin quintuplet cattle. (Rajakoski and Hafez, 1963.)

a. = antrum
c.o. = cumulus oophorus
i.c. = "interstitial cells"
l.c. = lutein cells
p.t.a. = primary tunica albuginea containing smooth muscle fibers
s.t.a. = secondary tunica albuginea
r. = rete

"Hormonal Theory"

Freemartinism has been ascribed by Lillie (1922) to hormones from the male twin which reach the female through vascular anastomoses between the fused placentae. The dominance of the male member of the pair was explained on histological evidence: the interstitial cells appear in the fetal testis much earlier than the fetal ovary, suggesting that the testis is active endocrinologically long before the ovary. In support of the "hormonal" theory are the results on androgen destruction by placental enzymes. The human and the marmoset are similar in that the female partner of unlike-sex twins are not freemartins (Benirschke and Brownhill, 1962). The female twins of these species are protected against the masculinizing effects from the male because their placentae contain an enzyme capable of converting androgen to estrogen (Ryan, *et al.*, 1961). Apparently this enzyme is absent in the bovine placenta.

All attempts to produce freemartinism experimentally in cattle (Mason, *et al.*, 1958) and other mammals (Moore, 1950) were unsuccessful. Also freemartins were not produced by injecting androgens from adult males into pregnant females (Jost, *et al.*, 1963). Intra-uterine injections of testosterone to cows in early pregnancy also failed to transform the embryonic ovaries to the male direction (Jainudeen and Hafez, 1965). Failure to induce freemartinism by injection of synthetic androgens may be due to physiological and structural differences between androgens secreted by the fetal and adult testes.

"Sex-Chromosome Mosaicism" Theory

The placentae of bovine multiple pregnancy have a common circulation as a result of anastomosis of allantoic blood vessels. Blood group studies (Stone, *et al.*, 1952) have shown that the vascular anastomosis between adjacent placentae may permit a reciprocal exchange of primordial hematopoietic tissues so that each twin carries more than one type of erythrocytes; erythrocytes formed by its own tissue and erythrocytes formed by alien tissue derived from its co-twin. Such twins are called *"chimeras"* and the phenomenon of mixed blood is called *"erythrocyte chimerism"* or *"mosaicism."* If allantoic fusion results in blood chimeras, cells other than erythrocytes may also be exchanged. Karyotype analysis of chromosomes show a definite chimerism in the bone marrow cells (Ohno, *et al.*, 1962) and leucocytes (Fechheimer, *et al.*, 1963) of bovine freemartins and their male partner, *i.e.*, they contain a mixture of XX and XY cells. These findings raise the possibility that freemartinism may be the function of germ cell chimerism, arising from physical transfer of germ cells from the male to female fetus.

Variability in the degree of intersexuality may be due to differences in the time of the introduction of the factor(s) from the male fetus, the duration of the action, and/or the stage of development at which vascular anastomosis is established. The location of the embryo *in utero* and the position of the female embryo in relation to the male may be important factors that determine the degree of transformation. An early and intimate vascular connection is very likely between the fetuses in the same horn, whereas a connection between fetuses in different horns would develop at later stages.

Diagnosis of Freemartinism

The diagnosis of freemartinism is based on: (1) genital abnormalities in the female member of heterosexual twins; (2) sex-chromatin test; (3) blood typing and skin grafts which provide indirect evidence for fetal vascular anastomosis; and (4) chromosomal analysis. Rectal palpation and vaginoscopy by a glass speculum are usually used. The genetic sex of a bovine freemartin can be determined by the sex chromatin test previously discussed, thereby differentiating it from male pseudohermaphroditism. Moore, *et al* (1957), showed that a typical sex chromatin was present in the nerve cell nuclei of the bovine freemartin.

The technique of "differential hemolysis" is used to detect mosaicism in twins. The proportions of the two cell types are determined spectrophotometrically as the fraction of cells not lysed by a given reagent (Stone, 1964). A heifer, born twin to a bull and possessing erythrocyte mosaicism, will be a freemartin; but a female, twin to a bull with a different blood type, will be as fertile as if born singly.

Erythrocyte mosaicism induces immunological tolerance, and therefore dizygotic twin calves of different sex are, with few exceptions, tolerant to grafts of each other's skin (Anderson, *et al.*, 1951). This is true of skin exchanges in monozygotic twins but is never found in other relationships. When skin grafts were exchanged between bull/heifer twins, it was found that all the females that were highly tolerant to grafts from their male twins were freemartins (Billingham and Lampkin, 1951).

Chromosome analysis is also conducted on the nuclei of cells undergoing active mitotic division. Chromosomes are morphologically most distinct during the metaphase of mitosis. Cells for such analysis may be obtained from bone marrow, skin, or peripheral blood samples.

2. True Hermaphrodites

True hermaphrodites are very common in pigs and rare in cattle and in horses. In the true hermaphrodite, one gonad may be an ovary and the other a testis or both gonads may be ovotestes. The testes may be located subcutaneously in the inguinal region or within the abdominal cavity. The pathogenesis of true hermaphroditism is the least understood. Chromosomal analysis in humans has revealed very few aberrations, *i.e.,* sex-chromosome mosaicism arising from non-disjunction that occurs after fertilization, either during the first mitosis of the zygote or at a later cleavage (Sohval, 1963). Such a mosaicism could explain the presence of genetically different cells and tissues in the same organism. However, no chromosomal imbalance is observed in the majority of true hermaphrodites. According to Beatty (1964) such intersex in goats, in swine and in cattle possesses the sex-chromosome constitution of the homogametic sex (XX).

Comparative Aspects

The syndrome is expressed differently in different species.

Cattle: The testes are either rudimentary or poorly developed and are surrounded by a tunica albuginea. The seminiferous tubules may be developed but spermatogenesis does not occur. Although the testicular tissue predominates in ovotestes, the ovarian tissue is more developed, showing Graafian follicles and sometimes even corpora lutea and corpora albicantia. Vestigial structures representing epididymis and vas deferens are often encountered. The accessory genitalia are variable, depending on the extent to which the Müllerian and Wolffian ducts are developed. The uterus may be normal or rudimentary. The vagina and cervix are usually underdeveloped.

Goats: Different types and degrees are encountered. In the female syndromes, the external appearance may resemble the normal female, but the animal may have seminal vesicles and vasa deferentia. In the male syndrome, the penis and descended testis are poorly developed and the Müllerian ducts survive along with the vas deferens. In some animals the clitoris and the penis both show conspicuous development, whereas in others the external genitalia are entirely of the female type. Older females tend to produce a slightly higher percentage of hermaphroditic kids (Haugen, 1960), probably as a result of the higher incidence of multiple births.

True hermaphroditism is more frequent in certain breeds, *e.g.,* Saanen, Toggenburg, Angora and some milk breeds (Eaton, 1943; Asdell, 1944). It is a hereditary trait due to the action of a recessive gene which apparently acts only on the female zygote. In the homozygous condition, the embryos bearing such genes develop simultaneously toward the male and female types. The sex ratios of the different crosses clearly place the hermaphrodites as genetic females. There seems to be a linkage between the dominant hornless and recessive hermaphroditism in goats (Brandsch, 1959).

Pigs: The reproductive tract is retarded to variable degrees or it may be lacking (Albertsen, 1951). Male organs are represented by testes, an epididymis, and a vas deferens. The testis, cryptorchid in type, is characterized by hyperplasia of the interstitial cells and absence of spermatogenic epithelium. The ovary may be rudimentary or normal. The fallopian tubes, uterine horns, cervix, vagina and labia are present. The clitoris is enlarged and the teats and mammary glands are lacking. In the unilateral type, the left gonad is consistently an ovary; testicular tissue is invariably present on the right gonad which is either a testis or an ovotestis (Johnston, *et al.*, 1958). In the ovotestis, the ovarian tissues contain normal oocytes, whereas the spermatic tubules lack any spermatogenic activity (Pond, *et al.*, 1961).

True hermaphrodites in pigs are chromatin positive (Cantwell, *et al.*, 1958) and therefore are genetic females. In addition they are capable of producing litters of apparently normal piglets (Hulland, 1964).

Horses: The female reproductive organs and the mammary glands are almost normal, whereas the male structures are represented by somewhat underdeveloped testes and a rudimentary penis (Senze, 1947; Culzoni, 1964). Other abnormalities may include penile clitoris or undescended testes.

3. Pseudohermaphroditism

Pseudohermaphroditism is more common in the male than in the female. The gonads in both sexes are undescended in the abdominal cavity. The Müllerian and Wolffian ducts may be at any stage of development, from almost complete degeneration to full development (Freudenberg, 1957).

Male Pseudohermaphrodite

Animals with this syndrome have undescended testes and varying combinations of male and female genital structures so that the external genitalia are often quite ambiguous.

Male pseudohermaphrodites in cattle resemble steers and lack any sexual desire (Collet and Lapeyre, 1955). The external genitalia which resemble that of a female, have a small normal vulva, clitoris and traces of male external genitalia. The scrotum, if present, does not contain testes and a poorly developed penis may be present. The internal genitalia may contain vas deferens, seminal vesicles, prostate gland, uterus, cervix, and/or vagina (Yapp, 1947). The gonads are invariably testes and are usually located in the subcutaneous tissue just dorsal to the udder. The testes are either vestigial or poorly developed. The seminiferous tubules, if present, do not show any spermatogenic activity.

Most male pseudohermaphrodites are born as singles or twins to normal calves (Paulsen, 1943). Results of castration experiments suggest that the fetal testes in these male intersexes failed to secrete sufficient hormones to produce full masculinization of the reproductive system (Jost, 1953).

Consequently, some of the segments of the Müllerian duct and bisexual external genitalia persisted.

In goats, the testis which lacks spermatogenic epithelium may be located in the scrotum or the abdominal cavity. A poorly developed vagina is usually present. The phallus, directed caudally, is intermediate in structure between a penis and a clitoris, and is terminated in a gland-like enlargement, but without any urethral canal (Halley and Baxter, 1953).

In the pig, the testes are located either intra-abdominally or sub-cutaneously just below the vulva and contain widely scattered seminiferous tubules devoid of germinal epithelium. The interstitial tissue contains a preponderance of Leydig cells. The epididymis, vas deferens, prostate glands, and seminal vesicles are present. The testes may connect with the uterus by epididymal tubules (Johnston, *et al.*, 1958; Delphia and Bolin, 1958). The uterus is well developed but the cervix is poorly developed and the vagina is either small or absent. The syndrome has been attributed to similar mechanisms as in bovine freemartinism, although the resulting phenotype may not be quite as extreme (Gowen, 1961). It is also more common in some breeds than in others, suggesting the hereditary background. This seems to be the result of a single recessive gene limited to one sex.

In horses, the external genitalia and sexual behavior are those of a female. The poorly developed testes are either undescended or located subcutaneously at the site of the udder. Segments of the female tract are usually found and a short penis capable of erection protrudes through the vulva (Franz and Widmaier, 1960; Runnels, *et al.*, 1960).

Female Pseudohermaphrodite

Female pseudohermaphrodites have essentially normal ovaries and female internal genitalia with definite evidence of masculinization of external genitalia. Such intersexes are extremely rare in cattle and horses. In pigs, the urogenital opening is located at the tip of a phallus-like structure below the anus; the vulva is absent, gonads and fallopian tubes are normal and the male excretory duct system is evident.

IX. Pathogens and Reproductive Failure

Specific and non-specific infections, usually in the reproductive duct system, play a major role in reproductive failure. Bacterial, mycotic, viral, rickettsial, protozoan, and non-specific infections may inhibit estrus, ovulation, fertilization, implantation, placentation, fetal survival, or normal parturition. The most common clinical sign of these infections is abortion or "repeat breeding" as a result of failure to conceive or early embryonic death. The effect of infection on reproduction depends upon the initial resistance to the organism and the rate of production of specific antibodies in the host in response to infection.

The physiological mechanisms of reproductive failure caused by pathological syndromes vary according to the type of organism, its location in the reproductive system, the mode of transmission, and the degree of outbreak. For the most part the mode of actions is unknown because of the lack of studies involving experimental infections. Some of the pathogens common to farm mammals are also known to infect man, *e.g.*, human brucellosis, leptospirosis, vibriosis, and listeriosis, and salmonellosis.

Bacterial Infections

Vibrio fetus invades the ruminant uterus particularly the caruncles; the number of affected caruncles increases gradually. The organism destroys the fetal-maternal vascular attachments causing abortion. Abortion occurs in the first three quarters of gestation in the cow and in the last quarter in the ewe. Abortion is probably due to interference with the placental circulation, destruction of the chorionic epithelium, with edema and cellular infiltrations in the subepithelial tissues. The infection is associated with a temporary infertility and revealed by an increase in the number of services required per pregnancy. A delayed estrus often follows the first service to an infected bull. The insemination with semen containing antibiotics increases pregnancy rate.

Brucellosis (contagious abortion) (*Brucella abortus, melitensis* and *suis*) occurs in several mammals including man. In the cow the organism invades the gravid uterus multiplying there particularly at the caruncles. It also attacks the fetal chorionic epithelium and sets up an inflammatory process, causing retention of the placenta or abortion. The fetal-maternal attachments are necrosed, yellowish-brown in color and covered with a thick sticky exudate. The fetus degenerates showing septicaemic lesions.

The leptospirae constitute a large group of microorganisms. At present they are grouped with the bacteria in the order *Spirochaetales* and are considered to be related to *Treponema pallidum,* the cause of syphilis in man. These organisms cause abortion.

Protozoan, Viral and Fungus Infections

Trichomonas fetus is transmitted by the bull and infects the prepuce, vagina, and uterus. In Trichomoniasis, fertilization or implantation may be hindered in a proportion of the infected animals. Other animals may abort at 1 to 7 months of gestation. Fetal degeneration is usually due to interference in placental circulation. In other cases infertility is caused by pyometra where the uterus becomes distended with fluid and pus.

Still in other cases the fetus is mummified and retained after death for several months. The mode of action is not clear but it may be due to mild metritis which interferes with embryonic nutrition or to disturbed cervical function.

Toxoplasma gondii causes toxoplasmosis in cattle, sheep, swine, dogs, and man. In animals acute toxoplasmosis is associated with fever, dyspnea, central nervous disturbances, abortion, premature birth, and stillbirth. Intra-uterine transmission of the organism has resulted in the high mortality in the neonate during the first few weeks of life. In man, the infection causes certain forms of congenital encephalomyelitis, chorioretinitis, and/or pneumonitis.

Several viruses interrupt fetal development, *e.g.*, blue tongue in sheep, hog cholera and rubella. Some fungi, such as *Aspergillus fumigatus* and *Absidia ramosa,* invade the placental tissues and also produce lesions in the fetus.

Non-specific Infections

Non-specific microorganisms, normally present in the female reproductive tract, may be activated under favorable circumstances and set up infections. The non-specific infections include *Alkigenes, Clostridium, Corynebacterium, Diplococcus, Hemophilus, Pasteurella, Staphylococcus,* and *Streptococcus.*

During the progestational and gestational phases, the uterus is quite susceptible to bacterial infection since it provides an environment for the nourishment of the embryo. This uterine environment is also favorable for organisms which may enter the uterus via the cervix, blood or lymph. In horses and swine, the anatomy and physiology of the cervix makes the uterus accessible to bacterial invasion. The male ejaculates a large volume of dilute semen directly into the cervix. The relaxed cervix allows the semen to gravitate into the uterus and carry contaminating infectious organisms which may cause embryonic death.

Unilateral and bilateral hydrosalpinx may be associated with pyosalpinx, the accumulation of pus resulting from inflammation by pyogenic organisms. The inflammation may cause adhesions of the fimbriae; this prevents ova pick up.

Copulation within 40 days after parturition in cattle is associated with high percentage of reproductive failure; but whether this is due to residual postpartum infection or to some other non-infectious causes has not yet been clarified (*e.g.*, Boyd and Reed, 1961). Inflammation of the endometrium and acute vaginitis may be associated with specific infections, parturient injuries, or retained placenta.

A variety of conditions resulting in prolonged pyrexia, especially those of infectious origin, may result in abortion or fetal atrophy, even after the recovery of the pregnant female. Survival from febrile infections is much more common at present with antibiotics and antisera therapy of affected animals.

Other Diseases

Diseases of general nature may also cause reproductive failure either indirectly through the general reduction in health caused by extensive infection, or directly by causing lesions which interfere with gamete production, inhibit implantation, or cause abortion. For example, tuberculosis may cause reproductive failure in cattle and pigs. Tuberculosis may also cause tubercles in the ovarian stroma, ovarian granular lesions, adhesions in the fimbriae, occlusion of the fallopian tube, pyometra, necrosis in fetal-maternal attachments, maceration of the fetus, lesions in the vagina, and enlargement of the labia.

X. FUTURE INVESTIGATIONS

Several factors have been shown to cause reproductive failure. In most instances, however, the mode of action and the physiological mechanisms involved are unknown. Comparative research should be conducted in animal reproduction using the modern techniques in developmental physiology, developmental genetics, microbiochemistry, histochemistry, immunology, and pathology.

Little information is available on reproductive failure as affected by aging, general and specific nutritional deficiency, climatic and social stress, toxins, drugs, poisons, and vaccines. For example, there is a high incidence of defective embryos and offspring in non-immune sows, vaccinated against swine fever with the modified virus vaccines 1 to 2 weeks after conception (*cf.* Roberts, 1956). Cholera vaccination during the first month of pregnancy with either virulent virus or attenuated live virus causes fetal anomalies with an increase in prenatal mortality (Young, *et al.,* 1955).

Studies are needed to understand the nature of implantation failure in relation to the biochemical changes of the endometrium and its secretions, *e.g.,* pH changes, enzyme content, metabolic activity with various substrates. Post-implantation mortality should be studied in association with the biochemical changes in the trophoblast, blastocoelic fluid, and placenta, as well as the maternal-fetal vascular attachment.

Blood groups and protein polymorphism should be considered in studies of prenatal and perinatal mortality. Similar studies have been conducted in chickens (Morton, *et al.,* 1965). Other studies should deal with experimental induction of prenatal and perinatal mortality using reliable methods for inducing embryonic loss, *e.g.,* overcrowding *in utero*

by superovulation or egg transfer, inter-specific egg transfer, interference
with the function of the corpus luteum, experimental infections with
V. fetus (*cf.* Frank *et al.*, 1962) or X-radiation, and the administration of
teratogenic substances.

The normal trophoblast could be considered as an anatomical barrier
between the mother and the embryo to protect the developing conceptus
from a homograft reaction. Studies on immunoreproduction will elucidate
aspects of fetal-maternal interactions. Such studies should deal with: (a)
the immunological mechanisms involved in sperm-ovum incompatibility
and fertilization failure, (b) the role of antigenic substances of reproductive tissue origin, particularly the absorption of the sperm by the tissue
of the female tract, and (c) the nature, frequency, quantity and routes
of fetal antigens reaching the mother and the maternal antigens reaching
the fetus.

Studies on intersexuality are needed in view of the recent advances in
micro-electronic instrumentation used in biological and biophysical
sciences. The use of tissue culture techniques and similar techniques in
cytogenetics should be useful for studies of karyotype analysis and
somatic cell matings. Bone marrow, skin and peripheral blood culture
techniques provide accurate information concerning the karyotypes.
Studies on cytodifferentiation and histochemical characteristics of fetal
gonads are of special interest since they may explain several phenomena
of academic and practical importance, *e.g.*, polyploidy and the mode by
which sex chromosomes and autosomes interact to determine sex phenotype. Karyotype analysis of tissue cultures should be done routinely on
aborted material.

The phenomenon of sex-chromosome mosaicism in cases of freemartinism requires further study. Knowledge of the levels of circulating and
excreted hormones and placental enzymes and hormones in the different
intersexual states is needed. More investigations involving the influence
of heredity coupled with physio-pathological studies of intersexual syndromes are clearly required to assess different syndromes of reproductive
failure in mammals.

References

Adams, C. E.: Prenatal mortality in the rabbit, *Oryctolagus cuniculus.* J. Reprod.
    Fertil. *1:*36, 1960.
————: Studies on prenatal mortality in the rabbit, *Oryctolagus cuniculus:* The
    effect of transferring varying numbers of eggs. J. Endocrin. *24:*471, 1962.
Adler, J. H., and D. Trainin: A hyperoestrogenic syndrome in cattle. Refuah
    vet. *17:*115, 1960.
Albertsen, K.: Hermafroditisme hos svin. Nord. Vet. Med. *3:*849, 1951.
Alexander, G., and J. E. Peterson: Neonatal mortality in lambs. Intensive

observations during lambing in a flock of maiden Merino ewes. Austr. Vet. J. *37*:371, 1961.

Alliston, C. W., B. Howarth, Jr., and L. C. Ulberg: Embryonic mortality following culture *in vitro* of one- and two-cell rabbit eggs at elevated temperatures. J. Reprod. Fertil. *9*:337, 1965.

Amoroso, E. C., and A. S. Parkes: Effects on embryonic development of X-irradiation of rabbit spermatoza in vitro. Proc. Roy. Soc. Ser. B: Biol. Sci. *134*:57, 1947.

Anderson, D., R. E. Billingham, G. H. Lampkin, and P. B. Medawar: The use of skin grafting to distinguish between monozygotic and dizygotic twins in cattle. Heredity *5*:379, 1951.

Asdell, S. A.: The genetic sex of intersexual goats and a probable linkage with the gene for hornlessness. Science *99*:124, 1944.

Asdell, S. A., and M. F. Crowell: Effect of retarded growth upon sexual development of rats. J. Nutrition *10*:13, 1935.

Ashton, G. C.: Beta-globulin type and fertility in artificially bred dairy cattle. J. Reprod. Fertil. *2*:117, 1961.

Ashton, G. C., and G. R. Fallon: β-globulin type, fertility and embryonic mortality in cattle. J. Reprod. Fertil. *3*:93, 1962.

Austin, C. R.: Effects of hypothermia and hyperthermia on fertilization in rat eggs. J. Exp. Biol. *33*:348, 1956.

————: "The Mammalian Egg." Oxford: Blackwell, 1961.

Austin, C. R., and A. W. H. Braden: An investigation of polyspermy in the rat and rabbit. Austr. J. Biol. Sci. *6*:674, 1953.

Baier, W.: Über den frühembryonalen Fruchttod als paternell bedingter Zuchtschaden. Züchtungskunde *27*:93, 1955.

Baker, L. N., A. B. Chapman, L. N. Grummer, and L. E. Casida: Some factors affecting litter size and fetal weight in purebred and reciprocal-cross matings of Chester White and Poland China swine. J. Anim. Sci. *17*:612, 1958.

Beatty, R. A.: "Parthenogenesis and Polyploidy in Mammalian Development." Cambridge: Cambridge University Press, 1957.

————: Chromosome deviations and sex in vertebrates. In "Intersexuality in Vertebrates including Man." Pp. 17–143, Ed. by C. N. Armstrong and A. J. Marshall. London: Academic Press, 1964.

Bellows, R. A., A. L. Pope, A. B. Chapman, and L. E. Casida: Effect of level and sequence of feeding and breed on ovulation rate, embryo survival and fetal growth in the mature ewe. J. Anim. Sci. *22*:101, 1963.

Benirschke, K., and L. E. Brownhill: Further observations on marrow chimerism in marmoset monkeys. Cytogenetics *1*:245, 1962.

Bennetts, H. W., E. J. Underwood, and F. L. Shier: A specific breeding problem of sheep on subterranean clover pastures in Western Australia. Austr. Vet. J. *22*:2, 1946.

Billingham, R. E., and G. H. Lampkin: Further studies in tissue homotransplantation in cattle. J. Embryol. Exp. Morph. *5*:351, 1951.

Bishop, M. W. H.: Paternal contribution to embryonic death. J. Reprod. Fertil. *7*:383, 1964.

Bishop, M. W. H., and A. Walton: Spermatogenesis and the structure of mammalian spermatozoa. Chapter 7 in "Marshall's Physiology of Reproduction." A. S. Parkes, ed. Vol. 1, Part 2. London: Longman's, 1960.

Black, D. L., and S. A. Asdell: Mechanism controlling entry of ova into rabbit uterus. Amer. J. Physiol. *197*:1275, 1959.

Blandau, R. J.: The female factor in fertility and infertility. I. Effects of delayed fertilization on the development of the pronuclei in rat ova. Fertil. and Steril. *3*:349, 1952.

————: The effects on development when eggs and sperm are aged before fertilization. Ann. N.Y. Acad. Sci. *57*:526, 1954.

Bomsel-Helmreich, O.: Experimental heteroploidy in sows. Proc. IVth Int. Congr. Anim. Reprod. The Hague, 1961. *3*:578, 1962.

Boyd, H., and H. C. B. Reed: Investigations into the incidence and causes of infertility in dairy cattle. II. Influence of some management factors affecting the semen and insemination conditions. Brit. Vet. J. *117*:74, 1961.

Bradbury, R. B., and D. E. White: Estrogens and related substances in plants. Vitamins and Hormones *12*:207, 1954.

Braden, A. W. H., and J. E. Peterson: The persistence of endometrial cysts induced by oestrogen in guinea pigs. Austr. J. Biol. Sci. *6*:520, 1953.

Brambell, F. W. R.: Intra-uterine mortality of the wild rabbit, *Oryctolagus cuniculus*. Proc. Roy. Soc. B. *130*:462, 1942.

Brandsch, H.: Die Vererbung geschlechtlicher Missbildung und des Hornes bei der Hausziege in ihrer gegenseitigen Beziehung. Versuch einer genetischen Analyse. Arch. Geflügelz. Kleintierk. *8*:310, 1959.

Bruce, H. M.: A block to pregnancy in the mouse caused by proximity of strange males. J. Reprod. Fertil. *1*:96, 1960.

Cantwell, G. E., E. F. Johnston, and J. H. Zeller: The sex chromatin of swine intersexes. J. Hered. *49*:199, 1958.

Caroli, J., and M. Bessis: Sur la cause et le traitement de l'ictère grave de Muletons nouveaunés. C. R. Acad. Sci. *224*:969, 1947.

Casida, L. E., and W. Wisnicky: Effects of diethyl stilbestrol-dipropionate upon postpartum changes in the cow. J. Anim. Sci. *9*:238, 1950.

Chang, M. C.: An experimental analysis of female sterility in the rabbit. Fertil. Steril. *3*:251, 1962.

Chang, M. C., and E. B. Harvey: Effects of ionizing radiation of gametes and zygotes on the embryonic development of rabbits and hamsters. In: "Effects of Ionizing Radiation on Reproductive System," Eds. W. D. Carlton and F. X. Gassner. New York: Macmillan, 1964.

Collet, P., and P. Lapeyre: Hermaphrodisme chez un bovin. Bull. Soc. Sci. Vet. Lyon. *57*:17, 1955.

Conneally, P. M., W. H. Stone, W. J. Tyler, L. E. Casida, and N. E. Morton: Genetic load expressed as fetal death in cattle. J. Dairy Sci. *46*:232, 1963.

Crichton, J. A., J. N. Aitken, A. W Boyne: The effect of plane of nutrition during rearing on growth, production, reproduction and health of dairy cattle. I. Growth to 24 months. Anim. Prod. *1*:145, 1959.

Cronin, M. T. I.: Haemolytic disease of newborn foals. Vet. Record. *67*:479, 1955.

Culzoni, V.: Aspetti strutturali di un cavallo intersessuato. Proc. 5th Congress of Anim. Reprod. Artif. Insem. (Trento, Italy) *3*:311, 1964.

Cupps, P. T., R. C. Laben, and S. W. Mead: Histology of the pituitaries, adrenals, ovaries, and uteri of dairy cattle associated with different reproductive conditions. J. Dairy Sci. *39*:155, 1956.

Curnow, D. H.: Oestrogenic activity of subterranean clover. 2. The isolation of genistein from subterranean clover and methods of quantitative estimation. Biochem. J. *58*:283, 1954.

Curnow, D. H., T. J. Robinson, and E. J. Underwood: Oestrogenic action of extracts of subterranean clover (*T. Subterranean,* L. var. Dwalganup). Austr. J. Exp. Biol. Med. Sci. *26*:171, 1948.

Davis, M. E., N. W. Fugo, and K. G. Lawrence: Effect of alloxan diabetes on reproduction in the rat. Proc. Soc. Exper. Biol. Med. *66*:638, 1947.

Delphia, J. M., and F. M. Bolin: An anomaly of the reproductive system of the pig. J. Amer. Vet. Med. Assn. *132*:281, 1958.

Dutt, R. H.: Fertility rate and embryonic death loss in ewes early in the breeding season. J. Anim. Sci. *13*:465, 1954.

———: Temperature and light as factors in reproduction among farm animals. J. Dairy Sci. *43* (Suppl. no. 123): 1960.

Dutt, R. H., E. F., Ellington, and W. W. Carlton: Fertilization rate and early embryo survival in sheared and unsheared ewes following exposure to elevated air temperature. J. Anim. Sci. *18*:1308, 1959.

Dutt, R. H., and E. C. Simpson: Environmental temperature and fertility of Southdown rams early in the breeding season. J. Anim. Sci. *16*:136, 1957.

Dziuk, P. J., and G. Henshaw: Fertility of boar semen artificially inseminated following *in vitro* storage. J. Anim. Sci. *17*:554, 1958.

Eaton, O. N.: An anatomical study of hermaphrodism in goats. Amer. J. Vet. Res. *4*:333, 1943.

Eckstein, P., T. McKeown, and R. G. Record: Variation in placental weight according to litter size in the guinea pig. J. Endocrin. *12*:108, 1955.

Edgar, D. G.: Studies on infertility in ewes. J. Reprod. Fertil. *3*:50, 1962.

Edgar, D. G., and S. A. Asdell: The valve-like action of the uterotubal function of the ewe. J. Endocrin. *21*:315, 1960.

El-Sheikh, A. S., C. V. Hulet, A. L. Pope, and L. E. Casida: The effect of level of feeding on the reproductive capacity of the ewe. J. Anim. Sci. *14*:919, 1955.

Erb, R. E., and E. W. Holtz: Factors associated with estimated fertilization and service efficiency of cows. J. Dairy Sci. *41*:1541, 1958.

Fechheimer, N. S., M. S. Herschler, and L. O. Gilmore: Sex chromosome mosaicism in unlike sexed cattle twins. In "Genetics Today." Proc. 11th Int. Congr. Gent. (The Hague), Vol. 1 (Abstr.): 265, 1963.

Finn, C. A.: Embryonic death in aged mice. Nature *194*:499, 1962.

First, N. L., F. W. Stratman, E. M. Rigor, and L. E. Casida: Factors affecting ovulation and follicular cyst formation in sows and gilts fed 6-methyl-17-acetoxyprogesterone. J. Anim. Sci. *22*:66, 1963.

Foley, R. C., and R. P. Reece: Histological studies of the bovine uterus, placenta, and corpus luteum. Mass. Agric. Exp. Sta. Bull. No. 468, 1953.

Foote, W. C., A. L. Pope, A. B. Chapman, and L. E. Casida: Reproduction in the yearling ewe as affected by breed and sequence of feeding levels. I. Effects on ovulation rate and embryo survival. J. Anim. Sci. *18*:453, 1959.

Foote, W. D., and J. E. Hunter: Post-partum intervals of beef cows treated with progesterone and estrogen. J. Anim. Sci. *23*:517, 1964.

Frank, A. H., W. T. Shalkop, J. H. Bryner, and P. A. O'Berry: Cellular changes in the endometrium of *Vibrio fetus*-infected and noninfected heifers. Amer. J. Vet. Res. *23*:1213, 1962.

Franz, W., and R. Widmaier: Ein intersexuelles, kernmorphologisch weibliches Pferd. Berl. Münch. tierärztl. Wschr. *73*:341, 1960.

Freudenberg, F.: Die Bedeutung der Intersexualität beim Schwein als erbliche Geschlechtsmissbildung. Mh. Vet. Med. *12*:608, 1957.

Garm, O.: A study on bovine nymphomania. Acta Endocrinol. *2*: Supplement *3*:5, 1949.

———: Investigations on cystic ovarian degeneration in the cow, with special regard to etiology and pathogenesis. Cornell Vet. *39*:39, 1949.

Goode, L., A. C. Warnick, and H. D. Wallace: The effect of energy level on reproductive performance of gilts. Proc. Assoc. Southern Agric. Workers *57*:82, 1960.

Gowen, J. W.: Cytologic and genetic basis of sex. In "Sex and Internal Secretions." Vol. 1. W. C. Young, ed. Baltimore: The Williams and Wilkins Company, 1961.

Greenstein, J. S., and R. C. Foley: Early embryology of the cow. I. Gastrula and primitive streak stages. J. Dairy Sci. *41*:409, 1958.

Grisebach, H., and W. D. Ollis: Biogenetic relationships between coumarins, flavonoids, isoflavonoids, and rotenoids. Experientia *17*:4, 1961.

Hadorn, E.: "Developmental Genetics and Lethal Factors." London: Methuen, 1961.

Hafez, E. S. E.: Effects of over-crowding *in utero* on implantation and fetal development in the rabbit. J. Exp. Zool. *156*:269, 1964.

———: Transuterine migration and spacing of bovine embryos during gonadotropin-induced multiple pregnancy. Anat. Rec. *148*:203, 1964.

———: "Reproduction in Farm Animals," 2nd edit. Lea & Febiger, Philadelphia, 1967.

Hafez, E. S. E., and M. R. Jainudeen: Intersexuality in farm mammals. Animal Breeding Abstracts. *34*:1, 1966.

Hafez, E. S. E., and Y. Tsutsumi: Changes in endometrial vascularity during implantation and pregnancy in the rabbit. Amer. J. Anat. *118*:249, 1966.

Haines, C. E., A. C. Warnick, and H. D. Wallace: Effect of energy level on puberty, ovulation, fertilization and embryonic survival (pig). Proc. Amer. Soc. Anim. Prod. in J. Anim. Sci. *14*:1174–1261, 1955.

Haines, C. E., A. C. Warnick, and H. D. Wallace: The effect of two levels of energy intake on reproductive phenomena in Duroc Jersey gilts. J. Anim. Sci. *18*:347, 1959.

Halley, G., and J. S. Baxter: Leucocyte invasion of the genital tract of an intersexual goat. J. Comp. Path. Therap. *63*:179, 1953.

Hammond, J.: The effects of high and low planes of nutrition on reproduction in rabbits. New Zeal. J. Agric. Res. *8*:708, 1965.

Hancock, J. L.: Polyspermy of pig ova. Anim. Prod. *1*:103, 1959.

———: Fertilization in the pig. J. Reprod. Fertil. *2*:307, 1961.

Hanly, S.: Prenatal mortality in farm animals. J. Reprod. Fertil. *2*:182, 1961.

Hanset, R.: La "maladie des genisses blanches": son aspect génétique. I. Terminologie, degrés d'expressivité, anomogenèse. Ann. Med. Vet. *103*:281, 1959–60.

II. L'hypothèse d'un gène récessif autosome complètement pénétrant lié au gène semidominant du blanc, N. *104*:3, 1959–60. III. Hypothèses en présence. Discussion générale. *104*:49, 1959–60.

Hartman, C. G.: Early death of mammalian ovum with special reference to the aplacental opossum. In "Mammalian Germ Cells," G. E. W. Wolstenholme, ed. London: Churchill, 1953.

Haugen, E.: Tvekjørn hos geit. Meld. Norg. LandbrHøgsk, *39*:1, 1960.

Hawk, H. W., W. J. Tayler, and L. E. Casida: Effect of sire and system of mating on estimated embryonic loss. J. Dairy Sci. *38*:420, 1955.

Heitman, H., Jr., E. H. Hughes, and C. F. Kelly: Effects of elevated ambient temperature on pregnant sows. J. Anim. Sci. *10*:907, 1951.

Henricson, B.: Genetical and statistical investigation into so-called cystic ovaries in cattle. Acta Agric. Scand. *7*:3, 1957.

Hertig, A. T.: In "The Placenta and Fetal Membranes," C. A. Villee, ed., p. 109. Baltimore: Williams & Wilkins, 1960.

Hillarp, N. A., H. Olivecrona, and W. Silfverskiöld: Evidence for the participation of the preoptic area in male mating behaviour. Experientia *10*:224, 1954.

Hulet, C. V., H. P. Voigtlander, Jr., A. L. Pope, and L. E. Casida: The nature of early-season infertility in sheep. J. Anim. Sci. *15*:607, 1956.

Hulland, T. J.: Pregnancy in a hermaphrodite sow. Canad. vet. J. *5*:39, 1964.

Jainudeen, M. R., and E. S. E. Hafez: Attempts to induce bovine freemartinism experimentally. J. Reprod. Fertil. *10*:281, 1965.

Jamieson, A.: Studies on the J-group system of cattle. Paper presented at the Immuno-genetics Conference of Blood Group Workers, Edinburgh. 1960 (unpublished).

Jennings, W. E.: Some common problems in horse breeding. Cornell Vet. *31*:197, 1941.

Johnsen, H. H. K., J. Moustgaard, and N. H. Olson: Am vitamin B_{12}'s betydning for søer og gyltes frugtbarked. (Importance of vitamin B_{12} for fertility of sows and gilts). Dansk Maanedsskr. Dyrlaeg. *63*:1, 1952.

Johnson, K. R., R. H. Ross, and D. L. Fourt: Effect of progesterone administration on reproductive efficiency. J. Anim. Sci. *17*:386, 1958.

Johnston, E. F., J. H. Zeller, and G. E. Cantwell: Sex anomalies in swine. J. Hered. *49*:254, 1958.

Jost, A.: Problems of fetal endocrinology, the gonadal and hypophyseal hormones. Hormone Research *8*:379, 1953.

Jost, A., M. Chodkiewicz, and P. Mauléon: Intersexualité du foetus de veau produite par des androgènes. Comparaison entre l'hormone foetale responsable du freemartinisme et l'hormone testiculaire adulte. C. R. Acad. Sci. *256*:274, 1963.

Joubert, D. M.: The influence of high and low nutritional planes on the oestrous cycle and conception rate of heifers. J. Agric. Sci. *45*:164, 1954.

King, J. W. B., and G. B. Young: Maternal influences on litter size in pigs. J. Agric. Sci. *48*:457, 1957.

Koch, P., H. Fischer, and H. Schumann: Erbpathologie der landwirtschaftlichen Haustiere. Berlin und Hamburg: Paul Parey. 1957.

Kristjansson, F. K.: Transferrin types and reproductive performance in the pig. J. Reprod. Fert. *8*:311, 1964.

Lasley, E. L.: Ovulation, prenatal mortality and litter size in swine. J. Anim. Sci. *16*:335, 1957.

Lawrence, A. M., and A. V. Contopoulos: Reproductive performance in the alloxan diabetic female rat. Acta Endocrin. *33*:175, 1960.

Leuchtenberger, C., I. Murmanis, L. Murmanis, S. Ito, and D. R. Weir: Interferometric dry mass and microspectrophotometric arginine determinations on bull sperm nuclei with normal and abnormal DNA content. Chromosoma *8*:73, 1956.

Lillie, F. R.: The etiology of the freemartin. Cornell Vet. *12*:332, 1922.

Lindan, O., and M. E. Morgan: Alloxan diabetes and pregnancy: a long-term observation. J. Endocrin. *6*:463, 1950.

McClure, T. J.: Temporary nutritional stress and infertility in female mice. J. Physiol. *147*:221, 1959.

———: Uterine pathology of temporarily-fasted pregnant mice. J. Comp. Pathol. Ther. *71*:16, 1961.

McLaren, A., and D. Michie: Control of prenatal growth in mammals. Nature *187*:363, 1960.

MacFarlane, W. V., P. R. Pennycuik, and E. Thrift: Resorption and loss of foetuses in rats living at 35°C. J. Physiol. *135*:451, 1957.

Mares, S. E., A. C. Menge, W. J. Tyler, and L. E. Casida: Some sources of variation in conception rate and pregnancy loss in parous Holstein cows. J. Anim. Sci. *17*:1217, 1958.

Mason, I. L.: Symptoms of muscular hypertrophy in heterozygous steers. Anim. Prod. *5*:57, 1963.

Mason, R. W., J. F. Bone, R. Bogart, and H. Krueger: Urogenital anomalies in calf born to beef cow treated with testosterone during pregnancy. J. Pharmacol. Exp. Ther. *122*:49A, 1958.

Miller, H. C.: The effect of pregnancy complicated by alloxan diabetes on the fetuses of dogs, rabbits and rats. Endocrinology *40*:251, 1947.

Moor, R. M., and L. E. A. Rowson: Influence of the embryo and uterus on luteal function in the sheep. Nature *201*:522, 1964.

Moore, C. R.: The role of the fetal endocrine glands in development. J. Clin. Endocrin. Metab. *10*:942, 1950.

Moore, K. L., M. A. Graham, and M. L. Barr: The sex chromatin of the bovine freemartin. J. exp. Zool. *135*:101, 1957.

Morton, J. R., D. G. Gilmour, E. M. McDermid, and A. L. Ogden: Association of blood-group and protein polymorphisms with embryonic mortality in the chicken. Genetics *51*:97, 1965.

Moule, G. R., A. W. H. Braden, and D. R. Lamond: The significance of oestrogens in pasture plants in relation to animal production. Anim. Breed. Abstr. *31*:139, 1963.

Myers, W. L., and D. Segre: The immunologic behavior of baby pigs. III. Transplacental transfer of antibody globulin in swine. J. Immunol. *91*:697, 1963.

Neimann-Sørensen, A., J. Rendel, and W. H. Stone: The J substance of cattle, II. A comparison of normal antibodies and antigens in sheep, cattle and man. J. Immunol. *73*:407, 1954.

Noyes, R. W., and Z. Dickman: Relationship of ovular age to endometrial development. J. Reprod. Fertil. *1:*186, 1960.

Odor, D. L., and R. J. Blandau: Incidence of polyspermy in normal and delayed matings in rats of the Wistar strain. Fertil. and Steril. *7:*456, 1956.

Ogden, A. L.: Biochemical polymorphism in farm animals. Proc. Roy. Soc. Med. *52:*955, 1959.

———: Biochemical polymorphism in farm animals. Anim. Breed. Abstr. *29:*127, 1961.

Ohno, S., J. M. Trujillo, C. Stenius, L. C. Christian, and R. L. Teplitz: Possible germ cell chimeras among newborn dizygotic twin calves (*Bos taurus*). Cytogenetics *1:*258, 1962.

Overzier, C.: True hermaphroditism. In "Intersexuality." C. Overzier, ed. London and New York: Academic Press, 1963.

Paulsen, C. L.: False masculine hermaphrodism in cattle. Vet. Med. *38:*274, 1943.

Perry, J. S.: Observations on reproduction in a pedigree herd of large white pigs. J. Agric. Sci. *47:*332, 1956.

———: The incidence of embryonic mortality as a characteristic of the individual sow. J. Reprod. Fertil. *1:*71, 1960.

Piko, L., and O. Bomsel-Helmreich: Triploid rat embryos and other chromosomal deviants after colchicine treatment and polyspermy. Nature *186:*737, 1960.

Pitkjanen, I. G.: Some mechanisms of ovulation, fertilization, and the first stages of embryonic growth in swine (translated title). Translation in: Referat. Zhur. Biol., 1956, no. 86160. Izv. Akad. Nauk S.S.S.R. Ser. Biol. No. 3, p. 120, 1965

———: K voprosu. ob oplodotvorenii i pervych stadijach razvitija zarodysa u korov. Izv. Akad. Nauk S.S.S.R. Ser. Biol. No. 3, 77, 1956.

———: Fertilization and early stages of embryonic development in the sheep (translated title). Izv. Akad. Nauk S.S.S.R. Ser. Biol. No. 3, p. 291, 1958.

Pond, W. G., S. J. Roberts, and K. R. Simmons: True and pseudohermaphroditism in a swine herd. Cornell Vet. *51:*394, 1961.

Race, R. R., and R. Sanger: "Blood Groups in Man," 3rd Edition, Chap. 20, p. 385. Oxford: Blackwell, 1958.

Rae, A. L.: The genetics of the sheep. In "Advances in Genetics," *8:*189, 1956. New York: Academic Press, 1956.

Ragsdale, A. C., S. Brody, H. J. Thompson, and D. M. Worstell: Influence of temperature, 50° to 105°F, on milk production and feed production in cattle. Res. Bull. Mo. Agric. Exp. Stn. No. 425, 1948.

———: Derivatives of cortical cords in adult freemartin gonads of bovine quintuplets. Anat. Rec. *147:*457, 1963.

Rajakoski, E., and E. S. E. Hafez: Cytological differentiation of fetal bovine gonads. Cytogenetics *3:*193, 1964.

Rathnasabapathy, V., J. F. Lasley, and D. T. Mayer: Genetic and environmental factors affecting litter size in swine. Res. Bull. Univ. Missouri Agric. Exp. Sta. No. 615, 1956.

Reddy, V. B., J. F. Lasley, and D. T. Mayer: Genetic aspects of reproduction in swine. Res. Bull. Mo. Agric. Exp. Stn. No. 666, 1958.

Rendel, J. M.: White heifer disease in a herd of dairy shorthorns. J. Genetics *51:*89, 1952.

Rigor, E. M., R. K. Meyer, N. L. First, and L. E. Casida: Endocrine differences associated with follicular development and ovulation rate in swine due to breed and energy intake. J. Anim. Sci. *22*:43, 1963.

Roberts, S. J.: "Veterinary Obstetrics and Genital Diseases." Ann Arbor: Edwards Brothers, Inc. 1956.

Robson, J. M., and A. A. Sharaf: Effect of adrenocorticotrophic hormone (ACTH) and cortisone on pregnancy. J. Physiol. *116*:236, 1952.

Runnels, R. A., W. S. Monlux, and A. W. Monlux: "Principles of Veterinary Pathology," 1st edition. Ames, Iowa: Iowa State University Press, 1960.

Ryan, K. J., K. Benirschke, and O. W. Smith: Conversion of androstenedione-4C14 to estrone by the marmoset placenta. Endocrin. *69*:613, 1961.

Ryle, M.: Studies on possible serological blocks to species hybridization in poultry. J. Exp. Biol. *34*:365, 1957.

————: Early reproductive failure of ewes in a hot environment. I. Ovulation rate and embryonic mortality. J. Agric. Sci. *57*:1, 1961.

————: Early reproductive failure of ewes in a hot environment. II. The uterus. J. Agric. Sci. *58*:137, 1962.

————: Early reproductive failure of ewes in a hot environment. III. The thyroid. J. Agric. Sci. *60*:95, 1963.

Schmid, D. O.: Fertility in cattle and blood groups. Mh. Tierheilk. *15*:302, 1963.

Self, H. L., R. H. Grummer, and L. E. Casida: The effects of various sequences of full and limited feeding on the reproductive phenomena in Chester White and Poland China gilts. J. Anim. Sci. *14*:573, 1955.

Senze, A.: Obojnactwo prawdziwe u konia. Med. wet. *3*:170, 1947.

Shah, M. K.: Reciprocal egg transplantations to study the embryo-uterine relationship in heat-induced failure of pregnancy in rabbits. Nature *177*:1134, 1956.

Shikata, A.: Influence of maternal age upon the development of foetuses in mice. (Japanese). Acta anat. Nippon *35*:109, 1960.

Shipley, E. G., and K. S. Danley: Pituitary and ovarian dysfunction in experimental diabetes. Am. J. Physiol. *150*:84, 1947.

Siddiqui, S.: Constituents of Chana. I. Isolation of three new crystalline products from Chana germ. J. Sci. Ind. Res. *4*:68, 1945.

Smidt, D.: Sexualpotenz und Fruchtbarkeitsvererbung beim Schwein. Munich: BLV Verlagsgesellschaft. 1962.

Sohval, A. R.: Chromosomes and sex chromatin in normal and anomalous sexual development. Physiol. Rev. *43*:306, 1963.

Sorensen, A. M., W. Hansel, W. H. Hough, D. T. Armstrong, K. McEntee, and R. W. Bratton: Causes and prevention of reproductive failures in dairy cattle. I. Influence of underfeeding and overfeeding on growth and development of Holstein heifers. Cornell Univ. Agric. Exp. Sta. Bull. No. 936, 1959.

Squiers, C. D., G. E. Dickerson, and D. T. Mayer: Influence of inbreeding, age and growth rate of sows on sexual maturity, rate of ovulation, fertilization and embryonic survival. Res. Bull. Mo. Agric. Exp. Stn. No. 494, 1952.

Stone, W. H.: The J substance in cattle. Ann. N.Y. Acad. Sci. *97*:269, 1962.

————: Significance of immunologic phenomena in animal reproduction. Proc. 5th Inter. Congr. Anim. Reprod. Artif. Insem. (Trento, Italy) *2*:89, 1964.

Stone, W. H., and M. R. Irwin: Blood groups in animals other than man. Adv. in Immunol. *3*:316. New York: Academic Press, 1963.

Stone, W. H., C. Stormont, and M. R. Irwin: Blood typing as a means of differentiating the potentially fertile from non-fertile heifer born twin with a bull. J. Anim. Sci. *11*:744, 1952.

Stratman, F. W., and H. L. Self: The effect of semen volumes and sperm concentrations on fertility and embryonic survival in artificially inseminated gilts. J. Anim. Sci. *17*:1238, 1958.

Sugiyama, T.: Morphological studies on the placenta of mice of various ages and strains. 4. Placental fusions in mice of colony bred hybrids. Acta Sch. Med. Univ. Kioto *37*:167, 1961.

Sykes, J. F., T. R. Wrenn, and S. R. Hall: The effect of inanition on mammary-gland development and lactation. J. Nutrit. *35*:467, 1948.

Tarkowski, A. K.: Mouse chimaeras developed from fused eggs. Nature *190*:857, 1961.

Teague, H. S.: The influence of alfalfa on ovulation rate and other reproductive phenomena in gilts. J. Anim. Sci. *14*:621, 1955.

Thibault, C.: Analyse de la fécondation de l'oeuf de la truie après accouplementou insémination artificielle. Ann. Inst. natn. de la Rech. agron. Paris: Ser. *D*. Ann de Zootech. *8*, (Suppl.) p. 165, 1959.

Trimberger, G. W.: Ovarian functions intervals between estrus, and conception rates of dairy cattle. J. Dairy Sci. *39*:448, 1956.

Ulberg, L. C.: The influence of high temperature on reproduction. J. Hered. *49*:62, 1958.

———: The reproduction of cattle. Chapter 13. In "Reproduction in Farm Animals," E. S. E. Hafez, ed. Philadelphia: Lea & Febiger, 1962.

Ulberg, L. C., and C. E. Lindley: Use of progesterone and estrogen in the control of reproductive activities in beef cattle. J. Anim. Sci. *19*:1132, 1960.

Waldorf, D. P., W. C. Foote, H. L. Self, A. B. Chapman, and L. E. Casida: Factors affecting fetal pig weight late in gestation. J. Anim. Sci. *16*:976, 1957.

Warnick, A. C., L. E. Casida, and R. H. Grummer: The occurrence of estrus and ovulation in postpartum sows. J. Anim. Sci. *9*:66, 1950.

Wiltbank, J. N., and A. C. Cook: The comparative reproductive performance of nursed cows and milked cows. J. Anim. Sci. *17*:640, 1958.

Woody, C. O., and L. C. Ulberg: Viability of one-cell sheep ova as affected by high environmental temperature. J. Reprod. Fertil. *7*:275, 1964.

Yapp, W. W.: A case of intersex in dairy cattle. J. Dairy Sci. *30*:552, 1947.

Yeates, N. T. M.: Foetal dwarfism in sheep—an effect of high atmospheric temperature during gestation. J. Agric. Sci. *51*:84, 1958.

Young, G. A., R. L. Kilchell, A. J. Luedke, and J. H. Sautter: The effect of viral and other infections of the dam on fetal development in swine. I. Modified live hog cholera viruses—immunological, virological, and gross pathological studies. J. Amer. Vet. Med. Assoc. *126*:165, 1955.

Young, W. C.: Gamete ages at the time of fertilization and the course of gestation in mammals. In "Conference on Pregnancy Wastage," E. T. Engle, ed. Springfield: Thomas, 1953.

CYTOGENETICS OF ABORTIONS

D. H. CARR

Health Sciences Centre, University of Western Ontario,
London, Ontario, Canada

The occurrence of chromosome anomalies in plants and invertebrate animals has been known for many years. In these organisms, various disorders, including the presence of extra chromosomes, have been shown to produce abnormal phenotypes, diminished fertility and increased pollen abortion (Burnham, 1962; Swanson, 1963). Early in 1959, the first chromosome anomalies in man were described. Each was associated with a clearly defined phenotype (Lejeune *et al.*, 1959; Jacobs and Strong, 1959; Ford *et al.*, 1959). In each of these three syndromes, affected individuals presented certain anomalies and diminished fertility but they did not always lead to early death. When lethal chromosome anomalies were demonstrated in newborn infants it was a logical step to study the chromosomes in cases of intrauterine death (Edwards *et al.*, 1960; Patau *et al.*, 1960). Further interest in this subject was stimulated by the description of two abortuses with triploidy (Penrose and Delhanty, 1961; Delhanty *et al.*, 1961). These findings provided the background which led to the study of a large series of spontaneous abortions in man.

MATERIALS AND METHOD

Due to excellent co-operation by physicians and hospital staff in the area, over 400 abortions were collected during two and a half years. An attempt was made, without selection, to culture all material received. Further information regarding the abortions was collected only after tissue culture had been attempted.

The technique of tissue culture was similar to that described by Lejeune *et al.* (1960). Modifications were introduced mainly to deal with the material for study.

All specimens received were washed in running tap water to remove bacteria from the surface. Strict precautions were taken to reduce the risk of contamination. Many specimens received consisted of an intact sac of the chorionic membrane. Sometimes it contained recognizable embryonic tissue but it was frequently empty. Tissue for culture was obtained from amnion, chorionic membrane or umbilical cord. The limbs of smaller embryos were sometimes removed for culture, but they frequently failed to grow. Late in the course of this study, it was discovered that the eyes of the embryo have good growth potential. In the case of larger fetuses, the best tissues are muscle sheath, dura mater and peritoneum.

The selected tissue was cut into fragments about 1 mm. square. Three of these pieces were placed in Leighton tubes on cover slips coated with a thin layer of chicken plasma (Difco dehydrated). The explants were set in place with a bent Pasteur pipette which was discarded after use. One drop of chicken embryo extract (Difco dehydrated—reconstituted to EE_{50}) was added to each tube and allowed to run over the explants. The explants became firmly adherent after setting aside the Leighton tubes, capped with metal covers, for about two hours at room temperature. After the addition of 20 drops of medium, the tubes were closed with silicone stoppers and placed in an incubator at 37°C.

The medium used consisted of 45 per cent fresh frozen, single donor, human AB serum; 45 per cent Hanks' solution and 10 per cent chicken embryo extract (EE_{50}). As a routine, the following antibiotics were added to give final concentrations of: penicillin G sodium 100 U./cc.; streptomycin or dihydrostreptomycin 50 μg./cc. and chloramphenicol 5 μg./cc. These are well below the level known to be toxic to cells in culture (Paul, 1960).

The first cellular outgrowth from the explants was seen within 12 hours to six days. If no growth was seen by the sixth day, experience showed that there was no hope of a growth of cells, adequate for chromosome study. The need to change medium was assessed by change in the color of indicator in the Hanks' solution from red to yellow. When the growing cells extended about half way from an explant to the edge of the cover slip, the culture could usually be prepared for chromosome study. With this amount of growth, the culture medium had become acid and the explants were transferred to new tubes to continue the culture. The explants were simply lifted off with a bent Pasteur pipette and the cellular outgrowth left undisturbed. The medium on these cells was then changed, and next morning, Colcemid (Ciba) was added to the cultures. To obtain the correct concentration, one drop of Colcemid was mixed with 2 cc. of Hanks' solution warmed to 37°C. One drop of this mixture was then added to each culture tube which was then agitated and quickly returned to the incubator. The cultures were processed 6 hours later.

After removing the culture medium, ½ cc. of 0.9 per cent sodium citrate was added to each tube and the cells exposed to this solution for two minutes. The cover slip was then removed from the Leighton tube with a bent metal spatula and placed, cells uppermost in 45 per cent acetic acid in water. After 10 minutes' fixation, the cover slip was inverted on to a siliconised slide and gently squashed with a metal bar weighing 2.5 kg., no additional pressure being required. The cover slip was removed after thorough freezing on a block of solid carbon dioxide, dipped in absolute alcohol and air dried. The cover slip could then be kept for staining without deterioration. The cells were stained with carbol fuchsin and the cover slips mounted on slides with DePeX to make permanent preparations (Carr and Walker, 1961).

Slides were scanned using a 10× objective of a Zeiss binocular photomicroscope. Good cells were studied with the oil immersion lens. An attempt was made to count the chromosomes of at least thirty cells. At least three cells of the best quality available were photographed using Kodak high contrast copy film. The film was printed on 8″ x 10″ Kodabromide No. 5 paper to give a final magnification of the chromosomes of 5,000 times. The chromosomes were cut from the photograph and stuck on to a card with Kodak mounting tissue to form a karyotype according to the London-modified Denver system (1964). According to this internationally agreed method of chromosome array, the 22 pairs of autosomes are arranged in descending order of size. They fall into 7 groups sometimes given the letters A to G, and a pair of sex chromosomes XX in the female and XY in the male.

In addition to the chromosome preparations, sex chromatin was studied in interphase nuclei. This examination was made by staining a whole mount of amnion, or cells grown in culture or both. Cells were stained with carbol fuchsin after fixation in acetic alcohol (3:1).

Additional information was sought for the first 200 successfully cultured abortions as well as for 149 abortions, obtained during the same period, which failed to grow in culture. This information included maternal age, period of gestation before abortion and number of previous pregnancies and their results.

During the course of the study, it became clear that stillborn infants were being sent for analysis only if they presented anomalies. It was therefore decided to restrict the study to previable abortions which were defined as fetuses under 500 gm., regardless of their menstrual age, in accordance with the recommendation of Javert (1957). In addition to 23 infants, stillborn or dying within two hours of birth, there were 38 induced abortions and ectopic pregnancies which were excluded from the series.

RESULTS

There were 227 abortions which grew successfully in culture. As far as it was possible to judge from records, they were aborted without outside interference. A history of the pregnancy was available for all but 5 of the abortions. Among the 227 spontaneous abortions for which chromosome study was possible, there were 50 with numerical anomalies, an incidence of 22 per cent. Though various types of structural chromosome anomalies have been described in living subjects, none were found among the abortions.

The mean length of time between setting up the culture and obtaining the first cells suitable for chromosome analysis was 8.7 days. This was about the same for chromosomally normal and abnormal abortions.

The results are summarized in Table 1. In just over half of the chromosomally abnormal abortions one extra chromosome was found. This anomaly is known as trisomy. Trisomies were found for each of the 7 groups of autosomes. In the case of certain trisomies, involving the groups C and G, it is possible to confuse an extra element with one of the sex chromosomes, X or Y respectively. Karyotypes and sex chromatin analysis of the trisomies in groups C and G did not suggest the presence of an extra X or Y chromosome. In other words, there were no specimens with an XXY or XYY sex chromosome complex. This is not surprising as living subjects with these chromosome anomalies do not usually suffer from severe systemic disorders.

One of the specimens consisted of an intact sac containing an umbilical cord but no embryo. On culture, cells were found to contain an extra

Table 1

ASSOCIATION BETWEEN CHROMOSOME ANOMALY AND ABORTUS

Anomaly	Number	Mean Maternal Age	Types of Specimen
Trisomy no. 3	1	25	Intact sac. Cord but no embryo.
Trisomy B	1	29	Intact sac. Nodular embryo.
Trisomy C	2	41	Intact sac. Ruptured sac.
Trisomy D	6	35	3 embryos. Others sacs.
Trisomy E (16)	8	25 ⎫	Intact or ruptured
(17–18)	1	33 ⎭	sacs.
Trisomy F	1	32	Intact empty sac.
Trisomy G	6	32.5	1 embryo. Others sacs.
Trisomy E and G	1	25	Intact empty sac.
Triploid	9	28	5 embryos. Varied from nodular to normal.
Tetraploid	2	23.5	1 nodular embryo. 1 intact empty sac.
X0	12	25.2	9 embryos or fetuses.

chromosome which fell in group A. The three pairs of autosomes in this group can be separated from one another as well as from other members of the complement and the extra chromosome was indistinguishable from chromosome no. 3. This chromosome can be mimicked by a translocation (Edwards *et al.*, 1962) or by an X-isochromosome (Fraccaro *et al.*, 1960). To exclude, as far as possible, a translocation product carried in the gametes, chromosome studies of blood and skin cells were made on the woman who produced the abortion and her husband. The chromosomes of both were normal and she has since borne a normal child. An autoradiographic study was made on several cultures from the abortion using tritiated thymidine. The results were not good and positive identification of the extra element could not be made. The data from the autoradiographic study may be summarized as follows: None of the elements resembling chromosome no. 3 were late labelling, which seems to exclude the possibility that the extra member was formed from an X chromosome. The labelling of the no. 3 chromosomes were never strikingly different from one another and in two or three cells they were identical. The autoradiographic studies, therefore, were not against the identification of the extra chromosome as no. 3.

One intact sac containing a small nodule of tissue produced cells with an extra chromosome in group B. The two chromosomal pairs which make up this group can be distinguished from all other members of the complement but not from each other. It is therefore impossible to tell if this was a trisomy 4 or 5. The woman who produced this abortion had had two previous abortions and two normal children. Such a history suggests the possibility of a translocation existing in a parent. The chromosomes in blood cultures of both parents were normal.

Although the chromosomes in group F are among the smallest members of the complement, only one abortion, consisting of a small intact empty sac, produced cells with an extra element in this group. The two pairs of chromosomes in group F, like those in group B, can be distinguished from all other members of the complement but not from each other. Therefore, it was not possible to say whether the extra element was chromosome 19 or 20. The woman who produced the abortion had normal chromosomes, as did her husband. They had one normal child and no history of previous abortion.

The presence of two extra chromosomes, double trisomy, was demonstrated in cells grown from an intact empty sac. Although only two cells were satisfactory for chromosome study, the karyotype was identical in each instance. There was an extra element in group E, but it was uncertain whether it was no. 16 or 17, and an extra member in group G indistinguishable from the normal members of that group.

The individual members of group C are so similar that an extra element can only tentatively be assigned a number. The X chromosome falls in this group but the presence of an extra X chromosome can be distinguished from an autosome by studying the sex chromatin in interphase nuclei. This was done, both in cultured cells and in a whole mount of amnion for the two specimens whose cells were trisomic for a member of the C group. It was concluded that the extra element was an autosome in each instance. A karyotype from one of the specimens tentatively assigns the extra member to the no. 10 position (Fig. 1). One of the specimens trisomic for a C group autosome was an intact empty sac. The other was a ruptured sac with no indication of embryonic remains.

It is impossible to distinguish the three chromosomal pairs in group D. The six abortions whose cells showed D trisomy did not present features which distinguished them chromosomally, either from one another, or from the trisomy found in the "Bartholin-Patau" syndrome to which reference will be made later. In one culture, autoradiographic studies were attempted but the result was uninformative.

Three of the abortions with D trisomy contained an embryo. In one instance the embryo was so macerated as to be useless for study. How-

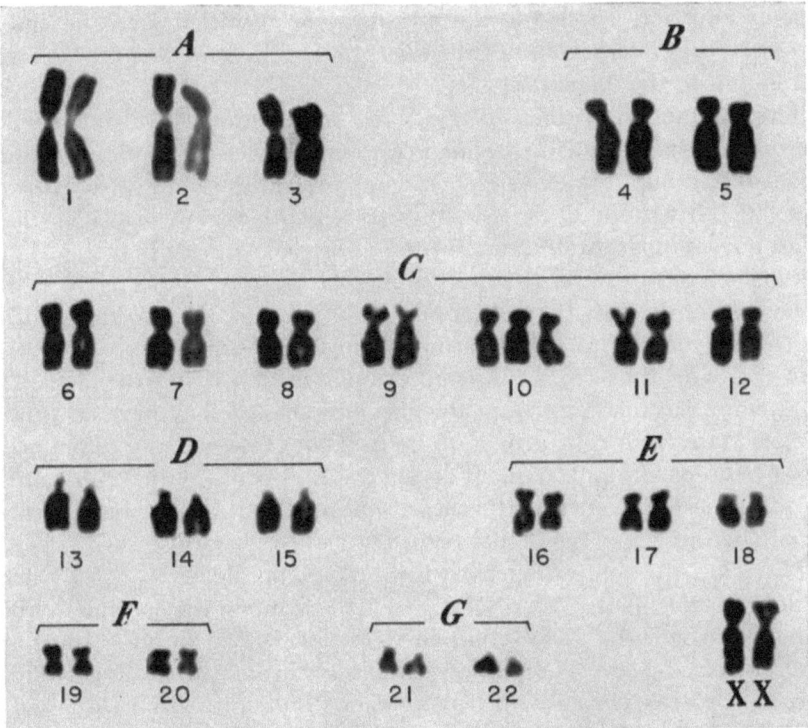

Fig. 1. Karyotype showing C group trisomy. Extra element tentatively placed as chromosome 10.

ever, it was noted that both eyes were developed normally. A 17 mm. embryo from an intact sac appeared to be normal. There was no indication of eye anomalies or polydactyly at that stage. These are areas commonly affected in the clinical syndrome first described by Patau *et al.* (1960). The third specimen, twin of a triploid embryo, was decapitated during abortion. The only striking defect of the trunk was that the lower limb development was retarded in relation to the upper limb. It seems likely that had such an embryo come to term it would have had phocomelia.

Although in occasional cells it is possible to distinguish a larger and a smaller pair in group G, there is no way to regularly separate them. However, it is almost always possible, in good preparations, to recognize the Y chromosome, which in some respects resembles the autosomes in group G. The extra element did not resemble the Y chromosome in any of the 6 cultures with trisomy G. In 5 of the 6 specimens with trisomy in group G, the extra element was indistinguishable from autosomes 21 and 22. In the cells derived from the other abortion, the extra G group chromosome appeared to lack short arms in which respect it differed from the normal members of the complement. This was the only G trisomy abortion which was not an empty or ruptured sac. The embryo was obviously microcephalic and microphthalmic and was very retarded in relation to the menstrual age.

The commonest trisomy in this series of abortions involved members of group E. In 8 of the 9 specimens whose cells were trisomic for a member of this group, the extra element was indistinguishable from chromosome 16. This is one of the 4 autosomes which is regarded as identifiable in all good preparations. None of the abortions with cells trisomic for a member of group E provided an embryo for study. Trisomy 16 proved to be the commonest type of trisomy in this series of abortions which is in striking contrast to the situation in the liveborn.

In studying cells from the 9 abortions which proved to be triploid, an attempt was made to count the chromosomes, if they were adequately spread. Many such cells proved to be broken. However, the difference in count between diploid, triploid and tetraploid cells (with 46, 69 and 92 chromosomes) is so striking that a triploid cell could easily be recognized visually without making actual counts.

There are three possible sex chromosome complexes in a cell with a triploid complement, XXY, XXX and XYY. Among the specimens about to be described, the XXY gonosomal complex was found 6 times, the XXX twice and the XYY once. The sex chromatin findings differed according to sex chromosome complement and have been discussed in detail elsewhere (Carr, 1965).

The triploid abortuses varied from intact empty sacs to an apparently normal embryo. One triploid embryo had an encephalocele and the gonosomes in this instance were XXX (Singh, 1966). A karyotype from this abortus is shown in Figure 2. Since this series was completed another triploid embryo has been received and it appeared to be normal. The chromosomes could not be counted accurately and owing to the poor quality of the preparation no karyotypes could be constructed.

There were two specimens which gave rise to cells all of which were tetraploid. As with the triploid specimens, accurate counts were only possible on a few well spread cells. Tetraploid cells are, however, easily distinguishable visually, both from normal diploid and from triploid cells. One specimen was an intact empty sac and the chromosome complex was 88 + XXXX. The other specimen, also an intact sac, contained a small nodule of embryonic tissue. In this instance, the sex chromosome complex was XXYY.

It is quite common for tetraploid cells to arise in an otherwise normal culture. It was most important to consider whether or not the tetraploidy in cells grown from these two specimens, was in fact representative of the tissue before culture. In deciding that these were not artefacts, the following criteria were observed. The tetraploid cells were seen in the first culture studied which was within 8 days of explanting the tissue. All cells seen were in the tetraploid range and none were diploid. The tetraploid cells were seen in more than one culture. Finally, in the case of the XXXX

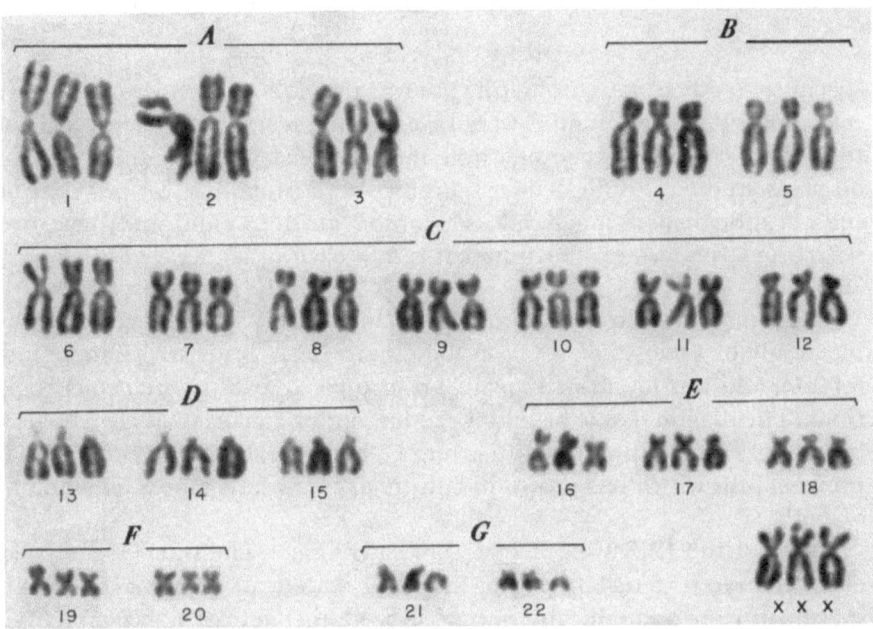

Fig. 2. Triploid karyotype with XXX sex-chromosome complex.

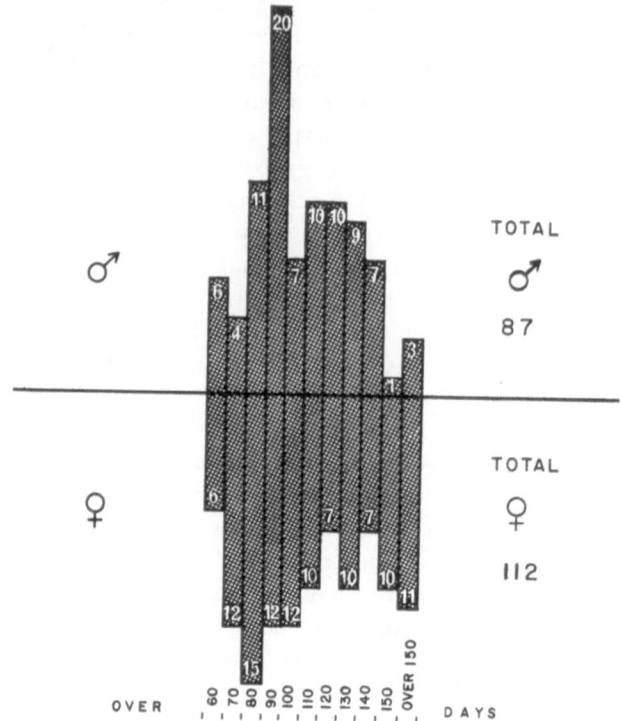

Fig. 3. Histogram showing sex ratio of abortions at different stages of pregnancy.

specimen, study of sex chromatin confirms that the amnion was uniformly tetraploid. Pieces of amnion were taken from several areas of the wall of the empty sac and stained as whole mounts for sex chromatin study. In all areas studied, duplicated sex chromatin was found in 95 per cent of cells. Although occasional cells in amnion are tetraploid and have two sex chromatin masses, the number rarely exceeds 2 per cent (Klinger, 1957).

The commonest chromosome anomaly among 227 successful cultures was the result of absence of a sex chromosome. This leads to X-monosomy (or X0), and was found in 12 or just over 5 per cent of spontaneous abortions. The diagnosis was based on the finding of 45 chromosomes, including only 15 in group C and 4 in group G, and the absence of sex chromatin from cells which had grown in culture or in a whole mount of amnion, or both.

Three of the 12 X0 specimens were empty sacs. The other 9 consisted of an embryo or fetus, in varying stages of maceration. Four of the 9 X0 specimens were anatomically normal. Two larger fetuses had lymphangiomata in the neck region. Two of the X0 embryos had horseshoe kidneys, an anomaly commonly found in living individuals with this chromosome disorder. In all of the X0 embryos and fetuses the gonads appeared normal and on section contained primitive germ cells. This is in contrast with the findings in most X0 adults who have streak ovaries without germinal elements (Singh and Carr, 1966).

If the polyploid specimens with a Y chromosome are included with the males and those without a Y with the females, the sex ratio of the 227 successfully cultured abortions was 116 female:99 male, the 12 X0 specimens being unassigned. Though most series of abortions contain an excess of males (Stevenson, 1966) there are exceptions (Moore and Hyrniuk, 1960; Wingate, 1966). There is some indication that there may be an excess of female abortions early in pregnancy (Stevenson, 1966). With this in mind, the sex of the 215 abortions were plotted in a histogram according to gestational age (Fig. 3). There did not appear to be any consistent pattern regarding the sex ratio at different stages of pregnancy.

The maternal age at the time of abortion, as well as the gestational age of the specimens, was analysed for the first 200 abortions. Though not included in this analysis, data for 6 previously unreported chromosomally abnormal specimens is shown in Table 2. The mean maternal age at the time of abortion for the first 44 abnormal specimens was 28.9 (±1.19) years. This was not significantly different from the mean maternal age of 151 women producing chromosomally normal abortions, which was 27.1 (±0.52) years. However, there is reason to believe that the trisomic specimens should be treated separately. It is known that trisomy in living subjects becomes increasingly common with increasing maternal age (Penrose, 1963; Taylor and Polani, 1964). On the other hand, Boyer *et al.* (1961) concluded that there was no increase in parental age in cases of chromatin-negative Turner's syndrome. It has even been suggested

Table 2

INFORMATION ON 6 CHROMOSOMALLY ABNORMAL SPECIMENS

Specimen No.	Last Menstrual Period	Date of Abortion	Length of Pregnancy (days)	Maternal Age	Previous Abortions	Live Births	Anomaly
202	22/9/64	17/12/64	86	30	0	0	Trisomy G ♂
211	20/11/64	1/2/65	73	20	0	0	Trisomy 16 ♀
213	4/11/64	6/2/65	94	32	0	1	Trisomy F ♀
214	15/11/64	18/2/65	95	34	0	2	Trisomy 16 ♂
216	15/1/65	6/4/65	81	17	0	0	X0
218	15/1/65	6/4/65	81	25	0	1	Trisomy E & G ♂

that infants with Turner's syndrome may be commoner as the offspring of younger mothers (Lamy *et al.*, 1965). Though the X0 abortuses in this series had a mean maternal age less than that of women producing abortions with normal chromosomes, the difference was not significant. Nothing is known regarding parental age in relation to triploidy because the anomaly has not been found to affect all the cells of the body in any living individual. The triploid abortions did not come from significantly older women.

For 22 trisomic specimens in this series the mean maternal age was 31.3 (±1.88) years. This was significantly different from the mean maternal age for the abortuses with normal chromosomes (p < 0.01). Polani (1966) has pointed out that the effect of maternal age on trisomy in man is not the same for all chromosomes. He noted that known abortuses with trisomy 16 came from younger women. As more cases are described, it will be interesting to see whether this is a difference in behaviour between acrocentric and metacentric chromosomes. So far, it appears that trisomy for the acrocentric chromosomes in groups D and G is commoner in abortions of older mothers. Lack of age effect may apply to all trisomies of metacentric chromosomes or be characteristic of trisomy 16. It is not possible to tell until more cases are available for study.

There was a striking difference between the mean gestational age of abortions with and without chromosome anomalies. The mean duration of pregnancy for 44 specimens with chromosome anomalies was 85.9 (±3.21) days from the first day of the last menstrual period to the day of abortion. The mean gestational age for 144 abortions with normal chromosomes was 106.7 (±8.97) days. The difference is highly significant statistically (p < 0.001). It should also be noted that the difference in gestational age for trisomic, polyploid (triploid and tetraploid) and X0 abortions is striking. The mean gestational age of these three groups was 74.3, 87.6 and 98.4 days respectively. These differences are illustrated in histogram form in Figure 4.

The finding of an extra chromosome resembling a normal member of the complement does not exclude other possible origins for the element. Vislie *et al.* (1962) described a family in which the mother proved to have a balanced translocation between D and E chromosomes. Of the two products of the translocation, one was the size of a chromosome in group C, whereas the other resembled members of the G group. The mother had had two living children and a miscarriage. Both children were retarded and had 47 chromosomes, the extra element being the translocation product from the mother which resembled the members of group G. It was recognized that this mechanism might be responsible for the extra chromosome in some of the abortions studied. Families were contacted in order to study their chromosomes. The woman producing the

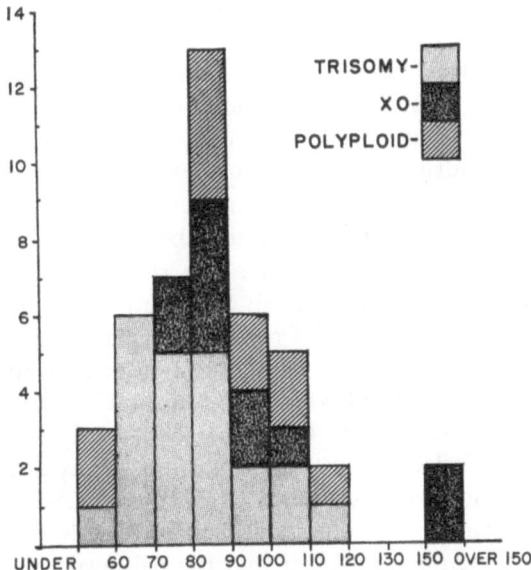

Fig. 4. Histogram of 44 chromosomally abnormal abortions in relation to gestational age.

abortion and her husband had leucocyte cultures performed for 14 of the 26 cases of trisomic abortion. The wife alone had chromosome studies in two other instances. In all cases, chromosome analysis was normal (Singh, Rubinoff and Carr, 1966).

During the period in which 200 specimens of spontaneous abortions were successfully cultured for chromosome analysis, a further 149 abortuses failed to grow in culture. This group of specimens which failed to grow may have affected the incidence of chromosome anomalies in the series as a whole. The chromosome constitution was unknown but it was possible to study three features; the duration of pregnancy and maternal age at the time of abortion and the previous obstetrical history. Among the successfully cultured specimens, the previous obstetrical history did not differ between women producing abortions with normal chromosomes and those aborting chromosomally abnormal specimens. This aspect was therefore of no value in assessing the significance of the failed cultures.

The maternal age of women producing trisomic abortuses was significantly greater than those whose abortions had normal chromosomes. If the maternal age of the women whose abortions failed to grow had been higher than normal, this would have suggested an excess of trisomic abortions in this group. However, it was found that the maternal age of women whose abortions failed to grow was not significantly different from that of the women whose abortions were of known chromosomal constitution.

It is known that the abortions with abnormal chromosomes were from shorter pregnancies than those with normal chromosomes, and that the difference was highly significant statistically. It could therefore be argued that, if the failed cultures contained an excess of chromosomally abnormal specimens, they should come from pregnancies of shorter duration. The mean duration of pregnancy for the abortions which failed to grow in culture was intermediate between the chromosomally normal and abnormal abortions and significantly different from both. Two comparisons were made: between the duration of pregnancy for the failed cultures and the pooled data for the successful cultures and also between the "failed" and "normal" plus twice "abnormal" to see if it improved the fit. In neither case was the difference statistically significant. It was concluded that the failed cultures did not differ in any measurable degree from those with known chromosomal constitution (Carr *et al.*, 1966).

It had been intended to include chromosome analysis of fetuses of all ages. However, it soon became clear that the stillborn infants we received were not unselected. The presence of abnormalities seemed to be the main reason that viable fetuses were sent for study. On account of this, all abortuses weighing more than 500 gm. were excluded from the original analysis. However, chromosome study was successfully performed on 17 of the 23 fetuses we received which were stillborn or died within two hours of birth. The chromosomes were normal in each case.

At the time abortions were received for study, the pregnancy history was usually unknown. Though induced abortions and ectopic pregnancies were not solicited, several such specimens were received. The 34 successfully cultured specimens in these two categories all had normal chromosomes.

DISCUSSION

The incidence of chromosome anomalies among 227 spontaneous abortions was 22 per cent and this figure did not vary appreciably throughout the study. The specimens were unselected and it is felt that 20–25 per cent is a valid incidence for chromosome anomalies among clinically recognizable previable human abortions. This is about 50 times as high as the incidence of all types of chromosome anomalies in the liveborn infant (Marden, *et al.*, 1964; Carr, 1965). The total "zygotic loss" in pregnancy is certainly higher than the 15 per cent occurrence of clinically recognizable abortion. Hertig *et al.* (1956) obtained 34 human ova within 17 days of fertilization. Of these specimens, 13 or 30 per cent were abnormal on microscopic examination. It is likely, in view of these findings, that man, like other mammals studied, suffers a zygotic loss of at least 30 per cent, which is about twice the incidence of clinical abortion. The high zygotic loss in mammals has been found in the horse, ferret (Robinson, 1921), pig (Corner, 1923; Casida, 1953), rhesus monkey (Corner and Bartelmez, 1953) and rabbit (Casida, 1953). Definite proof of chromosome anomalies in early zygotes is available for the pig (McFeely, 1966). Earlier, Vara and Peronen (1947) found abnormal chromosome numbers in unfertilized and in early fertilized ova in rabbit and cat.

The incidence and types of chromosome anomalies is comparable in most respects to studies by other workers. In three studies of chromosome anomalies in spontaneous abortions, the incidence varied from 17 to 30 per cent (Clendenin and Benirschke, 1963; Hall and Källen, 1964; Kerr and Rashad, 1966). However, in two other studies of spontaneous abortions the incidence of chromosome anomalies reached 65 per cent (Thiede and Salm, 1964; Szulman, 1965). Thiede and Salm (1964) selected their material for study mainly from "blighted ova." It is clear that such abortuses have a high incidence of chromosome anomalies (Carr, 1965). Szulman (1965) appears to have selected abortions of short gestational age but the figure of 65 per cent is higher than one would expect from the present study.

The types of anomalies found by various workers are similar and have been reviewed by Kerr and Rashad (1966). It is interesting to note that other investigators have found abortions trisomic for chromosome no. 2 (Hall and Källen, 1964; Kerr and Rashad, 1966). The absence of any example in this series suggests that cells trisomic for this chromosome may have special nutritional requirements not met by the medium used in this study. Anyone embarking on this type of study in the future would be well advised to set up cultures in different types of media.

As a group, trisomy, or the presence of one extra chromosome, was the commonest anomaly in the present series. With one exception, the groups containing larger chromosomes were less likely to have trisomic elements than those containing smaller chromosomes. The rarity of trisomy in group F was the striking exception. The two chromosomal pairs in this group are among the smallest members of the complement. Only one other example of an abortion with F group trisomy has been described in the literature (Inhorn *et al.,* 1964). It may be that certain genes on members of group F are vital to early embryogenesis and any disorder leads to maldevelopment and abortion. If this is so, we must presume that not one but both pairs of chromosomes in group F have this special importance.

No living individuals have been found to be trisomic for chromosome no. 3 or a member of group B. However, single abortuses with B group trisomy and trisomy no. 3 have been reported (Sato, 1965; Kerr and Rashad, 1966).

The presence of an extra autosome in group C has never been found in all the cells of a living subject though it has been described in abortions

(Szulman, 1965; Rashad and Kerr, 1966). The two abortions with an extra element in group C were believed to be trisomic for an autosome and not an X chromosome on the basis of sex chromatin studies. The X chromosome, which cannot be distinguished morphologically from the larger chromosomes in group C, behaves differently from the autosomes. It is believed that the additional X chromosome is relatively inactive genetically (Russell, 1964). It is this property which leads to the formation of the sex chromatin body. As a general rule, any X chromosomes above one in the cell, become contracted, relatively inactive and are visible as the sex chromatin body. It is this property of the X chromosome which makes trisomy for this element relatively innocuous to the individual (Barr, 1966).

Trisomy for chromosome no. 3 and for groups B, C and F were all associated with "blighted ova." The effect of the extra chromosome on the phenotype was so devastating that the embryo was virtually destroyed.

The commonest trisomy in liveborn infants involves one of the members of group G. The phenotype is readily recognized and known as Down's syndrome. The extra element is commonly called chromosome 21, though the two pairs of chromosomes in group G cannot usually be distinguished from each other. In any case, no constant phenotype has been recognized for trisomy of the other pair in group G and is perhaps recovered among aborted specimens. The only embryo recovered from 6 abortions with G trisomy was grossly abnormal.

The other clearly defined phenotypes associated with trisomy in man are both more lethal than Down's syndrome and in each instance a larger chromosome is involved. In the E_1 trisomy syndrome, it is not certain which member of group E is found in excess, 17 or 18 (Butler *et al.,* 1965). In any case infants with trisomy for this chromosome present a constant phenotype and usually die within weeks or months of birth (Smith, 1964). Only one specimen with trisomy for chromosome 17 or 18 was found in the present series. The commonest single trisomy among the abortions was due to the presence of an extra element indistinguishable from chromosome 16. The phenotype could not be assessed as all the abortions were empty sacs. However, Clendenin and Benirschke (1963) found an embryo with trisomy no. 16 and it appeared to be normal.

Trisomy for an extra member of group D produces multiple congenital anomalies in affected infants who may be stillborn or die within hours or days of birth (Smith, 1964). It is not possible to distinguish the three pairs in group D from one another. Therefore it is impossible to state whether or not the trisomic abortions in this series involved the same chromosome as that found in the D_1 trisomy syndrome. Three embryos were recovered from the 6 abortions with D trisomy. The eyes were well developed in two of them and the head was missing in the other. The eyes are usually small in the D_1 trisomy syndrome but even the presence of normal eyes in the early embryo does not guarantee their continued normal development. The one embryo without a head had lower limbs which were so retarded in development as to constitute an anomaly. No such defect has ever been described in the D_1 trisomy syndrome.

Trisomy in man could theoretically occur for each of the 22 pairs of autosomes as well as the X chromosome. Only three autosomal trisomies are known to be compatible with live birth and in two of these syndromes, as already mentioned, the affected infants soon die. Trisomy is now known to affect chromosomes of the 7 groups of autosomes but most of these have been recovered only in abortions.

We have little information regarding trisomy in other mammals. Trisomy has been described in the mouse and involved a smaller chromosome in the complement (Cattanach, 1964; Griffen and Bunker, 1964). The only apparent phenotypic effect was sterility. If further work confirms the relatively mild effect of autosomal trisomy in other mammals, it will indicate a striking contrast to the situation in man.

Only one example of double trisomy, the presence of two different extra chromosomes, was found in this series. The abortus was an intact empty sac. Double trisomy has been described, both in living individuals and in spontaneous abortions (Pergament and Kadotani, 1965; Kerr and Rashad, 1966).

Triploidy has now been described in a large number of aborted specimens (Penrose and Delhanty, 1961; Delhanty *et al.*, 1961; Atkin and Klinger, 1962; Aspillaga *et al.*, 1964; Inhorn *et al.*, 1964; Makino *et al.*, 1964; Thiede and Salm, 1964; Szulman, 1965; Rashad and Kerr, 1966). No living individual has yet been found to have a triploid chromosome complement in all the body cells. It may be the hemopoietic system which is particularly sensitive to this chromosomal disorder. Judging by the abortions in this series, the effect on the developing embryo is highly variable. Triploid specimens included intact empty sacs, nodular, cylindrical and stunted embryos and one well developed 18 mm. embryo with an encephalocele. Two other triploid embryos of about 7 weeks' developmental age appeared quite normal. In addition to the effect on the fetus, there is a possible relationship between triploidy and degenerative changes in the placental villi (Atkin and Klinger, 1962; Makino *et al.*, 1964).

Tetraploidy is less damaging to development than triploidy in plants and is usually associated with normal fertility (Swanson, 1963). However, in sexually reproducing organisms, tetraploidy and higher polyploids are rare. In man, the only description of tetraploidy has been in abortions (Inhorn *et al.*, 1964; Thiede and Salm, 1964). Thiede and Salm (1964) found tetraploid mosaicism in a 75 mm. fetus but both the tetraploid abortions reported here were in the category of "blighted ova."

Perhaps the most surprising finding in this cytogenetic study of abortions was the high incidence of the X0 anomaly. Twelve of the 227 abortions were found to be monosomic for the X chromosome, an incidence of over 5 per cent. This is in contrast to the incidence of 1 in 5,000 among liveborn infants (Maclean *et al.*, 1964). If we accept a 5 per cent figure for the true incidence of X0 abortions, and a figure of 15 per cent as the incidence of spontaneous abortion in man, it can be calculated that only 1 in 40 X0 zygotes survive to birth.

It can also be calculated from this data that at the time of fertilization, about 0.8 per cent of all zygotes are X0. This is very close to the average calculated incidence for the X0 anomaly in the mouse, 0.7 per cent (Russell and Saylors, 1961). The X0 mouse is a normal fertile female but may have a reduced prenatal viability of about 60 per cent (Russell *et al.*, 1959). Cattanach (1962) found no evidence that the shortage of X0 females in mice was due to death during embryogenesis. In any case, the prenatal loss would not approach the level found in man. It is tempting to suggest that the human X chromosome carries a lethal gene normally suppressed by the presence of a second X chromosome. If that is the case, it is necessary to propose the same protective function for the Y chromosome in the male. Lindsley *et al.* have shown that the Y chromosome suppresses the expression of sex-linked recessive lethals in Drosophila. Whatever the mechanism, it is clear that the Y chromosome in man is responsible for protecting the X-hemizygote against prenatal death. To complete the comparison with the mouse it may be said that laboratory strains, being highly in-bred, have selectively lost animals carrying X-linked lethal genes.

What is the cause of chromosome anomalies? We do not know, but there are indications that multiple factors may be involved. Radiation, maternal aging, genetic effects and virus infections have all been proposed, as well as a combination of these factors (Hecht *et al.*, 1964). This is one direction for future research in this field. We already have evidence that maternal age is a factor in the etiology of some human trisomies. Uchida (1962) was unable to demonstrate any effect of maternal age on the frequency of X0 and XXY progeny in Drosophila. However, the effect of radiation in producing abnormal progeny was enhanced by maternal aging. Goodlin (1965) attempted to study the effect of maternal age on non-disjunction in the mouse. He concluded "that in the stock of mice studied there was no evidence of increased aneuploidy with maternal age among the live newborn offspring."

Another guide to the effect of aging comes from work on amphibia. Witschi and Laguens (1963) have shown that aging the ova of amphibian eggs produces a high incidence of monosomy, trisomy and occasionally polyploidy.

The work of Witschi and Laguens (1963) also emphasizes the fact that aging of ova can be considered in two ways. This aspect is especially relevant in mammals. There is the effect due to an aging female in the reproductive sense whose ova have been in prophase (dictyotene) for more years than the postnatal age of the animal. On the other hand, the scarring of the ovary associated with repeated ovulation and corpus luteum formation in mammals could interfere with the bursting of mature follicles. In that case, the liberation of the oocyte could be delayed by hours and thus appreciably lengthen the time before fertilization and the completion of meiosis. This itself may well interfere with normal disjunction of chromosomes.

The final point of interest in this series of abortions is the sex ratio. It is known that there is an excess of males over females at birth though it varies geographically (Stern, 1960). The conflicting results in studies of sex ratio have already been mentioned. Although the majority of workers have found an excess of male embryos at all stages of pregnancy there is still a suspicion of early embryonic loss of female conceptuses (Stevenson, 1966). In sex chromatin studies the chromatin-negative specimens must be reduced by 5 per cent of the total to eliminate the X0 abortions. In addition, great care is required in the interpretation of sex chromatin studies in spontaneous abortions which often give poor cellular detail.

Summary

1. In a series of 227 unselected spontaneous human abortions, 50 were found to have numerical chromosome anomalies. None were found to have structural anomalies.

2. About half the chromosomally abnormal specimens had one extra chromosome, resulting in trisomy. One quarter were monosomic for the X chromosome (X0) and nearly a quarter were polyploid (triploid or tetraploid).

3. Abortions trisomic for chromosomes in groups A, B, C, E and F, as well as the two tetraploid specimens, were either "blighted ova" or ruptured chorionic sacs from which no embryonic remains could be recovered.

4. Half of the 6 abortuses with D trisomy had at least part of an embryo. One embryo was too macerated for study, one was apparently normal, while the other had gross underdevelopment of the lower limbs.

5. Only one of 6 abortions trisomic for a chromosome resembling the G group provided an embryo for study. This was grossly abnormal.

6. The 9 triploid abortions included intact empty sacs, nodular, cylindrical, stunted, abnormal and normal embryos. Thus the most variable phenotype was found in this group.

7. Nine of 12 X0 abortions had an embryo or fetus. Two had lymphangiomata, two had horseshoe kidneys and four were anatomically

normal. Germ cells were recognized in the gonads of all the X0 embryos and fetuses.

8. Increase in maternal age appears to be related to the incidence of certain trisomies. Other trisomies, triploidy and X0 abortions were not related to maternal age.

9. In those families which could be studied, no chromosome anomalies were found among women who had had a chromosomally abnormal abortion or among their husbands.

10. As far as could be determined, the abortions which failed to grow did not invalidate any conclusions regarding incidence of chromosome anomalies among successfully cultured specimens.

11. Though the series is small, the excess of female abortions is in conflict with most other reports based only on sex chromatin studies.

Acknowledgments

The author wishes to thank all those who co-operated in the collection of material for this research. This study was supported by grants from the Medical Research Council of Canada and the D. H. McDermid Research Fund.

References

Aspillaga, M. J., R. J. Schlegel, R. Neu, and L. I. Gardner: Triploid/diploid mosaicism in tissue from an early fetus. J. Pediat. (Abstr.) *65:*1098, 1964.

Atkin, N. B., and H. P. Klinger: The superfemale mole. Lancet *2:*727, 1962.

Barr, M. L.: Correlations between sex chromatin patterns and sex chromosome complexes in man. In K. L. Moore, ed. The Sex Chromatin. Philadelphia: W. B. Saunders Co. 129, 1966.

Boyer, S. H., M. A. Ferguson-Smith, and M. M. Grumbach: The lack of influence of parental age and birth order in the aetiology of nuclear sex chromatin-negative Turner's syndrome. Ann. Hum. Genet. *25:*215, 1961.

Burnham, C. R.: Discussions in Cytogenetics. Minneapolis: Burgess Publishing Co., 1962.

Butler, L. J., G. J. A. I. Snodgrass, N. E. France, L. Sinclair, and A. Russell: E (16–18) trisomy syndrome: Analysis of 13 cases. Arch. Dis. Child. *40:*600, 1965.

Carr, D. H., and J. E. Walker: Carbol fuchsin as a stain for human chromosomes. Stain Technol. *36:*233, 1961.

————: Chromosome studies in spontaneous abortions. Obstet. Gynec. *26:*308, 1965.

Carr, D. H., A. J. Bateman, and A. B. Murray: Analysis of data from abortions which failed to grow in culture. *In press.* Obstet. Gynec. 1966.

Casida, L. E.: Fertilization failure and embryonic death in domestic animals. In E. Engel, ed. Pregnancy Wastage. Springfield: C. C. Thomas, 27, 1953.

Cattanach, B. M.: X0 mice. Genet. Res. *3:*487, 1962.

————: Autosomal trisomy in the mouse. Cytogenetics (Basel) *3:*159, 1964.

Clendenin, T. M., and K. Benirschke: Chromosome studies on spontaneous abortions. Lab. Invest. *12:*1281, 1963.

Corner, G. W.: The problem of embryonic pathology in mammals, with observations upon intra-uterine mortality in the pig. Amer. J. Anat. *31:*523, 1923.

Corner, G. W., and G. W. Bartelmez: Early Embryos of the Rhesus Monkey. In E. Engel, ed. Pregnancy Wastage. Springfield: C. C. Thomas, 3, 1953.

Delhanty, J. D. A., J. R. Ellis, and P. T. Rowley: Triploid cells in a human embryo. Lancet *1:*1286, 1961.

Edwards, J. H., D. G. Harnden, A. H. Cameron, V. M. Crosse, and O. H. Wolff: A new trisomic syndrome. Lancet *1:*787, 1960.

Edwards, J. H., M. Fraccaro, P. Davies, and R. B. Young: Structural heterozygosis in man: analysis of two families. Ann. Hum. Genet. *26:*163, 1962.

Ford, C. E., K. W. Jones, P. H. Polani, J. C. de Almeida, and J. H. Briggs: A sex-chromosome anomaly in a case of gonadal dysgenesis (Turner's syndrome). Lancet *1:*711, 1959.

Fraccaro, M., D. Ikkos, J. Lindsten, R. Luft and K. Kaijser: A new type of chromosomal abnormality in gonadal dysgenesis. Lancet *2:*1144, 1960.

Goodlin, R. C.: Non-disjunction and maternal age in the mouse. J. Reprod. Fertil. *9:*355, 1965.

Griffen, A. B., and M. C. Bunker: Three cases of trisomy in the mouse. Proc. Nat. Acad. Sci. U.S.A. *52:*1194, 1964.

Hall, B., and B. Källen: Chromosome studies in abortuses and stillborn infants. Lancet *1:*110, 1964.

Hecht, F., J. S. Bryant, D. Gruber, and P. L. Townes: The nonrandomness of chromosomal abnormalities: association of trisomy 18 and Down's syndrome. New Eng. J. Med. *271:*1081, 1964.

Hertig, A. T., J. Rock, and E. C. Adams: A description of 34 human ova within the first 17 days of development. Amer. J. Anat. *98:*435, 1956.

Inhorn, S. L., E. Therman and K. Patau: Cytogentic studies in spontaneous human abortion. Amer. J. Clin. Path. (Abstr.) *42:*528, 1964.

Jacobs, P. A., and J. A. Strong: A case of human intersexuality having a possible XXY sex-determining mechanism. Nature (London) *183:*302, 1959.

Javert, C. T.: Spontaneous and Habitual Abortion. New York: McGraw-Hill. 1957.

Kerr, M., and M. N. Rashad: Chromosome studies on spontaneous abortions. Am. J. Obstet. Gynec. *94:*322, 1966.

Klinger, H. P.: The sex chromatin in fetal and maternal portions of the human placenta. Acta Anat. (Basel) *30:*371, 1957.

Lamy, M., N. Josso, J. de Grouchy, and A. Bitan: Anomalies des gonosomes. 20e Congres Pediat. Langue Franç. 1965.

Lejeune, J., M. Gautier, and R. Turpin: Les chromosomes en culture de tissue. C. R. Acad. Sci. (Paris) *248:*602, 1959.

————, ————, and ————: Étude des chromosomes somatiques humains. Technique de culture de fibroblastes in vitro. Rev. Franc. Étud. Clin. Biol. *5:*406, 1960.

Lindsley, D. L., C. W. Edington, and E. S. Von Halle: Sex-linked recessive lethals in Drosophila whose expression is suppressed by the Y chromosome. Genetics *45*:1649, 1960.

London Conference on "The normal human karyotype." Amer. J. Hum. Genet. *16*:156, 1964.

Maclean, N., D. G. Harnden, W. M. Court Brown, J. Bond, and D. J. Mantle: Sex-chromosome abnormalities in newborn babies. Lancet *1*:286, 1964.

Makino, S., M. S. Sasaki, and T. Fukuschima: Triploid chromosome constitution in human chorionic lesions. Lancet *2*:1273, 1964.

Marden, P. M., D. W. Smith, and M. J. McDonald: Congenital anomalies in the newborn infant, including minor variations. J. Pediat. *64*:357, 1964.

McFeely, R. A.: A direct method for the display of chromosomes from early pig embryos. J. Reprod. Fertil. *11*:161, 1966.

Moore, K. L., and W. Hyrniuk: Sex diagnosis of early human abortions by the chromatin method. Anat. Rec. (Abstr.) *136*:277, 1960.

Patau, K., D. W. Smith, E. Therman, S. L. Inhorn, and H. P. Wagner: Multiple congenital anomaly caused by an extra autosome. Lancet *1*:790, 1960.

Paul, J. R.: Cell and Tissue Culture. Edinburgh: E. and S. Livingstone, 1960.

Penrose, L. S., and J. D. A. Delhanty: Triploid cell cultures from a macerated foetus. Lancet *1*:1261, 1961.

————: The Biology of Mental Defect. London: Sidwick and Jackson Ltd. 1963.

Pergament, E., and T. Kadotani: A new double aneuploid: XXY D-trisomy. Lancet *2*:695, 1965.

Polani, P. E.: Chromosome anomalies and abortions. Develop. Med. Child Neurol. *8*:67, 1966.

Rashad, M. N., and M. G. Kerr: Observations on the so-called holocardius amorphus. J. Anat. (Abstr.) *100*:425, 1966.

————, ————: A human triploid embryo. Proceedings Anatomical Society of Great Britain and Ireland, p. 10, April, 1966.

Robinson, A.: Prenatal death. Edinburgh Med. J. *26*:137 and 209, 1921.

Russell, W. L., L. B. Russell, and J. S. Gower: Exceptional inheritance of a sex-linked gene in the mouse explained on the basis that the X0 sex-chromosome constitution is female. Proc. Nat. Acad. Sci. *45*:554, 1959.

Russell, L. B., and C. L. Saylors: Spontaneous and induced abnormal sex-chromosome number in the mouse. Genetics (Abstr.) *46*:894, 1961.

————: Another look at the single-active-X hypothesis. Trans. N.Y. Acad. Sci. *26*:726, 1964.

Sato, H.: Chromosome studies in abortuses. Lancet *1*:1280, 1965.

Singh, R. P.: Anatomical findings in human abortions of known chromosomal constitution. University of Western Ontario, Ph.D. Thesis. 1966.

Singh, R. P., and D. H. Carr: The anatomy and histology of X0 human embryos and fetuses. Anat. Rec. *155*:369, 1966.

Singh, R. P., A. Rubinoff, and D. H. Carr: Lancet. *2*:445, 1966.

Smith, D. W.: Autosomal abnormalities. Am. J. Obstet. Gynec. *90*:1055, 1964.

Stern, C.: Principles of Human Genetics. San Francisco: W. H. Freeman and Co. 1960.

Stevenson, A. C.: Sex chromatin and the sex ratio in man. In K. L. Moore, ed. The Sex Chromatin. Philadelphia: W. B. Saunders Co. 263, 1966.

Swanson, C. P.: Cytology and Cytogenetics. London: Macmillan and Co., 1963.

Szulman, A. E.: Chromosomal aberrations in spontaneous human abortions. New Eng. J. Med. *272:*811, 1964.

Taylor, A. I., and P. E. Polani: Autosomal trisomy syndromes excluding Down's. Guy Hosp. Rep. *113:*231, 1964.

Thiede, H. A., and S. B. Salm: Chromosome studies on human spontaneous abortions. Amer. J. Obstet. Gynec. *90:*205, 1964.

Uchida, I. A.: The effect of maternal age and radiation on the rate of non-disjunction in Drosophila melanogaster. Canad. J. Genet. Cytol. *4:*402, 1962.

Vara, P., and S. Personen: Uber Abortiveier. Acta Obst. Gynec. Scand. 27:215, 1947.

Vislie, H., M. Wehn, A. Brøgger, and J. Mohr: Chromosome abnormalities in a mother and two mentally retarded children. Lancet 2:76, 1962.

Wingate, M. B.: The sex ratio of mid-trimester abortions. J. Obstet. Gynaec. Brit. Comm. *73:*296, 1966.

Witschi, E., and R. Laguens: Chromosomal aberrations in embryos from overripe eggs. Develop. Biol. *7:*605, 1963.

GENETIC AND BIOCHEMICAL ASPECTS
OF REPRODUCTIVE FAILURE *

DAVID YI-YUNG HSIA

*Professor of Pediatrics, Northwestern University, Medical School,
Chicago, Illinois*

A few years ago, Wilson (1959) proposed a schematic representation of the growing embryo and some of its relationships to its environment (Fig. 1). It is apparent that genetic and biochemical factors are likely to play a role at various stages of development in this embryo. The early development is sustained by the genetic information stored in the egg during "oögenesis." Following fertilization, there is a sequential development of the various forms of RNA leading ultimately to the laying down of specific proteins. Certain enzyme systems are undoubtedly involved in the induction of chemical processes and others probably in differentiation and organ formation. Functional immaturity of these enzyme systems can extend into the neonatal period and will sometimes alter the future development of the liveborn infant.

Reproductive failure will occur if there is sufficient alteration of the normal biochemical processes of growth and differentiation throughout gestation. These can be caused by genetically determined mutations which can alter the DNA and RNA's very early and result in termination of the pregnancy when the embryo has not developed much beyond the 16 or 64 cell stage. They can alter enzyme systems during gestation so that chemical differentiation and organ formation would be altered in midstream. In the more severe cases, this would result in the death of the fetus *in utero*, and in the less severe cases, this might result in the birth of an infant with congenital malformations. Finally, they may cause the alteration of a single enzyme required for normal physiological processes after birth and result in an "inborn error of metabolism"

* These studies were aided by grants from the Chicago Community Trust, the Illinois Mental Health Fund, and the United States Public Health Service.

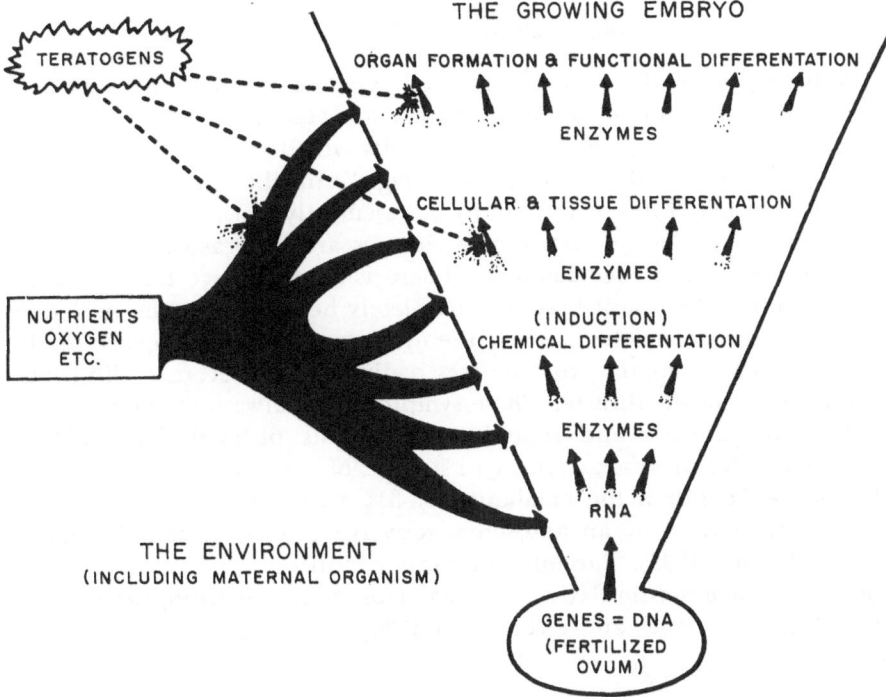

Fig. 1. Scheme illustrating the biochemical development of the growing embryo and the influence of environment at all levels of development. (After Wilson, 1959.)

(Hsia, 1966a). However, not all instances of reproductive failure in this context need be genetic in origin. Environmental factors such as a deficiency of oxygen or nutrients could result in alterations of enzyme systems during gestation, and certainly both "iatrogenic" as well as "teratogenic" agents have been shown to damage embryos.

As one views the genetic and biochemical aspects of reproductive failure as a whole, one finds that during the past few years, rather detailed knowledge has been gained regarding the human infant after birth and perhaps during the latter part of the third trimester of pregnancy. Similarly, some very useful information has been reported on the biochemical changes during the early part of embryonic development in sea urchin eggs and amphibia. Some very preliminary knowledge of control mechanisms has been described in mammalian organisms, but its application to chemical differentiation and organ formation is virtually unknown. The present paper will try to point out the kind of information which is available and whenever possible, I shall try to cite one or two concrete examples.

Inborn Errors of Metabolism

Perhaps one of the best documented types of reproductive failure is an "inborn error of metabolism" which fits Hadorn's (1949) definition of lethal factors as being "Mendelian units which cause the death of an organism prior to the reproductive stage." In the case of Tay-Sachs disease, affected children appear to be essentially normal up to the age of 4–6 months. The first noticeable changes are hyperacusis, irritability, and a leveling-off of development. There is a progressive loss of muscle strength until the infant becomes completely helpless. Also, he no longer notices or recognizes the mother; the eyes cease to fix on objects; gradually it becomes apparent that he sees badly, and he becomes blind later in the course of the disease. These symptoms are always accompanied by striking hyperacusis and sometimes by bursts of explosive laughter. Eventually the process advances to a state of complete idiocy. Most of the affected children die at about $2\frac{1}{2}$–$3\frac{1}{2}$ years of age. Tay-Sachs disease is transmitted by an autosomal recessive gene with complete penetrance (Slome, 1933). Patients with this condition have been found to show a massive accumulation of a galactosamine-containing ganglioside (Fig. 2) (Klenk, 1962) and recently a UDP-galactose: glycolipid galactosyl

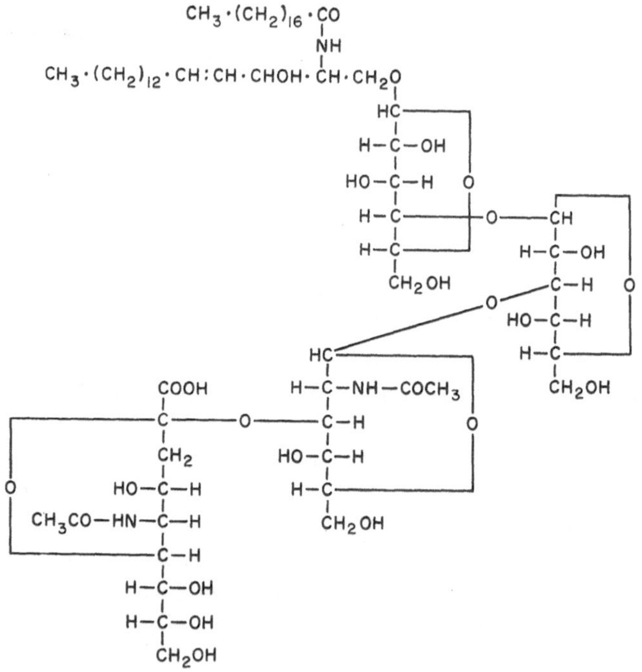

Fig. 2. Galactosamine-containing ganglioside from human brain.

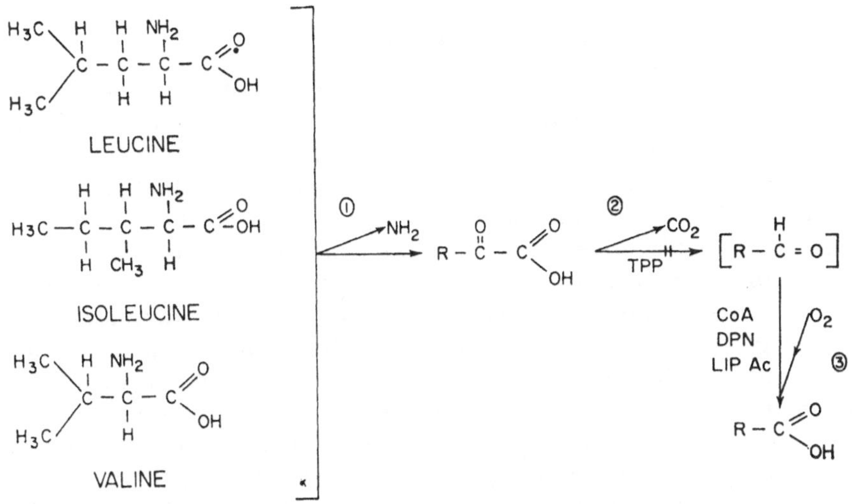

Fig. 3. Partial scheme of degradation of the branched-chain amino acids showing transamination of the keto acid (1) followed by oxidative decarboxylation (2) to the simple acid which is shorter by one carbon (3).

transferase has been detected in a particulate preparation from chicken embryonic brain that catalyzed the synthesis of a monosialoganglioside from Tay-Sachs ganglioside (Basu, *et al.*, 1965). Since no treatment is currently available for Tay-Sachs disease and all affected children die before the age of reproduction, this genetically determined condition may be regarded as an example of a pregnancy which came to term, but still a reproductive failure.

Before leaving the subject of "inborn errors of metabolism," I would like to cite one additional example where reproductive failure may ultimately be turned to success. About a decade ago, Menkes, *et al.* (1954) described a new syndrome characterized by cerebral symptoms and the passage of urine with an odor strikingly similar to that of maple sugar. All of the affected infants begin to show clinical symptoms between the third and fifth day of life. There is difficulty in feeding, an absence of the Moro reflex and the development of irregular jerky respirations. This is followed by signs of spasticity and opisthotonos and the infants go rapidly downhill and die within weeks or months. The condition is believed to be transmitted as an autosomal recessive trait and is caused by a deficiency of the decarboxylase of the keto acids of the branched-chain amino acids (Fig. 3) (Dancis, *et al.*, 1963). Recently, restricting the dietary intake of leucine, isoleucine, and valine has resulted in the prevention of the mental and neurological signs and normal development (Snyderman, *et al.*, 1964). Thus, there is a possibility that these children will eventually reach the age of reproduction themselves and not be regarded as examples of reproductive failure (Hsia, 1966b).

Homozygous Lethals in Man

Next we want to turn to a consideration of whether some of the lethals with "inborn errors of metabolism" die at an early stage of embryonic development and are not viable. In the upper half of Table 1 are listed four recessive lethal factors in animals which have been studied for segregation for monohybrid recessive lethal factors following matings of heterozygotes. In the first two examples, the numbers affected follow closely the quarter ratio expected for the recessive trait. In the second two examples, however, there is a significant decrease in the percentage of affected. This shortage of lethals can be explained either by the fetuses dying at an early stage of embryonic development in which case these animals would be missing if the litters are counted at full term, or the fetuses are destroyed by the mother at birth. In the case of the congenital hydrocephalus in the mouse, Grüneberg (1943) observed that the mother of hydrocephalic mice often eat the heads or the whole animals immediately after birth. This accounts for the disturbance of the Mendelian ratio. In the lower half of Table 1 are listed four "inborn errors of metabolism" corrected for ascertainment. It would appear that all four do not show any evidence of loss of lethals during gestation.

In point of fact, except for chromosomal aberrations, it is extremely difficult to demonstrate any clear-cut examples of loss of homozygous lethals among autosomal traits in man. One or two possibilities have been

Table 1

Data Comparing the Percentage Affected in Four Recessive "Inborn Errors of Metabolism" in Man with Four Recessive "Lethal" Factors in Animals (In the latter group, two of the conditions show the expected frequency of homozygous lethals while the other two show a significant decrease)

Species	Condition	No. Studied	No. Affected	Per Cent Affected	Reference
Pig	Paralysis of hind legs	28	7	25	Berge (1941)
Pig	Hydrocephalus	178	42	24	Blunn & Hughes (1938)
Mouse	Congenital hydrocephalus	2474	405	16	Grüneberg (1952)
Pig	Muscle contraction	266	46	17	Hallqvist (1934)
Man	Tay-Sachs	247	61	25	Slome (1933)
Man	Phenylketonuria	1571	430	27	Jervis (1954)
Man	Galactosemia	72	16	22	Walker et al. (1962)
Man	Cystic fibrosis	690	168	24	Danks et al. (1965)

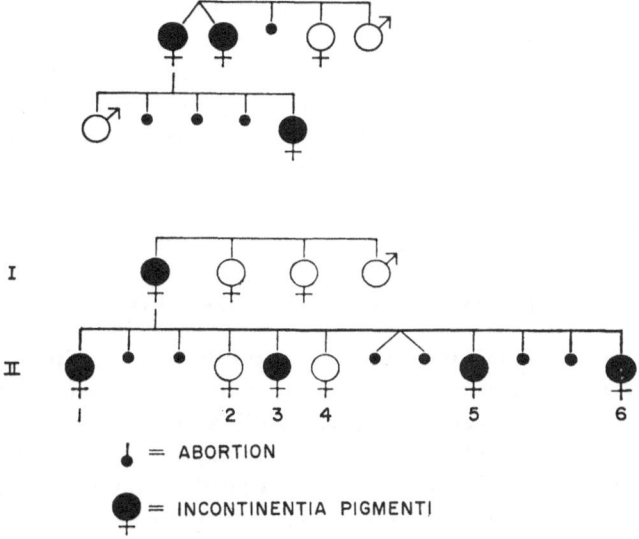

I = ABORTION

= INCONTINENTIA PIGMENTI

Fig. 4. Families with incontinentia pigmenti. (After Lenz, 1963.)

suggested for X-linked recessive traits. The sex ratio at fertilization is unknown in man, but may be at least 1.22 (male/female) (Szontagh, *et al.,* 1961). At birth, this has been reduced to 1.06 in white populations in the United States. The preferential loss of male zygotes, embryos, and fetuses may be in part of lethal X-linked recessive genes. Two examples may be cited. Two families with incontinentia pigmenti are shown in Figure 4. In this dominant trait, the hemizygous male may be lethal prenatally resulting in only females with the defect being seen. Another example would be the specific human intersex type consisting of XY individuals. Externally, these individuals are typical females and regard themselves as such. Internally, except for a small vagina, they lack all female structures but possess a pair of testes and derivatives of the Wolffian ducts. In the pedigree shown in Figure 5, the intersexuality is trans-

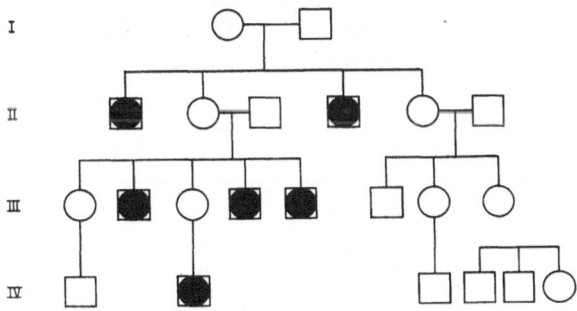

Fig. 5. Pedigree of intersexuality. (After Petterson and Bonnier, 1937.)

mitted by normal females and that, among the sibs of all six intersexes, no normal males occurred. Combining data from four similar pedigrees with this one, there were 28 normal females, 22 intersexes, and 8 normal males. Since the carrier women are heterozygous for an allele which, in an XY individual, results in the XY individual, one would expect equality among the children of carrier females between the number of intersexes and of normal males. The ratio of 22:8 suggests an excessive loss of males.

The Development of Enzyme Systems

During recent years, data have been published showing that the activity of certain enzymes, which are low in early periods of life, will increase sharply at a certain age (Driscoll and Hsia, 1958; Sereni and Principi, 1965). Most of these enzymes belong to the "adaptive" group whose activity can be induced *in vivo* in adult animals by changing environmental factors such as substrate, corticosteroid concentration, cofactors, etc. In Table 2 are listed some data which we have recently collected on the development of the α-disaccharidases in rat intestinal mucosa (Hsia, 1966c). It can be seen that sucrase and isomaltase are almost completely absent up to 15 days of life and then abruptly increase to almost adult levels. Doell and Kretchmer (1964) have shown that if 9 day old rats are given hydrocortisone, sucrase activity becomes detectable in the intestine within 24 hours and the enzyme is fully active within 2 to 3 days. After 5 days the activity began to decrease and remained low until about the 9th day after injection when it increased again during the usual physiological increase. When hydrocortisone was injected daily, the enzyme was fully active at least until the animals were 21 days of age. In contrast to sucrase and isomaltase, small but significant maltase activity can be detected from birth. This is caused

Table 2

Development of α-Disaccharidases in Whole Homogenate of Rat Intestinal Mucosa During Neonatal Period (All values expressed as mean $\pm$ standard deviation) (After Hsia, 1966c)

Age (Days)	Sucrase (I.U./Gm P)	Isomaltase (I.U./Gm P)	Maltase (I.U./Gm P)
1–5	0	0	113 $\pm$ 10
6–10	0	0	95 $\pm$ 10
11–15	2 $\pm$ 3	0	106 $\pm$ 11
16–20	80 $\pm$ 8	79 $\pm$ 8	345 $\pm$ 37
Adult	100 $\pm$ 10	125 $\pm$ 13	415 $\pm$ 42

by the differing development of isoenzymes for maltase. Semenza and Auricchio (1962) have shown in the human, there are five different intestinal maltases. Maltase 1 and 2 have only maltase activity, maltase 3 and 4 have also sucrase activity and maltase 5 has also isomaltase and palatinase activity. By determining maltase 1 and 2 separately, Auricchio, *et al.* (1965) have shown that maltase 1 is still very low in the full-term newborn, but maltase 2 is at maximal adult values by the sixth month of intrauterine life.

A second and perhaps better example of differences in the development of isoenzymes is that of lactic acid dehydrogenase (LDH). There is general agreement that LDH is a tetramer of four equal-sized subunits existing in two distinct electrophoretic varieties, which may be designated A and B (Markert, 1964; Dawson, *et al.*, 1964). Assortment of these two kinds of subunits in all combinations of four yields the five isoenzymes, which may be viewed as follows:

$$LDH\text{-}1 = A_0B_4 \ (H_4)$$
$$LDH\text{-}2 = A_1B_3 \ (H_3M_1)$$
$$LDH\text{-}3 = A_2B_2 \ (H_2M_2)$$
$$LDH\text{-}4 = A_3B_1 \ (H_1M_3)$$
$$LDH\text{-}5 = A_4B_0 \ (M_4)$$

In mammals, there is a preponderance of isoenzymes in the LDH-5 end of the spectrum which permits anaerobic metabolism during early fetal life. As the embryo develops, LDH-1 and LDH-2 gradually becomes predominant in the heart whereas LDH-5 remains prodominant in skeletal muscle. In the chick embryo, which relies more on aerobic rather than anaerobic metabolism, the opposite occurs. Early in gestation, there is a predominance of LDH-1. After hatching, LDH-5 becomes dominant, particularly in the muscles. Very little is known about the factors which control the shift of LDH isoenzymes. Goodfriend and Kaplan (1964) have found that injection of hormones, particularly estradiol induces a preferential synthesis of LDH-5 units. The LDH isoenzyme story may be taken as a reasonable example of how autogenesis could be considered a shortened repetition of phylogenesis!

One of the techniques used for the study of the development of enzyme systems in the developing embryo is the use of the technique for hormonal deprivation in the fetal rat (Jost and Jacquot, 1958). This is carried out by performing bilateral adrenalectomy of the mother rat at 14 days post coitum since maternal steroids traverse the placenta; and decapitation *in utero* of two to four fetuses of a litter at 18 days post coitum. By this procedure production of fetal steroids is eliminated, and there is a resultant atrophy of the adrenal glands. The pregnant rat is killed at 20 to 21 days post coitum and the fetuses quickly removed. Jacquot and Kretchmer (1964) have undertaken a study on the effect of hormonal deprivation on the enzymes concerned with the synthesis of hepatic glycogen. Decapitation caused a failure in the physiological rise of glucose-6-phosphatase (Fig. 6), uridine diphosphoglucose-glycogen transglucosylase, and to a lesser extent of phosphoglucomutase, and a lack of the usual fall of glucose-6-phosphate dehydrogenase (Fig. 7) at term. These enzymatic changes were associated with a negligible deposition of glycogen and thus the decapitated fetus at 20 and 21 days of gestation remains roughly in the biochemical status corresponding to 18 days at the time of decapitation.

Finally, we should discuss the effect of external agents upon the development of enzyme systems in the developing fetus. There is general agreement that the conjugation of bilirubin is carried out by means of the following reaction:

$$\text{Bilirubin} \quad + \quad \text{UDPGA} \quad \xrightarrow{\text{Glucuronyl Transferase}}$$

(indirect-reacting)   (uridine diphosphate
                         glucuronic acid)

$$\text{Bilirubin Glucuronide} + \quad \text{UDP}$$

(direct-reacting)   (uridine diphosphate)

Hyperbilirubinemia in the newborn infant results either from an excessive breakdown of red cells such as in erythroblastosis fetalis, or from a functional immaturity of glucuronyl transferase, or both. This results in

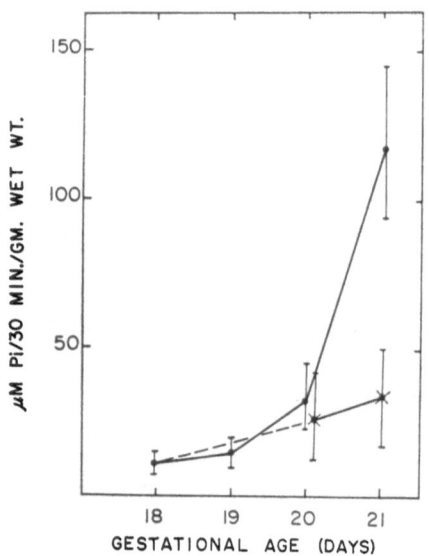

Fig. 6. Effect of decapitation *in utero* on activity of glucose-6-phosphatase in fetal rat liver. Decapitated rats are shown by X. (After Jacquot and Kretchmer, 1964.)

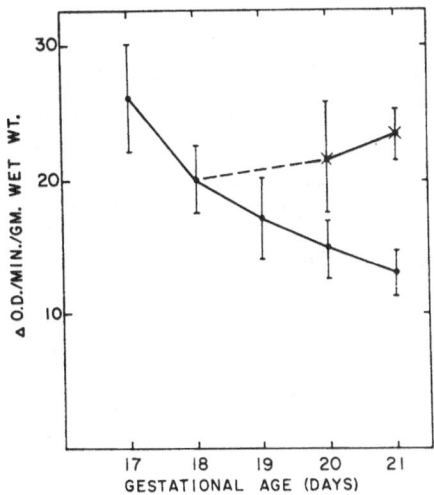

Fig. 7. Effect of decapitation *in utero* on activity of glucose-6-phosphate dehydrogenase in fetal rat liver. Decapitated rats are shown by X. (After Jacquot and Kretchmer, 1964.)

the accumulation of indirect-reacting bilirubin in the serum, and if the levels become excessive, kernicterus will result.

A few years ago, Sutherland and Keller (1961) observed that the serum bilirubin levels were increased threefold when a group of newborn infants were given novobiocin to control a staphylococcal infection. A similar, though less intense, jaundice was also produced in newborn rats who were injected with novobiocin. In an investigation carried out a few years ago, we found that novobiocin is a potent inhibitor of glucuronyl transferase (Lokietz, *et al.*, 1963). As shown in Table 3, when 500 mg/kg of novobiocin was injected into newborn rats, enzyme activity in animals sacrificed 2 to 8 hours later was found to be less than half that observed in litter mates who were injected with saline solution. This *in vivo* effect becomes less marked at 24 hours, when bilirubin levels were at their peak. These findings were confirmed in *in vitro* systems using rat liver microsomes, and novobiocin was found to be a noncompetitive inhibitor using Lineweaver-Burk plots as shown in Figure 8. The administration of certain drugs to mothers will also cause hyperbilirubinemia to the newborn infant. Lucey and Dolan (1959) reported seven cases of marked hyperbilirubinemia occurring among premature infants following the administration of 72 mg of a vitamin K analogue (Hykinone) prior to delivery. This would suggest that this analogue would pass through the placenta and exert a hepatotoxic effect upon the newborn infant.

According to the 1961 census figures released by the Children's Bureau,

Table 3

GLUCURONYL TRANSFERASE ACTIVITY AND INHIBITION LEVELS OF YOUNG RATS GIVEN 500 MG/KG BODY WEIGHT OF NOVOBIOCIN (After Lokietz, *et. al.*, 1963)

Hours After Injection	Litters (no.)	Serum Bilirubin Levels		Activity * (%)
		Control (mg/100 ml)	Experimental (mg/100 ml)	
2–4	4	0.3 ± 0.1	1.1 ± 0.9	44 ± 19
4–8	5	0.3 ± 0.1	0.9 ± 0.5	39 ± 11
8–18	4	0.5 ± 0.3	5.1 ± 1.8	69 ± 25
24	8	0.6 ± 0.3	4.0 ± 0.9	85 ± 15
48	5	0.4 ± 0.1	1.4 ± 1.0	115 ± 15
72	4	0.4 ± 0.1	0.7 ± 0.2	109 ± 21

* Using rat liver homogenate and o-aminophenol as substrate and expressed as percentage of activity over controls. Mean and standard deviation are given.

there are 25 deaths per 1000 during the neontal period compared with only 1.5 deaths per 1000 for the ages of one to 14 years. Thus, over 15 times as many children are lost during the first few days of life as during the next 14 years combined. Except for a small group of infants with congenital anomalies, incompatible with life, and other specific pathology, the cause of death for the vast majority of full-term and premature infants remains unknown. It appears not unlikely that a lot of reproductive failure at this stage including probably the respiratory distress syn-

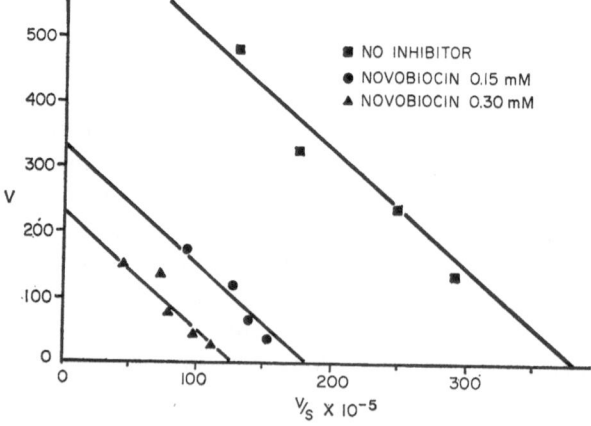

Fig. 8. Lineweaver-Burk plot of kinetic studies showing values without inhibitor and with two concentrations of novobiocin. The data are plotted with V/S against V. (After Lokietz *et al.*, 1963.)

drome represent examples of "metabolic" failure, where the fetus was not able to make a suitable adaptation to extrauterine existence. We are only just beginning to acquire some information on the normal processes of metabolism during this transitional stage. As we learn more about these mechanisms, it appears not unlikely that we will learn some of the causes of fetal failure at this stage of development.

EMBRYOGENESIS AND ORGANOGENESIS

We shall now turn to the much more difficult problems of embryogenesis and organogenesis (Monroy, 1965; DeHaan and Ursprung, 1964). Since relatively little is known in this area with certainty in man, we shall consider several of the classical experiments which have been performed in lower organisms. If one assumes that many biological mechanisms are universal, some of this information may help explain the success or failure of embryogenesis in man.

We shall first turn to the work of Boveri. In the early part of the present century before the rediscovery of Mendel, this remarkable biologist demonstrated the genetic diversity of chromosomes by a relatively simple experiment. Sea urchin eggs can be artificially fertilized. If one uses much sperm, an egg can be simultaneously fertilized by two sperms resulting in a tetrapolar or sometimes tripolar mitosis with unequal distribution of the chromosome content (Fig. 9). As development proceeds, the triaster cells cleave and form three blastomeres while the tetraster cells separate into four cells. If different chromosomes are carriers of different genetic information, then defects will arise because of this unequal distribution among the cells. Boveri (1907) postulated that

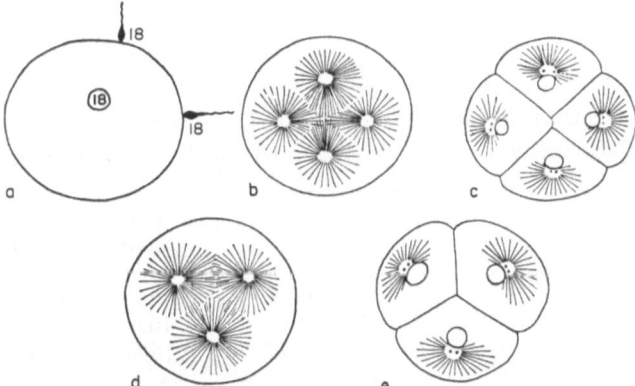

Fig. 9. Cleavage in dispermic sea urchin eggs. (a) Fertilization of an egg with two sperm, shown schematically. (b and c) Tetraster and resulting simultaneous cleavage into four cells ("Simultanvierer"). (d and e) Triaster and resulting simultaneous cleavage into three cells ("Simultandreier"). (After Baltzer, 1964.)

if the sea urchin egg has a haploid chromosome number of 18, one would expect 11 per cent of the triaster eggs to be provided with each kind of chromosome and develop normally, and that a favorable distribution of chromosomes would hardly ever occur among the tetraster eggs. With this in mind, he isolated 719 triaster eggs and obtained 79 pluteus larvae—just about the predicted 11 per cent. On the other hand, he got only one pluteus in 1500 tetraster eggs cultivated.

Half a century later, Briggs and King (1952) introduced the technique of nuclear transplantation and opened up the important question of nuclear specificity. The single cell zygote of a frog with its diploid nucleus, divides about 35 times to form the adult organism. The first dozen or so of these divisions, which lead to an early gastrula, are accompanied by little obvious cyto-differentiation. In the postgastrula stages, however, there is increasing specialization of the different lineages of the embryonic cells leading to the cytologically distinct types of the mature organism. If one takes the cell of a blastula and crushes it in a certain way so that the membrane is broken but the nucleus is not damaged, the nucleus can be injected into an enucleated egg. In 40 to 60 per cent of the cases, the egg with its blastula nucleus will develop normally through the larval stages. This would suggest that the blastula nucleus had not undergone any irreversible differentiation up to that point and possessed genetic capabilities equal to that of the zygote nucleus. In older embryos, Briggs and King (1957) found evidence that the nuclei of endodermal cells become changed. The change is reflected in part by the decrease in the frequency of the experimental embryos developing to the blastula stage and then continuing on to form larvae as shown in Table 4. In an effort to learn something of the genetic capabilities of these nuclei that replicate in a foreign cytoplasm, Moore (1958) has

Table 4

THE DEVELOPMENT OF ENUCLEATED OVA INJECTED WITH ENDODERM NUCLEI OF PROGRESSIVELY OLDER STAGES (After Briggs and King, 1957)

Type of Donor Nucleus	No. of Ova Injected	Per Cent Forming Blastulae	Per Cent Forming Larvae
Early gastrula	92	42	36
Late gastrula	155	51	10
Endoderm of neurula	98	16	1
Endoderm of tailbud	130	7	0

transferred a blastula nucleus of *R. sylvatica* to an enucleated ovum of *R. pipiens*. He found that normal development will only prosper to the late blastula stage whether the nucleus is kept in the foreign cytoplasm or transferred back to the cytoplasm of the original species. On the basis of these observations, Moore (1962) has suggested that when the nucleus of one species replicates in the cytoplasm of another, the cytoplasmic DNA is used in part for the newly-formed chromosome. Because of differences in the base sequence of DNA between species, copy errors are made, and even if the nucleus is returned to the original cytoplasm, the scrambled code cannot recover and if anything might become more scrambled after the back-transfer. Essentially the same conclusions have been reached by Markert and Ursprung (1963) who injected very small amounts of various protein fractions from adult *R. pipiens* liver cells into zygotes of the same species. The albumin fraction and, to a lesser extent, the histones when injected cause a highly reproducible cessation of cell division and arrested development in the late blastula stage. Nuclei from the arrested blastulae were serially transplanted for seven generations without showing any recovery. They also concluded that the effect is primarily in the chromosomes and is replicated during cell division.

More complete information on the biochemical changes during early development may be derived from studies on the sea urchin egg (Harvey, 1956). Speaking in general terms, it appears clear from the behavior of parthenogenetic merogones that all the machinery needed for segmentation of the egg is laid down in the cytoplasm before fertilization. Normal fertilization evokes an outburst of protein synthesis in a system which makes essentially no protein beforehand. Artificial activation of the egg produces, for a time, an almost identical effect.

There is general agreement that polypeptide synthesis is dependent upon the interaction of messenger ribonucleotides (m-RNA) and ribosomes. The absence of RNA synthesis before fertilization could be caused either by an absence of m-RNA in the unfertilized egg or by a process which prevents the interaction between the ribosomes and m-RNA prior to fertilization. Let us consider the first of these possibilities. The argument in favor of this possibility was advanced by the experiments of Nemer (1962) and Wilt and Hultin (1962) who both showed that polyadenylic acid can stimulate incorporation of phenylalanine both before and after fertilization as shown in Table 5. This suggested that the increase in protein synthesis upon fertilization is due to the synthesis and attachment of endogenous m-RNA. Recently, several pieces of evidence have been published showing that this is probably not the case. The biological activity of the actinomycins depends upon their specific binding to DNA and its subsequent effects on DNA-primed RNA synthesis. The selective suppression of this mechanism, while gene replication continues, offers

Table 5

STIMULATION BY POLY U OF L-PHENYLALANINE-C[14] INCORPORATION WITH
MICROSOMES FROM EMBRYOS OF VARIOUS STAGES OF DEVELOPMENT (After
Nemer, 1962)

Microsomal Source	Additions	L-Phenylalanine Incorporation (C.p.m./mg protein)	Ratio of Incorporations: Stimulated / Control
Unfertilized Egg	None	21	
	Poly U	3615	173
2-Cell Stage (90 min.)	None	113	
	Poly U	3010	27
Blastula (12 hr.)	None	117	
	Poly U	1600	14

a unique kind of chemical enucleation where there is a restriction of
the flow of gene products into the cytoplasm. With this in mind, Gross
and Cousineau (1963) showed that when sea urchin eggs are fertilized
and permitted to develop in the presence of Actinomycin D, cell divisions
continue for many hours, at normal or slightly subnormal rates. Cellular
differentiation is completely inhibited, so that at the end of a period
during which controls have progressed through gastrulation, the actino-
mycin-treated embryos are inert, multicellular masses. This suggested
that the unfertilized egg contained a store of marked template material
(m-RNA), whose information concerns in part the protein which must
be made for cell division. These templates, which are probably not
attached to the ribosome of the unfertilized egg, become functional and
attached upon activation, and are responsible for the actinomycin-
resistant protein synthesis. Cellular differentiation, however, appears to
be under direct genomic control and to depend upon the synthesis after
fertilization of appropriate messengers. This synthesis is stopped by
actinomycin. Using another approach, Brachet, *et al.* (1963) demonstrated
that anucleate fragments of unfertilized eggs respond to parthenogenetic
activation by increased incorporation of amino acids into their proteins.
In these experiments, unfertilized eggs are divided into light (nucleate)
and heavy (anucleate) halves by centrifugation. After washing with
sea water, the fragments are activated by treatment with hypertonic
water. Amino acid incorporation is studied by C[14] leucine incubation.
As shown in Table 6, the anucleate halves respond to parthenogenetic

Table 6

Sample	Blackening of the emulsion
1. Unfertilized whole eggs	+ +
2. Activated whole eggs	+ + + +
3. Fertilized whole eggs	+ + + + +
4. Unfertilized, centrifuged whole eggs	+
5. Activated, centrifuged whole eggs	+ + + +
6. Light (nucleate) halves	+ +
7. Same as 6, after activation	+ + + +
8. Heavy (anucleate) halves	0
9. Same as 8, after activation	+ + +

activating in essentially the same way as whole eggs or nucleate halves. Finally, Maggio, *et al.* (1964) have shown that RNA extracted from unfertilized eggs can act *in vitro* as a template for protein synthesis. As shown in Figure 10, total RNA prepared from unfertilized eggs stimulates incorporation of labelled amino acids into proteins in the rat liver test system. The stimulation is proportional to the amount

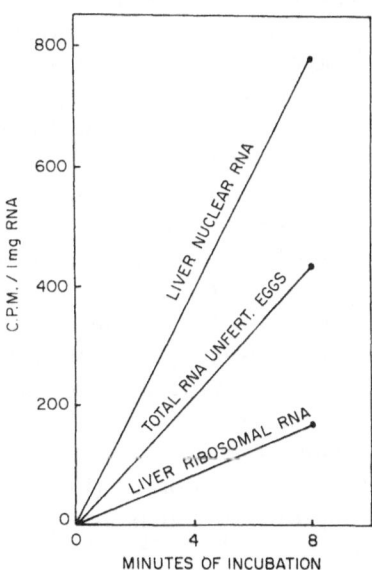

Fig. 10. Incorporation into proteins of a mixture of 0.71 μC L-alanine + 1.42 μC L-phenylalanine + 0.71 μC L-lysine + 0.14 μC L-valine stimulated by total RNA from unfertilized Paracentrotus eggs and by nuclear and ribosomal rat liver RNA. (After Maggio *et al.*, 1964.)

Table 7

EFFECT OF TREATMENT WITH TRYPSIN ON THE ABILITY OF RIBOSOMES
FROM UNFERTILIZED EGGS TO INCORPORATE AMINO ACIDS INTO PROTEINS
(After Monroy, *et al.*, 1965)

Expt. No.	Amino Acid(s)	Addition	C.p.m/mg Ribosomal RNA	
			Control	Trypsin-treated
1	Phenylalanine	None	0	122
	Phenylalanine	Poly U	270	340
2	Phenylalanine	None	0	72
	Phenylalanine	Poly U	360	2200
	AH	None	25	360
	AH	RNA blastula	0	475
3	AA	None	18	448
	AA	RNA unfert.	9	644

of RNA added and is better than RNA from liver ribosomes, but not quite as good as RNA from liver nuclear RNA.

These observations have led investigators to turn away from the concept of a missing m-RNA more to the possibility of a process which prevents the interaction between the ribosomes and m-RNA prior to fertilization. Recently, Monroy, *et al.* (1965) have shown that trypsin is able to activate the ribosome of the unfertilized egg. As shown in Table 7, the ribosomes of unfertilized eggs following treatment with trypsin can be stimulated by RNA extracted from sea urchin embryos or from unfertilized eggs, show a markedly increased response to polyuridylic acid, and become capable of carrying out *in vitro* incorporation of amino acids into proteins in the absence of any exogenous RNA or polyuridylic acid. On the basis of these findings, Monroy (1965) has concluded that m-RNA which is synthesized in the cause of oögenesis actually reaches the ribosome. The ribosome-m-RNA complex is, however, rendered inoperable by a protein coat. This may form after the m-RNA has reached the ribosome, or the m-RNA may travel from the nucleus to the ribosome in the form of nucleoprotein particles. Activation of the ribosome would result from the removal of the shielding protein. At fertilization, there is a transient activation of a protease (Lundblad, 1949) which may break down this shielding protein and permit protein synthesis to take place.

Using a somewhat different approach, Brown and Littna (1964) have described the pattern of RNA synthesis during the development of *Xenopus laevis,* the South African Clawed Toad. In this technique, p^{32} labelled phosphate at pH 7 is injected into gravid *X. laevis* females and

the RNA synthesized is separated by sucrose density gradient zonal centrifugation. As shown in Figure 11, the sedimentation pattern of RNA
divides into three peaks—the 28S, the 18S, and the 4S fractions. By comparing the changes of the protein, which are shown in the white circles
and radioactivity which are shown in black circles, one can see the
developmental pattern all the way from the unfertilized egg to the tailbud stage. It would appear that there is no detectable difference between
the content of newly synthesized RNA associated with ribosomes in ovulated unfertilized eggs and early cleavage embryos. Both contain an
exceedingly small quantity of radioactive RNA sediment at a lower rate
than the 18S ribosomal RNA. By late cleavage, an increase of RNA with
the same approximate sedimentation coefficient has occurred. However,
there is still no radioactivity in the 28S RNA. At gastrulation, there
are signs of radioactive 28S RNA being synthesized, but the low molecular form still predominates. Even at neurulation, the 28S form is less
than 18S form. Only at the tailbud stage, one finds that the newly
synthesized ribosomal RNA exceeds the slower sedimentary component.
These data may be interpreted to mean that the synthesis of new ribosomal RNA begins with the onset of gastrulation at the same stage that

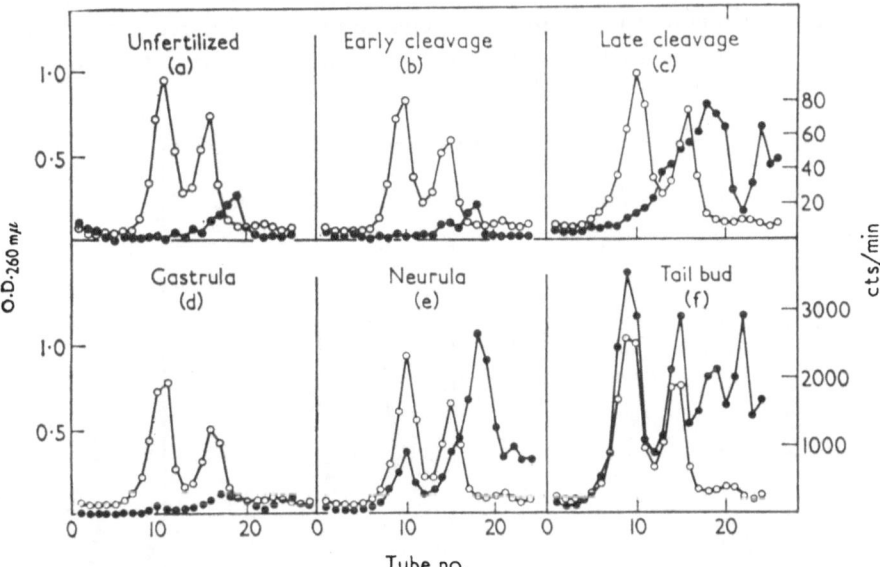

Fig. 11. Sedimentation pattern of RNA purified from the isolated ribosomes of
developing *X. laevis* embryos. Sibling embryos were used for the entire experiment
and each pattern represents RNA from 200 embryos. The stages and hours after
fertilization of each group of embryos are: (a) unfertilized egg; (b) early cleavage 2 to 3,
3 hr; (c) late cleavage 8 to 9, 5 hr; (d) gastrula 10 to 12, 28 hr; (e) neurula 14 to 16,
35 hr; (f) tail bud 21 to 24, 52 hr. (After Brown and Littna, 1964.)

the nucleolus first appears, while during the cleavage stages, DNA-like RNA, soluble RNA, and DNA are synthesized to the exclusion of ribosomal RNA. Brown and Gurdon (1964) have repeated these studies on the lethal anucleolate mutant of *X. laevis* first described by Elsdale, *et al.* (1958). The mutant has numerous small nucleolar "blobs" instead of typical nucleoli. These embryos show retarded development shortly after hatching, become microcephalic and edematous and die as swimming tadpoles before feeding. As shown in Figure 12, at stages 26–28 (muscular response) when the mutant embryos were still indistinguishable grossly from the control embryos, the anucleolate mutant showed no synthesis of 28S and 18S ribosomal RNA while DNA, 4S RNA and rapidly labelled heterogenous RNA is synthesized in the usual manner. It is of interest to note that the heterozygous embryos with one normal and one abnormal nucleolus, the wild type genes regulate to produce twice as much 28S and 18S ribosomal RNA as do the same genes when present in homozygous wild type individuals.

Finally, we turn to a consideration of the question of regulation of protein synthesis (Jacob and Monod, 1961). While relatively little is known about genetic regulation in higher organisms, Weber, *et al.* (1965, 1966) have suggested that there may be a coordinated biosynthetic pattern of the key glycolytic enzymes in the liver. As shown in Figure 13, there are probably three functional genome units in the liver. One group of enzymes are related to gluconeogenesis, another related to glycolysis,

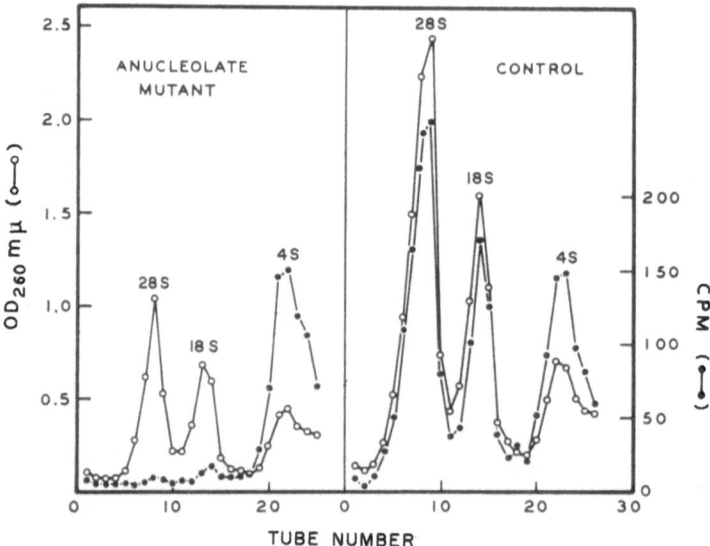

Fig. 12. Sucrose density gradient centrifugation of total RNA isolated from 0-nu and control embryos. (After Brown and Gurdon, 1964.)

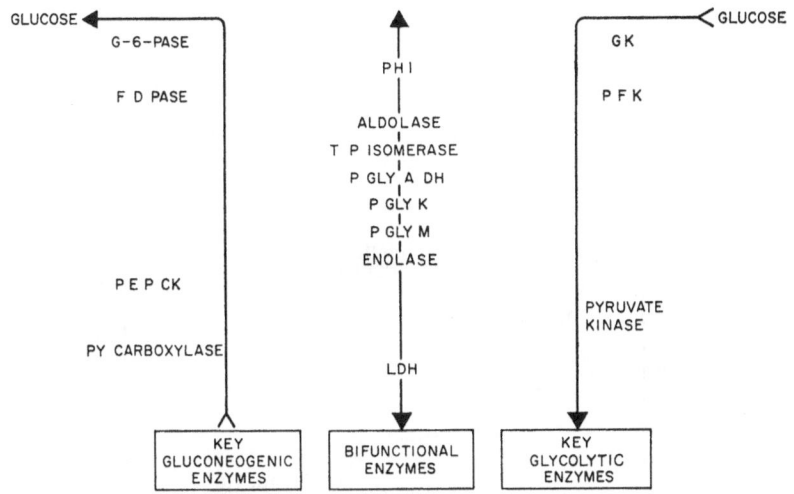

Fig. 13. The three functional genome units. G-6-Pase = glucose 6-phosphatase; FDPase = fructose 1,6-diphosphatase; PEP CK = phosphoenolpyruvate carboxykinase; Py carboxylase = pyruvate carboxylase; PHI = phosphohexose isomerase; TP isomerase = triosephosphate isomerase; P Gly A DH = phosphoglyceraldehyde dehydrogenase; P Gly K = phosphoglycerokinase; P Gly M = phosphoglyceromutase; LDH = lactic dehydrogenase; GK = glucokinase; PFK = phosphofructokinase.

and a third which are bifunctional. Weber (1966) has suggested that the three enzymes concerned with glycolysis—glucokinase, phosphofructo-kinase, and pyruvate kinase—are products of the same functional unit. Evidence for this are: (1) In starvation and diabetes mellitus, where there is a low insulin level, there is a decrease of all three enzymes. When insulin is replaced by feeding in starvation and by injection in diabetes mellitus, there is a rise of enzyme activity to the normal range as shown in Figure 14; (2) This insulin-induced rise of enzyme activity is blocked by Actinomycin D or ethionine; (3) The effect of insulin upon enzyme activities is dependent on the hormone dosages and unrelated to the blood glucose level, and (4) There is a simultaneous increase of all three enzymes during the developmental period of 18–51 days postpartum, a rise which is blocked by Actinomycin D. In a final experiment, Weber (1965) has shown that pyruvate kinase activity becomes fully active when the enzyme is saturated with DPNH in a certain concentration. However, if the DPNH concentration is further increased, pyruvate kinase activity is inhibited and finally approaches zero as shown in Figure 15. This might represent a mechanism for switching off pyruvate kinase activity when the lactate is connected to pyruvate through gluconeogenesis, and increasing amounts of DPNH are produced.

To summarize, although we know virtually nothing about the factors

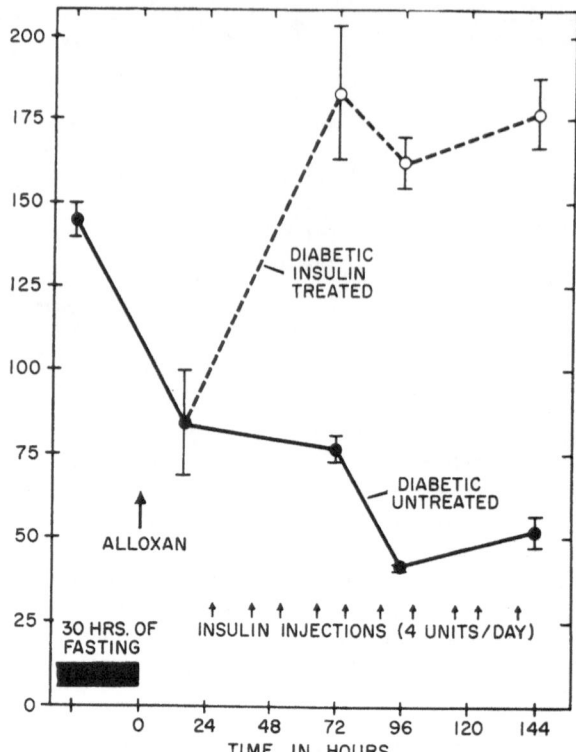

Fig. 14. Effect of insulin administration on hepatic pyruvate kinase activity in alloxan diabetic rats. In the untreated diabetic rats pyruvate kinase activity decreased to 40%. Insulin administration in diabetic rats increased the low enzyme activity to normal range. (After Weber *et al.*, 1965.) In ordinate: PYRUVATE KINASE ACTIVITY PER CELL (μ moles of substrate metabolized/hr $\times$ 107 at 37°C).

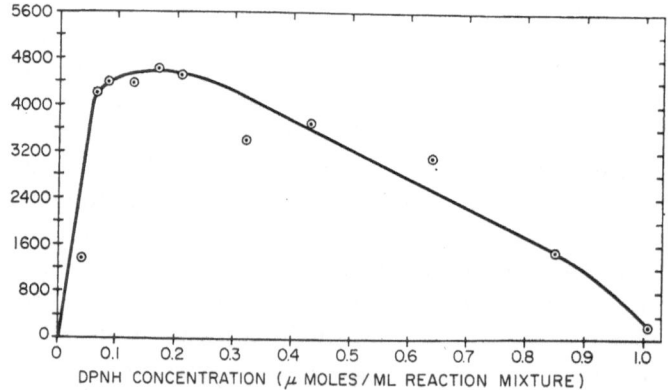

Fig. 15. Inhibition of hepatic pyruvate kinase activity by DPNH. (After Weber *et al.*, 1965.) In ordinate: PYRUVATE KINASE ACTIVITY (μ moles/g/hr at 37°C).

that lead to reproductive failure during early embryogenesis and organogenesis, we are beginning to acquire some basic information on the major biochemical and metabolic changes that take place at various stages of gestation. This will undoubtedly lead to a better understanding of pathological processes that are likely to occur at the critical stages of fertilization, implantation and differentiation.

References

Auricchio, S., A. Rubino, and G. Mürset: Intestinal glycosidase activities in the human embryo, fetus, and newborn. Pediatrics *35*:944, 1965.

Basu, S., B. Kaufman, and S. Roseman: Conversion of Tay-Sachs ganglioside to monosialoganglioside by brain uridine diphosphate D-galactose: glycolipid galactosyl-transferase. J. Biol. Chem. *240*:PC4115, 1965.

Berg, S.: The inheritance of paralysed hind legs, scrotal hernia and atresia ani in pigs. J. Hered. *32*:271, 1941.

Blunn, C. T., and E. H. Hughes: Hydrocephalus in swine, a new lethal defect. J. Hered. *29*:203, 1938.

Boveri, T., cited in F. Baltzer, Theodor Boveri: Science *144*:809, 1964.

Brachet, J., A. Ficg, and R. Tencer: Amino acid incorporation into proteins of nucleate and anucleate fragments of sea urchin eggs: effect of parthenogenetic activation. Exp. Cell Res. *32*:168, 1963.

Briggs, R., and T. J. King: Transplantation of living nuclei from blastula cells into enucleated frogs' eggs. Proc. Nat. Acad. Sc. *38*:455, 1952.

———, ———: Changes in the nuclei of differentially endoderm cells as revealed by nuclear transplantation. J. Morph. *100*:269, 1957.

Brown, D. D., and E. Littna: RNA synthesis during the development of xenopus laevis, the South African clawed toad. J. Mol. Biol. *8*:669, 1964.

Brown, D. D., and J. B. Gurdon: Absence of ribosomal RNA synthesis in the anucleolate mutant of xenopus laevis. Proc. Nat. Acad. Sc. *51*:139, 1964.

Dancis, J., *et al.*: The metabolism of leucine in tissue culture of skin fibroblasts of maple syrup urine disease. Biochim. et Biophys. Acta 77:523, 1963.

Danks, D. M., J. Allan, and C. M. Anderson: A genetic study of fibrocystic disease of the pancreas. Ann. Human Genetics *28*:323, 1965.

Dawson, D. M., T. L. Goodfriend, and N. O. Kaplan: Lactic dehydrogenases: functions of the two types. Science *143*:929, 1964.

DeHaan, R. L., and H. Ursprung, editors: Organogenesis. New York: Holt, Reinhart and Winston, 1964, 804 pp.

Doell, R. G., and N. Kretchmer: Intestinal invertase: Precocious development of activity after injection of hydrocortisone. Science *143*:42, 1964.

Driscoll, S. G., and D. Y. Y. Hsia: The development of enzyme systems during early infancy. Pediatrics *22*:785, 1958.

Elsdale, T. R., M. Fischberg, and S. Smith: A mutation that reduces nucleolar number in xenopus laevis. Exp. Cell Res. *14*:642, 1958.

Goodfriend, T. L., and N. O. Kaplan: Effect of hormone administration on lactic dehydrogenase. J. Biol. Chem. *239*:130, 1964.

Gross, P. R., and G. H. Cousineau: Effects of actinomycin D and macromolecule synthesis and early development in sea urchin eggs. Biochem. & Biophys. Res. Comm. *10:*321, 1963.

Grüneberg, H.: Congenital hydrocephalus in the mouse, a case of spurious pleiotropism. J. Genet. *45:*1, 1943.

———: The Genetics of the Mouse. 2nd edition. The Hague: Nijhoff, pp. 1–650, 1952.

Hadorn, E.: Begriffe und Termini zur Systematik der Letalfaktoren. Arch. Jul. Klaus-Stiftg. *24:*105, 1949.

Hallqvist, C.: Ein Fall von Letalfaktoren beim Schwein. Hereditas *18:*215, 1934.

Harvey, E. B.: The American Arbacia and Other Sea Urchins. Princeton: Princeton University Press, 1956.

Hsia, D. Y. Y.: Inborn Errors of Metabolism. 2nd Edition. Chicago: Year Book Publishers, 396 pp., 1966a.

———: The detection and treatment of inborn errors of metabolism associated with mental deficiency. In Enzymes in Mental Health. Philadelphia: Lippincott, 1966b, p. 121.

———: The development of intestinal β-galactosidase and β-glucuronidase in the newborn. Biochim. et Biophys. Acta *122:*550, 1966c.

Jacob, F., and J. Monod: Genetic regulatory mechanisms in the synthesis of protein, J. Molec. Biol. *3:*318, 1961.

Jacquot, R., and N. Kretchmer: Effect of fetal decapitation on enzymes of glycogen metabolism. J. Biol. Chem. *239:*1301, 1964.

Jervis, G. A.: Phenylpyruvic oligophrenia (phenylketonuria). Res. Publ. Asso. Nerv. and Ment. Dis. *33:*259, 1954.

Jost, A., and R. Jacquot: Sur le rôle de l'hypophyse, des surrenales et du placenta dans le synthese de glycogene par le foie foetal du lapin et du rat. Comp. Rend. *247:*2459, 1958.

Klenk, E., W. Gielen, and G. Padberg: The structure of the gangliosides. In S. M. Aronson, and B. W. Volk, editors. Cerebral Sphingolipidoses. New York: Academic Press, 1962, p. 301.

Lenz, W.: Medical Genetics. Chicago: University of Chicago Press, 1963, p. 87.

Lokietz, H., R. M. Dowben, and D. Y. Y. Hsia: Studies on the effect of novobiocin on glucuronyl transferase. Pediatrics *32:*47, 1963.

Lucey, J. F., and R. G. Dolan: Hyperbilirubinemia of newborn infants associated with the parental administration of a vitamin K analogue to the mothers. Pediatrics *23:*553, 1959.

Markert, C. L.: Developmental genetics. Harvey Lectures *59:*187, 1964.

Markert, C. L., and H. Ursprung: Production of replicable persistent changes in zygote chromosomes of *Rana pipiens* by injected proteins from adult liver nuclei. Dev. Biol. *7:*560, 1963.

Menkes, J. H., P. L. Hurst, and J. M. Craig. A new syndrome: Progressive familial infantile cerebral dysfunction associated with an unusual urinary substance. Pediatrics *14:*462, 1954.

Monroy, A.: Certain embryological considerations. Birth Defects Original Article Series *1:*15, 1965.

Monroy, A., R. Maggio, and A. M. Rinaldi: Experimentally induced activation of the ribosomes of the unfertilized sea urchin egg. Proc. Nat. Acad. Sc. *54:*107, 1965.

Moore, J. A.: Transplantation of nuclei between *Rana pipiens* and *Rana sylvatica*. Exp. Cell. Res. *14*:532, 1958.

———: Nuclear transplantation and problems of specificity in developing embryos. J. Cell and Comp. Phys. *60* and Suppl. (1): p. 19, 1962.

Nemer, M.: Interrelation of messenger polyribonucleotides and ribosomes in the sea urchin egg during embryonic development. Biochem. and Biophys. Res. Comm. *8*:511, 1962.

Petterson, G., and G. Bonnier: Inherited sex-mosaic in man. Hereditas *23*:49, 1937.

Semenza, G., and S. Auricchio: Chromatographic separation of human intestinal disaccharidases. Biochim. et Biophys. Acta *65*:173, 1962.

Sereni, F., and N. Principi: The development of enzyme systems. Ped. Clin. N. A. *12*:515, 1965.

Slome, D.: The genetic basis of amaurotic family idiocy. J. Genetics *27*:363, 1933.

Snyderman, S. E., *et al.:* Maple syrup urine disease with particular reference to dietotherapy. Pediatrics *34*:454, 1964.

Sutherland, J. M., and W. H. Keller: Novobiocin and neonatal hyperbilirubinemia. Am. J. Dis. Child. *101*:447, 1961.

Szontagh, F. E., A. Jakobovits, and C. Méhes: Primary embryonal sex ratio in normal pregnancies determined by nuclear chromatin. Nature *192*:476, 1961.

Walker, F. A., D. Y. Y. Hsia, H. M. Slatis, and A. G. Steinberg: Galactosemia: A study of 27 kindreds in North America. Am. J. Human Genetics *25*:287, 1962.

Weber, G., R. L. Singhal, N. B. Stamm, and S. V. Scrivastava: Hormonal induction and suppression of liver enzyme biosynthesis. Fed. Proc. *24*:745, 1965.

Weber, G., M. A. Lea, R. L. Singhal, N. B. Stamm, and E. A. Fisher: Coordinated biosynthetic pattern in liver: synchronous behavioral pattern of key glycolytic enzymes. Fed. Proc. *25*:378, 1966.

Wilson, J. G.: Experimental studies on congenital malformations. J. Chronic Dis. *10*:111, 1959.

Wilt, F. H., and T. Hultin: Stimulation of phenylalanine incorporations by polyuridylic acid in homogenates of sea urchin eggs. Biochem. and Biophys. Res. Comm. *9*:313, 1962.

CHEMO-MECHANICS OF IMPLANTATION

BENT G. BÖVING

*Department of Embryology, Carnegie Institution of Washington,
Baltimore, Maryland*

The Darwinian revolution established phylogeny as a principal perspective in biology. During the same period, supremacy of the experimental method in physiology was advocated by Claude Bernard. From union of those prominent ideas derive the sciences of comparative physiology, comparative physiology of reproduction among them. Within that science and in the next generation of subsciences, such as comparative physiology of implantation, the Bernardian genes for experimentation have been dominant, and, as if by orthogenesis, have generated unassimilated accumulations of spottily consistent information; whereas the genes for theoretical synthesis have been recessive and have generated no comprehension with broad Darwinian sweep. That failure of comparative physiology to attain the ancestral perspective achieved over a century ago by comparative anatomy deserves question of its causes. Attention is required too by species differences having been used too rarely for systematic and revealing comparisons and too often as excuses for isolation and misunderstanding between veterinarians and clinicians and even colleagues in research.

About half a century ago, Loeb (1908a, b) performed the classical experiments of stimulating a decidual reaction by uterine trauma. Onto those valid experiments was tied the dubious interpretation that the decidual reaction was equivalent to the maternal part of the placenta (in spite of lacking its principal functional part: the maternal blood circulating in the intervillous space) and upon that fallacy a school of thought developed with something akin to a value judgement that, since the maternal part of the placenta could be developed without the benefit of an implanting ovum, the ovum could be assumed to be unimportant and perhaps even passive or nearly so in the process of implantation and placenta formation (Fig. 1, top line). That point of view has been encouraged by the number and success of studies revealing many details of endocrine and other maternal chemical conditions necessary to permit experimental decidualization in one or another species and by the experience that decidualization can be induced by a great variety of chemical or physical stimuli—even models of ova made from inert substances. Accordingly, decidualization has been considered the reaction of a specifically conditioned uterus to a non-specific stimulus. That is not to be mistaken as an explanation of placentation, for there is no placenta; nor is the normal action by the ovum on the uterus informatively or accurately described by the term "non-specific." All it really means is that inappropriate applications of the experimental method of simple substitution have not identified the normal stimulus. Were it not for the meticulously physiological studies of Blandau (1949a, b)

Maternal	
	Fetal
Maternal	Fetal
Maternal ⇄	Fetal

Fig. 1. The diagram lists viewpoints of components responsible for implantation.

showing the stimulus to be far from non-specific and clearly related to ovum model size (in the rat) and proteolysis (in the guinea pig), one might be tempted to conclude that the normal process by which ova become implanted in uteri is unlikely to be explained by studies that begin by excluding ova.

While Loeb (1908a, b) was thinking about uteri developing placental components without the benefit of ova, Brachet (1913) was studying the development of mammalian ova without the benefit of uteri. Having carried a series of rabbit ova up to and perhaps slightly beyond the normal time of implantation by overlapping two day periods of culture, he concluded that the mammalian ovum does not require a uterus for its development (Fig. 1, line 2). The same conclusion could be supported by observations, in several species, of fetal development in ectopic sites, either spontaneously or as result of experiment.

Ectopic implantation is, however, not only uncommon but rather unlikely to succeed. Nicholas (1934) found only 2 per cent of rat ova shed into the peritoneal cavity developed. Fawcett, Wislocki and Waldo (1947), after growing mouse ova in various ectopic locations, drew the tempered conclusion that neither uterus nor ovum can be considered "chiefly" responsible for normal implantation; both have important roles (Fig. 1, line 3).

Since reproduction is more efficient when uteri and ova are associated normally than when they are not, I turned my attention to the normal interaction between them (Fig. 1, bottom line) and adopted the strategy of at least beginning various branches of investigation by observation of ovum and uterus together in as near normal relationship as achievable. Initial steps avoided experimental procedures because of the danger that they might alter as yet uncomprehended relationships. It is the development of that strategy and some insights gained through it that I will stress—at the expense of repeating details and data most of which have been not only published but summarized already (Böving, 1963).

The attachment of the rabbit ovum, or blastocyst, to the uterus may be described as having three stages or families of mechanisms: muscular, adhesive and invasive. In the muscular stage, waves of uterine contraction transport blastocysts along the cylindrical uterine horn containing them, then space them equidistantly. The impression of equal spacing is an old one. Spacing has been attributed to preformed implantation sites in a number of species and, in the rabbit, to waves of uterine contraction traveling the full length of the uterus and described as peristalsis milking the uterine fluid to the cervical end of the uterus and recoil when the contraction eases. Such contractions were observed in experiments on artificially ovulated rabbits into whose uteri were injected fixed sea urchin ova which were discovered to be scattered throughout the length of the uterine horn within 2 hours. Unfortunately, preformed sites, ova and the objects scattered throughout the uterus have all been described as evenly spaced without a distinction having been made between random spacing and equidistant spacing. The distinction is vital, because random spacing implies passive disposition of objects by the uterus, whereas even spacing implies that the uterus has reacted to each one individually—unless the unlikely situation occurs that the number of blastocysts that varies from horn to horn is consistently matched by an equal number of preformed implantation sites.

Simply opening a series of rabbit uteri of exactly known gestation age covering the period of blastocyst transport and measuring the location of each blastocyst permitted a description of normal transport. The mean location of blastocysts on each day served to mark their typical progress along the uterus, and the equality of distances separating them was expressed by a simple coefficient of variation: the mean of the separations divided into the standard deviation. With perfectly even spacing, standard deviation is 0 and the coefficient is 0; random spacings give coefficients whose average is 1. The blastocysts required over 2 days to fill out the uterus to the degree that the fixed sea urchin ova did in 2 hours, suggesting inappropriateness of the experimental technique. Blastocyst spacing was not distinguishable from random during the first 2 days that blastocysts are in the uterus but became significantly more even than random during the next 2 days. (Blastocysts enter at 3 days after mating and attach at 7.) Measurements of blastocyst diameter that had been made together with the location measurements added the information that the shift from random to even spacing began at the same time that blastocysts began a rapid increase in diameter. On the basis of these observations of the normal situation without experimentation, it was concluded that increasing blastocyst size or something associated with it was the stimulus for equidistant spacing, and it was guessed that the spacing was accomplished by waves of contraction originating from each end of each uterine horn and wherever a blastocyst within caused distention. Opening normally pregnant rabbits and just looking, albeit helped by a time lapse movie camera, has supported the conclusion.

Experimental confirmation was attempted by putting into empty horns of contralaterally pregnant uteri a number of blastocyst sized glass spheres. A few horns did indeed space glass beads as evenly as the blastocysts in the control horn, but when enough experiments were done for statistical evaluation, the usual glass bead spacing could not be distinguished from random, whereas the spacing of blastocysts in the sham operated control horns was as near even as normal. Does the experiment disprove the conclusions from observation? I think rather it proves that I, as well as others, am capable of experiments that are insufficiently physiological to fool the living system.

The muscular stage of attachment just discussed appears explainable mainly in terms of mechanical interplay between uterus and blastocysts, although it should be added that both the blastocyst expansion and the characteristic behavior of the rabbit's uterine muscle during spacing is progesterone dependent.

The adhesive and invasive stages of implantation introduce chemical factors with a more direct role. Again, they were approached by the strategy of beginning with observation of mother and offspring in as near normal relationship as possible and looking for relations presumably established before and obviously not entirely removed by the only procedure applied: histological preparation. A rabbit blastocyst just implanted and fixed at 7 days after mating was studied in serial sections by using the numerous discrete points of attachment between uterus and blastocyst as tracers to indicate sites where all essentials for attachment could be presumed to have been present. A method of quantitative analysis was introduced to test whether various structures suspected of promoting attachment were in fact consistently closely associated with attachment, randomly related or inversely related. The width of attachments was also used as a clue to the dimension or range of activity of the stimulus. Mechanical aspects considered included the crests and depressions of the folded endometrium. With respect to them the only conclusion was the obvious one that attachment can occur only where the maternal and fetal tissues touch. Chemical aspects were looked for in three categories: sites of stimulus storage (glycogen), sites of stimulus production (uterine glands) and sites of transfer for conceivably helpful substances of remote origin, such as hormones, nutrients and gas exchange (capillaries at the base of uterine epithelium). In brief, no relation was found between attachments and storage or local production sites, but a statistically significant relation was found between trophoblast attachments and uterine epithelium with a blood vessel at its base. There was agreement also between the mean width of attachments ($16 \pm 2\mu$) and the mean dimension of contact between the vessels and involved epithelium ($18 \pm 2\mu$). In anatomical terms, the blastocyst had attached to the uterus only where it touched epithelium with a vessel at its base, and that complex was thereby identified as supplying the necessary and sufficient stimulus for attachment.

Two stages of attachment were distinguished: adhesions and penetrations, according to whether the trophoblast touched only the surface of the uterine epithelium or had extended through it to the underlying stroma. Presumably adhesion precedes penetration, yet adhesions are as accurately related to vessels as are penetrations. Thus, the site for trophoblastic attachment either must be chosen by the trophoblast's somehow detecting the maternal vessel some 50μ away on the other side of the uterine epithelium, or else the proximity of vessel and trophoblast attachment is caused by vessels growing into the site of attachment even at the early adhesive stage. Ingrowth of vessels can be denied responsibility for the association, because vessels are not significantly closer together at attachment sites than elsewhere, but how the trophoblast can "see" through the epithelium remained to be explained. A blood borne agent passing through the epithelium by unhindered diffusion would reach the surface almost uniformly distributed; its point of origin could not be detected from the lumen. To find the presumably narrow epivascular pathway presumably followed by the presumed chemical stimulus presumed to pass from the maternal blood to the implanting ovum, the security of observation was forsaken for an experimental procedure. A horribly but necessarily unphysiological solution of silver nitrate was perfused through the vessels of a pregnant rabbit at 7 days after mating with the hope that silver would precipitate wherever it diffused and thereby mark its path. In the uterus, that path was straight through epithelial cells with a vessel at their base. The consequence was that the epithelial surface next to the blastocyst had a chemical image of the vessel at the base—exactly the location and extent that the anatomical analysis had defined as providing the necessary and sufficient stimulus for attachment.

By 8 days after mating, trophoblast attachments spread in the plane of the epithelium. That seemed inconsistent with confined exchange.

The silver perfusion was applied, and it revealed that the pathway of exchange had also spread. The anatomical basis for it is a synchronous disappearance of intercellular membranes from the epithelium—a change known to be progesterone dependent. In more diagrammatic terms, a chain of consequences goes from a maternal chemical factor to a maternal structural factor that changes the pathway of chemical exchange and alters the mechanical relations of both mother and offspring at their point of contact.

As I may have hinted, experimentation is fraught with the danger of misinterpretation. An incidental observation that silver perfusion usually failed to reveal epivascular pathways deep in endometrial depressions or other regions where there were no blastocysts suggested that there might be an error in the basic assumption that the technique would mark a pathway by which chemical substances might travel from maternal circulation to blastocyst. What was observed might rather be a reaction product of the silver solution and something coming from the blastocyst and being removed by the first available maternal circulation. Association with normally developed blastocysts and absence near a dead one characterized the chemical agent from the blastocyst as not only precipitable by silver but the product of a living system rather than something passively stored. A heavier concentration near the abembryonic pole of blastocysts, known to develop a higher pH on exposure to air and a higher concentration of calcium carbonate crystals upon drying suggested that the agent was in the family of carbon dioxide, bicarbonate or carbonate.

The discoveries by Lutwak-Mann and Laser (1954) that rabbit blastocysts have a bicarbonate concentration 3 and 2 times that of mother's blood at 6 and 7 days after mating and have equilibrated by 8 days, when attachment is intimate, seem consistent as do discoveries of high carbonic anhydrase activity in the rabbit endometrium near the time of implantation and as a result of progesterone (Lutwak-Mann, 1955; Adams and Lutwak-Mann, 1955; Lutwak-Mann and Adams, 1957).

Actually, the silver perfusion method appears to be an *in vivo* histochemical test that reveals where the enzyme carbonic anhydrase is operating—more particularly where bicarbonate has diffused from the blastocyst into epivascular epithelium as result of the concentration gradient and where the equilibrium with alkaline carbonates and carbonic acid is continually pushed the same way as carbon dioxide is released from carbonic acid with the enzyme's assistance and removed by the maternal circulation, leaving an alkaline residue and consequent epivascular rise of pH.

So local a rise of pH is difficult to demonstrate by conventional methods, but the adhesions seen histologically are considered a physical consequence or evidence of high pH, just as indicator color change or potential change across a glass membrane are so considered. The surface of blastocysts become mushy and sticky when subjected to solutions of high pH, and acidification restores firmness. Looked at in reverse, it is considered that trophoblast picks epivascular epithelium for invasion by a preliminary adhesion to such epithelium because just that epithelium and apposed region of the blastocyst get sticky from carbon dioxide blowoff to the maternal circulation.

Since trophoblast penetration of uterine epithelium is localized epivascularly just as adhesion is, one may suppose that it could result from the same epivascular pH rise. To test that idea, solutions of bicarbonate or carbonate comparable in concentration to those of a blastocyst 6 days after mating were placed in the lumen of rabbit uteri at various pertinent times. At the time of implantation, dissociation of epivascular epithelium was observed; apparently it provides a path through the epithelium that the trophoblast may travel. Two days later, when attachment has spread maximally and become arrested, the epithelium is not dissociable —presumably because progesterone has converted it from a cellular to syncytial state. Thus, endocrine control of uterine epithelial cell carbonic anhydrase and intercellular membranes turns on bicarbonate transfer from blastocyst to mother, keeping it local near vessels for adhesion and early penetration, then allowing it to spread for the spreading stage of attachment, and finally helping to turn off trophoblastic invasion.

The propulsive force that actually drives trophoblast through the epivascular gap in uterine epithelium remains to be explained. Again, the basis for explanation lies in observation. Again, over half a century ago, it was noticed that rabbit trophoblast penetrates uterine epithelium only where there is a knob or aggregate of what has been called trophoblastic syncytium. There is current controversy on the degree to which it is cellular or syncytial, but so far as function is concerned the crux of the matter seems rather to be what happens when it is in the epivascular microenvironment where a pH rise is considered to occur. A few unusually favorable sites in a specimen fixed promptly by injection of fixative into the uterine lumen in vivo show that trophoblast knobs are not necessarily just syncytium or something like it differentiated on the outer surface of the cellular trophoblast covering the blastocyst. Some knobs consist of syncytiotrophoblast encapsulated between endoderm and a cover of cytotrophoblast. Moreover, other stages show the contents of knobs being extruded through uterine epithelium through a duct formed by the cellular covering of the knob adhering to the epithelium and, like the adjacent epithelium, developing a gap or separation of cells epivascularly, presumably as result of pH rise there. To test that interpretation, and in a sense all of the foregoing tentative and somewhat more secure interpretations from which it has been derived, micromanipulated jets of alkaline solution (sodium carbonate 0.15N) were

DAYS AFTER MATING	STIMULUS (P = progesterone dependent)	EFFECTOR	ACTION	RESULT
3	"spontaneous"	uterine muscle	well-propagated contraction	blastocyst transport, rapid with random arrangement
4–5	progesterone / P 2-5mm blastocyst distention	uterine muscle	decreasingly propagated contraction	blastocyst transport slow with increasingly even spacing
6	progesterone / P over 5mm blastocyst distention	uterine muscle	dome & grasp	transport stopped, endometrium flattened, blastocyst held antimesometrially
7	HCO⁻ accumulated in blastocyst	maternal circulation	$2HCO_3 \rightleftharpoons H_2CO_3 + CO_3$ (alkaline/residue); carbonic anhydrase; $H_2O + CO_2 \longrightarrow$ (to maternal circulation)	epivascular alkalinity
	P carbonic anhydrase increase	uterine epithelium		
	epivascular alkalinity	lemmas	adhesion	first attachment (fixes blastocyst orientation, always abembryonic & antimesometrial)
	?	trophoblast knobs	lysis	lemmas disintegrated abembryonically (aids chemical exchange, uncovers trophoblast)
	? pressure within knob	trophoblast knobs	adhesion	epivascularly localized attachments (aim subsequent penetrations at vessels)
	penetration to vessel	trophoblast knob	cell dissociation	gap in cytotrophoblast — path for penetration to vessel
		uterine epithelium	cell dissociation	gap in epithelium
		trophoblast knob	extrusion of syncytiotrophoblast	penetration to vessel
8	maternal circulation	early yolk-sac placenta	anchor	holds blastocyst while lemma remnants are shed
	progesterone		chemical exchange	alkaline reaction becomes diffuse / attachment spreads
	presence of trophoblast	uterine epithelium	becomes syncytiol	undissociable / penetration
9	continued & improved chemical exchange	late yolk-sac placenta	HCO_3 gradient eliminated	basis for pH rise eliminated / penetration stops

Fig. 2. The table presents the process of antimesometrial placentation in the rabbit as an orderly sequence of related mechanisms. (By courtesy of Carnegie Institution of Washington.)

directed at the knobs on blastocysts removed shortly before they would have attached to the uterus. In a few cases the knobs were caused to pop. It was concluded that they have an internal pressure that under favorable circumstances can be released, as if to drive the contents through uterine epithelium, by the stimulus of high pH of a carbonate that for a variety of reasons has been thought responsible for epivascular trophoblast penetration.

The preceding interpretations of antimesometrial attachment of rabbit blastocysts may be brought together to illustrate the process as a whole by listing the sequence of reactions described in the usual terms of stimulus, effector and result (Fig. 2). It is hoped that doing so conveys the sense that implantation is an orderly process. It is also intended to replace the idea that the orderly sequence of anatomical stages implies a corresponding sequence of remote and presumably chemical controls; mechanical changes may have consequences on how chemical factors are distributed and operate to cause subsequent mechanical consequences. There seems to be, as suggested in the left column, a feedback or oscillation at least to the extent that steps or mechanisms of predominantly mechanical character tend to be followed by changes of more chemical nature—although less pronouncedly so toward the end. One may perhaps speculate that such oscillation is in fact necessary to maintain a progressive process and stave off a steady state.

At the very least, it is clear that the process of implantation cannot be described realistically in terms of maternal factors alone or fetal factors alone or even both in interaction, as diagrammed in Figure 1. Since both mother and offspring have both mechanical and chemical aspects, and they all interact with one another, Figure 3 is offered as a better diagram. However, in order to indicate that changes occur with time, it should have an added dimension, as in Figure 4. Although not complete, it is nevertheless troublesomely complex; an abstraction from it may be worthwhile (Fig. 5). That brings us back to Mossman's (1937) classical definition of a placenta as an apposition of fetal and maternal parts for purposes of physiological exchange. The balance and completeness of the definition can be appreciated. It regards placentation equally from the

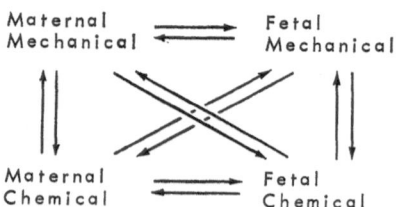

Fig. 3. The diagram expands upon the last line of Fig. 1 to suggest the components and interactions involved in implantation.

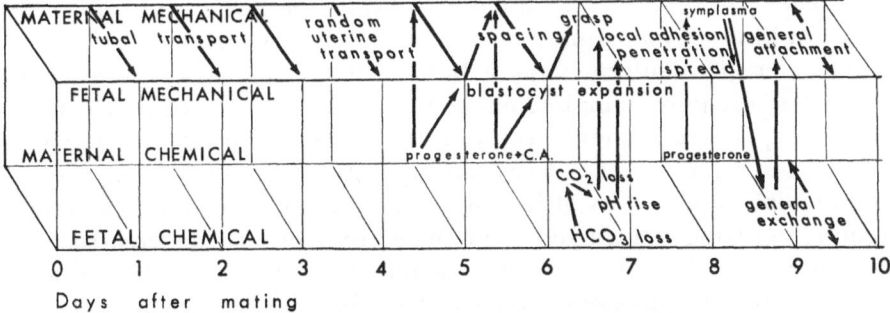

Fig. 4. The diagram expands Fig. 3 by adding a dimension for time to the components and interactions involved in rabbit blastocyst implantation.

maternal and fetal points of view and gives equal stress to the mechanical and chemical aspects of both. There is also a hint that, if the purpose of placentation is physiological exchange, then—since the circulatory systems of mother and offspring provide the principal structures for exchange—that fraction of the implantation process most promising for comparative study is likely to be the mechanisms by which maternal and fetal circulations are brought into association. That thought not only reflects the observation that trophoblastic attachment exhibits hemotropism in the macaque as well as in the rabbit, but it helps relate observations that might otherwise seem irrelevant. For example, at the time before implantation, capillaries grow into the previously avascular, subepithelial region (Bartelmez' Zone I) of the human endometrium, whereas the rabbit endometrium acquires increased carbonic anhydrase activity. Both apparently result from progesterone stimulation, and both promote gas

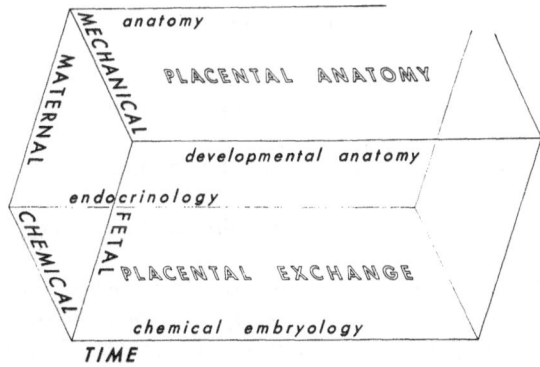

Fig. 5. An abstraction of Fig. 4 removes some of the unfamiliarity and complexity yet shows that many viewpoints must be considered for a complete and balanced explanation of implantation.

exchange across uterine epithelium in spite of being very obviously different kinds of mechanisms.

The fact that adult anatomy is but one line or corner or subspecialty (Fig. 5) and requires sampling only once along the time axis may explain how comparative anatomy could survey enough species to derive a phylogenetic picture. To derive an equivalent comparative picture of the physiological process of implantation and placenta formation would require applying at least the diagrammed four of our sundered sciences at a number of points in developmental time and to as many orders, families and perhaps even species as have been considered necessary to draw phylogenetic conclusions from just anatomical evidence. It has taken many years to understand just the process of implantation in just one species with what is hoped to be reasonable balance and completeness. To do so for a significant number of species will take many men and many years, and when it is reflected that the physiology of implantation is but a small part of the physiology of reproduction, it may be appreciated that comparative physiology of reproduction is not so much a well developed body of existing science as it is an objective that we share.

Summary

1. Explanations of implantation may be regarded as incomplete if they do not consider mother and offspring, the mechanical and chemical aspects of each, and the interplay among them.

2. The supremacy of experimental methods in physiology is questioned with respect to rabbit blastocyst implantation, and it is demonstrated that observational methods may have commanding advantages over experimental ones, particularly in early stages of exploration.

References

Adams, C. E., and C. Lutwak-Mann: Endometrial carbonic anhydrase as hormone indicator for progesterone and related steroids. J. Endocrinol. *13*:xix, 1955.

Blandau, R. J.: Observations on implantation of the guinea pig ovum. Anat. Rec. *103*:19, 1949a.

————: Embryo-endometrial interrelationships in the rat and guinea pig. Anat. Rec. *104*:331, 1949b.

Böving, B. G.: Implantation Mechanisms. In Mechanisms Concerned with Conception, C. G. Hartman, ed. New York: Pergamon Press, 1963.

Brachet, A.: Recherches sur le déterminisme héréditaire de l'oeuf des Mammifères. Dévelopement *in vitro* de jeunes vesicules blastodermiques de lapin. Arch. Biol., Paris, *28*:447, 1913.

Fawcett, D. W., G. B. Wislocki, and C. M. Waldo: The development of mouse ova in the anterior chamber of the eye and in the abdominal cavity. Am. J. Anat. *81*:413, 1947.

Loeb, L.: The experimental production of the maternal part of the placenta in the rabbit. Proc. Soc. Exper. Biol. and Med. *5*:102, 1908a.

———: The production of deciduomata and the relation between the ovaries and the formation of the decidua. J.A.M.A. *50*:1897, 1908b.

Lutwak-Mann, C.: Carbonic anhydrase in the female reproductive tract. Occurrence, distribution and hormonal dependence. J. Endocrinol. *13*:26, 1955.

Lutwak-Mann, C., and C. E. Adams: Carbonic anhydrase in the female reproductive tract. II. Endometrial carbonic anhydrase as indicator of luteoid potency: Correlation with progestational proliferation. J. Endocrinol. *15*: 43, 1957.

Lutwak-Mann, C., and H. Laser: Bicarbonate content of the blastocyst fluid and carbonic anhydrase in the pregnant rabbit uterus. Nature *173*:268, 1954.

Mossman, H. W.: Comparative morphogenesis of the fetal membranes and accessory uterine structures. Contrib. Embryol. *26*:129, 1937.

Nicholas, J. S.: Experiments on developing rats. I. Limits of fetal regeneration; behavior of embryonic material in abnormal environments. Anat. Rec. *58*: 387, 1934.

COMPARATIVE ASPECTS OF STEROID HORMONES IN REPRODUCTION

K. J. RYAN AND L. AINSWORTH

Department of Obstetrics and Gynecology,
Western Reserve University School of Medicine, Cleveland

In considering the role of steroid hormones in the comparative aspects of reproductive failure in animals, it is important to bear in mind that the evolution of reproduction, in general, and of viviparity, in particular, has involved predominantly endocrine mechanisms. Within these endocrinological adaptations, the steroid hormones have appeared to be the major mediating agents in the control of pregnancy. Hence, the generality has been offered that progestins and estrogens are essential for reproduction in all higher forms of animal life. However, the roles that these two types of hormones play once gestation is established have not yet been completely defined nor has their essentiality in an absolute or relative sense been incontrovertibly established for all species. Our knowledge of the mechanisms involved in comparative reproductive failure is incomplete, and clear cut pathophysiological bases for endocrine related pregnancy failures have yet to be defined even in the human where it has been studied extensively. As will be elaborated upon in the following section, the normal physiological roles of steroid hormones in the course of pregnancy still require extensive study.

I. POSSIBLE ROLES FOR THE STEROID HORMONES IN CONTROL OF GESTATION

Before one can possibly implicate steroid hormones in reproductive failure, their normal role must be considered. It is surprising how little is known and how much is assumed in this area. The requirement for estrogens and progestins for the development of the reproductive tract, cyclic function, estrus receptivity and all of the events leading up to conception are well documented and will not be elaborated further. Review

154

articles on synthesis of hormones in the ovary (Ryan, 1963a), and the biogenesis of estrogens (Breuer, 1962; Ryan, 1963b) have been published. The role of these hormones after conception is still incompletely described and several areas where comparative data might be of assistance are noted below. It is generally assumed that where the length of gestation exceeds the luteal phase of the cycle new endocrine mechanisms must be invoked to maintain the pregnancy (Amoroso, 1955, 1960).

1. Implantation

In certain species, the armadillo and guinea pig specifically, implantation will occur after oophorectomy. To the extent that other tissues assume the ovary's endocrine function, this implantation may still require the presence of estrogen and progesterone, but certainly a placenta is not present at this pre-implantation period and such hormones may not be essential in these animals for nidation. On the other hand it is assumed that both hormones are needed for implantation in most other species (Amoroso and Finn, 1962; Buchanan, *et al.*, 1956).

2. Pituitary Inhibition

It is generally assumed that the infertility, anovulation, lack of cyclic pituitary function during gestation are a reflection of selective inhibitory effects of estrogen and progesterone on the pituitary or hypothalamus. Except in species (marsupials) where hypophysectomy always results in abortion, such a mechanism would seem to be a generalized phenomenon indicating that gonadotrophic hormones are produced by an extra pituitary source in most instances (Amoroso, 1955).

3. Maintenance of Decidua and the Endometrial Implantation Site

In those species where oophorectomy always results in failure to implant or abortion, it would be logical to predict that steroid hormones would be necessary for the maintenance of the implantation site. Where only replacement of progesterone is required after oophorectomy, it has been assumed that estrogen is provided by an extra ovarian source but this has not been demonstrated directly in such species. There is, in fact, no evidence that both estrogen and progesterone are needed for maintenance of the endometrium in all species during the course of pregnancy.

4. Myometrial Tone and Contractility

An elaborate theory concerning the effects of estrogen and progesterone on the myometrium has been advanced in which progesterone in some way modulates the tone and contractility of the uterus keeping it quiescent during gestation. Estrogen has an effect antagonistic to that of progesterone, the net result being dependent on their respective concentrations. Attempts to relate the initiation of labor to blood or tissue levels of hormones in varying species have not been successful. On the contrary, no good correlation with the relative levels of these steroids has as yet been made during normal pregnancy. The so-called progesterone or estrogen dominated uterine muscle with a predictable activity and response to oxytocin has however been well established, but the extension of this to the intact uterus in all species requires further study (Csapo, 1956; Short, 1960).

5. Length of Gestation

It appears that the normal length of gestation varies widely among various animal species but is reasonably constant within a given species or strain. Where interbreeding has occurred, the genotype of the fetus seems to determine the length of the gestation rather than a purely maternal factor. How this genotypic information is imparted is of course not known. Within a given species or strain, the length of gestation appears to be inversely proportional to litter size. Whether the relative size of the uterine contents determines this for different species remains to be evaluated. That there is primarily an endocrine basis for determining the onset of parturition is by no means established in spite of the attention paid to this concept (Short, 1960).

6. Mammary Function

Although estrogens and progestins are needed for mammary development and preparation for lactation, there is evidence that they are not essential for the maintenance of this mammary function (Cowie and Folley, 1961).

7. Metabolic Effects

Under this heading can be included all of the poorly understood changes which characterize the pregnant state which may or may not be due to the estrogens and progestins and which may or may not be important in the course of normal pregnancy. Immune-suppressive phenomena, alterations in the function of other endocrine glands, changes in the metabolism of their hormonal products and changes in cardiopulmonary and renal function are all examples of these possible effects. The requirement of these changes for the normal course of gestation is poorly understood.

8. Summary on the Role of Steroid Hormones in Gestation

In short, although estrogens and progestins appear to be needed in all higher species for reproductive functions up to conception, their essentiality and role in the further development and maintenance of pregnancy are by no means established. By attention to the comparative aspects, our deficiencies in knowledge become much more apparent. The impossibility of making generalizations on the basis of a single species (usually the human or rat) is dramatically exemplified in this area of the comparative physiology of pregnancy.

II. The Role of the Corpus Luteum in Pregnancy

Studies on the general role of the corpus luteum in gestation have been extensively reviewed (Hisaw, 1932). A possible endocrine role for the corpus luteum in pregnancy has been determined by either checking its anatomic integrity during gestation or studying the effect of its removal on the subsequent course of the pregnancy. Since anatomic appearance cannot always be used as a guide to function, greater credence has been placed on the latter type of study. It is interesting, however, to note how well the two types of data can be correlated (Tables 1 and 2).

Table 1

SPECIES IN WHICH PREGNANCY MAY CONTINUE IF OOPHORECTOMY IS PERFORMED AFTER A CERTAIN GESTATIONAL AGE

Species	Gestational Age for Oophorectomy in which Abortion not inevitable (Days)	Approx. Total Length of Gestation (Days)	Regression of Corpora Before Term	Presence of Progesterone Demonstrated in Placenta
Guinea Pig	40	67–68	No	Yes
Cat	49	63	Yes	—
Sheep	55	143–159	Yes	Yes
Horse	170–270	330	Yes	Yes
Monkey	25	165	—	Yes
Human	40	267	Yes	Yes
Cow	207–230	280	No	No
Armadillo *	Approx. 40	150	No	—

* Not allowed to proceed to term.

Table 2

SPECIES IN WHICH PREGNANCY IS NOT MAINTAINED
AFTER OOPHORECTOMY

Species	Length of Gestation (Days)	Regression of Corpora Before Term	Presence of Progesterone Demonstrated in Placenta
Opossum	12.5	No	—
Hamster	16–19	No	—
Mouse	20	No	—
Rabbit	30–32	No	No
Goat	151	—	No
Sow	112–115	No	No
Rat	21	No	No
Dog *	58–63	Yes	No

* Not tested in second half of gestation.

There are generally two types of response to ovariectomy; either abortion at any stage or maintenance of the pregnancy if the procedure is performed at some specific time for the given species (Amoroso and Finn, 1962).

In those instances where oophorectomy does not cause abortion, it is generally assumed that the placenta provides a source of progesterone and/or estrogen. When abortion after oophorectomy can be circumvented by exogenous progesterone alone, it has been assumed that the placenta can synthesize estrogens but as previously mentioned this has not been proven directly. Some species might require only progesterone for maintenance of pregnancy at that point. Where progesterone and estrogen are both required after oophorectomy to maintain gestation, it has been assumed that the placenta does not make either progesterone or estrogen or that the amounts made are too small to be effective unless most of the fetuses are removed.

It can be seen in Table 1 that maintenance of pregnancy after oophorectomy generally seems to occur in animals with prolonged gestational periods (Zarrow, 1961), in whom regression of corpora lutea ordinarily is completed long before parturition (Amoroso and Finn, 1962) and in which progesterone can be isolated from the placentae. The cow is a notable exception but the presence of progesterone in cow placenta has been reported in one of six samples tested (Bowerman and Melampy, 1962) although it was not isolated by other investigators (Short, 1961). Progesterone has been isolated from guinea pig placentae (Heap and Deanesly, 1966).

Species in which pregnancy is not maintained after oophorectomy seem to have relatively short gestational periods (Zarrow, 1961), a prolonged life of the corpora lutea until parturition (Amoroso and Finn, 1962) and have placentae in which progesterone has not been demonstrated (Short, 1961; Amoroso and Finn, 1962).

III. Steroid Endocrine Role of the Placenta

In those species listed in Table 1 and others where ovariectomy does not result in termination of the pregnancy it has been postulated that an extra ovarian source, the fetal placenta, has assumed the endocrine control of gestation. Direct evidence for this has been obtained only in the human where isolation from placental effluent and peripheral blood, isolation from the placenta, perfusion studies and in vitro enzymatic formation of progesterone and estrogens have been documented (Ryan, 1962; Diczfalusy and Troen, 1961; Ryan, *et al.*, 1966). More detailed discussion of placental enzymatic function will be documented below.

IV. Progesterone

Progesterone has been isolated from the placentae of several species (Table 3) and peripheral blood levels have often been obtained as well (Short, 1957, 1961). In many of these species as noted previously abortion does not occur following oophorectomy (Table 3) suggesting that placen-

Table 3

SPECIES IN WHICH PROGESTERONE WAS ISOLATED FROM PLACENTAE

Species	Placental Level (μg/kg)	Peripheral Blood (μg/100 ml)	Abortion Inevitable After Oophorectomy
Human	500–10,000	14.2	No
Chimpanzee	*	—	—
Monkey	35–72	2.4 †	No
Horse	60–370	0.5–1.3 ‡	No
Sheep	4–9	0.23–0.39 §	No
Common Seal	46	—	—
Indian Rhinoceros	41	—	—
Giraffe	10–12	—	—

* Values comparable to human.
† Uterine vein blood—lower in periphery.
‡ Detected up to day 70 of gestation. Not detected from day 120 to 310.
§ Ovarian vein levels 40-215 μg/100 ml.

Table 4

SPECIES IN WHICH PROGESTERONE WAS NOT ISOLATED FROM PLACENTAE

| Species | Blood or Sera Levels | | Abortion Inevitable After Oophorectomy |
	Peripheral (μg/100 ml)	Ovarian (μg/100 ml)	
Sow	0.34	—	Yes
Goat	0.71	230	Yes
Dog	—	—	Yes *
Rabbit	100 †	—	Yes
Cow ‡	0.24–0.28	140–264	No
Rat	7	—	Yes

* Not tested in second half of gestation.

† Hooker-Forbes Bioassay.

‡ Progesterone (4.4 μg/kg) found in one of six placentae tested.

tal production of progesterone is adequate. Only in the case of the human, however, has direct data been obtained (Ryan, 1962; Diczfalusy and Troen, 1961).

In Table 4, the species in which progesterone was not isolated from placentae are listed. In most such cases, abortion follows oophorectomy, peripheral blood concentrations of progesterone are generally at non-pregnancy levels (Short, 1961; Zarrow and Neher, 1955; Wiest, 1958) and ovarian vein levels (Gomes and Erb, 1965) are high relative to the periphery.

The distribution of 20α- and 20β-hydroxy-Δ^4-pregnen-3-ones are listed in Table 5 (Short, 1957; Short, 1961; Wiest, 1958; Zander, 1961). These compounds have progestational activity and the importance of their relative amounts in relationship to progesterone has not yet been defined.

Progesterone is not excreted in the urine without chemical modi-

Table 5

ISOLATION OF 20α-HYDROXY AND 20β-HYDROXY-Δ^4-PREGNEN-3-ONES FROM PLACENTA AND BLOOD

Species	20α	20β	Blood (μg/100 ml)
Horse	—	+	—
Human	+	+	—
Sheep	+	—	3.7
Rat	+	—	3.0

fication and there is as yet no comprehensive evaluation of the comparative aspects of the types of excretion products in various species. The metabolite found predominantly in the human, pregnanediol, has not been recognized in many species and more reliance has been placed on progesterone blood levels in following endocrine changes than on urinary or fecal excretory products.

V. Estrogens

In animals that do not abort after oophorectomy (Table 1) or in those where only progesterone is needed for maintenance of pregnancy after such surgery, it is assumed that the placenta produces estrogens. The isolation of estrogenic substances from many species has been reported (Newton, 1938; Harkness, *et al.,* 1964) and as was the case with progesterone, there is a rough inverse correlation between the presence of estrogens in the placenta and the need for the corpus luteum (Table 6).

Chemical identification of specific estrogens in the placenta has also been accomplished (Diczfalusy and Troen, 1961; Velle, 1963) and an interesting species difference in the epimeric estradiols is apparent (Table 7).

In contrast to studies on progesterone metabolism, work on the estrogens has involved assay of urinary excretory products rather than blood levels (Newton, 1938; Velle, 1963). This is not too surprising in view of the lower blood estrogen levels and difficulties in assay. Surprisingly, estrogens could not be demonstrated in the urine of the pregnant dog (Kristoffersen and Velle, 1961), although it is present in pregnancy plasma of this species (Metzler, *et al.,* 1966).

Table 6

ISOLATION OF ESTROGENS FROM PLACENTA

Species Where Estogens Demonstrated	Abortion Inevitable Following Oophorectomy	Species Where Estrogens Not Demonstrated	Abortion Inevitable Following Oophorectomy
Cow	No	Rat	Yes
Sheep	No	Dog	Yes *
Mare	No	Cat	No
Monkey	No	Mouse	Yes
Chimpanzee	—		
Sow	Yes		
Human	No		
Goat	Yes		

* Not tested in second half of gestation.

Table 7

ISOLATION OF SPECIFIC ESTROGENS FROM PLACENTA

Species	Steroid Isolated
Human	Estrone
	Estradiol-17β
	16-Keto Estradiol-17β
	Estriol
	Epiestriol
Cow	Estrone
	Estradiol-17α
	Estradiol-17β
Sheep	Estradiol-17α
Goat	Estradiol-17α
Pig	Estrone

Urinary excretion patterns during pregnancy in the pig (Raeside, 1963) and horse (Amoroso, 1955) have been correlated with various changes in the source of the estrogens (ovary to placenta) as gestation progresses. Studies on blood and placental estrogen levels and the factors which control them in the human have been reviewed (Ryan, 1965) but as with progesterone, no correlation of blood levels with the onset of parturition has been obtained.

VI. ESTROGEN SYNTHESIS BY THE PLACENTA

Progesterone synthesis by the human placenta has been well established in several studies (cf. Ryan, Meigs and Petro, 1966) but there is as yet little comparative data. In contrast, studies on estrogen biosynthesis have been carried out in several species (Table 8). The series of enzymatic steps from androgen precursors to estrogens (aromatization) by the human placenta has been reported (Bolté, *et al.*, 1964a; Bolté, *et al.*, 1964b; Cedard, 1962; Ryan, 1959a; Ryan, 1959b; Ryan, 1963). Work with the marmoset (Ryan, *et al.*, 1961), horse (Starka, *et al.*, 1965; Ainsworth and Ryan, 1966), cow, sheep, pig, guinea pig, rabbit (Ainsworth and Ryan, 1966) and mouse (Vinson and Jones, 1964) placentae has also been performed. There is a good correlation of enzymatic capacity and isolation of estrogens from the placenta (Table 8) with the mouse work being a notable exception.

Table 8

Species	Precursor	Estrogens Formed	Estrogen Found by Extraction of Placenta
Human	Androstenedione Testosterone Dehydroepiandrosterone	Estrone Estradiol-17β	Yes
	16-Hydroxy- androstenedione 16-Hydroxytestosterone 16-Hydroxydehydro- epiandrosterone Androstenetriol	16α-Hydroxyestrone Estriol	Yes
Horse	Testosterone Androstenedione Dehydroepiandrosterone	Estrone Estradiol-17β	Yes
Marmoset	Androstenedione	Estrone	—
Cow Sheep	Androstenedione Dehydroepiandrosterone	Estrone Estradiol-17α Estradiol-17β	Yes
Sow	Androstenedione Dehydroepiandrosterone	Estrone Estradiol-17β	Yes
Mouse *	Progesterone	16-Oxoestrone 17-Epiestriol	No
Guinea Pig Rabbit	Androstenedione Dehydroepiandrosterone	None	No

* Presumptive evidence.

VII. STEROID BIOSYNTHESIS BY THE HUMAN PLACENTA—THE FETO-PLACENTAL UNIT

Only in the human has sufficient work been performed to outline a scheme for placental endocrine synthesis during pregnancy. The concept has evolved that the human placenta is an incomplete endocrine organ that does not synthesize significant quantities of any steroid from acetate. It depends largely on blood borne precursors for estrogen and progesterone formation (Baulieu and Dray, 1963; Siiteri and MacDonald,

1963). In the case of estriol, it has been demonstrated that fetal integrity, and specifically the fetal adrenal, is of paramount importance (Frandsen and Stakemann, 1963, 1964; Greene and Touchstone, 1963). In fact precursors for estriol have been isolated from fetal blood in high concentrations (Colas *et al.,* 1964; Magendantz and Ryan, 1964).

Circulating dehydroepiandrosterone sulfate (Baulieu and Dray, 1963; Bolté, *et al.,* 1964a, b; Siiteri and MacDonald, 1963) appears to be the precursor for estrone and estradiol formation by the placenta and circulating 16-hydroxy-dehydroepiandrosterone sulfate appears to be the precursor of estriol (Magendantz and Ryan, 1964).

Decline of urinary estrogens in cases of fetal demise and anencephaly stress the importance of the fetal contribution.

With progesterone, the blood borne precursors appear to be pregnenolone sulfate and cholesterol (cf. Ryan, Meigs and Petro, 1966).

Such an outline for endocrine synthesis replaces the concept of placental autonomy with that of a feto-placental unit, makes the maternal and fetal adrenal source of steroid precursors an essential part of the unit and increases the importance of blood flow and placental transfer as parameters of endocrine control. The remarkable similarity of the aromatization step in all of the placentae tested in Table 8 does raise the question of whether the situation existing in the human pertains for these other species as well. Although failure of *de novo* synthesis of steroids from acetate has not been reported in any placenta other than the human, Ainsworth and Ryan (1966) could not find evidence for conversion of C_{21} steroids to estrogens in the horse, cow, sheep, sow or rabbit. This would suggest that conventional biosynthetic pathways for estrogens (Ryan, 1963) are not being followed by the placentae of any of these species as is the case for the human (Fig. 1).

VIII. Relationship of Steroid Hormones to Pregnancy Failures

It is obvious from the foregoing that steroid biosynthesis during pregnancy is complex and varies widely among different species. The similarities of such endocrine function among mammals may be more helpful than the differences since a common thread might pinpoint essential features of the endocrine control of viviparity.

The participating units in steroid endocrinology for continuation of pregnancy after implantation are:

1. A corpus luteum that can function for the time required, either to parturition or until placental activity can supervene.

2. A source of gonadotrophic principle that can maintain corpus luteum function for its required life span.

3. Where the corpora lutea regress prior to parturition, an extra-ovarian source of hormones (feto-placental unit) is necessary. This unit

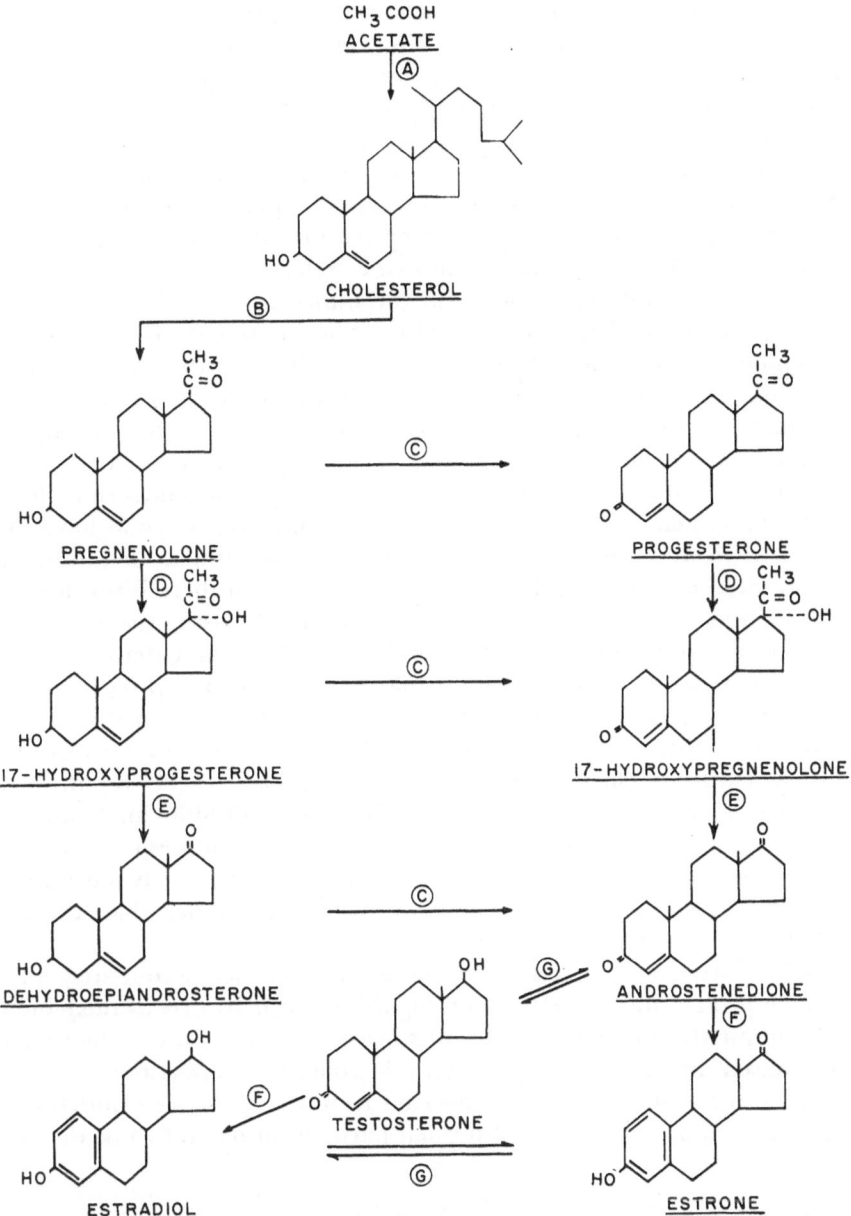

Fig. 1. Conventional biosynthetic pathway for estrogens in the human ovary. This total biosynthetic scheme does not apparently function in the placenta and the estrogen and progesterone precursors must be brought to the placenta via the fetal and maternal blood streams. In the human placenta, the conversion of dehydroepiandrosterone to the estrogens readily takes place and the conversion of pregnenolone and cholesterol to progesterone has been demonstrated. (From Smith and Ryan: Am J. Obstet. Gynec. *84:*141, 1962, The C. V. Mosby Company, St. Louis.)

is not autonomous as originally believed, but requires maternal and fetal sources of hormone precursors. The maternal adrenal is not essential since pregnancy proceeds after adrenalectomy and the fetal adrenal is not essential since even in anencephalics where the fetal adrenal is essentially absent, the pregnancy proceeds normally. Pregnancy can proceed after fetal demise or removal and hence the fetus is not an essential component. Only a viable placenta is required. The placenta in turn requires an intact implantation site and a receptive uterine musculature.

Two examples of possible endocrine related reproductive failure might be cited from experience with the human.

Early abortion which is not related to germ plasm defects or to specific environmental effects might be caused by a failure of the corpus luteum of pregnancy. The corpus luteum in turn would be dependent on its own integrity and a placental supply of chorionic gonadotrophin. Attempts have been made to measure urinary pregnanediol or gonadotrophin excretion to determine whether this specific endocrine defect occurs. Treatment with exogenous estrogen and progesterone have also been tried to avert abortion. To date there is no clinical or laboratory method for diagnosing this hypothetical situation and there is no clear-cut evidence that therapy is effective (Shearman and Garrett, 1963).

In the late pregnancy failures of human toxemia and diabetes, there have been reports of decline in urinary estrogens and progestins and elevation in gonadotrophins (Smith and Smith, 1948). As in the cases of reproductive loss early in pregnancy, exogenous hormone therapy has been used but its efficacy has been widely challenged (cf. Ryan, 1962; White, *et al.*, 1956; Given, *et al.*, 1950; Pedowitz and Shlevin, 1953).

It is apparent that there are no well established examples of pregnancy failure based solely on defects in steroid formation, and it is more likely that aberrations in steroid synthesis are the result rather than the sole immediate cause of pregnancy failures.

The importance of determining those factors which secondarily affect hormone production cannot be too highly stressed. In this context blood flow and nutrition may be significant since estriol excretion in the human is depressed in vascular disease and is roughly proportional to fetal weight or maturity. Whether endocrine factors are the determinants of premature labor or the onset of normal parturition has yet to be decided.

SUMMARY

A review of the bases for our current concepts of steroid synthesis and function during pregnancy has been presented.

The limitations in our knowledge of the specific roles that the hormones play in pregnancy after implantation have been stressed, including the void in such important areas as the control of myometrial activity.

Extensive comparative data on the effect of oophorectomy upon the course of gestation and the isolation of hormones from body fluids and tissues have provided notable indirect evidence for the relative contributions of the ovary and feto-placental unit to steroid formation.

Direct biochemical data on the enzymatic capacity of the placenta is becoming increasingly available and has allowed additional insight into the factors which might control hormone synthesis.

Only when knowledge of the normal physiological control mechanism for hormone synthesis and hormone action in pregnancy is reasonably complete will it be possible to approach reproductive failure in a meaningful way.

References

Ainsworth, L., and K. J. Ryan: Steroid hormone transformations by endocrine organs from pregnant mammals. Endocrinology *79:*875, 1966.

Amoroso, E. C.: Endocrinology of Pregnancy. Brit. Med. Bull. *11:*117, 1955.

——: Comparative Aspects of the Hormonal Functions in the Placenta and Fetal Membranes, ed. C. A. Villee, p. 3, Baltimore: Williams and Wilkins, 1960.

Amoroso, E. C., and C. A. Finn: Ovarian Activity During Gestation, Ovum Transport and Implantation. In The Ovary, ed. S. Zuckerman, Vol. 1, Chap. 9, p. 451, New York: Academic Press, 1962.

Baulieu, E.-E., and F. Dray: Conversion of H^3-Dehydroisoandrosterone Sulfate to H^3-Estrogens in Normal Pregnant Women. J. Clin. Endocrinol. and Metab. *23:*1298, 1963.

Bolté, E., S. Mancuso, G. Eriksson, N. Wiqvist, and E. Diczfalusy: Studies on the Aromatisation of Neutral Steroids in Pregnant Women. Acta Endocrinol. *45:*535, 1964a.

——, ——, ——, ——, and ——: Studies on the Aromatisation of Neutral Steroids in Pregnant Women. Acta Endocrinol. *45:*576, 1964b.

Bowerman, A. M., and R. M. Melampy: Progesterone and Δ^4-Pregnen-20α-ol-3-one in Bovine Reproductive Organs and Body Fluids. Proc. Soc. Exp. Biol. Med. *109:*45, 1962.

Breuer, H.: The Metabolism of the Natural Estrogens. Vitamins and Hormones *20:*285, 1962.

Buchanan, G. D., A. C. Enders, and R. V. Talmage: Implantation in Armadillos Ovariectomized During the Period of Delayed Implantation. J. Endocrinol. *14:*121, 1956.

Cedard, L., J. Varangot, and S. Yannotti: Biosynthese de l'oestriol dans les placentas humains perfuses *in vitro*. Comptes rendus *254:*3896, 1962.

Colas, A., W. L. Heinrichs, and H. J. Tatum: Pettenkoffer Chromogens in the Maternal and Fetal Circulations. Steroids *3:*417, 1964.

Cowie, A. T., and S. J. Folley: The Mammary Gland and Lactation. In Sex and Internal Secretions, ed. W. C. Young and G. W. Corner, Vol. I, p. 590, Baltimore: Williams and Wilkins Co., 1961.

Csapo, A.: The Mechanism of Effect of the Ovarian Steroids. Recent Progr. Horm. Res. *12*:405, 1956.

Diczfalusy, E., and P. Troen: Endocrine Functions of the Human Placenta. Vitamins and Hormones *19*:229, 1961.

Frandsen, V. A., and G. Stakemann: The Site of Production of Oestrogenic Hormones in Human Pregnancy. Acta Endocrinol. *43*:184, 1963.

———, ———: The Site of Production of Oestrogenic Hormones in Human Pregnancy. Acta Endocrinol. *47*:265, 1964.

Given, W. P., R. G. Douglas, and E. Tolstoi: Pregnancy and Diabetes. Am. J. Obstet. Gynec. *59*:729, 1950.

Gomes, W. R., and R. E. Erb: Progesterone in Bovine Reproduction: A Review. J. Dairy Science *48*:314, 1965.

Greene, J. W., and J. C. Touchstone: Urinary Estriol as an Index of Placental Function. Am. J. Obstet. Gynec. *85*:1, 1963.

Harkness, R. A., A. McLaren, and E.-J. Roy: Oestrogens in Mouse Placentae. J. Reprod. Fertil. *8*:411, 1964.

Heap, R. B., and R. Deanesly: Progesterone in Systemic Blood and Placentae of Intact and Ovariectomized Pregnant Guinea Pigs. J. Endocrinol. *34*:417, 1966.

Hisaw, F. L.: Physiology of the Corpus Luteum. In: Sex and Internal Secretions, ed. E. Allen, p. 499, Baltimore: Williams and Wilkins Co., 1932.

Kristoffersen, J., and W. Velle: Urinary Oestrogens of the Dog. Nature *185*:253, 1960.

Magendantz, H., and K. J. Ryan: Isolation of an Estriol Precursor, 16α-Hydroxydehydroepiandrosterone from Human Umbilical Sera. J. Clin. Endocrinol. and Metab. *24*:1155, 1964.

Mellin, T. N., and R. E. Erb: Estrogens in the Bovine—a Review. J. Dairy Science *48*:687, 1965.

Metzler, F., Jr., B. E. Eleftheriou, and M. Fox: Free Estrogens in Dog Plasma During the Estrous Cycle and Pregnancy. Proc. Soc. Exper. Biol. and Med. *121*:374, 1966.

Newton, W. H.: Hormones and the Placenta. Physiological Reviews *18*:419, 1938.

Pedowitz, P., and E. L. Shlevin: The Pregnant Diabetic. Obst. and Gynec. Survey *8*:54, 1953.

Raeside, J. I.: Urinary Oestrogen Excretion in the Pig During Pregnancy and Parturition. J. Reprod. Fertil. *6*:427, 1963.

Ryan, K. J.: Biological Aromatization of Steroids. J. Biol. Chem. *234*:268, 1959a.

———: Metabolism of C-16-Oxygenated Steroids by Human Placenta, The Formation of Estriol. J. Biol. Chem. *234*:2006, 1959b.

———: Hormones of the Placenta. Am. J. Obstet. Gynec. *84*:1695, 1962.

———: Synthesis of Hormones in the Ovary. In: The Ovary, ed. H. G. Grady and D. E. Smith, p. 69, Baltimore: Williams and Wilkins Co., 1963a.

———: Biogenesis of Estrogens. In: Biosynthesis of Lipids, ed. G. Popjak, p. 381, Proc. Vth Int. Congress Biochem., New York: Macmillan Co., 1963b.

———: Estrogens: Blood and Placental Levels and the Factors Which Control Them. Excerpta Medica International Congress Series No. 83, Proceedings of the 2nd International Congress of Endocrinology, Part II, p. 727, Amsterdam: Excerpta Medica Foundation, 1965.

Ryan, K. J., K. Benirschke, and O. W. Smith: Conversion of Androstenedione-4-C14 to Estrone by the Marmoset Placenta. Endocrinology 69:613, 1961.

Ryan, K. J., R. A. Meigs, and Z. Petro: Biosynthesis of Progesterone in the Human Placenta. Am. J. Obstet. Gynec. 96:676, 1966.

Shearman, R. P., and W. J. Garrett: Double-blind Study of the Effect of 17-Hydroxyprogesterone Caproate on Abortion Rate. Brit. M. J. 1:292, 1963.

Short, R. V.: Progesterone and Related Steroids in the Blood of Domestic Animals. Ciba Foundation Colloquia on Endocrinol. 11:362, 1957.

————: Blood Progesterone Levels in Relation to Parturition. J. Reprod. Fertil. 1:61, 1960.

————: Progesterone. In: Hormones in Blood, ed. C. H. Gray and A. L. Bacharach, Chapter 13, p. 379, New York: Academic Press, 1961.

Siiteri, P. K., and P. C. MacDonald: The Utilization of Circulating Dehydroisoandrosterone Sulfate for Estrogen Synthesis During Pregnancy. Steroids 2:713, 1963.

Smith, G. V. S., and O. W. Smith: Internal Secretions and Toxemia of Late Pregnancy. Physiol. Rev. 28:1, 1948.

Starka, L., J. Breuer, and H. Breuer: Biogenese von Oestrogenen in der Placenta des Pferdes. Naturwiss. 52:540, 1965.

Velle, W.: Metabolism of Estrogenic Hormones in Domestic Animals. General and Comp. Endocrinol. 3:621, 1963.

Vinson, G. P., and I. C. Jones: The Capacity of Mouse Fetus and Placenta to Synthesize Steroids from Progesterone in Vitro. General and Comp. Endocrinol. 4:415, 1964.

White, P., L. Gillespie, and L. Sexton: Use of Female Sex Hormone Therapy in Pregnant Diabetic Patients. Am. J. Obstet. Gynec. 71:57, 1956.

Wiest, W. G.: Isolation of Progesterone and 4-Pregnen-20α-ol-3-one from Pregnant Rat Tissues. Fed. Proc. 17:335, 1958.

Zander, J.: The Chemical Estimation of Progesterone and Its Metabolites in Body Fluids and Target Organs. In: Progesterone, p. 77, Brook Lodge Symposium, Augusta: Michigan, Brook Lodge Press, 1961.

Zarrow, M.: Gestation. In: Sex and Internal Secretions, Vol. II, ed. W. C. Young, p. 958, Baltimore: Williams and Wilkins Co., 1961.

Zarrow, M. X., and G. M. Neher: Concentration of Progestin in the Serum of the Rabbit During Pregnancy, the Puerperium and Following Castration. Endocrinol. 56:1, 1955.

PROTEIN HORMONES AND GESTATION

J. B. JOSIMOVICH

Department of Obstetrics and Gynecology,
The University of Pittsburgh School of Medicine,
Pittsburgh, Pennsylvania

The frequency of mammalian fetal loss caused by a deficiency of endocrine hormones is not known. This ignorance is attributable to a lack of knowledge concerning the number of hormones required for maintenance of gestation at various stages, and the specific biologic roles of such hormones. It is hoped that a brief review of our knowledge about certain protein gonadotrophic hormones associated with human pregnancy may induce those present at this conference to search for such hormones in the mammals with which they are most familiar. Methods of bioassay, immunologic assay and protein separation have been used in studying two human gestational gonadotrophins: human chorionic gonadotrophin (HCG) and human placental lactogen (HPL). It seems appropriate, therefore, to describe and criticize in some detail the techniques used in studying these two hormones.

Dr. Ryan (1962) has previously reviewed the evidence for the secretion of human gestational protein hormones other than the gonadotrophins. Furthermore, the review of Hisaw and Zarrow (1951) makes it unnecessary to recount here the studies performed on the protein hormone relaxin, which appears to induce relaxation of the pelvic interosseous ligaments in certain mammals. Protein gonadotrophins arising from sites other than the pituitary gland during pregnancy in subprimate species will also be discussed briefly.

It is important to note, however, that the human and rhesus monkey represent an extreme degree of fetal autonomy in that the trophoblastic syncytium produces gonadotrophins in these species which may maintain gestation in the absence of the pituitary gland after implantation of the fertilized ovum. The pituitary gonadotrophins continue to be required until mid-gestation in the rat; or throughout pregnancy in the rabbit, as seen in Table 1. The availability of techniques for detecting minute quantities of gonadotrophins now permits us to search systematically for them in other mammals whose pregnancies are not terminated by hypophysectomy.

Table 1

DIFFERING MAMMALIAN PATTERNS OF PITUITARY VS. PLACENTAL
MAINTENANCE OF GESTATION REVEALED BY HYPOPHYSECTOMY
(*AP* studies) [1]

RAT—Abortion if *AP* before 12th day of 22 day gestation.

Exogenous replacement after AP 8 days	*Placental hormones demonstrated*
Sheep LTH not effective [2]	Mammotropin after day 12 [4]
Sheep LTH + estrone effective [2]	? Second luteotrophin [5]
HCG effective [3]	

RABBIT—Abortion if *AP* at any time in 31 day gestation.

Exogenous replacement after AP day 5	*Placental hormones demonstrated*
Sheep LTH not effective [6A]	None
Sheep LH effective in pseudo-pregnancy [6B]	

MONKEY + HUMAN—No abortion if *AP* before day 35 of 155 day gestation [7] (rhesus). Abortion may not occur in *AP* humans becoming pregnant after induction of ovulation with gonadotrophins.[8, 9]

Exogenous replacement after AP	*Placental hormones demonstrated*
Luteotrophins not needed	HCG [10]
	HPL [11]

---

[1] *AP* signifies hypophysectomized
[2] Lyons *et al.*, 1943
[3] Josimovich, unpublished data
[4] Averill *et al.*, 1950
[5] Astwood and Greep, 1938
[6] Rennie *et al.*, 1964A, and Kilpatrick *et al.*, 1964B
[7] Smith, 1954
[8] Gemzell, 1966
[9] Bettendorf, 1966
[10] Philipp, 1930
[11] Josimovich *et al.*, 1963

HUMAN CHORIONIC GONADOTROPHIN

Aschheim and Zondek (1928) discovered HCG by finding that extracts of pregnancy urine, when injected six times over a two-day period into immature mice, would result in hemorrhagic ovarian follicles, signifying ovulation. Philipp (1930) first suggested the placental origin of this hormone. Final proof of the placental ability to synthesize and secrete HCG was the production demonstrated in tissue culture by Jones, Gey and Gey (1943). Further localization of storage in the syncytial tropho-blast was provided by the fluorescein-labeled antibody studies of Thiede (1963).

The biologic effects of HCG extracted from pregnancy urine are qualitatively similar in primates and in rats. Hisaw (1944, 1959) demon-strated that HCG given to normal cycling female rhesus monkeys will pro-long the life of the corpus luteum as evidenced by a delay in menstruation for approximately two weeks, as can be seen in Table 2. As shown there, ovulation takes place at day 14, and menstruation at day 28, as in humans. Monkeys treated daily with HCG from the time of supposed ovulation suffered a delay in menstruation for an average of two weeks. Menstruation did occur following this delay, despite continued HCG administration. During the following two to three weeks, there was evidence of continued estrogen secretion by the corpus luteum: At laparotomy, corpora lutea were still visible; endometrial biopsies showed an estrogenic effect alone (endometrial hyperplasia); and the sexual skin of the perineum of the monkeys was reddened and edematous. An end to this prolonged estrogen-dominated phase was signified by a final episode of vaginal bleeding, per-haps brought about by formation of antibodies in the monkeys to HCG. Brown and Bradbury (1947) were also able to prolong the life of the human menstrual cycle corpus luteum for two weeks, after which break-through bleeding occurred. They did not carry out studies, however, to determine whether estrogen alone was secreted for a further period in response to continued HCG therapy.

Table 2 also shows the analogous experimental situation created by HCG administration to the rat. When electrical or mechanical stimula-tion of the cervix uteri is performed in female rats during estrus, just before ovulation, a state of pseudopregnancy may be induced as evidenced by the ability to produce a decidual cell reaction in the endometrial stroma by scratching the uterine horn four days later. Yochim and De Feo (1962) showed that maintenance of the decidual reaction and pseudopregnancy depend on the secretion of balanced amounts of estrogen and progesterone. If hypophysectomy is performed at the same time as induction of the decidual reaction, as seen in Table 2, administration of HCG from the time of this uterine traumatization will only permit formation and maintenance of the decidual reaction for four to five days. After that time, the deciduoma gradually disappears and an estrogen-dominated phase ensues, as demonstrated by vaginal cornification and endometrial hypertrophy. In summary, in humans, monkeys, and rats, HCG administration appears to cause a short period of balanced estrogen and progesterone secretion from the corpus luteum of the normal cycle or of pseudopregnancy. After a critical time has passed (about one half

Table 2

EFFECTS OF HCG ON CORPUS LUTEUM FUNCTION IN THE RHESUS MONKEY
(Hisaw, 1944) AND RAT (Josimovich and Cato, 1966)

MONKEY (28 day cycle)					
Day of cycle	28	14	28	42	56
Ovulation	XX	X		XX	XX
Menstruation		x			x
HCG administration					
Endometrium	proliferative	secretory	.. proliferative hyperplasia .. atrophy		
Sex skin color	red	purple	red	.. regression	
Presumed ovarian steroids*	E	E & P	.. excess E		?

RAT (5 day cycle)				
Day of cycle	5	5	10	15
Ovulation	X			
Procedure	cervical trauma	uterine trauma + hypophysectomy		
HCG administration			x	
Endometrium		Decidua	.. endometrial hypertrophy	
Vaginal mucosa	cornified	mucified	cornified	
Presumed ovarian steroids*	E	E & P	.. excess E	

* E—estrogens; P—progesterone.

Table 3

VARIOUS BIOASSAYS EMPLOYED FOR HCG DETERMINATION

Animal	End point of assay	References
Prepubertal mouse	Ovulation points	Aschheim and Zondek, 1928
Rabbit	Ovulation points	Friedman and Lapham, 1931
Xenopus laevis	Release of ova	Shapiro and Zwarenstein, 1934
Prepubertal rat	Ovarian hyperemia	Frank and Berman, 1941
Bufo arenarum	Release of sperm	Galli-Mainini, 1947
Prepubertal mouse	Uterine weight	Diczfalusy and Loraine, 1955

of an ovulatory cycle) continued administration of HCG results in secretion of an excess of estrogen such that the progestational state of the reproductive tract can no longer be maintained.

Using a variety of bioassays listed in Table 3, variation in HCG excretion rate with age of gestation was found to be remarkably similar. Figure 1 shows this variation with weeks of human gestation, as obtained from the data of Loraine and Schmidt-Elmendorff (1963), using the increase in immature mouse uterine weight as an assay. A marked rise in HCG excretion (30,000-fold) occurs in urine at about 60 days of pregnancy. After the 60th day of gestation, there is a decrease in urinary excretion of HCG as determined by bioassay to levels of approximately 5000 international units (i.u.) per day, followed by a slight rise near

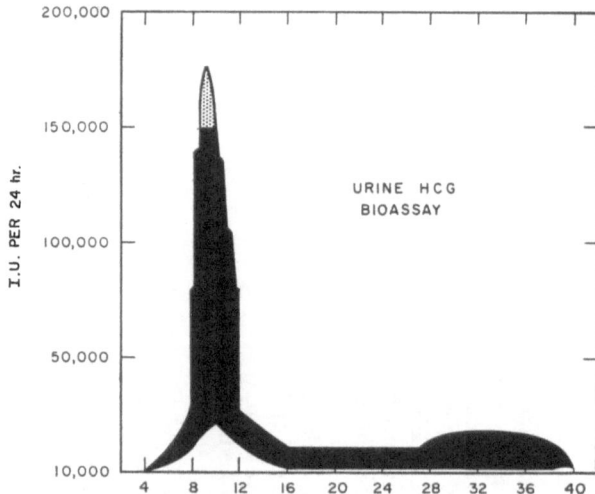

Fig. 1. Urinary HCG excretion by bioassay. (After Loraine and Schmidt-Elmendorff, 1963.) Abscissa: weeks gestation.

term. Despite unanimity of such methods in the data they provide, the biologic specificity of the assays was called into question by many workers, including Lyon and colleagues (1953). They noted that the degree of follicle development seen in histologic sections of ovaries of assay animals treated with urinary HCG was different when the HCG was obtained at various times during pregnancy. The magnitude of increase in ovarian and uterine weight or in ovarian hyperemia seen in animals given urinary protein obtained from patients excreting maximal amounts of gonadotrophic activity (10–15 weeks' gestation) was thought to represent synergism between follicle-stimulating hormone (FSH) and HCG in the urine. Albert and co-workers (1960), however, were of the opinion that the large peak in urinary and serum activity of HCG seen at the end of the first trimester of human pregnancy was due to the biologic activity of HCG alone; and that HCG itself, when given in high doses, inherently possesses follicle-stimulating activity. Recent evaluation of the follicle-stimulating activity of HCG, however, shows that as purification has progressed, there is less and less FSH activity at any dose. Figures 2 to 4 from the unpublished work of Dr. Donald Goss of Harvard Medical School show the essential lack of follicle-stimulating activity in high doses of partially purified chorionic gonadotrophin prepared directly from placentas. If immature 15-day old female albino rats are hypophysectomized and one allows two weeks to pass, one may observe that there are no graafian follicles present (Fig. 2). Many small primordial follicles are visible, however. There is little cytoplasm in the interstitial cells. No corpora lutea are present since the rats were too immature to have ovulated prior to surgery.

Figure 3 shows that the ovaries of rats treated for four days with 1800 i.u. daily of a neutral buffered extract of homogenized placentas respond with hyperplasia and hypertrophy of the interstitial cells. The primordial follicles are unstimulated.

Figure 4 shows the ovarian effects of a kaolin-adsorbed protein fraction obtained from women during the peak of gonadotrophic activity (8–15 weeks' gestation) when tested in immature hypophysectomized rats. 1800 i.u. were given daily for four days. Here we see marked follicle stimulation resulting in a swollen preovulatory graafian follicle, in addition to interstitial cell stimulation. The doses of 1800 i.u. per day correspond to the higher doses used by Albert and co-workers (1960) with their pregnancy urine extracts. Thus, follicle stimulation (Fig. 4) is apparently not an inherent property of HCG, but must depend upon the simultaneous presence of a follicle-stimulating substance in the urine of pregnant women, presumably pituitary FSH, acting in conjunction with HCG which, when administered by itself, is solely interstitial cell stimulating (Fig. 3). In confirmation of these histological observations, Goss also showed that trophoblast extract caused no significant changes in the

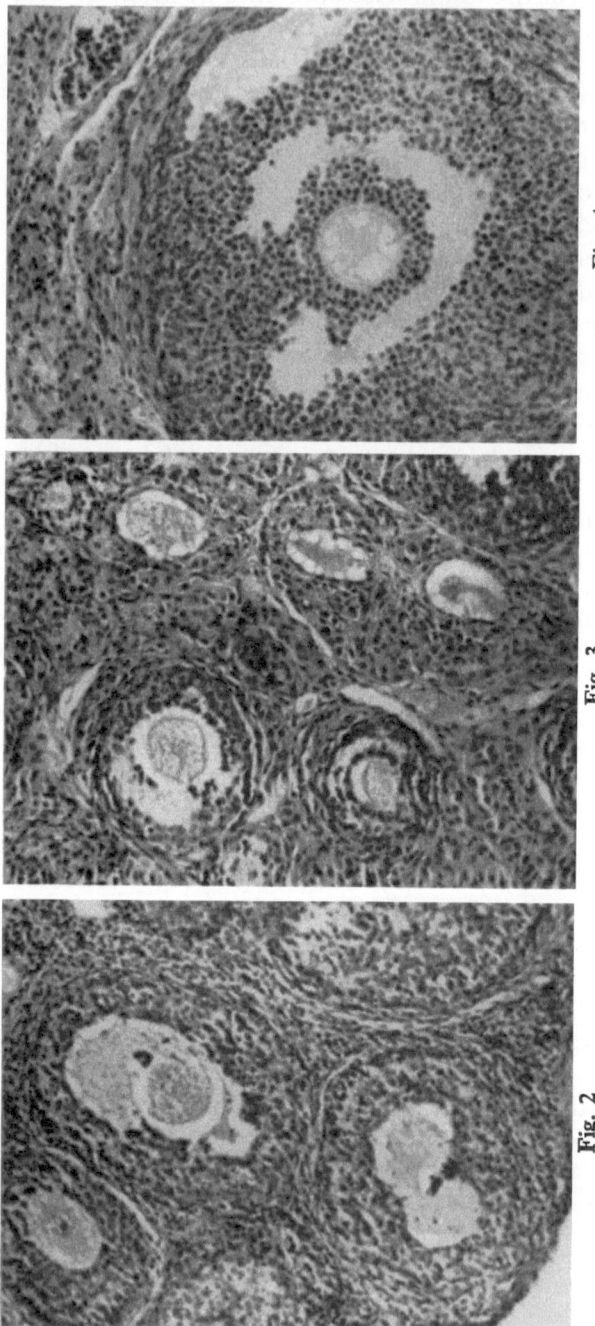

Fig. 2

Fig. 3

Fig. 4

Figs. 2–4. Differentiation of the gonadotrophic effects of human pregnancy urine extracts and placental HCG from the work of D. A. Goss. See text for details. Microphotographs all taken at same magnification from sections stained with hematoxylin and eosin. **Fig. 2.** Immature rat ovary two weeks after hypophysectomy. **Fig. 3.** Interstitial cell stimulation with placental HCG. **Fig. 4.** Marked follicle growth and maturation in addition to interstitial cell stimulation induced with pregnancy urine extract (at same dose used in rats depicted in Fig. 3, in terms of i.u. HCG).

histology or weight of the uterus of treated hypophysectomized rats; and that there was only a small, although significant, increase in ovarian weight. In contrast, extracts of pregnancy urine resulted in sizable increases in both uterine and ovarian weight in the assay rats.

A less distorted picture of the variation in rate of excretion of HCG has been afforded by recent immunologic assays of urine. Such immunoassays have demonstrated that there is a two to three-fold greater amount of HCG excreted in the urine at the end of the first trimester of pregnancy than that found later. Brody and Carlstrom (1965), and Mishell, Wide and Gemzell (1963), have utilized an assay for HCG which measures the inhibition of agglutination of red cells coated with HCG in the presence of rabbit HCG-antiserum by added amounts of HCG. The shape of both urine excretion and serum concentration curves are similar. Hamashige and Arquilla (1964), in an extensive series of immunologic studies, have pointed out, however, that it is unlikely that any of the immunologic assays for chorionic gonadotrophin are actually measuring biologically active HCG. They preferentially adsorbed HCG antisera and tested the remaining antibodies for biologic neutralizing activity in rats, microimmunoelectrophoretic behavior, and in standard precipitating and hemagglutination-inhibition immunoassays. They found that pregnancy-specific urinary antigens were not the same as the antigens whose antibodies neutralized biologic activity of HCG in rats. They admit that the two pregnancy-specific urinary antigens may be breakdown products normally produced by the human body to trophoblast HCG. Nevertheless, the various immunologic assays which have been developed for HCG have proved to be clinically useful even though they may not be measuring biologically active HCG. Immunologic methods which have been used include assays measuring inhibition of HCG antibody-induced agglutination of HCG-coated red cells or latex particles with added HCG. Another method is the radioimmunoassay in which I^{131}-labeled HCG is bound by rabbit antibodies against HCG, and unlabeled HCG in test samples competes with the radioactively-labeled HCG for the binding to antibody (Midgley, 1966). It has been shown that antibodies prepared to human chorionic gonadotrophin will cross-react with pituitary luteinizing hormone (LH) as seen by precipitin reactions in agar gel, thus showing similarities in the structure between the two hormones and the need for differentiating the two in immunoassays (Goss and Lewis, 1964).

The natural role of HCG during pregnancy remains obscure although speculation has often been made that it might stimulate steroid secretion in the placenta as it can in the ovary. Speculation about possible physiologic roles of HCG in controlling fetal endocrine function has been engendered by the finding of HCG in human fetal serum, as shown in Table 4, which gives the estimated concentrations of HCG and human placental lactogen in various maternal and fetal compartments. Chorionic

Table 4

COMPARISON OF HPL AND HCG CONCENTRATIONS IN VARIOUS
MATERNAL AND FETAL COMPARTMENTS NEAR TERM IN HUMANS

Compartment	HPL [1,2]	HCG [3]
Maternal blood	2 μg/ml	10 i.u./ml
Maternal urine	100 mμg/ml	12 i.u./ml
Fetal cord blood	Less than 0.2 mμg/ml	3 i.u./ml
Fetal urine	Less than 1 mμg/ml	??
Amniotic fluid	200 mμg/ml	6 i.u./ml

[1] Grumbach and Kaplan, 1964.
[2] Josimovich, unpublished.
[3] Brody and Carlstrom's data for maternal blood (1965) using an immunoassay, with conversion of Bruner's data obtained by bioassay (1951) to same unitage. Note that fetal serum HCG data were obtained by bioassay which cannot distinguish placental from fetal pituitary gonadotrophins.

gonadotrophin is present in large amounts in maternal serum and urine, particularly in the first trimester, while it is found in smaller quantities in the fetal serum and amniotic fluid. One effect of HCG present in the fetal circulation might be to induce the marked Leydig cell hyperplasia seen in the testes of the fetal rhesus monkey in the latter half of gestation (Knobil and Josimovich, 1961), even while spermatogenesis had not commenced. We noted a comparable state of Leydig cell stimulation after 14 days of HCG administration to immature hypophysectomized rhesus monkeys. In addition to possible effects of HCG on the fetal testis, Lauritzen (1966) has suggested that HCG may stimulate the fetal adrenal secretion of precursors for placental estriol synthesis, since he was able to induce a marked increase in dehydroepiandrosterone (DHEA) excretion by the administration of HCG to newborn human infants. DHEA is thought to be an early progenitor of estriol in human pregnancy.

HUMAN PLACENTAL LACTOGEN

Ito and Higashi first discovered the presence of a lactogenic protein fraction in the human placenta (see Higashi, 1961). In our laboratory, we isolated this lactogenic protein and found it to be closely related immunologically to human pituitary growth hormone (Josimovich and MacLaren, 1962). Evidence was given that the hormone was produced by the placenta, but proof of trophoblast secretion of this hormone was later obtained by Grumbach and Kaplan (1964). They demonstrated placental synthesis by tissue culture techniques; and their colleague, Sciarra, established the site of HPL production as the syncytial trophoblast (1963). This latter group of workers calls the hormone "chorionic growth hormone-prolactin!" because of ambiguity which exists concerning the relative potency of HPL as a lactogenic and somatotrophic agent. This confusion in terminology is not surprising in view of the well known lactogenic properties of human pituitary growth hormone or electrophoretically homogeneous fractions thereof (Ferguson and Wallace, 1961). Various human placental lactogen preparations possess 20–80% of the potency of the sheep pituitary prolactin (LTH) from the National Institutes of Health, as evidenced by the local and systemic pigeon crop assays. HPL was found to induce milk production in the mammary glands of pseudopregnant rabbits (Josimovich and MacLaren, 1962; Friesen, 1965); as well as promotion of casein and nucleic acid synthesis by mouse mammary tissue *in vitro* (Turkington and Topper, 1966). It is not surprising that HPL, an immunologic relative of human growth hormone, is lactogenic, since growth hormone itself was found to be highly lactogenic in the rabbit and somewhat less so in the pigeon crop growth assay (Forsyth *et al.*, 1965). One anticipates that more purified placental lactogen will be more potent than pituitary growth hormone in such lactogenic hormone assays.

The luteotrophic effects of placental lactogen, first described in our laboratory (Josimovich *et al.*, 1963), were demonstrated by maintenance of the decidual reaction in pseudopregnant rats with pituitaries removed at the time of uterine traumatization. We have found that combined treatment with HPL and HCG will prolong balanced estrogen and progesterone secretion by the corpora lutea of such rats for several days longer than HCG can alone (Josimovich and Cato, 1966). Such hormonal balance is required for rat decidual maintenance, as mentioned earlier. Figure 5 demonstrates that, after gradual progression of necrosis of the deciduomata in such rats between the fourth and seventh day of HCG administration, only hyperplastic endometrium surrounding a coagulum of necrotic decidua remains. In many instances, we observed expulsion of the bloody coagulum *per vaginam* on the sixth or seventh day. In contrast to the results in the HCG-treated rats, seven days of combined HCG and HPL therapy maintained a healthy decidual reaction, as seen in Figure 6. The vaginal mucosa of such rats was still well mucified in all but one case even after ten days of HCG-HPL treatment, although decidual necrosis had begun. The necrotic decidua seen in rats treated for ten days with HCG and HPL was not expelled by the myometrium even after 14 days HCG-HPL therapy. These data imply a continuation of progesterone secretion when HPL is added to the administered HCG. Perhaps HPL acts as a brake on certain biochemical pathways of estrogen synthesis, thus preventing total conversion of progesterone to estrogens.

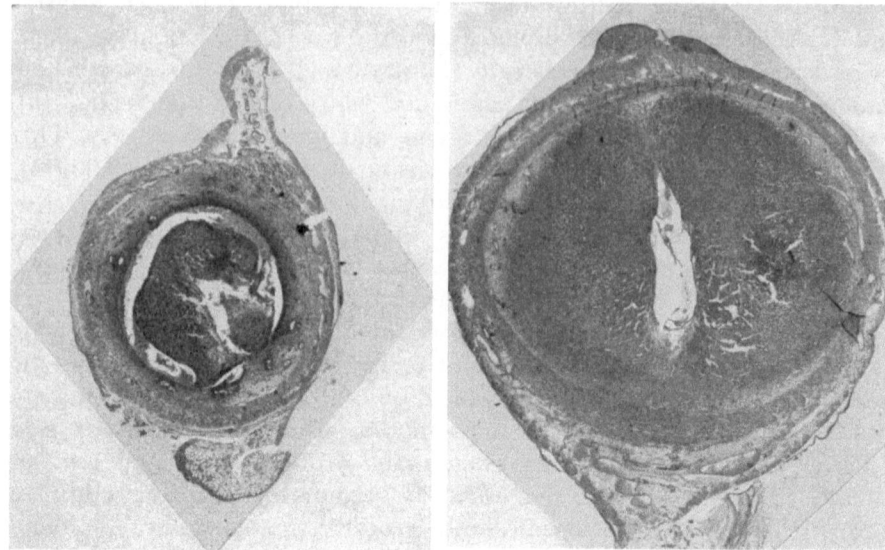

Fig. 5   Fig. 6

Figs. 5–6. Ability of HPL to prolong the effects of HCG in maintenance of corpus luteum of pseudopregnancy in the rat. Microphotographs in both figures taken at same magnification of cross-sections of uterine horns in which a decidual reaction had been induced by scratching on the same day as hypophysectomy. Fig. 5. Necrotic decidua in rather small horn after seven days HCG treatment. Fig. 6. Healthy decidua in large horn after seven days HCG-HPL treatment.

It is of considerable interest in this regard that Fried and Rakoff (1952) found that administration of small doses of HCG and sheep pituitary prolactin will prolong the life of the corpus luteum of the menstrual cycle (as measured by estrogen and pregnanediol excretion and a delay in menstruation), while the same dose of HCG, given without prolactin, failed to produce this effect.

The somatotrophic potency of HPL is still the subject of controversy, as reviewed elsewhere (Josimovich, 1966). Bioassays which were used by Kaplan and Grumbach (1964) and by Franchimont (1965) to demonstrate somatotrophic effects of HPL were not specific. The only direct growth hormone-like effect of HPL demonstrated to date was a marked increase in plasma free fatty acid concentration after administration of HPL to hypopituitary children whose plasma contained no detectable growth hormone (Grumbach *et al.*, 1966).

In contrast to the limited nature of the direct somatotrophic effects of HPL (fatty acid mobilization in hypopituitary subjects after injection of large doses of the hormone), small doses of HPL will potentiate the effects of human growth hormone in certain rat assays (Josimovich, 1966). It remains to be determined, therefore, whether the large amounts of HPL present in late pregnancy potentiate one or more of the biologic effects of pituitary growth hormone also present in the mother's serum.

Plasma immunoassays with rabbit antisera to HPL were introduced by Kaplan and Grumbach (1965) and have been successfully carried out in other laboratories (Beck *et al.,* 1966) with the finding that serum concentrations increase in a rectlinear manner from the twelfth to fortieth weeks of gestation (term), with smaller amounts being detectable between 6 and 12 weeks. HPL extracted from the fetal placenta from 5½ weeks (3½ weeks after fertilization) to term increases in amount in direct proportion to that found in the mother's serum (Josimovich and MacLaren, 1962, and unpublished observations). By term, all groups have found between one and thirty micrograms per milliliter of serum, an amount one to ten thousand times the serum growth hormone concentration. In contrast to the high rate of excretion of HCG, less than 100 micrograms of HPL are excreted per day, although 200–300 micrograms may be excreted during labor (Table 4). Of interest was the finding of Kaplan and Grumbach (1965) and of Greenwood *et al.,* 1964, that the newborn infant umbilical cord serum contains little or no placental lactogen. Tallberg *et al.* (1965), on the other hand, have found large amounts of this hormone in the amniotic fluid. This finding has been confirmed and quantitated in our own laboratory. The mechanism by which such large amounts of HPL gain entrance to the amniotic fluid is still unknown.

In summary, the human trophoblast secretes a second gonadotrophin, human placental lactogen, in increasing amounts throughout gestation. The primary role is presumed to be the final preparation of the breasts for milk secretion, while the secondary roles may be to synergize with luteotrophic effects of HCG and with certain metabolic effects of pituitary growth hormone.

Extrapituitary Gestational Gonadotrophins in Other Mammals

Studies on luteinizing hormones like HCG are less complete in mammals other than in man. Zuckerman (1935) found that the chimpanzee excretes a luteinizing hormone into the urine between the 25th and 130th days of gestation; and Hamlett (1937) found that the rhesus monkey excretes a similar material between the 18th and 25th days of a 155 day gestation. Pregnant mare serum hormone (PMS) is a gonadotrophin which appears to be produced by maternal, rather than by fetal tissue, located in the endometrial cups of the uterus. PMS is detectable between the 40th and 150th days of a 360 day gestation (Cole and Saunders, 1935), and in the endometrial cup secretion during early pregnancy (Clegg *et al.,* 1954). PMS differs from HCG not only in its site of production but also in molecular weight, in its lack of urinary excretion, and by its greater follicle-stimulating effects in laboratory animals.

W. R. Lyons and colleagues (1943) have detailed the presence of a lactogenic protein in the rat placenta and serum midway through gestation, which is termed "chorionic mammotropin" (Averill *et al.,* 1950; Matthies, 1966). It is not known whether another rat placental gonadotrophin described by Astwood and Greep (1938) differs from the mammotropin. A mouse placental luteotrophin has also been described (Deanesly and Newton, 1940).

It is clear that a more systematic search for extrapituitary gonadotrophins can now be made because the techniques of immunologic tagging permit the tracing of minute quantities of protein hormones during fractionation of tissue extracts. It is unlikely that such gonadotrophins are restricted to the few species cited herein. Inspiration for this quest will long be provided by F. L. Hisaw's remark at a symposium on comparative endocrinology eight years ago: "It is not the hormones which have evolved, but the uses to which they are put" (Hisaw, 1959).

References

Albert, A., and I. Derner: Studies on the biologic characterization of human gonadotrophins. VI. Nature and number of gonadotrophins in human pregnancy urine. J. Clin. Endocrinol. *20:*1225, 1960.

Aschheim, S., and B. Zondek: Die Schwangerschaftsdiagnose aus dem Harn durch Nachweis des Hypophysenvorderlappenhormons. Klin. Wschr. 7:1404, 1928.

Astwood, E. B., and R. O. Greep: A corpus luteum-stimulating substance in the rat placenta. Proc. Soc. Exp. Biol. and Med. *38:*713, 1938.

Averill, S. C., E. W. Ray, and W. R. Lyons: Maintenance of pregnancy in hypophysectomized rats with placental implants. Proc. Soc. Exp. Biol. and Med. *75:*3, 1950.

Beck, P., M. L. Parker, and W. H. Daughaday: Radioimmunologic measurement of human placental lactogen in plasma by a double antibody method during normal and diabetic pregnancies. J. Clin. Endocrinol. *25:*1457, 1965.

Bettendorf, A.: Transactions of the Fifth International Congress on Fertility and Sterility. Stockholm, June, 1960.

Brody, S., and G. Carlstrom: Human chorionic gonadotrophin in serum and its relation to the sex of the fetus. J. Clin. Endocrinol. *25:*792, 1965.

Brown, W. E., and J. T. Bradbury: A study of the physiologic action of human chorionic gonadotrophin. Am. J. Obstet. Gynec. *53:*749, 1947.

Bruner, J. A.: Distribution of chorionic gonadotrophin in mother and fetus at various stages of pregnancy. J. Clin. Endocrinol. *11:*360, 1951.

Clegg, M. T., J. Boda, and H. H. Cole: The endometrial cups and allantochorionic pouches in the mare with emphasis on the sources of equine gonadotrophin. Endocrinology *54:*448, 1954.

Cole, H. H., and F. J. Saunders: The concentration of gonad-stimulating hormone in the blood serum and of estrin in the urine throughout pregnancy in the mare. Endocrinology *61*:765, 1935.

Deanesly, R., and W. H. Newton: The influence of the placenta on the corpus luteum of pregnancy in the mouse. J. Endocrinol. *2*:317, 1940.

Diczfalusy, E., and J. A. Loraine: Sources of error in clinical bioassay of serum chorionic gonadotrophin. J. Clin. Endocrinol. *15*:424, 1955.

Ferguson, K. A., and A. L. C. Wallace: Prolactin activity of human growth hormone. Nature *190*:632, 1961.

Forsyth, I. A., S. J. Folley, and A. Chadwick: Lactogenic and pigeon crop-stimulating activities of human pituitary growth hormone preparations. J. Endocrinol. *31*:115, 1965.

Franchimont, P.: Presence d'une hormone de croissance dans le placenta. Ann. d'endocrinol, *26*:346, 1965.

Frank, R. T., and R. L. Berman: A twenty-four hour pregnancy test. Am. J. Obstet. Gynec. *42*:492, 1941.

Fried, P. N., and A. E. Rakoff: The effects of chorionic gonadotrophin and prolactin on the maintenance of corpus luteum function. J. Clin. Endocrinol. *21*:321, 1952.

Friedman, M. H., and M. E. Lapham: A simple, rapid procedure for the laboratory diagnosis of early pregnancies. Am. J. Obstet. Gynec. *21*:405, 1931.

Friesen, H.: Purification of a placental factor with immunologic and chemical similarity to human growth hormone. Endocrinology *76*:369, 1965.

Galli-Mainini, C.: Pregnancy test using the male toad. J. Clin. Endocrinol. *7*:653, 1947.

Gemzell, C. A.: Transactions of the Fifth International Congress on Fertility and Sterility. Stockholm: June, 1966.

Goss, D. A., and J. Lewis, Jr.: Immunologic differentiation of luteinizing hormone and human chorionic gonadotrophin in compounds of high purity. Endocrinology *74*:83, 1964.

Greenwood, F. C., W. M. Hunter, and A. Klopper: Assay of human growth hormone in pregnancy, at parturition and in lactation. Detection of a growth hormone-like substance from the placenta. Brit. Med. J. *1*:22, 1964.

Grumbach, M. M., and S. L. Kaplan: On the placental origin and purification of chorionic "growth hormone-prolactin" and its immunoassay in pregnancy. Trans. N.Y. Acad. Sci. 27:167, 1964.

Grumbach, M. M., S. L. Kaplan, C. L. Abrams, J. J. Bell, and F. A. Conte: Plasma free fatty acid response to the administration of chorionic "growth hormone-prolactin." J. Clin. Endocrinol. *26*:478, 1966.

Hamashige, S., and E. R. Arquilla: Immunologic and serologic study of human chorionic gonadotrophin. J. Clin. Invest. *43*:1163, 1964.

Hamlett, G. W.: Positive Friedman test in the pregnant rhesus monkey, *Macaca mulatta*. Amer. J. Physiol. *118*:664, 1937.

Higashi, K.: Studies on the prolactin-like substance in human placenta. III. Endocrinol. Japan. *8*:288, 1961.

Hisaw, F. L.: The placental gonadotrophin and luteal function in monkeys, *Macaca mulatta*. Yale J. Biol. Med. *17*:119, 1944.

————: Endocrine adaptations of the mammalian estrous cycle and gestation.

In: Comparative Endocrinology, A. Gorbman, ed. New York: John Wiley and Sons, 1959.

Hisaw, F. L., and M. X. Zarrow: The physiology of relaxin. In: Vitamins and Hormones, New York: Academic Press, Inc., *8*:151, 1951.

Jones, G. E. S., G. O. Gey, and M. K. Gey: Hormone production by placental cells maintained in continuous culture. Bull. Johns Hopkins Hosp. *72*:26, 1943.

Josimovich, J. B.: Potentiation of somatotrophic and diabetogenic effects of growth hormone by human placental lactogen (HPL). Endocrinology *78:* 707, 1966.

Josimovich, J. B., B. L. Atwood, and D. A. Goss: The luteotrophic, immunologic and electrophoretic properties of human placental lactogen. Endocrinology *73:*410, 1963.

Josimovich, J. B., and L. Cato: Unique luteotrophic properties of combinations of human placental lactogen and chorionic gonadotrophin. Transactions of the Fifth International Congress on Fertility and Sterility. Stockholm: June, 1966. *In press.* Excerpta Medica.

Josimovich, J. B., and J. A. MacLaren: Presence in the human placenta and term serum of a highly lactogenic substance immunologically related to pituitary growth hormone. Endocrinology *71:*209, 1962.

Kaplan, S. L., and M. M. Grumbach: Studies of a human and simian placental hormone with growth hormone-like and prolactin-like activities. J. Clin. Endocrinol. *24:*80, 1964.

————, ————: Immunoassay for human chorionic "growth hormone-prolactin" in serum and urine. Science *147:*751, 1965.

Kilpatrick, R., D. T. Armstrong, and R. O. Greep: Maintenance of the corpus luteum by gonadotrophins in the hypophysectomized rabbit. Endocrinology *74:*453, 1964.

Knobil, E., and J. B. Josimovich: The interstitial cell stimulating activity of ovine, equine and human luteinizing hormone preparations in the hypophysectomized male rhesus monkey. Endocrinology *69:*139, 1961.

Lauritzen, C.: Action of HCG and steroids on physiological processes in the neonatal period. In: Research on Steroids, Rome: I. L. Pensiero Scientifico *2:*109, 1966.

Loraine, J. A., and H. W. Schmidt-Elmendorff: Human gonadotrophins. In: Human Reproductive Physiology, H. M. Carey, ed. Washington: Butterworths, 1963.

Lyon, R. A., M. E. Simpson, and H. M. Evans: Qualitative changes in urinary gonadotrophins in human pregnancy during the period of rapid increase in hormone titer. Endocrinology *53:*674, 1953.

Lyons, W. R., M. E. Simpson, and H. M. Evans: Hormonal requirements for pregnancy and mammary development in hypophysectomized rats. Proc. Soc. Exp. Biol. and Med. *52:* 134, 1943.

Matthies, D. L.: Organs responsive to rat chorionic mammatropin. (Abstract) Anat. Rec. *154:*384, 1966.

Midgley, A. R., Jr.: Radioimmunoassay: A method for human chorionic gonadotrophin and human luteinizing hormone. Endocrinology *79:*10, 1966.

Mishell, D. R., Jr., L. Wide, and C. A. Gemzell: Immunologic determination of human chorionic gonadotrophin in serum. J. Clin. Endocrinol. *23*:125, 1963.

Philipp, E.: Die innere Sekretion der Plazenta. Zbl. Gynäk. *54*:2754, 1930.

Rennie, P., J. Davies, and E. Friedrich: Failure of ovine prolactin to show luteotrophic or luteolytic effects in the rabbit. Endocrinology *75*:622, 1964.

Ryan, K. J.: Hormones of the placenta. Am. J. Obstet. Gynec. *84*:1695, 1962.

Sciarra, J. J., S. L. Kaplan, and M. M. Grumbach: Localization of anti-human growth hormone serum within the human placenta: evidence for a human chorionic "growth hormone-prolactin." Nature *199*:1005, 1963.

Shapiro, H. A., and H. A. Zwarenstein: A rapid test for pregnancy of *Xenopus laevis*. Trans. Royal Soc. South Africa *22*:75, 1934.

Smith, P. E.: Continuation of pregnancy in Rhesus monkey, *Macaca mulatta,* following hypophysectomy. Endocrinology *55*:655, 1954.

Tallberg, H., E. Ruoslahti, and C. Ehnholm: Immunologic studies in human placental proteins and the purification of human placental lactogen. Ann. med. exp. fenn. *43*:67, 1965.

Thiede, H. A., and J. W. Choate: Chorionic gonadotrophin localization in the human placenta by immunofluorescent staining. II. Demonstration of HCG in the trophoblast and amnion epithelium of immature and mature placentas. Obstet. Gynec. *22*:433, 1963.

Turkington, R. W., and Y. J. Topper: Stimulation of casein synthesis and mammary gland development by human placental lactogen *in vitro*. (Abstract) Program of the Endocrine Society, forty-eighth meeting. Chicago: June, 1966.

Yochim, J. M., and V. J. DeFeo: Control of decidual growth in the rat by steroid hormones of the ovary. Endocrinology *71*:134, 1962.

Zuckerman, S.: The Aschheim-Zondek diagnosis of pregnancy in the chimpanzee. Amer. J. Physiol. *110*:597, 1935.

PROLONGED GESTATION

P. C. Kennedy

*Department of Pathology, School of Veterinary Medicine,
University of California, Davis, California*

G. C. Liggins

*Postgraduate School of Obstetrics and Gynaecology,
National Women's Hospital, Auckland, New Zealand*

L. W. Holm

*Department of Physiological Sciences, University of California,
Davis, California*

Abnormally long gestation periods due to fetal abnormalities have been reported in man (Comerford, 1965), cattle and sheep (Holm, 1966). The unique quality of the hereditary forms of prolonged gestation in cattle has tended to obscure the broad biological significance of this phenomenon. It has, however, focused attention on the fact that the fetus does make a critical contribution to his timely delivery. Recent experimental data coupled with data derived from naturally occurring cases of prolonged gestation, both of genetic and non-genetic origin, make it possible to discuss some of the mechanisms involved. Our attempt here will be to review this material and define some of the links in the chain of events which lead to parturition.

Abnormally long gestation periods were recognized in cattle since the latter part of the last century (Servatius, 1894). Even the early reports indicated an awareness that some of these syndromes had a genetic basis, but it was not until Gregory *et al.* (1951) and Stormont *et al.* in 1956 that the details of inheritance were reported. Both reports, although dealing with different breeds and, as we will discuss later, different diseases, indicated that in each case prolonged gestation was controlled by a single autosomal recessive gene. A fetus suffered an abnormally long gestation period if it was homozygous for the trait. Since the trait is lethal, the parents must both be heterozygotes. Herds which are closely inbred create the conditions under which this defect can become manifest. It is interesting that heterozygosity with respect to these genes is as widespread within the breeds of dairy cattle as it is, as judged by the fact that prolonged gestation, probably of genetic origin, has been reported in: Swedish Red and White (Hallgren, 1951), Guernsey (McEntee *et al.*, 1959; Kennedy *et al.*, 1956), Jersey (Blood *et al.*, 1957) Ayrshire (Wilson and Young, 1958), and Holstein-Friesian breeds (Jasper, 1950; Holm, 1958).

Although the patterns of inheritance are similar, there are at least two distinct genetically controlled syndromes in cattle which have prolonged gestation as their hallmark. In the first type, which is seen in the Holstein and Ayrshire breeds, the calf is large; there are no facial abnormalities, and polyhydramnios has not been reported. The pituitary is present although the acidophils are poorly granulated. The adrenal cortices are hypoplastic. These calves are usually weak when delivered by Caesarian section, lack ability to nurse, and die in six to eight hours in severe hypoglycemia (Holm, 1958).

Holm *et al.* (1961) have demonstrated that these postmature animals exhibit hypoadrenocorticism as measured by fasting blood sugar and response to exogenous ACTH. The situation which exists in these calves with regard to their pituitary-adrenal relationship is far from clear because, although after Caesarian section hypofunction and hypoplasia were demonstrated, the only calf which has been homozygous for this trait and survived to breeding age died with bilateral adrenal cortical hyperplasia and signs of hyperadrenocorticism.

Holm *et al.* (1961) reported that the animals also had elevated protein-bound iodine levels and that these correlated with signs of hyperthyroidism.

In the second type, which occurs in the Guernsey and Jersey breeds, the affected fetuses are small, many have facial abnormalities, lack an adenohypophysis, and have varying degrees of hypotrichosis. Many of these fetuses have hydramnios. These fetuses are incapable of initiating parturition but can survive for remarkably long periods *in utero*. Several affected fetuses survived over 200 days past term, but if delivered by Caesarian section live only a few minutes (Kennedy *et al.*, 1957).

Not all types of abnormally long gestation in cattle can be fitted into these two categories and in many of the reports fetuses have not been examined carefully enough to characterize precisely. Also it is unlikely that all prolonged gestations in cattle are of genetic origin. The report of Huston and Gier (1958) of a large hydrocephalic calf which had a pituitary and suffered prolonged gestation clearly dealt with a different morphological type and one that was probably not of genetic origin. The reports of prolonged gestation in sheep in which the ewe's ingestion of the plant *Veratrum californicum* damages the conceptus producing severe cranial malformation (Binns *et al.,* 1959, 1960, 1963) suggest other mechanisms which may also operate in cattle. Fetuses damaged by this plant do resemble in some respects the Guernsey fetuses and they are reported to lack a pituitary. Detailed morphological studies have not yet been reported; however, the size of these fetuses contrasts sharply with that of calves with adenohypophyseal aplasia. Some of the affected lamb fetuses weigh up to 26 pounds. Polyhydramnios was present in many.

Another pattern of prolonged gestation which is not thought to be hereditary was reported in Karakul ewes in South-West Africa (De Lange, 1961). These fetuses shared many features with the Holstein type of postmaturity in that the fetuses were large and had no gross cranial abnormalities. They had an adenohypophysis which, like the calf fetuses, had few granules in the acidophils. Also like their bovine counterparts, if delivered, they lacked vitality, made no attempt to nurse, developed incoordination and tremors, and died in a few hours.

Both cows and ewes carrying fetuses destined to fail to be delivered at term do develop prior to calculated term temporary udder filling and some enlargement and flaccidity of the vulva. After the expected date of term these signs regress and the animals appear to be in mid-pregnancy. Cows carrying fetuses beyond term maintain blood progesterone at the same levels as found in normal cows during mid-pregnancy (Holm and Short, 1962). There is not the normal fall in the progesterone just prior to the expected date of delivery characteristic of normal cows (Short, 1958).

Recently pituitary lesions have been made by electrocautery with a probe placed in the fetal pituitary fossa (Liggins *et al.,* 1966). In 11 ewes in which the fetal pituitary destruction was more than 50 per cent complete, parturition failed to occur and pregnancy was prolonged beyond 160 days in ten (normal gestation in ewes is 150 days) until interrupted electively by Caesarian section. In the one exception surgery was performed on the 143rd day of pregnancy and spontaneous delivery occurred within 24 hours. These animals appear to represent the experimental counterparts of inherited prolonged gestation due to adenohypophyseal aplasia seen in cattle. As in calves suffering from pituitary aplasia, there were multiple secondary endocrinopathies. The severity of these changes was related to age of gestation at which the pituitary was destroyed. The earlier during pregnancy that surgery was done the more severe were the secondary changes and the more closely the histological changes approximated those seen in the calves which had pituitary aplasia (Kennedy *et al.,* 1957). In a group of four ewes (three twin and one triplet pregnancies) one fetus in each instance was left intact. These ewes delivered spontaneously at term. This suggests a single intact fetus

Table 1

BODY WEIGHT

Ewe	Lamb		Maturity at Operation (days)	Body Wt. (g)
2358	A	(Intact)	95	4800
	B	(Operated)		2750
191	A	(Intact)	122	5100
	B	(Operated)		3500
2387	A	(Operated)	124	3200
	B	(Intact)		3900
	C	(Operated)		3600
2257	A	(Intact)	135	3400
	B	(Operated)		3400

can supply the necessary triggering for normal delivery in multiple births. It is also possible that this phenomenon explains at least in part the rarity of prolonged gestation in polytocous species.

Pituitary destruction in the fetus had a profound effect on body weight and weights of other endocrine glands at delivery (Tables 1, 2, 3, 4) (Liggins *et al., in press a, b*). This again is in keeping with the experience in pituitary aplasia in calves in which all fetuses were small, but it contrasts with the situation reported to exist in *Veratrum* intoxication in which gestations are prolonged but the fetuses continue to grow and the pituitaries are thought to be absent. It is very possible that in the badly distorted heads of these lambs an intact pituitary remained. This would also be in keeping with the experimental cases in which in three fetuses the pituitary of each remained intact, but the hypothalamus or stalk was destroyed, gestation was prolonged and the fetuses continued to grow.

It appears from this that certain pituitary functions are maintained

Table 2

THYROID WEIGHT

	No. of Lambs	Mean Wt. (mg)
With pituitary lesion	12	494
Intact	9	728
Stalk section	2	963
Hypothalamic destruction	1	651

Table 3

TESTIS WEIGHT

	No. of Lambs	Mean Wt. (mg)
With pituitary lesion	8	972
Intact	4	1735
Hypothalamic destruction	1	1910

in the fetus at substantial levels in the absence of neural control. It is difficult to be precise about the specific hormones involved, but fetal growth and bone maturity are two parameters that show differences in fetuses with pituitary lesion contrasted with fetuses with neural damage. It is interesting in this connection that there is little difference with regard to adrenal weight; that is, pituitary destruction and hypothalamic damage produced comparable levels of adrenal hypoplasia.

This evidence with regard to neural control or independence of fetal pituitary is obviously incomplete but it suggests each facet of pituitary function will have to be examined separately. It also suggests that anencephaly in human infants may not be a single endocrinological entity but that each case may vary according to the amount of functional pituitary tissue present. Observations on the fetal testis in anencephaly in which considerable variation was found in numbers of Leydig cells (Zondek and Zondek, 1965) might be interpreted in this way.

In the original studies of pituitary aplasia in Guernsey calves it was felt that the fetuses failed to grow because of deficiencies of trophic hypophyseal hormones, but since these animals were genetically defective and very possibly could have been suffering other defects which interfered with growth, they did not provide satisfactory evidence on this point. The evidence, on the other hand, provided by decapitation experi-

Table 4

ADRENAL WEIGHT

	No. of Lambs	Mean Wt. (mg)
With pituitary lesion	13	317
Intact	9	716
Normal	6	719
Stalk section	2	439
Hypothalamic destruction	1	294

ments on mice, rats and rabbits, summarized by Deanesly (1961), appeared to indicate fetal pituitary and thyroid hormones were not important in the regulation of intrauterine growth. These experiments represent the first unequivocal demonstration of the role of the pituitary in regulating fetal growth as well as in the initiation of parturition. The discrepancy between the data obtained from lambs and calves and that from mice, rats and rabbits can best be explained in terms of stage of maturity at birth. It is probable that all species have a certain intrinsic growth capacity which in the case of species that are well-developed at birth, such as lambs and calves, is exhausted *in utero* while in other species, which are delivered while still immature, it may still be present at birth. Birth obviously does not represent a specific stage in development and those species which are mature at delivery are apt to have fetal growth regulated in part by the fetal pituitary and species which are immature at birth probably will not.

Since prolonged gestation in all species and in all syndromes has as one of its characteristic features adrenal hypoplasia, this was one of the obvious points to be studied experimentally; *i.e.*, will fetal adrenalectomy prolong pregnancy? Unfortunately technical problems make this an easier question to ask than to answer. So far, mortality after surgery has been high and this coupled with the remarkable regenerative capacity of any adrenal tissue left at surgery has prevented us from satisfactorily examining this point. Additional evidence suggestive of adrenal involvement has been provided in one preliminary experiment. Infusion of ACTH into one of twins, both with pituitary lesions and beyond term, was followed by parturition on the sixth day of infusion; in the case of a single post-term fetus which had been subjected to adrenalectomy as well as pituitary destruction, ACTH had no effect during nine days' infusion (Liggins *et al.*, 1966).

These experiments, supported by our interpretations of the naturally occurring syndromes, suggest that the fetal triggering of parturition either originates in the hypothalamus, or is mediated through it, and via the hypothalamic-hypophyseal portal system stimulates pituitary release of ACTH (?). Delivery follows rapid growth of the fetal adrenal cortex and increase in its hormone production.

Acknowledgments

Dr. Holm's investigations reported herein have been supported by grants GMS-005784 and HD02299 from the National Institutes of Health and by a grant from the Association for the Aid of Crippled Children.

Dr. Liggins has been supported by a fellowship from the Lalor Foundation.

References

Binns, W., E. G. Thacker, L. F. James, and W. T. Muffman: A congenital cyclopian-type malformation in lambs. J. Am. Vet. Med. Assoc. *134*:180, 1959.

Binns, W., W. A. Anderson, and D. J. Sullivan: Further observations on a congenital cyclopian-type malformation in lambs. J. Am. Vet. Med. Assoc. *137*:515, 1960.

Binns, W., J. F. Lynn, J. L. Shupe, and G. A. Everett: Congenital cyclopian-type malformation in lambs induced by maternal ingestion of a range plant, *Veratrum californicum*. Am. J. Vet. Res. *24*:1164, 1963.

Blood, D. C., D. R. Hutchins, K. V. Jubb, and J. H. Whitten: Prolonged gestation of Jersey cows. Aust. Vet. J. *33*:329, 1957.

Comerford, J. B.: Pregnancy with anencephaly. Lancet *1*:679, 1965.

Deanesly, R.: Foetal endocrinology. Br. Med. Bull. *17*:19, 1961.

De Lange, M.: Prolonged Gestation in Karakul Ewes in South-West Africa. Proc. IVth Int. Congr. Anim. Reprod., The Hague, Vol. *3*:590, 1961.

Gregory, P. W., S. W. Mead, and W. M. Regan: A genetic analysis of prolonged gestation in cattle. Port. Acta Biol. Series A, RB Goldschmidt: *861*, 1949–51.

Hallgren, W.: Abnormt lang draktighet hos ko. (graviditas prolongata cum/sine partu serotino). Nord. Vet. Med. *3*:1043, 1951.

Holm, L. W.: Some aspects of glucose metabolism in calves after prolonged gestation. Am. J. Vet. Res. *19*:842, 1958.

———: The gestation period of mammals. In: Comparative Biology of Reproduction in Mammals. I. W. Rowlands, ed. London: Academic Press Inc., 1966.

Holm, L. W., H. R. Parker, and S. J. Galligan: Adrenal insufficiency in postmature Holstein calves. Am. J. Obstet. Gynec. *81*:1000, 1961.

Holm, L. W., and R. V. Short: Progesterone in the peripheral blood of Guernsey and Friesian cows during prolonged gestation. J. Reprod. Fert. *4*:137, 1962.

Huston, K., and H. T. Gier: An anatomical description of a hydrocephalic calf from prolonged gestation and the possible relationship of these conditions. Cornell Vet. *48*:45, 1958.

Jasper, D. E.: Prolonged gestation in the bovine. Cornell Vet. *40*:165, 1950.

Kennedy, P. C., J. W. Kendrick, and C. Stormont: Adenohypophyseal aplasia, an inherited defect associated with abnormal gestation in Guernsey cattle. Cornell Vet. *47*:160, 1957.

Liggins, G. C., L. W. Holm, and P. C. Kennedy: Prolonged pregnancy following surgical lesions of the foetal lamb pituitary. Proceedings of Soc. for Study of Fertility in J. Reprod. Fert. *12*:419, 1966.

———, ———, ———: Failure of the initiation of parturition following surgical lesions of the fetal lamb pituitary. Am. J. Obstet. Gynec.: *in press* (a).

———, ———, ———: Effects on growth and development of surgical lesions in the pituitary of the fetal lamb. Acta Endocrinol.: *in press* (b).

McEntee, K., S. J. Roberts, and R. M. Sears: Prolonged gestation in two Guernsey cows. Cornell Vet. *42*:355, 1952.

Servatius: Abnorme Trächtigkeitsdauer bei Kühen. Dt. Tierärztl. Wschr. *2*:117, 1894.

Short, R. V.: Progesterone in blood. II. Progesterone in the peripheral blood of pregnant cows. J. Endocrin. *16:*426, 1958.

Stormont, C., J. W. Kendrick, and P. C. Kennedy: A "new" syndrome of inherited lethal defects associated with abnormal gestation in Guernsey cattle. Genetics *41:*663 (Abst.) 1956.

Wilson, A. L., and G. B. Young: Prolonged gestation in an Ayrshire herd. Vet. Rec. *70:*73, 1958.

Zondek, L. H., and T. Zondek: Observations on the testis in anencephaly with special reference to the Leydig cells. Biol. Neonat. *8:*329, 1965.

OÖGENESIS—OVULATION AND EGG TRANSPORT

R. J. Blandau

School of Medicine, University of Washington, Seattle, Washington

Basic knowledge of the biology of germ cells, ovulation and egg transport are essential if proper assessments of the problems of excessive population growth and infertility in man and animals are to be made. There are very large hiatuses in our knowledge of any of these phenomena. It is fair to state that we would have difficulty in describing the natural history of reproductive processes with accuracy for any animal much less the physical and chemical changes occurring at the cellular level.

Many of the procedures recommended for human infertility problems are basically empirical because of our uncertainties of physiological mechanisms and because fundamental research in reproduction has not kept pace with what become urgent human problems. For example, tuboplasty is often performed in the human with remarkably little success. Because of its anatomical location and general morphology it is assumed that, if the tube is patent, it functions as a simple conduit for the fertilized eggs. Egg transport through the oviduct is dependent largely upon the proper location and function of the ciliated cells, the secretory cells and the complex activity of its musculature. Research is revealing that oviducal physiology is very complex indeed. Its propulsive and secretory activities are delicately balanced by the waxing and waning and the specific actions of various hormones. Egg nutrition and transport can be altered dramatically by injections of steroid hormones with the result that the eggs are either "trapped" within the oviducts or transported at an accelerated rate resulting in the death of the egg.

Oögenesis

A voluminous literature and comprehensive reviews debate the concept of the somatic versus the germinal origin of oöcytes of mammals.

194

The primary question as to whether or not the germinal epithelium at any time in the life of the animal serves as a source of definitive oögonia or whether the embryonic gonads contain the full complement of germ cells is being rapidly resolved.

It is becoming increasingly clear that the primordial germ cells segregate early during embryonic life and that females of mammals enter the reproductive period with ovaries that contain a finite number of oöcytes (Franchi, Mandl and Zuckerman, 1962; Mintz and Russell, 1957; Ohno, Klinger and Atkin, 1962).

The following facts related to the origin of the primordial germ cells have gone beyond the state of conjecture:

1. Germ cells of all birds and mammals have an extragonadal origin (Witschi, 1948; Brambell, 1927; Chiquoine, 1954; Mintz, 1957 a and b; Mintz and Russell, 1957; Blandau, White and Rumery, 1963; Meyer, 1964).

2. In mammals the primordial germ cells arise in the yolk sac from where they migrate by innate ameboid movements to the developing gonads (Witschi, 1948; Blandau, White and Rumery, 1963).

3. In birds the germ cells arise in the area of the anterior crescent. They are transported to the embryonic gonads by the blood stream. They show no ameboid movements during their migratory phase (Meyer, 1964; Simon, 1957 a and b; Meyer and Blandau, unpublished observations).

4. New germ cells are formed by mitotic divisions of pre-existing ones. These divisions occur either while the germ cells are migrating from the yolk sac to the germinal ridges or during a restricted period of time after having seeded the embryonic gonads (Witschi, 1948; Blandau, White and Rumery, 1963).

5. All oögonia have completed their mitotic divisions and have entered the prophase of meiosis before birth in mammals or hatching in birds (Mandl, 1964; Beumont and Mandl, 1962; Baker, 1963; Ohno, Klinger and Atkin, 1962; Hughes, 1963; Monataya and Patten, 1963).

6. Oöcytes remain in the dictyotene stage of meiosis until shortly before ovulation (Ohno, Makino and Kaplan, 1961; Winiwarter, 1901; Ohno, Klinger and Atkin, 1962; Odor, 1955 and 1960).

7. Segregation of homologous chromosomes occurs during the period between birth and sexual maturity (Ohno, Klinger and Atkin, 1962).

8. Tenuous evidence permits the conjecture that if the oöcytes are to survive they must be surrounded by a layer of follicular cells which constitute the primordial follicles. If individual bivalents are free from one another when entering the diplotene stage they proceed to diakinesis only to undergo atresia (Ohno, Christian and Stenius, 1963; Ohno and Smith, 1964; Blandau, Warrick and Rumery, 1965).

An host of intriguing problems concerning oögenesis remain to be investigated. Among these may be mentioned the origin of the precursors of the germ cells seen in the entodermal yolk sac, the significance of the chemo-differentiation that renders them so conspicuous during the early stages of development, further evaluation of their selective migration and factors that control it, reasons for the selective location of oögonia within the embryonic gonads, factors that effect the synchronous onset of meiotic prophase in oöcytes, the role of the somatic cells in inhibiting the completion of meiosis, the origin of the cells which encompass the ovum to form the primary follicle, and the interrelationship of these cells with the ovum and their role in atresia.

OVULATION

Reproductively various animals may be classified as either seasonal breeders, reflex or cyclic ovulators. Ovulation occurs spontaneously in most mammals. In animals such as the cat, ferret, ground squirrel, martin, mink, rabbit, shrew and others, external stimuli must be applied to evoke LH release (Asdell, 1946; Critchlow, 1958). It is firmly established that mature ovulatory follicles rupture in response to the action of the luteinizing hormone (Everett, 1956; Fee and Parkes, 1929; Rowlands and Williams, 1943). The role that this hormone plays in initiating the specific changes that occur within the follicular wall is still almost completely unknown. Morphological study, either with the light or electron microscope has thus far told us very little about the dynamics of the process of ovulation. Chemical study is perhaps the more fruitful approach. However, neither chemistry nor pure morphology in themselves can fully explain mechanisms of as complicated a system as living tissues. Even though no two cell-types or tissues are just alike it is fortunate that there is an underlying similarity in the morphology and biochemistry of living material. Studies of ovulation and egg transport in mammals other than man can provide the factual knowledge so that at least the proper questions can be raised as to the nature of these vital functions in man. It is becoming clear that successful ovulation involves a sequence of events which are exceedingly complex and still shrouded in mystery. The blister-like appearance of the ovulatory follicle and the general thinning of its wall as ovulation approaches naturally led to the conclusion that increasing internal fluid pressures are primarily responsible for its rupture (Rouget, 1858). Rising intrafollicular pressure cannot be the basic cause for ovulation when it is recognized (1) that there are considerable variations in the size and shape of ovulatory follicles; (2) that the point of rupture does not always appear at the apex of the follicle; (3) that conditions may exist where two follicles of identical size develop side by side—one ovulating and the other never rupturing—despite the fact that they may have identical intrafollicular fluid pressures; (4) that Graafian follicles may continue to grow far beyond their preovulatory size and become cystic, yet never ovulate; (5) that ovulatory follicles may actually become flaccid or folded shortly before ovulation; and (6) that ovulation occurs normally in certain animals in which antra are not formed.

Ovulation is preceded by a series of events which affect the orderly growth and development of an ovarian follicle. The essential role of the hypothalamus and pituitary is recognized. What determines the choice of follicles which are destined to become vesicular remains an intriguing problem. When the developmental pattern of a group of follicles has been determined they must either rupture or eventually undergo atresia and become a functional part of the interstitial gland. The temporal and morphologic features of follicular growth during the estrus or menstrual cycle have been described in detail for many animals (Allen, 1932, 1939). The morphologic changes that occur within a growing preovulatory follicle are effected by (1) the rapidly increasing number of follicular cells; (2) the formation of an antrum by the coalescence of fluid-filled spaces; (3) the displacement of the egg to an eccentric position encompassed by a variable number of cell layers comprising the cumulus oöphorus; (4) the elaboration and secretion of a specific fluid by the follicular cells which becomes the rather highly polymerized liquor folliculi; (5) the straight line increase in the size of the follicle related to its overall growth and the secretion of increasing quantities of follicular fluid, a growth leading to a bulging of the follicle above the surface of the ovary; and (6) the hypertrophy and increasing vascularity of the theca.

A significant event during the period of preovulatory growth is the depolymerization of the intercellular cement substance of the cells forming the cumulus oöphorus. This allows the cells to become partially separated one from another and from the underlying stratum granulosum. The extent of this dispersal varies significantly in different mammals. In most animals the cells forming the corona radiata are much less affected than those of the cumulus. In the hamster and mouse the cells of the corona radiata may be removed easily by gently brushing them from the zona. In contrast the corona cells of the freshly ovulated rabbit egg can be removed only by vigorous shaking in a test tube. Although the cells comprising the cumulus oöphorus are of the same lineage as those of the stratum granulosum they respond quite differently in the presence of the luteinizing hormone. In the rat, if cells from both the stratum granulosum and cumulus are recovered from growing follicles which have been under the influence of follicle stimulating hormone primarily and placed in tissue culture, they grow rapidly. In contrast, similar preparations from ovulatory follicles which have come under the influence of the luteinizing hormone show significant difference in growth response. The cells from the stratum granulosum grow very well indeed while those comprising the cumulus, though remaining alive for several days, fail to multiply. Thus it appears that, at least in the rat, the follicular cells comprising the cumulus oöphorus respond to the action of the luteinizing hormone in a manner quite different than those of the stratum granulosum (Blandau and Rumery, 1962).

Retrospective studies of large follicles which do not ovulate show that their cumuli oöphori may not undergo the usual dispersion. Cumuli recovered from these follicles, several hours after expected ovulation, usually grow reasonably well in culture (Blandau, unpublished observations). It would be interesting to explore the reciprocal relationship that seems to exist between the ovum and the cumulus oöphorus which leads to the dispersal of these cells, while those of the stratum granulosum are not similarly affected. It appears that, in the rat, cells of the cumulus oöphorus of ovulatory follicles have a definite life span once having come under the influence of the luteinizing hormone.

As emphasized previously the growth of the Graafian follicle is gradual but consistent throughout the preovulatory period. From 2 to 12 hours before the discharge of the ovum the ovulatory follicles undergo a brief period of accelerated growth which includes the accumulation of the secondary follicular fluid which is somewhat less polymerized than the original fluid. There is also a generalized increase of vascularity within the thin follicular walls.

The process of ovulation has now been observed in the living mouse, rat, rabbit, sheep and man (Blandau, 1955; Hill, Allen and Kramer, 1935; Markee and Hinsey, 1936; McKenzie and Terrill, 1937; and Doyle, 1951). In animals in which this event has been observed in some detail a circumscribed avascular area appears usually at the apex of the follicle. The over-all size of the macula pellucida or stigma is usually very small when one considers the total exposed surface area of the follicle. The avascular macula pellucida thins out until a thin, transparent membrane bulges above the surface. This transparent membrane has been identified as the basement membrane separating the stratum granulosum from the theca (Blandau, 1966). The transparent membrane gradually increases in height until it breaks and the follicular fluid flows from the rupture point. A significant feature of the stigma is that it is round in contour and quite smooth so that a tear is not suggested. There is considerable variation also in the size of stigmas in different follicles.

Descriptions of observed ovulations vary somewhat in regard to the rate of expulsion of the follicular contents. Observers emphasize that usually the follicular contents ooze from the follicle. Occasionally one may observe the leading follicular fluid escaping with a spurt. Careful observations of the thin walls of the ovulatory follicles reveal that their collapse does not occur simultaneously with rupture but usually begins after the cumulus has been ovulated. Once the stigma is open the time required to complete ovulation is related directly to the diameter of the stigma and the location of the cumulus oöphorus and egg within the follicle. If the cumulus oöphorus is located near the base of the follicle there is a significant stream-lining as it moves toward the stigma. In these cases ovulation may be completed in a matter of seconds or several minutes. If, on the other hand, the cumulus is located near the stigma it is moved into the opening without the opportunity of stream-lining, thus forming a plug. Under these circumstances it requires a considerably longer time for expulsion. Rouget as early as 1858 wrote that increasing intrafollicular pressure was the principal cause of follicular rupture. Various factors which have been suggested as placing stress upon the stigma causing it to rupture are (1) contraction of smooth muscles located either within the ovarian stroma or theca (Kölliker, 1848; Rouget, 1858; Aeby, 1861; Rugh, 1935; Phillips and Warren, 1937; Guttmacher and Guttmacher, 1921; Strauss, 1938; Lipner and Maxwell, 1960); (2) hypertrophy of the theca interna exerting cellular pressure on the stratum granulosum (Waldeyer, 1870); (3) increasing vaso-dilatation and hyperemia of the whole ovary but particularly the follicular wall (Heape, 1905; Keller, 1943; Pearson, 1944); (4) hypertrophy of the follicular cells as ovulation approaches (Strauss, 1938); (5) rapid preovulatory secretion of a thinner secondary follicular fluid (Smith and Ketteringham, 1938); (6) liberation of osmotically active substances into preovulatory follicles (Zachariae, 1959; Christiansen, Jensen and Zachariae, 1958; Zachariae and Jensen, 1958) ; and (7) secretion of proteolytic or cytolytic enzymes into the antrum effecting the digestion or destruction of the tissues of the macula pellucida (Schochet, 1916; Moricard and Gothie, 1946; Jung and Held, 1959; Rugh, 1935).

It is fair to state that the trend of opinion has been away from the pressure theory. Some of the reasons for lack of enthusiasm for the mechanical theory are (1) the limited portion of the follicular wall that is occupied by the macula pellucida (Blandau, 1955); (2) the smooth form and contour of the edges of the stigma and the orderly manner in which it appears; (3) the decrease in turgescence of the follicle just prior to ovulation as observed in the dog (Evans and Cole, 1931); in the bat (Wimsatt, 1944); and in the ungulates (Hansel, 1958); (4) the fact that most ovulations that have been observed in living animals are not violent eruptions but rather a "steady flowing of the contents from the opening" (Walton and Hammond, 1928; Markee and Hinsey, 1936; Doyle, 1951; and Blandau, 1955); (5) the evidence for the presence of contracting smooth muscle within the follicular wall has been lacking (Claesson, 1947); (6) the finding that intrafollicular pressure measurements in several hundred ovulatory follicles do not show an increase in mechanical pressure at the time of ovulation (Blandau and Rumery, 1963; Espey and Lipner, 1963).

It may be concluded that ovulation is the end result of an orderly progression of a series of integrated events that lead to the growth and development of an ovarian follicle, the formation of the macula pellucida, its rupture and the discharge of the cumulus oöphorus and egg. At the present time there is no theory that can explain adequately the various morphological features of ovulation that have been observed.

EGG TRANSPORT

Even though the oviducts of mammals are basically similar in their physiologic functions, there is considerable variation in their anatomical configuration and arrangement. In the Muridae and Mustelidae the ovaries are almost completely enclosed by a thin membranous periovarial sac (Alden, 1942; Wimsatt and Waldo, 1945). The relatively small, ciliated infundibulum of the oviduct pierces this sac and occupies only a small area of the periovarial space. In contrast, the fimbriated ends of the oviducts of the guinea pig, rabbit and primates are usually rather flattened and fluted and so extensively cover the ovaries as to form effective ovarian bursae.

Oviducts may be either rather straight or slightly folded tubes as in the guinea pig, rabbit, and man, or highly coiled as in the mouse, rat and hamster.

The length of the ampullar and isthmic subdivisions of the oviduct varies greatly in different animals. In the mouse and rat the first few loops form the ampulla. In the rabbit almost 50 per cent of the total length of the oviduct is ampulla (Harper, 1961).

There is considerable variation also in regard to the extent and arrangement of the smooth musculature within the membranes comprising the mesosalpinx. Specialized smooth muscle membranes such as the mesotubarium superius in the rabbit may span the antimesosalpingal border.

The specific method whereby the freshly ovulated eggs are moved from the site of rupture of the follicle into the ostium of the oviduct has been observed only for the rat and rabbit (Blandau, 1961; Clewe and Mastroianni, 1958).

As mentioned earlier the infundibulum of the rat oviduct occupies only a small portion of the periovarial space and it makes only minimal contact with the surface of the ovary. At ovulation the eggs and cumuli are shed completely from the follicle so that they come to lie free within the periovarial space. Contractions of the extensive smooth musculature in the mesovarium effect longitudinal movements of the ovary so that the eggs are constantly changing their positions within the periovarial space. The infundibulum of the oviduct is highly ciliated. All of the cilia beat in the direction of the ostium effectively establishing fluid currents which direct the cumulus to the ostium. If the fimbriated end of the oviduct of the rabbit is removed surgically and the animal ovulated, the cumuli with their eggs invariably adhere to the stigma as cellular strands. Thus in the rabbit the eggs are not cast free from the follicles. It is necessary for the ciliated fimbria to be in intimate contact with the surface of the ruptured follicle if the egg is to be brushed away and carried into the ampulla.

Obviously there are distinct species differences in the mechanism of transport of eggs from the surface of the ovary. This transport is effected by (1) the direct action of the cilia of the infundibulum or fimbria on the cumulus oöphorus; (2) contractions of the smooth muscles of the mesovarium; (3) rhythmic contractions that change the position of the ovary in relationship to the infundibulum and the fimbria, particularly in those animals in which it is in intimate contact with the ovarian surface and (4) currents of fluid directed toward the ostium of the oviduct by the action of cilia (Fredricsson and Björkman, 1962).

Once having entered the ostium the eggs are transported rather quickly through the ampulla to the ampullar-isthmic junction where sperm penetration and fertilization take place. In the rat, ampullar transport requires 2 to 5 minutes; in the rabbit 8 to 10 minutes (Harper, 1965; Blandau, unpublished observations). A physiologic stricture at the ampullar-isthmic junction retains the eggs within the ampulla from 15 to 20 hours. Nothing is known of the hormonal mechanisms effecting these valvular constrictions (Burdick, Whitney and Emerson, 1942; Odor and Blandau, 1951; Strauss, 1956; Noyes, Adams and Walton, 1959, Brundin, 1964).

The first phase of egg transport then is relatively rapid and is the result of both ciliary and muscular activity. The cilia play their role primarily in the movements of the egg from the surface of the ovary or from the periovarial space and perhaps through the first few millimeters of the ampulla. Once having been carried into the ostium, transport to the ampullar-isthmic junction is effected primarily by coordinated peristaltic contractions of the ampulla.

There is less information as to transport of eggs through the isthmic portions of oviducts. The walls of the isthmus are much thicker and eggs or objects that simulate eggs are much more difficult to visualize within its lumen. In addition to transporting eggs the fallopian tubes in certain animals secrete substances which when deposited on the eggs comprise the tertiary membranes. Greenwald (1958) showed that progesterone initiated the secretion of mucoprotein in the rabbit isthmus.

Attempts to unravel the endocrine basis for the transport of ova through oviducts have been proceeding since 1939 (Whitney and Burdick, 1939).

The results and interpretations of these investigations are contradictory and confusing. The removal of ovaries following ovulation in the mouse (Whitney and Burdick, 1939), the rat (Alden, 1942), and the rabbit (Adams, 1958) results in what is considered to be normal transport time. When donor eggs were placed into the ampullae of rabbits ovariectomized one month previously, ova transport was significantly accelerated (Noyes, Adams and Walton, 1959). When 1 to 4 mg. estradiol benzoate were injected daily for 5 to 10 days the majority of the ova were retained in the oviduct. Burdick and Pincus (1935) reported that estrogens prevented egg transport through the oviducts in intact mice and rabbits. Large doses accelerated passage. Greenwald (1959) suggested that small doses of estrogens accelerated transport. On the contrary, he injected 250 μg. depo-estradiol into rabbits at the time of mating and invariably found eggs in the oviducts 4 days later. Harper (1966) found that estradiol benzoate injections into spayed rabbits had a definite stimulatory effect on the rate of egg transport. Adams (1958) injected progesterone into recently ovariectomized rabbits and reported a delay in transport. When Chang injected intact rabbits with progesterone over a period of 3 days before ovulation was induced with gonadotrophins, egg transport was greatly accelerated. It is generally assumed that increasing progesterone levels result in retardation of egg transport through the ampulla due to its depressing effect on muscular activity (Harper, 1965 and 1966).

There is no doubt but that the normal function of the oviduct is delicately controlled by the actions of estrogens and progesterone.

The problem that confronts the investigator is that the physiologic levels of these hormones have not been adequately established. The injection of single or multiple doses of hormone may result in pharmacologic reactions which may be quite abnormal. Little attention has been paid so far to the physiologic effects of withdrawal of hormones and it is quite possible that reactions attributed to the hormones may be the result of its lack rather than its presence.

Acknowledgment

Unpublished observations referred to in this chapter were supported by grants from the National Institutes of Health, United States Public Health Service.

References

Adams, C. E.: Egg development in the rabbit: The influence of post-coital ligation of the uterine tube and of ovariectomy. J. Endocrinol. *16:*283, 1958.

Aeby, C.: Die glatten Muskelfasern in den Eierstöcken der Wirbeltiere. Arch. Anat. Physiol., p. 635, 1861.

Alden, R. H.: Aspects of the egg-ovary-oviduct relationship in the albino rat. J. Exper. Zool. *90:*159, 1942.

Allen, E. (Editor): Sex and Internal Secretions. Baltimore: The Williams & Wilkins Co., 1932. 1st Ed.

Allen, E. (Editor): Sex and Internal Secretions. Baltimore: The Williams & Wilkins Co., 1939. 2nd Ed.

Asdell, S. A.: Patterns of Mammalian Reproduction. New York: Comstock, 1946.

Blandau, R. J.: Ovulation in the living albino rat. Fertil. and Steril. *6:*391, 1955.

————: Biology of Eggs and Implantation. In: Sex and Internal Secretions, W. C. Young, ed., p. 797. Baltimore: Williams & Wilkins Co., 1961.

————: The Mechanisms of Ovulation. In: Ovulation. Philadelphia: J. B. Lippincott Co., 1966.

Blandau, R. J., and R. E. Rumery: Observations on cultured granulosa cells from ovarian follicles and ovulated ova of the rat. Fertil. and Steril. *13:*335, 1962.

Blandau, R. J., E. Warrick, and R. E. Rumery: *In vitro* cultivation of fetal mouse ovaries. Fertil. and Steril. *16:*705, 1965.

Blandau, R. J., B. J. White, and R. E. Rumery: Observations on the movements of the living primordial germ cells in the mouse. Fertil. and Steril. *14:*482, 1963.

Brambell, R. F. W.: The development and morphology of the gonads of the mouse. I. The morphogenesis of the indifferent gonad and of the ovary. Proc. Roy. Soc. London Ser. B. *101:*391, 1927.

Brundin, J.: An occlusive mechanism in the fallopian tube of the rabbit. Acta Physiol. Scand. *61:*219, 1964.

Burdick, H. O., and G. Pincus: The effect of oestrin injections upon the developing ova of mice and rabbits. Am. J. Physiol. *111:*201, 1935.

Chiquoine, D. A.: The identification, origin and migration of the primordial germ cells in the mouse embryo. Anat. Rec. *118:*135, 1954

Christiansen, J. A., C. E. Jensen, and F. Zachariae: Studies on the mechanism of ovulation. Some remarks on the effect of depolymerization of high-polymers on the preovulatory growth of follicles. Acta Endocrinol. *29:*115, 1958.

Claesson, L.: Is there any smooth musculature in the wall of the graafian follicle? Acta Anat. (Basel) *3:*295, 1947.

Clewe, T. H., and L. Mastroianni, Jr.: Mechanisms of ovum pickup. I. Functional capacity of rabbit oviducts ligated near the fimbria. Fertil. and Steril. *9:*13, 1958.

Critchlow, V.: Ovulation induced by hypothalamic stimulation in the anesthetized rat. Am. J. Physiol. *195:*171, 1958.

Doyle, J. B.: Exploratory culdotomy for observation of tubo-ovarian physiology at ovulation time. Fertil. and Steril. *2:*474, 1951.

Espey, L. L., and H. Lipner: Measurements of intrafollicular pressures in the rabbit ovary. Amer. J. Physiol. *205:*1067, 1963.

Everett, J. W.: The time of release of ovulating hormone from the rat hypophysis. Endocrinology *59:*580, 1956.

Fee, A. R., and A. S. Parkes: Studies on ovulation. I. The relation of the anterior pituitary body to ovulation in the rabbit. J. Physiol. (London) *67:*383, 1929.

Fredricsson, B., and N. Björkman: Studies on the ultrastructure of the human oviduct epithelium in different functional states. Z. Zellforsch. *58*:387, 1962.

Greenwald, G. S.: Endocrine regulation of the secretion of mucin in the tubal epithelium of the rabbit. Anat. Rec. *130*:477, 1958.

———: Tubal transport of ova in the rabbit. Anat. Rec. *133*:386, 1959.

Guttmacher, M. S., and A. F. Guttmacher: Morphological and physiological studies on the musculature of the mature graafian follicle of the sow. Bull. Johns Hopkins Hosp. *370*:394, 1921.

Hansel, W.: Personal communication, 1958.

Harper, M. J. K.: The mechanisms involved in the movement of newly ovulated eggs through the ampulla of the rabbit Fallopian tube. J. Reprod. and Fertil. *2*:522, 1961.

———: Transport of eggs in cumulus through the ampulla of the rabbit oviduct in relation to day of pseudopregnancy. Endocrinology *77*:114, 1965.

———: Hormonal control of transport of eggs in cumulus through the ampulla of the rabbit oviduct. Endocrinology *78*:568, 1966.

Heape, W.: Ovulation and degenerating ova in the rabbit. Proc. Roy. Soc. (Biol.) *70*:260, 1905.

Hill, R. T., E. Allen, and T. C. Kramer: Cinemicrographic studies of rabbit ovulation. Anat. Rec. *63*:239, 1935.

Keller, K.: Das Bindegewebegerüst des Eierstockes und seine funktionelle Bedeutung. Jahrb. Morph. Mikrosk. Anat. *88*:351, 1943.

Kölliker, A. von: Beiträge zur Kenntniss der glatten Muskeln. Z. wissen. Zool. *1*:48, 1848.

Lipner, H. J., and B. A. Maxwell: Hypothesis concerning the role of follicular contractions in ovulation. Science *131*:1737, 1960.

Mandl, A. M.: The radiosensitivity of germ cells. Biol. Rev. *39*:288, 1964.

Markee, J. E., and J. C. Hinsey: Observations on ovulation in the rabbit. Anat. Rec. *64*:309, 1936.

McKenzie, F. F., and C. E. Terrill: Estrus, ovulation, and related phenomena in the ewe. Res. Bull. Mo. Agric. Exp. Sta. No. 264, 1937.

Meyer, D. B.: The migration of primordial germ cells in the chick embryo. Develop. Biol. *10*:154, 1964.

Mintz, B.: Germ cell origin and history in the mouse: Genetic and histochemical evidence. Anat. Rec. *127*:335, 1957a.

———: Embryological development of primordial germ cells in the mouse: Influence of a new mutation, Wj. J. Embryol. Exp. Morph. *5*:396, 1957b.

Moricard, R., and S. Gothie: Dissociation des cellules de la granulosa et probleme d'un mecanisme diastasique dans la rupture du follicle ovarien de lapine. C. R. Soc. Biol. (Paris) *140*:249, 1946.

Noyes, R. W., C. E. Adams, and A. Walton: The transport of ova in relation to the dosage of oestrogen in ovariectomized rabbits. J. Endocrinol. *18*:108, 1959.

Odor, D. L.: The temporal relationship of the first maturation division of rat ova to the onset of heat. Am. J. Anat. *97*:461, 1955.

Odor, D. L., and R. J. Blandau: Observations on fertilization and the first segmentation division in rat ova. Am. J. Anat. *89*:29, 1951.

Odor, D. L., and D. F. Renninger: Polar body formation in the rat oocyte as observed with the electron microscope. Anat. Rec. *137*:13, 1960.

Ohno, S., L. C. Christian, and C. Stenius: Significance in mammalian oögenesis of the non-homologous association of bivalents. Exp. Cell Res. *32*:590, 1963.

Ohno, S., H. P. Klinger, and N. B. Atkin: Human oögenesis. Cytogenetics *1*:42, 1962.

Ohno, S., and J. B. Smith: Role of fetal follicular cells in meiosis of mammalian oöcytes. Cytogenetics *3*:324, 1964.

Pearson, O. P.: Reproduction in the shrew (*Blarina brevicauda,* Say). Amer. J. Anat. *75*:39, 1944.

Rouget, C.: I. Recherches sur les organes erectiles de la femme et sur l'appareil musculaire tubo-ovarien dans leurs rapports avec l'ovulation et la menstruation. J. Physiol. (Paris) *1*:320, 1858.

Rowlands, I. W., and P. C. Williams: Production of ovulation in hypophysectomized rats. J. Endocrinol. *3*:310, 1943.

Rugh, R.: Ovulation in the frog. II. Follicular rupture to fertilization. J. Exp. Zool. *71*:163, 1935.

Schochet, S. S.: A suggestion as to the process of ovulation and ovarian cyst formation. Anat. Rec. *10*:447, 1916.

Simon, D.: La localisation primaire des cellules germinales de l'embryon de poulet: preuves expérimentales. C. R. Soc. Biol. *151*:1010, 1957a.

———: Sur la localisation des cellules germinales primordiales chez l'embryon de poulet et leur mode de migration vers les ébauches gonadiques. C. R. Acad. Sci. *244*:1541, 1957b.

Smith, J. T., and R. C. Ketteringham: Rupture of the graafian follicle. Am. J. Obstet. Gynec. *36*:453, 1938.

Strauss, F.: Die Befruchtung und der Vorgang der Ovulation bei *Ericulus* aus der Familie der Centetiden. Biomorphosis *1*:281, 1938.

———: The time and place of fertilization of the golden hamster egg. J. Embryol. Exper. Morphol. *4*:42, 1956.

Waldeyer, W.: Eierstock und Ei. Leipzig: W. Englemann, 1870.

Walton, A., and J. Hammond: Observations on ovulation in the rabbit. Brit. J. Exp. Biol. *6*:190, 1928.

Whitney, R., and H. O. Burdick: Effect of massive doses of an estrogen on ova transport in ovariectomized mice. Endocrinology *24*:45, 1939.

Wimsatt, W. A.: Growth of the ovarian follicle and ovulation in *Myotis lucifugus*. Am. J. Anat. *74*:129, 1944.

Winiwarter, H. von: Recherches sur l'ovogenese et l'organogenese de l'ovaire des mammiferes (lapin et homme). Arch. Biol. *17*:33, 1901.

Zachariae, F., and C. E. Jensen: Studies on the mechanism of ovulation. Histochemical and physico-chemical investigations on genuine follicular fluids. Acta Endocrinol. *27*:343, 1958.

EXPERIMENTAL HYBRIDIZATION

M. C. CHANG *

*Worcester Foundation for Experimental Biology,
Shrewsbury, Massachusetts*

J. L. HANCOCK

ARC Animal Breeding Research Organization, Edinburgh, Scotland

This communication is primarily concerned with the results of some experiments in mammalian cross fertilization which appear to be relevant to problems of reproductive failure in man and animals. It is not within the scope of this paper to review the extensive literature on mammalian hybrids; those interested are referred to the comprehensive check list by Gray (1957). Nor is it the intention here to examine the problems of hybridization in relation to evolution and genetics; reference should be made to White (1954), Stebbins (1959), and Mayr (1963) for a full discussion of these aspects of the subject.

In nature there exist isolating mechanisms which effectively prevent the merging of the identities of separate species. The barriers to successful reproduction between species arise from incompatibilities affecting mating, fertilization, and embryonic development and survival. Hybrid sterility which is also one of the isolating mechanisms will be dealt with by other contributors to this meeting.

Artificial insemination provides an effective method for circumventing the incompatibilities which result in the failure of males and females of different species to mate. The technique was used for this purpose by Yamane and Egashira (1924) and by Hammond and Walton (1929) who showed that insemination of domestic rabbits (*Oryctolagus cuniculus*) with hare (*Lepus timidus* and *L. europaeus*) sperm failed to result in the birth of young. However, fertilization may fail from one of a number of causes when artificial insemination is substituted for natural

* Research Career Awardee of the National Institute of Child Health and Human Development (5K6 HD 18,334).

mating even if semen is deposited successfully at the normal site. Foreign spermatozoa may fail to reach the site of fertilization, or may fail to penetrate ova if they reach them, or penetration may occur without subsequent fertilization.

We are concerned here with the observations regarding fertilization and embryonic development following artificial insemination of a number of species with foreign spermatozoa.

(a) Cottontail rabbit (*Sylvilagus floridanus* and *S. transitionalis*) × Domestic rabbit (*Oryctolagus cuniculus*). Table 1 summarizes data from two series of experiments (Chang and McDonough, 1955; Chang, 1960). It shows that not more than 30 per cent of rabbit eggs were truly fertilized following insemination with cottontail spermatozoa. The rate

Table 1

INSEMINATION OF DOMESTIC RABBITS WITH COTTONTAIL RABBIT SPERM

Method of Insemination and No. of Sperm/Rabbit	Time of Examination (hrs. after insemination)	No. of Eggs Examined		Stages of Fertilized Eggs	No. of Extra-Sperm on the Fertilized Eggs
		Total	Fert.		
Vaginal, about 150 million	14–24	36	11	11: from the enlargement of sperm head to pronuclear chromosomes	0–2
	27	21	7	7: 2-cell	0–1
	48	32	12	3: 4-cell 5: 8-cell 4: 12-cell	
	168	24	8 (34%)	7: 4-cell to morulae 1: deg. early blastocyst	
Uterine, 60 million	14–26	48	12 (25%)	1: enlarged sperm head 10: pronuclear 1: 2-cell	0–100
Tubal, 10 million	14–26	39	10 (26%)	1: enlarged sperm head 8: pronuclear 1: 2-cell	1–13
Total		200	60 (30%)		

Combination of data from Chang and McDonough (1955) and Chang (1960).

of fertilization was not improved either by uterine or tubal insemination suggesting that the low fertilization rate is not necessarily due to failure of sperm transport. In a few eggs the number of sperm attached to the eggs was greatest when spermatozoa were deposited in the uterus but even here fertilization was not improved. First cleavage did not occur until 27 hours after insemination. The first cleavage of rabbit eggs occurs at 22.5 hours after mating (Gregory, 1930) and about 24 hours in the rabbit × hare (Table 2).

(b) Hare (*Lepus europaeus*) × rabbit. Adams (1957) recorded 97.7 per cent of eggs fertilized in rabbits inseminated with hare sperm. In a more detailed study, Chang and Adams (1962) examined eggs recovered 14 to 168 hours after insemination and found virtually all eggs fertilized. The fertilized eggs divided twice but development failed subsequently at various stages. Of 54 eggs examined, 14 were blastocysts when recovered 4 to 7 days after insemination but none developed further (Table 2).

Treus and Steklenjov (1961) found only 2-celled eggs following insemination of European hares with rabbit sperm.

Table 2

INSEMINATION OF DOMESTIC RABBITS WITH EUROPEAN HARE SPERM

Time of Examination (hrs. after insemination)	No. of Eggs Examined	Stages of Eggs	No. of Extra-Sperm on the Egg
14–22	31	31: from the enlargement of sperm head to cleavage	0–20
24–31	26	2: 1-cell (Unfer. ?) 12: 2-cell 12: 4-cell	1–18
49–73	55	7: 4-cell 13: 6–8-cell 10: 12–16-cell 25: Morulae	0–21
96–168	58	1: 6-cell 1: Morula 42: Smooth morulae 14: Early blastocysts	

Combination of data from Adams (1957) and Chang and Adams (1962).

(c) Snowshoe hare (*Lepus americanus*) × rabbit. Chang, Marston, and Hunt (1964) found a very high proportion (96%) of rabbit eggs fertilized by Snowshoe hare spermatozoa. Ten of 20 eggs from rabbits inseminated with hare semen reached the blastocyst stage; only 10 per cent of Snowshoe hare eggs were fertilized in the reciprocal cross (Chang, 1965c).

When rabbits were inseminated with a mixture containing equal numbers of Snowshoe hare and rabbit spermatozoa, the evidence was that in five rabbits all the ova were fertilized by hare sperm. Two rabbits had ova fertilized only by rabbit spermatoza and in five rabbits some ova were fertilized by rabbit spermatozoa and some by hare spermatozoa. There was here, therefore, no evidence of preferential fertilization of rabbit eggs by rabbit spermatozoa. Transportation and capacitation of hare spermatozoa occurred without delay in the rabbit tract.

(d) Ferret (*Mustela furo*) × mink (*Mustela vison*). Ferrets cannot be inseminated successfully by the vaginal route (Hammond and Walton, 1934) but intra-uterine insemination is effective and ovulation can be induced by the injection of HCG (Chang, 1965a). With this technique, ferrets were inseminated with mink epididymal sperm at varying times before and after ovulation. The animals were examined $2\frac{1}{2}$ to 42 days after the ovulating injection (Chang, 1965b). The morphological features of the egg during fertilization and the time sequence of sperm penetration and first cleave were not different from those observed in eggs from pure ferret matings.

However, the proportion of ferret eggs fertilized by mink sperm was low (12% to 23%). The fertilization rate was higher (41%) following insemination 12 hours after ovulation, than when the ferrets were inseminated at the time of ovulation or 12 or 30 hours previously. This may be evidence that the block to the entry of heterologous spermatozoa is less effective in older eggs as in the case of homologous spermatozoa (Austin and Braden, 1953). The percentage of eggs fertilized was also higher following insemination with 38 million spermatozoa (38% of eggs fertilized) than that recorded following insemination with 9 to 10 million spermatozoa (17% eggs fertilized). The cleavage rate of ferret × mink eggs was found to be lower than either the pure ferret or pure mink eggs. Delayed implantation, which occurs in mink, was not observed.

Increasing the number of spermatozoa inseminated appears to increase the proportion of eggs fertilized (Table 3). In one experiment (Chang, *unpublished*) different males were used to inseminate different uterine horns within the same ferret; there was no evidence of any difference in fertilization rate between the two horns.

With the reciprocal insemination, none of 96 mink eggs was fertilized (Chang, *unpublished*), but the result is not regarded as conclusive in view of the difficulties experienced in inseminating mink and in view of the poor quality of ferret sperm at the onset of their breeding season.

Table 3

DEVELOPMENT OF FERRET EGGS FERTILIZED BY MINK SPERM

Seasons	No. of Ferrets Used	No. of Sperm Insemin. in Million/ferret (Date of Insem.)	Time of Examination, Days After Ovulation Injection (Stages of development)				
			2½ (Fertilized)	4 (Cleaved)	6-13 (Morulae Blastocysts)	22-23 (Implantation Site)	38-50 (Degenerate Embryos)
1963	8	Low (4/15–4/22)	4/20 [1] (20%)	—	0/18 [2] (0%)	—	1/54 [2] (2%)
1965	18	9–38 (3/23–4/7)	16/69 [1] (23%)	5/29 [1] (17%)	6/51 [1] (14%)	0/30 [2] (0%)	6/24 [2] (25%)
1966	6	28–58 (3/25)	—	—	23/36 [2] (64%)	11/24 [2,3] (46%)	16 [2]/36 [2] (51%)

[1] Based on number of eggs examined.
[2] Based on the estimated number of corpora lutea.
[3] Some embryos are degenerating.
Combination of data from Chang (1965b) and unpublished results.

In these experiments, cleavage of eggs was shown to be due neither to parthenogenetic activation nor to gynogenetic development resulting from stimulation by a foreign spermatozoon. Evidence of normal fertilization was provided by the presence of spermatozoa in the perivitelline space, enlargement of the sperm head in the vitellus, presence of a male pronucleus and of a second polar body in many instances.

(e) Goat (*Capra hircus*) × sheep (*Ovis aries*). Warwick and his collaborators (Warwick, Berry and Horlacher, 1932, 1933, 1934) showed thirty years ago that conception readily occurs when goats are inseminated with sheep semen but they found virtually no evidence of fertilization with the reciprocal mating. Lopyrin and Loginova (1963) found no eggs fertilized in 6 ewes inseminated with goat semen into the Fallopian tube although inseminations of goats with ram semen by the same route resulted in fertilization. Following observations by Bratonov and Dikov (1962) which will be referred to again, the problem was reexamined by Bowerman and Hancock (1963) and it was confirmed that fertilization readily occurs in goats inseminated with ram semen. Of 18 ova recovered from goats inseminated with the semen of one of four different breeds of rams, 12 were fertilized. Eggs were fertilized by all four breeds of ram. Further evidence of the efficiency of fertilization is provided by data on the pregnancy rate in goats following insemination with ram semen. In a recent experiment (Hancock, *unpublished*) 29 goats were inseminated and 27 did not return to service within 24 days. Thirteen were examined at autopsy or at laparotomy and all were found to be pregnant.

Results on the sheep differed slightly from those recorded by Warwick and Berry (1949), a small but significant proportion of eggs fertilized were found in sheep inseminated with goat semen (Bowerman and Hancock, 1963). No attempt was made to isolate the reason for the failure of fertilization in sheep.

It is of interest to note the high proportion of rabbit eggs fertilized by hare sperm and that of sheep eggs by goat sperm in contrast to the low proportion of eggs fertilized in reciprocal insemination. This shows that the failure of fertilization between different species is not mainly due to the difference in the genetic constitution of the sperm or the egg. Apparently the failure is due rather to genetically determined differences in the physiological characteristics of the sperm and the egg, or of the genital tract.

Bratanov and his collaborators (1962) have claimed that in the sheep the chances of fertilization are increased by repeated insemination of the anoestrous ewe with goat semen before the effective insemination or by pretreatment of the goat spermatozoa with ram seminal plasma. Recent experiments have failed to confirm this observation (Buttle and Hancock, *unpublished*).

The survival to term of the horse × donkey hybrid is a notable exception to what is probably the more usual fate of hybrid embryos. In many species even when fertilization occurs, the embryo consistently fails to develop to term and in this situation the hybrid embryo can serve as a most valuable reproducible model of embryonic death and is, therefore, of particular interest in relation to the problems of reproductive failure in man and domestic animals.

The specific requirements for successful implantation are still unknown, but interspecies transfer of embryos shows that although the foreign uterus may provide an adequate environment before implantation, the conditions for implantation and successful development are much more exacting.

Tarkowski (1962) has reviewed some of the information available about the fate of embryos transferred to mothers of a different species. Briones and Beatty (1954) transferred eggs of 9 of the possible 12 combinations between mouse, rat, guinea pig, and rabbit but did not study the whole of the pre-implantation period. Successful development of sheep and goat embryos following reciprocal transfer between the two species was recorded by Warwick and Berry (1949). Similar findings have been recorded by Lopyrin, Loginova and Karpov (1951), and by Buttle and Hancock (1966). Averill, Adams and Rowson (1955) showed that 8-cell sheep eggs will develop to the blastocyst stage in the rabbit uterus and that they will subsequently develop to term if transferred back to sheep. Tarkowski (1962) has studied in detail the fate of rat and mouse embryos after reciprocal transfer between the two species. Both species of embryo evoked a typical decidual reaction in the foreign mother but neither was able to make normal contact with the mucosa. Degeneration of the epithelium was irregular and when it did occur, it appeared to do so as a result of spontaneous degeneration of the epithelium and not as a result of activity of the trophoblast. Reciprocal transfer of eggs between rabbit and ferret, however, has shown that ferret eggs can survive and develop in the tube but not in the uterus of rabbit while rabbit eggs cannot survive in the tract of ferret (Chang, 1966).

Chang and his collaborators found that rabbit hybrids invariably died before implantation. Chang (1965b), however, found evidence of considerable variation in the time of death of ferret × mink hybrids. Living embryos were recovered up to 23 days' gestation age (Fig. 1) and the maximum survival time was not determined (Chang, *unpublished*).

Warwick and Berry (1949) found that goat hybrids invariably died in the first half of gestation but they recorded finding one living hybrid 62 days' gestation age. In experiments by Hancock and his collaborators it has been found that death consistently occurs about the fortieth day of gestation. All of 13 embryos recovered 24 to 35 days after insemination

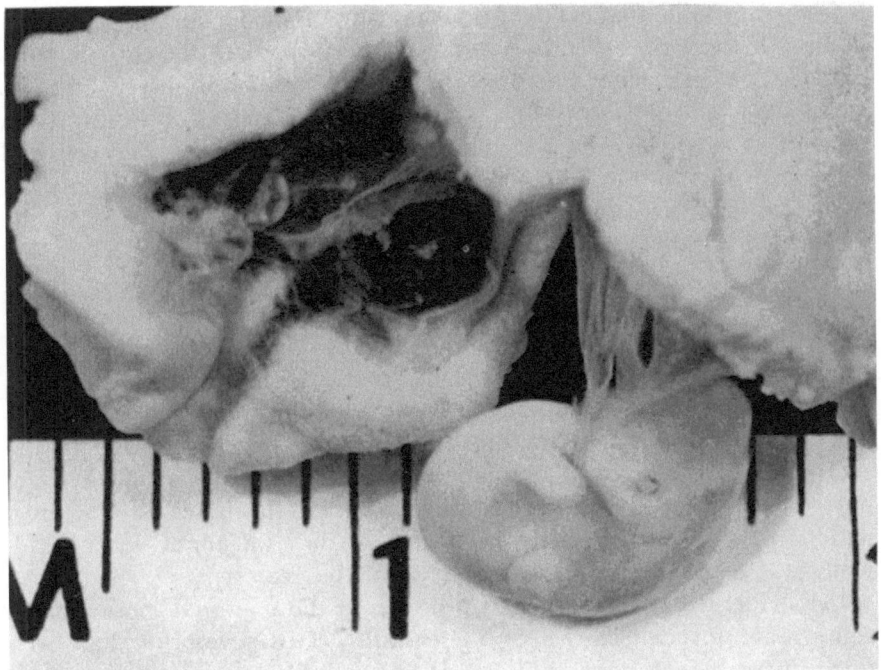

Fig. 1. A 23-day embryo developed from a ferret egg fertilized by mink sperm, showing hemorrhagic placenta.

were alive and all of 7 embryos recovered 39 to 42 days after insemination were dead.

In considering the cause of death of the hybrid embryo, it is evident that it may fail to survive for one of two main reasons. Either because of intrinsic defects of the embryo or because of incompatibilities between mother and foetus.

In view of the damaging effects of variations from the normal karyotype it is an obvious step to seek reasons for the poor survival of hybrid embryos in their cytological constitution. Chang and Adams (1962) noticed irregular metaphase plates in the hybrid blastocysts of hare × rabbit. Berry (1938) concluded that the death of the goat × sheep embryo was unlikely to be explained on cytological grounds. He found the chromosome number of the hybrid (57) to be intermediate between that of the goat (60) and sheep (54). He found no evidence of non disjunction. However, it seemed important to re-examine his conclusion using modern techniques.

Some uncertainty was expressed by Buttle and Hancock (1966) about the significance of the variation in the number of chromosomes recorded by them because these were counted in tissue-cultured hybrid cells, and it seemed possible that some of the variation might be due to artifacts of tissue culture. More recently counts have been made of the chromosomes of goat × sheep hybrid in fresh uncultured preparations of embryos 25–35 days old and the results clearly support Berry's conclusion. Of a total of 135 hybrid cells, 125 had 57 chromosomes. Only one cell had more than this number (Hancock and Jacobs, 1966).

So far, no information is available as regards the karyotype of the rabbit or ferret hybrids.

There is a growing body of evidence of the importance of foetal genotype in the maintenance of the foetal-maternal relationship. Of particular interest here is the finding (Bielanski *et al.*, 1955; Clegg *et al.*, 1962) that serum gonadotrophin fails to reach normal levels in mares carrying mule foetuses. An obvious source of incompatibility exists in the different mechanisms for the hormonal control of gestation in the sheep and goat. The corpus luteum is an active secretory organ throughout gestation in the goat but is unnecessary for the latter part of pregnancy in the ewe. Thus a hybrid foetus in the ewe might be at disadvantage if its placenta is predominantly of the goat type.

It is evident that the existence of a twin of the maternal species might compensate for deficiencies of these kinds. This possibility has been investigated by egg transfer and the results, so far, show no gain in survival of hybrids from twinning them with young of the maternal species. Ewes receiving sheep and hybrid eggs by transfer have produced at full term only lambs (Hancock, 1964) and goats inseminated with ram semen and receiving goat eggs by transfer have produced only goats (Hancock, *unpublished*).

We have, therefore, failed to confirm the findings of Bratanov and Dikov (1962) that hybrids will develop to term if twinned with one of the maternal species: they claimed to have produced mixed sets of twins by mating the mother to males of both species. Incidentally, Ohno (1966) has pointed out that chimaeras might occur in this sort of situation and the Bulgarian workers' observations apparently failed to exclude the possibility that the progeny of their mixed matings were sheep, goat chimaeras.

There is a growing volume of published work which has been prompted by the need to explain the privileged position occupied by the foetus, differing, as it does, in its genotype from its mother. Namely, to explain why the foetus should not suffer the usual fate of a homograft (Woodruff, 1957; Simmons and Russell, 1962). It seems possible that the privileged position enjoyed by the normal foetus is not extended to the hybrid foetus. Recent evidence suggests that the normal foetus owes its survival to an anatomical separation of trophoblast and maternal tissue (Bradbury *et al.*, 1965); clearly there is a need for precise information about the structure of the hybrid placenta.

In ferret uteri containing ferret × mink embryos, there is evidence of massive hemorrhage at the placental sites (Fig. 1). Preliminary findings from an experiment designed to compare the histopathology of hybrid and goat placentae also show that massive hemorrhages are a regular feature of hybrid placentae examined at the end of the sixth week of gestation (Hancock and Stamp, *unpublished data*). It is hoped that closer examination will provide decisive evidence for or against the view that the observed lesions are those which characterize the rejected homograft.

SUMMARY

This paper summarizes data from several experiments on cross fertilization between several mammalian species. When rabbits are inseminated with the semen of either the European hare or of the Snowshoe hare, virtually all the eggs are fertilized whereas reciprocal insemination results in few fertilized eggs. Similarly, fertilization commonly occurs in goats inseminated with ram semen but rarely occurs with the reciprocal insemination.

The hybrid rabbit eggs do not survive beyond the blastocyst stage. The proportion of eggs developed into blastocysts is higher in the Snowshoe hare × rabbit than in the European hare × rabbit and the lowest in the cottontail rabbit × rabbit.

Ferret × mink embryos implanted and were found to survive to the 23rd day of gestation.

Goat × sheep embryos were found to die about the end of the sixth week of gestation. There was no effect on survival of twinning hybrids by egg transfer with sheep embryos in sheep or with goat embryos in goats. The chromosome number of the goat × sheep hybrid is intermediate between that of the parent species. The cytological findings provide no explanation for the cause of embryonic death. The reasons for reproductive failure in attempts to cross between different mammalian species are briefly discussed.

References

Adams, C. E.: An attempt to cross the domestic rabbit (*Oryctolagus cuniculus*) and hare (*Lepus europaeus*). Nature *180*:853, 1957.

Austin, C. R., and A. W. H. Braden: An investigation of polyspermy in the rat and rabbit. Aust. J. Biol. Sci. *6*:674, 1953.

Averill, R. L. W., C. E. Adams, and L. E. A. Rowson: Transfer of mammalian ova between species. Nature (London) *176*:167, 1955.

Berry, R. O.: Comparative studies on the chromosome number in sheep, goats and sheep × goat-hybrids. J. Hered. *29*:343, 1938.

Bielanski, W., Z. Ewy, and H. Pigoniowa: Differences in endocrine secretion of mares pregnant with stallion or jack. Bull. Acad. Polonaise Sci. *111*:37, 1955.

Bowerman, H. R. L., and J. L. Hancock: Sheep ✕ goat hybrids. J. Reprod. Fert. *6*:326, 1963. (Abstr.)

Bradbury, S., W. D. Billington, and D. R. S. Kirby: Histochemical and electron microscopical study of the fibrinoid of the mouse placenta. J. roy, micr. Soc. *84*:199, 1965.

Bratanov, K. V., and D. Dikov: Fécondation entre les espèces brebis et chèvres et obtention d'hybride inter-espèces. Proc. IV Int. Congr. Anim. Reprod. The Hague IV: 744, 1962.

Briones, H., and R. A. Beatty: Interspecific transfer of rodent eggs. J. exp. Zool. *125*:99, 1954.

Buttle, H. L., and J. L. Hancock: The chromosomes of sheep, goats, and their hybrids. Res. Vet. Sci. *7*:230, 1966.

Chang, M. C.: Fertilization of domestic rabbit (*Oryctolagus cuniculus*) ova by cottontail rabbit (*Sylvilagus transitionalis*) sperm. J. exp. Zool. *144*:1, 1960.

————: Fertilizing life of ferret sperm in the female tract. J. exp. Zool. *158*:87, 1965a.

————: Implantation of ferret ova fertilized by mink sperm. J. exp. Zool. *160*:67, 1965b.

————: Artificial insemination of Snowshoe hares (*Lepus americanus*) and the transfer of their fertilized eggs to the rabbit (*Oryctolagus cuniculus*). J. Reprod. Fertil. *10*:447, 1965c.

————: Reciprocal transplantation of eggs between rabbit and ferret. J. exp. Zool. *161*:297, 1966.

Chang, M. C., and C. E. Adams: Fate of rabbit ova fertilized by hare spermatozoa. XXV Anno dalla Fondazione dello Instituto Sperimentale Italiano L. Spallanzani. 186, 1962.

Chang, M. C., and J. J. McDonough: An experiment to cross the cottontail and the domestic rabbit. J. Hered. *46*:41, 1955.

Chang, M. C., J. H. Marston, and D. M. Hunt: Reciprocal fertilization between the domesticated rabbit and the Snowshoe hare with special reference to insemination of rabbits with an equal number of hare and rabbit spermatozoa. J. exp. Zool. *155*:437, 1964.

Clegg, M. T., H. H. Cole, C. B. Howard, and H. Pigon: The influence of foetal genotype on equine gonadotrophin secretion. J. Endocrinol. *25*:245, 1962.

Gray, Annie P.: Mammalian hybrids—a check-list with bibliography. Commonwealth Agricultural Bureaux, Farnham, Royal, Bucks, England. 1957.

Gregory, P. W.: The early embryology of the rabbit. Contr. Embryol. No. 125. Carnegie Institution of Washington *21*:141, 1930.

Hammond, J., and A. Walton: Notes on ovulation and fertilization in the ferret. J. exp. Biol. *11*:307, 1934.

————: An attempt to cross hare and rabbit. J. Genet. *20*:401, 1929.

Hancock, J. L.: Attempted hybridization of sheep and goats. Proc. V. Int. Congr. Anim. Reprod. Trento. II, 445, 1964.

Hancock, J. L., and P. A. Jacobs: The chromosomes of goat ✕ sheep hybrids. J. Reprod. Fertil. *12*:591, 1966.

Lopyrin, A. I., and N. V. Loginova: Remote hybridization of animals. Anim. Br. Abstr. *22*:227, 1963.

Lopyrin, A. I., N. V. Loginova, and P. C. Karpov: The effect of changed conditions during embryogenesis on growth and development of lambs. Soviet Zootech. *6*:83, 1951.

Mayr, E.: Animal species and evolution. Cambridge: Harvard University Press, 1963.

Ohno, S.: Cited by K. Benirschke, and M. M. Sullivan: Corpora lutea in proven mules. Fertil. and Steril. *17*:24, 1966.

Simmons, P. L., and P. S. Russell: Antigenicity of mouse trophoblast. Ann. N.Y. Acad. Soc. *99*:717, 1962.

Stebbins, G. L.: The role of hybridization in evolution. Proc. Amer. Philosoph. Soc. *103*:231, 1959.

Tarkowski, A. K.: Interspecific transfers of eggs between rat and mouse. J. Embryol. Exp. Morph. *10*:476, 1962.

Treus, V. D., and O. P. Steklenjov: The hybridization of hare and domestic rabbits, 1961. Anim. Br. Abstr. *31*:510, 1963.

Warwick, B. L., and R. O. Berry: Intergeneric and intraspecific embryo transfers in sheep and goats. J. Hered. *40*:297, 1949.

Warwick, B. L., R. O. Berry, and W. R. Horlacher: Cytological and hybridization studies with sheep and goats. Rep. Texas Agric. Exp. Sta. p. 24, 1932.

———, ———, ———: Cytological and hybridization studies with sheep and goats. Rep. Texas Agric. Exp. Sta. p. 29, 1933.

———, ———, ———: Cytological and hybridization studies with sheep and goats. Rep. Texas Agric. Exp. Sta. p. 30, 1934.

White, M. J. D.: Animal Cytology and Evolution. Cambridge, England: Cambridge University Press, 1954.

Woodruff, M. F. A.: Transplantation immunity and the immunological problem of pregnancy. Proc. Roy. Soc. (London) Ser. B. *148*:68, 1957.

Yamane, J., and T. Egashira: Über Kreuzungsversuche zwischen Kaninchen und Schneehasen durch natürliche Paarung und künstliche Befruchtung. Zoolog. Mag. (Tokyo) *36*:1924.

STERILITY AND FERTILITY OF INTERSPECIFIC MAMMALIAN HYBRIDS

K. Benirschke *

*Department of Pathology, Dartmouth Medical School,
Hanover, New Hampshire*

The term Hybrid, according to the Oxford dictionary, was first used in 1601 to describe the offspring of a tame sow and a wild boar, hence also a half-breed. More specifically, hybrids are the results of crosses between different animal species and best known for their utility are the mule and the hinny, the offspring of horse × donkey matings. Hybridization between different mammalian species has been described to occur in nature but much more commonly it is the result of intentional breeding programs or it has occurred in zoos where sexual experience of various species living in the same enclosure is limited to members of other species. Descriptions of numerous hybrids have been accumulated in the admirable collective review by Gray (1954). In addition, this monograph gathers some information on attempted but unsuccessful hybridization and, also, it gives the limited information available on the fertility of these crosses.

While heterosis, *i.e.*, hybrid vigor, is expressed in some such animals, this is by no means universally true. Indeed, many crosses have produced inferior animals and a disturbed sex ratio has often been reported. What factual information on this point is available from the literature and his own experience has been summarized by Craft (1938). He finds evidence of a reduction of the heterogametic sex (in mammals this is the male) but no sound explanation for this inconstant phenomenon could be given. So far as heterosis is concerned, the best known hybrid is the mule whose hardiness and usefulness has led to the production of literally millions which have served mankind before the age of the motor industry. The beneficial results derived from judicious interbreeding of cattle have been discussed extensively by Rife (1965) and wool production can

* Supported by a grant from the National Institutes of Health, GM 10210.

be enhanced by hybridization of the American camelidae. Other hybrids (zebroids, tiger × lion, dog × wolf, etc.) have served only as curiosity items in zoos or they were exploited for scientific purposes.

It has long been of interest that some such interspecific hybrids fail to reproduce despite vigorous breeding attempts and no apparent defect in mating behavior. This is true particularly for the equine hybrids and it has been a frustration to those who wished to establish "a line" of mules. In other hybrids, no barrier to reproduction exists and it has been noted that the phenotypic disparity of the parental species does not obviously govern whether or not a hybrid is capable of reproduction. Numerous mechanisms can be envisaged on theoretic grounds which could account for these differences in performance; however, only few of these have been explored systematically in recent years. Potentially, such studies are of considerable interest. They may enhance our understanding of the taxonomic relationship among various species, give clues regarding evolutionary mechanisms and, perhaps they will lead to a better understanding of gestational and placental physiology.

One possible reason for sterility of interspecific mammalian hybrids is chromosomal and this has been well explored in lower species (Stebbins, 1958). If the parental chromosome structure is too much at variance, it is likely that synapsis at first meiotic division is incomplete and gametes fail to form. Similarly, too great a divergence of chromosomal structure and, more importantly, function (*i.e.*, enzyme and protein determination) may be an important limitation of the feasibility of the production of hybrids, let alone their gonadal function. It must be cautioned, however, that direct evidence for this is difficult to obtain in mammals, as placental and other gestational functions may also be incapable of sustaining growth of the hybrid. In this connection the controversial sheep × goat hybridization experiments (Craft, 1938) are of great interest. More recently, Berry (1938) found that 45 per cent of his goats became pregnant when inseminated by a ram; however, they all aborted before 145 days, while 5 living embryos were recovered between 30–65 days. From these embryos and sheep and goat amnions this author and Shiwago (1931) correctly determined the chromosome number of the species: goat (*Capra hircus*) 2n = 60, sheep (*Ovis aries*) 2n = 54, hybrids 2n = 57. Hancock (1965) has confirmed these findings recently and discusses the alleged success in Bulgaria of such crosses. He envisages, *i.a.*, that immunological recognition by the mother of the foreign paternal genotype may be the barrier. While trophoblast can be "recognized" if it is transplanted to a foreign host (*e.g.*, human to hamster cheek pouch), experience with equine hybrids does not lead to placental rejection and a more detailed study is needed of sheep × goat hybrids before such generalizations can be applied. Dalimier (1959) described a hybrid (ram × goat) whose intermediate characters and history favor acceptance of at least occasional success of this cross.

In a similar way, the hybridization between hares and rabbits has been of considerable interest. Two species of European hares (*Lepus europaeus* and *L. timidus*) both have 48 chromosomes and male and female hybrids are fertile (Ohno *et al.*, 1965). Dave *et al.* (1965) also find the karyotypes of two hares to be 48 and to differ appreciably from the rabbit (*Oryctolagus cuniculus*) with 44 elements. Chang *et al.* (1964) reported the first unequivocal success of hybridization between these two species (and genera) employing snowshoe hares (*L. americanus*) and the domestic rabbit. However, during cleavage and blastocyst formation the eggs degenerated and no such crosses have yet led to viability, for unknown reasons.

The purpose of this report is to review our own studies of interspecific mammalian hybrids, to present findings on some species in which hybrids have been reported frequently and to draw attention to some potentially valuable models for future study. The studies here described are principally chromosome studies on various mammals obtained from the Catskill Game Farm, from local breeders, the Washington Zoological Garden, the Antwerp Zoo and Drs. Thuline and Hard. Both short term lymphocyte cultures and fibroblast cultures from skin biopsies were employed. Occasionally, kidney tissue was available for culture. The techniques used for tissue culture have been described in some previous publications to which reference will be made.

I. Equidae

A systematic study of all available equine species (Benirschke and Malouf, 1966) has shown a remarkable diversity of chromosome numbers (Fig. 1). The Mongolian Przewalski's horse (*Equus przewalskii*), extant only in zoological collections now, possesses 66 chromosomes and the Hartmann mountain zebra (*E. zebra hartmannae*) has only 32 elements. Despite this enormous divergence, numerous hybrids among many of the equine species have been reported both by Gray (1954) and many authors since (*e.g.*, King *et al.*, 1966). We have studied mules and hinnies (2n = 63), a grant zebra (*E. burchelli boehmi*) × donkey (*E. asinus*) hybrid (2n = 53), a donkey × *E. zebra* hybrid (2n = 48, *v.i.*), the hybrid between a Przewalski's horse × domestic horse (*E. caballus*) (2n = 65) reported by Koulischer and Frechkop (1966) and a fertile male "Tarpan" (2n = 65), an example of selective rebreeding of this extinct species. In all of these animals the diploid number of the hybrid was the expected sum of the haploid numbers of the parental species, with the notable exception of the animal erroneously referred to as donkey × Grevy zebra (*E. grevyi*) hybrid (Benirschke *et al.*, 1964; Benirschke, 1964). This hybrid from

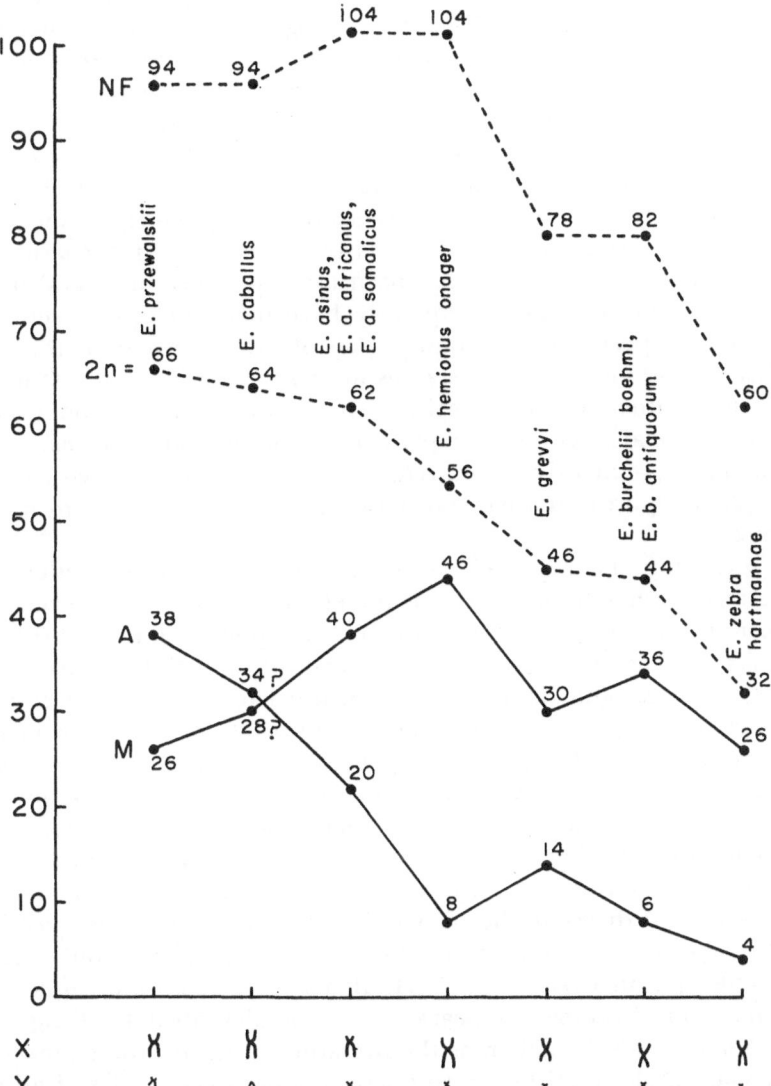

Fig. 1. Graphic representation of chromosome findings in various species of equidae. NF = nombre fondamental, *i.e.*, the total number of major chromosome arms, presumably an indicator of taxonomic relationship; 2n = diploid number of chromosome set; A = acrocentric elements; M = metacentric elements. The connecting lines are not to indicate direct evolutionary relations. At least two specimens of each species were studied; the results of the Onager are those kindly supplied by Dr. J. L. Hamerton (pers. comm., 1966). Inasmuch as the NF is not constant, it is inferred that complex chromosomal rearrangements have taken place during the evolution of this family. (Modified from Benirschke *et al.*, 1966.)

the Manila Zoo had (on repeated study) 48 elements instead of the expected 47. It is inferred from the photograph that the mother zebra is an *E. zebra hartmannae* for which the diploid number is 2n = 32 (confirmed by J. L. Hamerton, pers. comm., 1965). These parents could not be studied directly and the two possibilities still exist, namely (a) that the zebra mother (or, less likely the donkey father) has an aberrant karyotype or, (b) that it is an *E. zebra zebra,* a species closely similar to Hartmann's zebra (considered a subspecies) and one which is now so rare that it has not been studied directly as yet. This point is of considerable general importance. Firstly, it has been the experience of several investigators that hybrid animals frequently have more aneuploid cells than the parental species and, secondly, possible variations of chromosome number among members of a species must be borne in mind. Often the karyotypic delineation of a species is the result of study of only one or very few animals. As more experience is accumulated, translocations become more frequently recognized, as is of course best known from the many phenotypically normal translocation carriers in man with only 45 chromosomes.

In reviewing the reported equine hybrids and, in particular those whose fertility has been tested adequately, it appears that chromosome number and structure is the most likely factor governing the hybrid's performance at gametogenesis. Crossbreeding of the Nubian ass (*Equus asinus africanus*) and donkey with identical karyotypes produces fertile offspring and the same may be said of the Burchell zebras. Unfortunately, only one of the Asiatic hemiones has been studied and more experience with this group is desirable. In the absence of any apparent physiological or other barrier in the production of a large variety of other interspecific equine hybrids, their gametogenesis is often significantly disturbed (King *et al.,* 1966). So far as mules are concerned, this point has been reviewed by us in several direct studies. Two alleged female mules with offspring could be shown to be odd-looking donkeys by chromosome studies rather than mules (Benirschke *et al.,* 1964; Benirschke, 1965) and mule testes have the same histologic appearance as that described by King *et al.* (1966) for zebroids. Whether, as the literature contends, there are in fact fertile mare mules is still open to question and direct studies of all such animals are highly desirable. There are, of course, several theoretic possibilities for such an occurrence and to prove any one of these would be of great interest. As of the moment, I am skeptical of the older reports, despite the recent publication by Bratanov *et al.* (1964) of a few mule spermatozoa which is not supported by chromosome studies. Phenotypically, this cross is so diverse that reliance cannot solely be placed on this description. It is also pertinent to mention here that mules have an apparently undisturbed ovarian endocrine function and produce normal corpora lutea frequently (Benirschke and Sullivan, 1966). From this finding and the generally azoospermic semen found in mule ejaculates it might be inferred that mare mule fertility is a greater possibility than spermatogenesis. It must be pointed out, however, that primordial ova were extremely uncommon in the 94 ovaries examined. In this study we suggested that the female germ cells may have incited the development of granulosa cells which later responded to pituitary gonadotropins normally, while the germ cell itself died in meiosis. Unfortunately, ovaries from fetal or young mules have never been described, a study which should be undertaken by those having access to such relatively common material.

The only equine hybrid with attested fertility whose parental chromosome number and structure differed is the Przewalski's × domestic horse cross (Frechkop, 1964). This animal had the expected 65 chromosomes (Koulischer and Frechkop, 1966) and produced a viable quarter hybrid. Spermatogenesis is readily identified in the testicular biopsy (Fig. 2) and can be explained if one accepts the hypothesis that a one-step translocational event is the basic difference in the chromosome sets of the parental species. At present not enough studies have been conducted to define

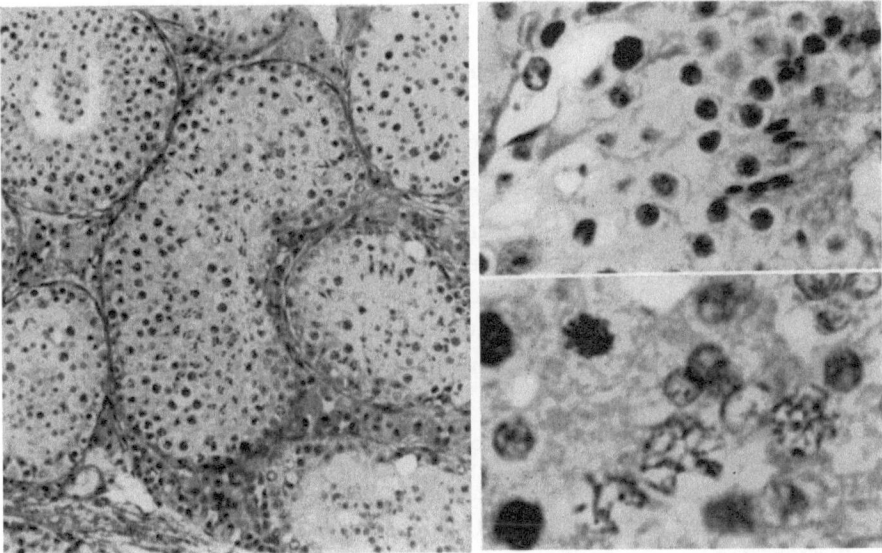

Fig. 2. Sections of testis of hybrid *E. przewalskii* × *E. caballus* (Koulischer & Frechkop, 1966) showing active spermatogenesis. Quarter hybrids have been produced by this animal with *E. caballus*. This hybrid has 65 chromosomes. Normal tubules and interstitial cells at left (H & E × 160). At right top secondary spermatocytes and spermatozoa are apparent; lumen is to the right (H & E ×640). At right bottom several pachytene figures are seen (H & E ×960). We are grateful to the Antwerp Zoo for allowing our study.

the chromosomes involved but in analogy with other and similar hybrids reported (*e.g.*, the pig × boar, McFee *et al.*, 1966) this assumption is reasonable.

II. Ursidae

Several hybrids in this family are listed by Gray (1954) and we have had occasion to study one hybrid whose parents are now listed as belonging to different genera: *Thalarctos maritimus* Phipps (Polar bear) × *Ursus arctos middendorffi* Merriam (Alaskan brown bear). This female animal (Wash. Zoo #20013) died at age 6 months, two litter mates surviving and producing cubs. This cross has been shown to be capable of reproduction. This hybrid bear and a polar bear studied had 74 chromosomes as their diploid sets (Table 1) and, despite marked phenotypic differences, the karyotypes are virtually indistinguishable one from another. Indeed, no significant differences can be detected from karyotypes prepared from skin cultures of a Tibetan bear (*Selenarctos thibetanus*) which we received from the Antwerp Zoo (Figs. 4–7). This morphologic homology among bears suggests perhaps a closer relationship than may be inferred from their taxonomic position, a point also suggested to us in various letters received after publication of the initial findings (Low *et al.*, 1964). Parenthetically, we performed autoradiographic studies in three of these tissue cultures which delineates the

Table 1

DISTRIBUTION OF CHROMOSOME COUNTS OF RANDOMLY CHOSEN METAPHASES FROM FIBROUS TISSUE CULTURES OF FOUR DIFFERENT BEARS [There is a steep mode of 74 chromosomes which is considered representative of the diploid number for these five species (counting the hybrid as representing two species)]

	Total Number of Chromosomes per Metaphase				
	71	72	73	74	75
Ursus americanus (Black bear ♂)	1	2	10	34	1
Thalarctos maritimus (Polar bear ♀)	4	3	5	23	1
Selenarctos thibetanus (Tibetan bear ♀)	1	3	6	28	0
Hybrid bear (*Thalarctos maritimus* × *Ursus middendorffi* ♀)	1	3	6	28	0

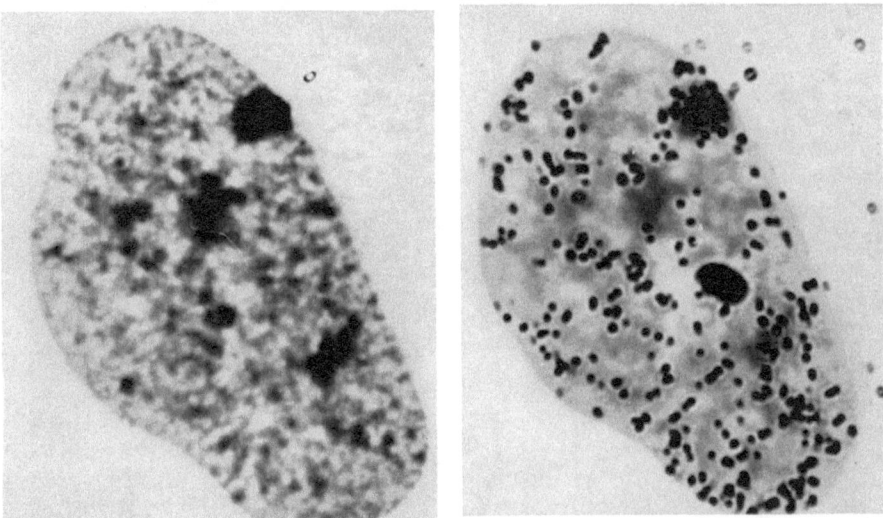

Fig. 3. Nucleus of fibrous tissue cell culture of female bear hybrid in Fig. 6 showing sex chromatin (Barr) body before (left) and after (right) autoradiography employing H_3-thymidine as DNA precursor. The Barr body is "late-replicating" in analogy with the one late-labeling X-chromosome. Aceto-orcein, ×1500.

X-chromosome by late labelling of one element (Figs. 5, 6), and shows heavy grain density also over the Y of one species (Fig. 4). Moreover, the Barr body is readily identified in sections of the three female animals and shows late replication (Fig. 3) as expected in analogy to other mammals.

III. CAMELIDAE

Our experience with various members of the Asiatic and South American camelidae is given in Table 2. As will be seen, all have 74 chromosomes and by present techniques it is not possible to differentiate the karyotypes of these animals morphologically. Hungerford and Snyder (1966) find similar karyotypes of guanaco (*Lama huanacus*) and Bactrian camel (*Camelus bactrianus*), while Capanna and Civitelli (1965) report a 2n = 72 for vicuna (*Lama vicugna*) and guanaco but 2n = 74 for the llama (*Lama glama*). The reason for these discrepancies is not immediately apparent but warrants more extensive study of animals of different origin. A representative karyotype of this family is shown in Figure 8, that of a male guanaco. It is interchangeable with the other members of the family.

This family is of interest because many crosses of the South American members have been reported, they are usually fertile and some are

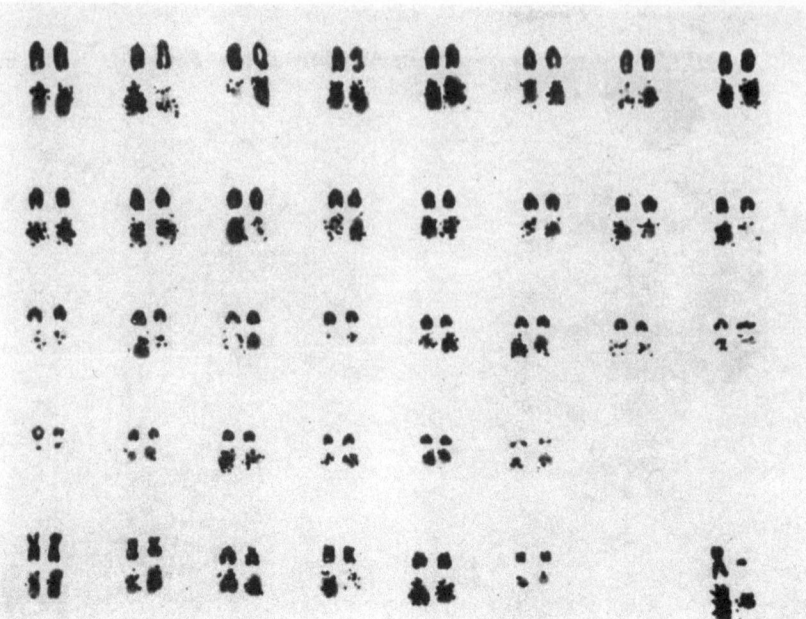

Fig. 4

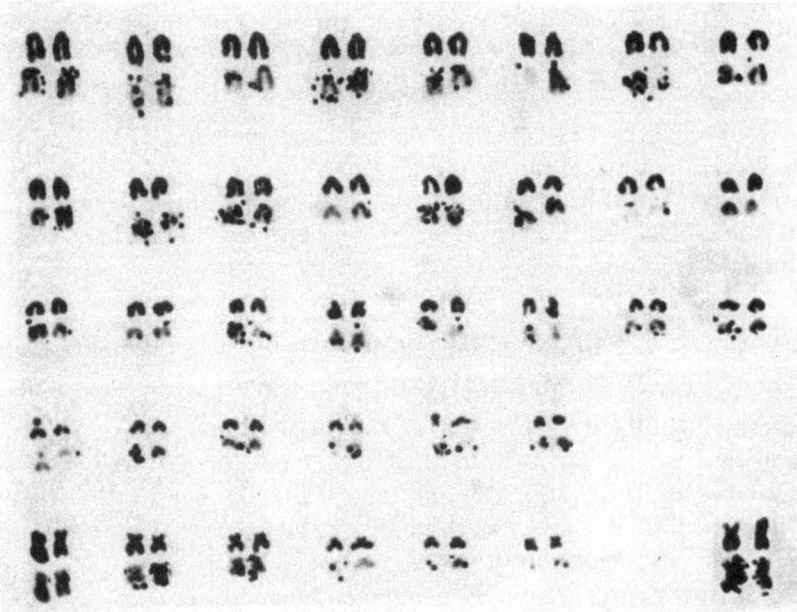

Fig. 5

Figs. 4–7. Karyotypes of black bear male (*Ursus americanus*) **Fig. 4**, polar bear female (*Thalarctos maritimus*) **Fig. 5**, hybrid female of Polar bear and Kodiak bear (*Ursus middendorffi*) **Fig. 6**, and female Tibetan bear (*Selenarctos thibetanus*) **Fig. 7**. All have 74 chromosomes and, by thymidine late replication studies (bottom line of first three species), the X-chromosome is defined. By visual inspection it is not possible to distinguish these karyotypes, even at higher magnification.

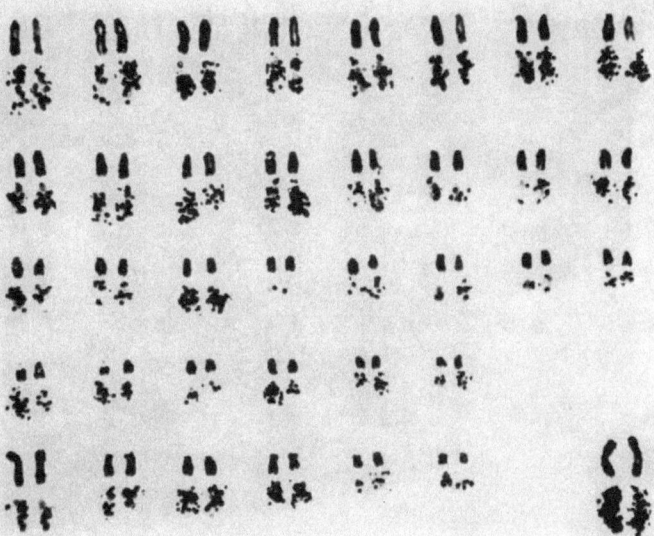

Fig. 6

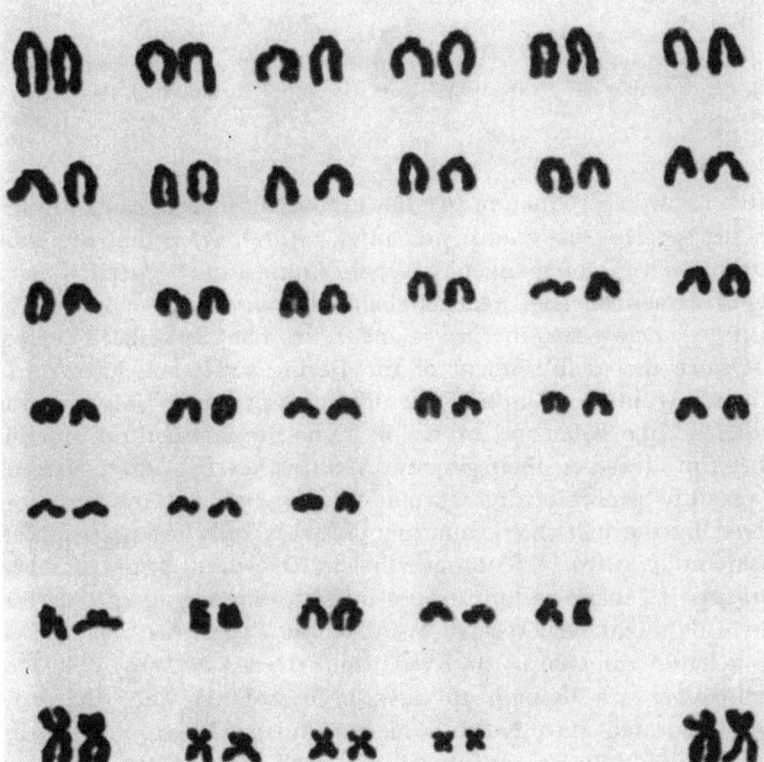

Fig. 7

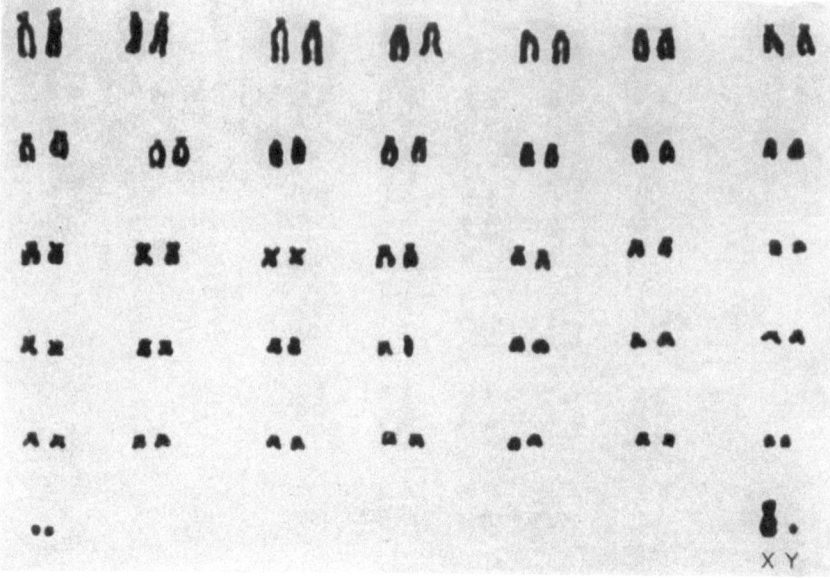

Fig. 8. Karyotype of guanaco (*Lama huanacus*) with 74 chromosomes. Morphologically these elements cannot be distinguished from the other members of this family and the bactrian camel.

exploited for wool production (see however Krumbiegel, 1952). Similarly, the Asiatic species cross readily and most authors agree that the hybrids are fertile. In view of many physiologic similarities between these two groups of camelidae (*e.g.,* red cell behavior, structure, etc., see Krumbiegel, 1952) a close relationship is apparent. They have been separated at least since the establishment of the Bering strait and have pursued their own evolution, adapting to different pressures (Moody, 1962). Nevertheless, like polar and brown bear, no gross structural differences have been produced in their genome. Would they hybridize? Size differences possibly prohibit direct attempts and artificial insemination is hampered by the fact that ovulation probably only follows copulation (Shalash and Nawito, 1965); nevertheless, it would be a challenging experiment. If generalizations are permissible one would speculate that such hybrids might be fertile, not unlike the experience reported with Wapiti (*Cervus canadensis*) and red deer (*Cervus elaphus*). These two Eurasiatic deer are thought to have been isolated since the Wapiti ancestors migrated into North America during Pleistocene (Colbert, 1955). When both were introduced to New Zealand the two species became sympatric and now hybridize freely, producing fertile offspring (Howard, 1965).

Of parenthetic interest in this connection is the distinct similarity of

Table 2

DISTRIBUTION OF CHROMOSOME COUNTS OF RANDOM METAPHASES FROM LYMPHOCYTE CULTURES OF THE SOUTH AMERICAN CAMELIDAE AND THE BACTRIAN CAMEL [All have 74 chromosomes and the karyotypes appear indentical and do not differ from those described by Hungerford and Snyder (1966). The methodology of blood culture in this family requires at least twice the volume of hypotonic solution]

	Total Number of Chromosomes per Metaphase				
	72	73	74	75	76
Alpaca *(Lama pacos)*	2	1	14		
Guanaco *(Lama huanacus)*	1	7	102	2	1
Vicuna ♂ *(Lama vicugna)*	1	1	5		
Vicuna ♀ *(Lama vicugna)*	1	2	10		
Llama *(Lama glama)*	2	3	25		
Alpaca × vicuna hybrid		1	3		
Bactrian camel *(Camelus bactrianus)*	1		9		

type and mobility of the glucose-6-phospate dehydrogenase of the American camelidae while the dromedary differs in one band (Fig. 9) when studied by starch gel electrophoretic techniques. In other species this technique affords an easy means to trace interspecific hybrids (Ohno *et al.,* 1965; Mathai *et al.,* 1966).

IV. CANIDAE

Dog *(Canis familiaris)* and coyote *(Canis latrans)* have identical karyotypes (Benirschke and Low, 1965), 2n = 78, and the chromosome structure of the following animals cannot be differentiated from the above: Jackal *(Canis aureus,* Ranjini, 1966), wolves *(Canis lupus,* Hungerford and Snyder, 1966; *Canis niger,* T. C. Hsu, pers. comm., 1966), dingo *(Canis dingo,* Valenti and Levy, 1965).

Hybridization among these various species has often been described (Gray, 1954) and Herre (1965) discusses the fertility of Jackal × coyote hybrids, among others. He doubts dog × coyote hybrids; however, a good description of a litter of three such bastards is given by Dice (1942). There still is doubt regarding the fertility of the latter cross (not that of the others) and Dice discusses the histologic picture of his male hybrid's testis. No spermatozoa were present at age 6½ months and a first meiotic divisional disturbance was found which, on description, is similar to that

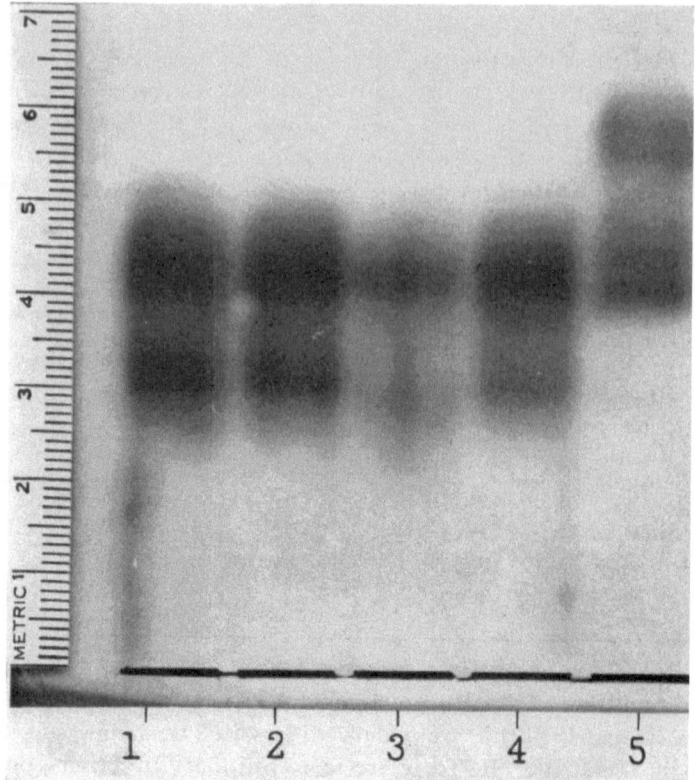

Fig. 9. Ascending starch-gel electrophoresis of red blood cell homogenate using the buffer system of Shows *et al.* (1964). Starting line at bottom. Development for G-6-PD, pH 7.6, 16 hours at 3.5–5v/cm. 1 = male alpaca (*Lama pacos*), 2 = male llama (*Lama glama*), 3 = male vicuna (*Lama vicugna*), 4 = male guanaco (*Lama huanacus*), 5 = female dromedary (*Camelus dromedarius*).

of mules. While there is then general agreement between the fertility of hybrids in canidae and chromosome structure, the dog × coyote cross needs further detailed study.

V. CERVIDAE

Of the two American deer, the white-tailed deer (*Odocoileus virginianus borealis*) and the mule deer (*O. hemionus*), we have recently identified identical chromosome sets (2n = 70) and fertile interspecific hybrids have been reported (Wurster and Benirschke, 1966). The diploid number of 68 is characteristic for fallow deer (*Dama dama*), the elk (*Cervus canadensis*), the red deer (*Cervus elaphus*) and sika deer (*Cervus nippon nippon*) with similarity of karyotypes (Gustavsson and Sundt,

1965; T. C. Hsu, pers. comm., 1966). Elk and red deer produce fertile hybrids (Gray, 1954). Other hybrids in this group are referred to but their fertility has not been tested, and further, there are deer (*Axis axis* and Père David) which cross with elk and red deer respectively, producing fertile offspring whose karyotype has not been determined. Here is a large number of species, represented in many zoological gardens, whose karyotypes need to be studied and in which experimental hybridization could readily be investigated.

VI. Other Families

In this field, burgeoning with new information, many other studies could be cited from the literature of species in which interspecific crosses are relatively common. Thus, Hsu *et al.* (1963) describe nine species of felidae, seven having 38, two 36 chromosomes. All are very similar, also, to other species of this family more recently reported. An enormous mass of information has accumulated with recent techniques in rodents, in primates, in cattle and their relatives, to name a few. These findings are beyond the scope of this brief consideration of our own studies but they merit attention in attempts at generalizations concerning interspecific hybrid reproductive performance.

Conclusions and Outlook

Aside from the various barriers which exist in the production of interspecific hybrids, their eventual reproductive behavior has been of considerable interest. Some investigators have suggested that such tests (fertility or infertility) should be used ultimately to delineate species when taxonomic problems arise in this regard (see discussion by Craft, 1938). That species designation is a difficult and arbitrary choice at best has been discussed in lucid detail by Mayr (1963) to which the interested reader is referred.

From this review of chromosome studies in various large mammals few general conclusions can be drawn. It is apparent that, in general, infertility of the hybrids is a rule when significant differences of parental chromosome sets exist, while minor rearrangements are capable of being overcome. The failure of gametogenesis appears to be largely due to asynapsis during first meiotic division and is best explored in equine hybrids. In some interspecific crosses the morphological similarity (or identity) of the parental chromosomes does not necessarily forecast hybrid fertility, particularly of males (*viz.* tiger × lion, dog × coyote; cattle hybrids with bison, yak, gayal) and this aspect of hybrid sterility is at present very poorly understood. Of all these studies it must be said that far too few well-documented hybrids have been examined by critical and modern techniques and a large field of investigation has been opened by technical advances.

The remarkable findings of Clegg *et al.* (1962) of a lowered PMSG-secretion in mares pregnant with a mule and other effects of fetal geno-type on pregnancy have been discussed by Holm (1966). The obscure mechanism underlying this effect and, *i.a.*, that of goat × sheep and rabbit × hare hybrid abortion, undoubtedly will stimulate detailed inquiries from which much might be learned about normal gestational physiology.

SUMMARY

Numerous interspecific mammalian hybrids have been reported. Many attempts at hybridization have been unsuccessful, others have led to abortion and many, but not all, hybrids are sterile. Sterility is more common in male (heterogametic) hybrids.

These factors are examined in the light of personal experimental studies of the chromosome complement in Equidae, Ursidae, Camelidae, Canidae and Cervidae. In many cases, asynapsis at meiosis due to grossly divergent parental chromosome structure can be held responsible for the sterility. There are some notable exceptions, however, which are discussed. Remarkable difference in chromosome numbers, as in members of the equidae, does not prevent hybridization, while in other families a much more similar karyotype of parental species is found, yet successful pregnancy does not take place. The karyotypes of several mammalian species known to hybridize are presented for the first time and it is suggested that detailed study of such mammals may yield important insight into reproductive pathophysiologic events.

References

Benirschke, K.: Corrigendum. Chromosoma *15*:300, 1964.

———: A second alleged fertile mare mule. In Fetal Homeostasis. Vol. I: Proc. First Conf., R. M. Wynn, ed. New York: N.Y. Acad. Sc., 1965.

Benirschke, K., and R. J. Low: Chromosome complement of the coyote, *Canis latrans*. Mammal. Chrom. Newsl. *15* (Febr.): 102, 1965.

Benirschke, K., and N. Malouf: Chromosome studies of equidae. Zool. Garten, *in press,* 1966.

Benirschke, K., and M. M. Sullivan: Corpora lutea in proven mules. Fertil. and Steril. *17*:24, 1966.

Benirschke, K., R. J. Low, L. E. Brownhill, L. B. Caday, and J. DeVenecia-Fernandez: Chromosome studies of a donkey-Grevy zebra hybrid. Chromosoma *15*:1, 1964.

Benirschke, K., R. J. Low, M. M. Sullivan, and R. M. Carter: Chromosome study of an alleged fertile mare mule. J. Hered. *55*:31, 1964.

Berry, R. O.: Comparative studies on the chromosome numbers in sheep, goat, sheep-goat hybrids. J. Hered. *29*:343, 1938.

Bratanov, K., V. Dikov, and V. K. Dokov: Recherches sur l'infecondité chez l'hybride male. 5th Intern. Congr. Anim. Reprod. and Artif. Insem. Trento, Italy. *3*:560, 1964.

Capanna, E., and M. V. Civitelli: The chromosomes of three species of Neotropical Camelidae. Mammal. Chrom. Newsl. *17* (August): 75, 1965.

Chang, M. C., J. H. Marston, and D. M. Hunt: Reciprocal fertilization between the domesticated rabbit and the Snowshoe hare with special reference to insemination of rabbits with an equal number of hare and rabbit spermatozoa. J. Exp. Zool. *155*:437, 1964.

Clegg, M. T., H. H. Cole, C. B. Howard, and H. Pigon: The influence of foetal genotype on equine gonadotrophin secretion. J. Endocrinol. *25*:245, 1962.

Colbert, E. H.: Evolution of the Vertebrates. New York: Wiley, 1955.

Craft, W. A.: The sex ratio in mules and other hybrid mammals. Quart. Rev. Biol. *13*:19, 1938.

Dalimier, P.: Les chabins, hybrides chèvres-moutons. Säugetierk. Mitt. *7*:49, 1959.

Dave, M. J., N. Takagi, H. Oishi, and Y. Kikuchi: Chromosome studies on the hare and the rabbit. Proc. Japan Acad. *41*:244, 1966.

Dice, L. R.: A family of dog-coyote hybrids. J. Mammal. *23*:186, 1942.

Frechkop, S.: Remarques concernant l'histoire et la genetique du cheval. Bull. Inst. Roy. Sc. Nat. Belg. *40*: #13, 1, 1964.

Gray, A. P.: Mammalian Hybrids. Techn. Commun. #10. Commonwealth Bureau of Animal Breeding and Genetics, Edinburgh. Bucks, England: Farnham Royal, 1954.

Gustavsson, I., and C. O. Sundt: Chromosome studies in five species of deer representing the four genera Alces, Capreolus, Cervus and Dama. Mammal. Chrom. Newsl. *18* (Nov.): 149, 1965.

Hancock, J. L.: Attempted hybridization of sheep and goats. 5th Intern. Congr. Anim. Reprod. and Artif. Insem. Trento, Italy. *3*:445, 1964.

Herre, W.: Demonstration im Tiergarten des Institutes für Haustierkunde der Universität, insbesondere von Wildcaniden und Canidenkreuzungen (Schakal/Coyoten F_1- und F_2 Bastarde sowie Pudel/Wolf Kreuzungen). Zool. Anz. *28* (Suppl.): 622, 1965.

Holm, L. W.: The gestation period of mammals. In: Comparative Biology of Reproduction in Mammals. I. W. Rowlands, ed. London: Academic Press, 1966.

Howard, W. E.: Interactions of behavior, ecology, and genetics of introduced mammals. In: The Genetics of Colonizing Species. H. G. Baker and G. L. Stebbins, eds. New York: Academic Press, 1965.

Hsu, T. C., H. H. Rearden, and G. F. Luquette: Karyological studies of nine species of felidae. Amer. Naturalist *97*:225, 1963.

Hungerford, D. A., and R. L. Snyder: Chromosomes of a European wolf (*Canis lupus*) and of bactrian camel (*Camelus bactrianus*). Mammal. Chrom. Newsl. *20* (April): 72, 1966.

King, J. M., R. V. Short, D. E. Mutton, and J. L. Hamerton: The reproductive physiology of male zebra-horse and zebra-donkey hybrids. In: Comparative

Biology of Reproduction in Mammals, I. W. Rowlands, ed. London: Academic Press, 1966.

Koulischer, L., and S. Frechkop: Chromosome complement: A fertile hybrid between *Equus prjewalskii* and *Equus caballus*. Science *151*:93, 1966.

Krumbiegel, I.: Lamas. Neue Brehm Bücherei. Heft 54. Leipzig: Geest & Portig, 1952.

Low, R. J., K. Benirschke, J. L. Grimmer, and T. G. Schneider: The chromosomes of three bears. Mammal. Chrom. Newsl. *13* (May): 3, 1964.

Mathai, C. K., S. Ohno, and E. Beutler: Sex-linkage of the glucose-6-phosphate dehydrogenase gene in equidae. Nature *210*:115, 1966.

Mayr, E.: Animal Species and Evolution. Cambridge, Mass.: Harvard University Press, 1963.

McFee, A. F., M. W. Banner, and J. M. Rary: Variation in chromosome number among European wild pigs. Cytogenetics *5*:75, 1966.

Moody, P. A.: Introduction to Evolution. 2nd ed. New York: Harper & Row, 1962.

Ohno, S., J. Poole, and I. Gustavsson: Sex-linkage of erythrocyte glucose-6-phosphate dehydrogenase in two species of wild hares. Science *150*:1737, 1965.

Ranjini, P. V.: The chromosomes of the Indian Jackal (*Canis aureus*). Mammal. Chrom. Newsl. *19* (Jan.): 5, 1966.

Rife, D. C.: Hybrids. Washington: Public Affairs Press, 1965.

Shalash, M. R., and M. Nawito: Some reproductive aspects in the female camel. 5th Intern. Congr. Anim. Reprod. and Artif. Insem. Trento, Italy. *2*:263, 1964.

Shiwago, P. I.: Karyotypische Studien an Ungulaten. I. Ueber die Chromosomenkomplexe der Schafe und Ziegen. Z. Zellf. Mikr. Anat. *13*:511, 1931.

Shows, T. B., R. E. Tashian, G. J. Brewer, and R. J. Dern: Erythrocyte glucose-6-phosphate dehydrogenase in Caucasians: New inherited variant. Science *145*:1056, 1964.

Stebbins, G. L.: The inviability, weakness, and sterility of interspecific hybrids. Advances in Genetics *9*:147, 1958.

Valenti, C., and C. Levy: The karyotype of *Canis dingo*. Mammal. Chrom. Newsl. *18* (Nov.): 147, 1965.

Wurster, D. H., and K. Benirschke: Chromosome studies in some deer, the springbok, and the pronghorn, with notes on placentation in deer. Cytologia, *in press*, 1967.

THE STERILITY OF TWO RARE EQUINE
HYBRIDS

J. M. KING

Game Department, Nairobi, Kenya

INTRODUCTION

Equine hybrids, other than the mule (ass ♂ × horse ♀) and hinny (horse ♂ × ass ♀), have not been bred in any numbers nor received much attention. The first record of a zebroid (ass ♂ × mountain zebra ♀) appears to be from Europe in 1782 (cited by Ewart, 1898). Then in the nineteenth century Captain Lugard recommended that "an attempt should be made to obtain zebra mules by horse or donkey mares, because he believed that such mules "would be found excessively hardy, and impervious to the (tsetse) fly, and to climatic diseases" (Ewart, 1896). However, the potential of the hybrids remained unexplored and they did not achieve any prominence until Ewart's telegony experiments at the turn of the century. He coined the terms "zebrule," "zebrinny," "zebryle," and "zebret," in an effort to distinguish the zebroids he produced. Now many other terms would have to be added to cover the hybrids which have been recorded between three zebra species, two horse species, and the African and Asian asses (Antonius, 1951).

Our knowledge of the reproductive physiology of the equine hybrid has been confined to the mule. The sterility of the male is notorious but there are a number of reports of mule mares that have foaled (Bonadonna, 1957). These animals could not be positively identified before the advent of modern chromosome techniques, and the only two fertile, alleged mule mares to be examined in this way have been shown to possess donkey chromosome complements (Benirschke *et al.*, 1964; Benirschke and Malouf, 1965).

It has been suggested that the sterility of the male mule is caused by inadequate pairing of the chromosomes at the meiotic stage of sperma-

Fig. 1. Hybrid (Grant's zebra ♂ × ass ♀) in the company of asses.

togenesis. This may be an acceptable hypothesis but it has been based on very little evidence, since the correct karyotype of the mule was only published in 1962 (Trujillo *et al.*), and there do not appear to be any accounts of the stages of equine spermatogenesis in the relevant literature (Gunn, 1898; Iwanoff, 1905; Wodsedalek, 1916; Kupfer, 1928; Nishikawa and Sugie, 1952; Horie and Nishikawa, 1954; Bielánski, 1955; Robert, 1956; Benirschke *et al.*, 1964).

The availability of the equine hybrids: Grant's zebra ♂ × ass ♀ (Fig. 1) and Grevy's zebra ♂ × horse ♀ (Fig. 2) presented an opportunity to test this hypothesis. Therefore, in the present study I have investigated spermatogenesis, testicular steroid production in collaboration with Dr. Short, and chromosome complement in collaboration with Mr. Mutton and Mr. Hamerton.

Material and Methods

The equine animals examined were either shot, or dart immobilized, or roped and cast by sidelines. Material was then collected from four male hybrids and the parent species for three lines of investigation: chromosome complement, studies of testicular histology, and steroid content. The experimental work has already been described in detail (King *et al.*, 1965; King, 1966).

Fig. 2. Hybrid (Grevy's zebra ♂ × horse ♀) in a herd of horses.

RESULTS

Histology of the testis

The first requirement was to characterize the normal stages of equine spermatogenesis. These are shown in Figure 3 which consists of two photographs of adjacent cell populations of the seminiferous epithelium, and is probably equivalent to a slightly tangential section of one tubule. The conspicuous stages of meiosis are particularly relevant to this study. It is also worth noting the well developed Sertoli cells with their large, elongated nuclei at right angles to the basement membrane.

The seminiferous epithelium of the hybrid (Fig. 4) does not show complete spermatogenesis. Meiosis fails at the pachytene II to diplotene stage of the primary spermatocyte. The Sertoli cell nuclei are small and only occasionally orientated at right angles to the basement membrane. The interstitial tissue is similar in the hybrids and parent species, with well developed Leydig cells (Fig. 4).

The lipid in the testis of the hybrid appears to be concentrated in the Leydig cells in the interstitial tissue. It was assumed that the method of fixation and storage resulted in loss of lipid, but by using acetone-treated control sections (washed in acetone for 15 minutes prior to staining) it was possible to show that some lipid was still present. In the testis of pure species the lipid is concentrated in the residual body debris surrounding

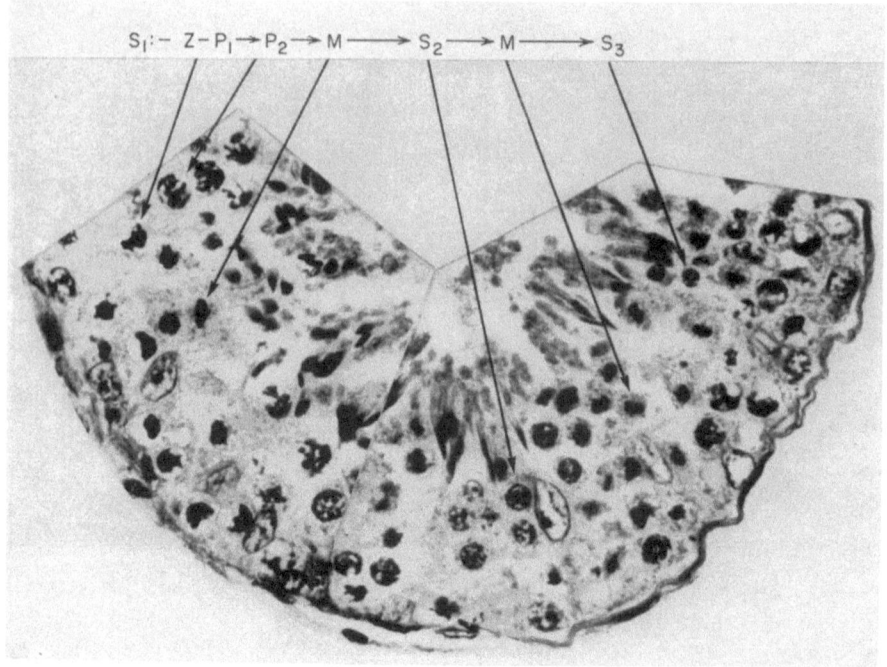

Fig. 3. The seminiferous epithelium of the Grevy's zebra showing normal equine spermatogenesis. The most conspicuous stages of meiosis are shown in series by apposing two sections from different tubules. S_1, primary spermatocyte; Z-P_1, zygotene-pachytene I; P_2, pachytene II; M, metaphase; S_2, secondary spermatocyte; S_3, spermatid. ×1143.

the spermatozoa, and in the Leydig cells of the interstitial tissue. None of the testes show lipid droplets in Sertoli cells.

Chromosome Studies

There is a wide range of chromosome numbers amongst the Equidae, from a diploid number of 32 for Hartmann's mountain zebra (Benirschke *et al.*, 1964a) to 66 for Przewalski's horse (Benirschke and Malouf, 1965). In the present study it was confirmed that the horse has a diploid number of 64, the ass 62 (Trujillo *et al.*, 1962), the Grant's zebra 44 (Benirschke and Brownhill, 1963), and the Grevy's zebra 46 (Mutton *et al.*, 1964). The hybrids possess a chromosome number that is the sum of the haploid sets of the two parent species, and so far as the morphology of these parental haploid sets is distinctive they can be separated in the hybrid karyotype. The Grevy's zebra × horse has a modal number of 55 (**Fig. 5**) and the Grant's zebra × ass a number of 53 (52–55) (**Fig. 6**). The later is a particularly interesting animal, being chromatin-positive and a presumptive sex chromosome mosaic with cell lines having X0, XX, XY, XYY, and XXYY sex chromosome constitutions (Mutton *et al.*, 1964).

It can be seen from Figures 5 and 6 that the two parental sets in each hybrid karyotype are markedly dissimilar.

Steroid Determinations in Testicular Tissue

The results are shown in the table. Testosterone was the principal steroid present in all the testes examined. The concentrations in the two mature hybrids exceeds those found in the horse and Grevy's zebra. There is, therefore, no suggestion of impaired androgen production in the zebroids, which is in agreement with the behaviour of these animals. The oestrogen production by the hybrid testis is also normal.

DISCUSSION

The relation of the three separate studies presents a fairly clear picture of the reproductive physiology of the male zebroid. There is a marked disturbance in the exocrine activity of the hybrid testis, shown by the small diameter of the seminiferous tubules and the failure of spermatogenesis. In contrast, the libido and endocrine activity, as judged by

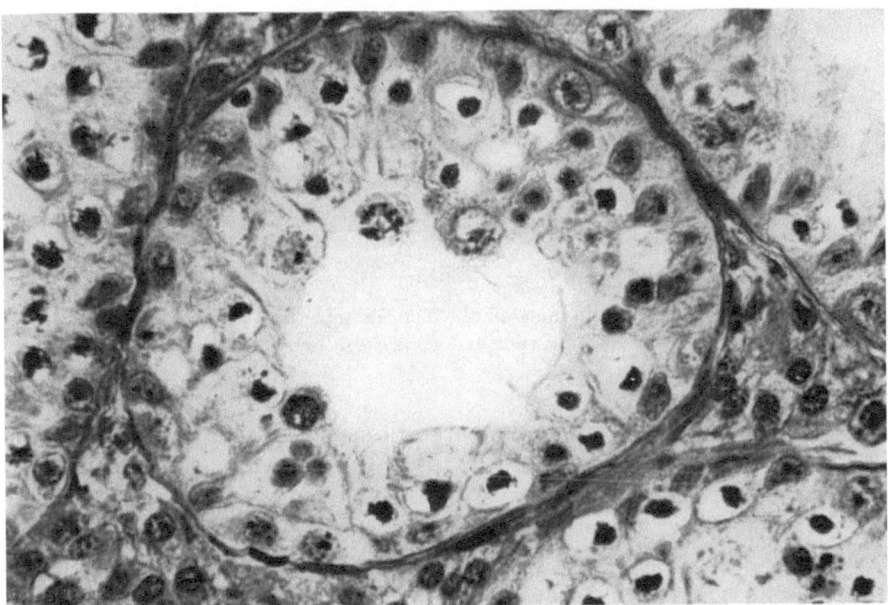

Fig. 4. Hybrid (Grant's zebra ♂ × ass ♀) testis showing normal Leydig cells in interstitial tissue, but failure of meiosis at the late pachytene stage of the primary spermatocyte. ×920.

Fig. 5

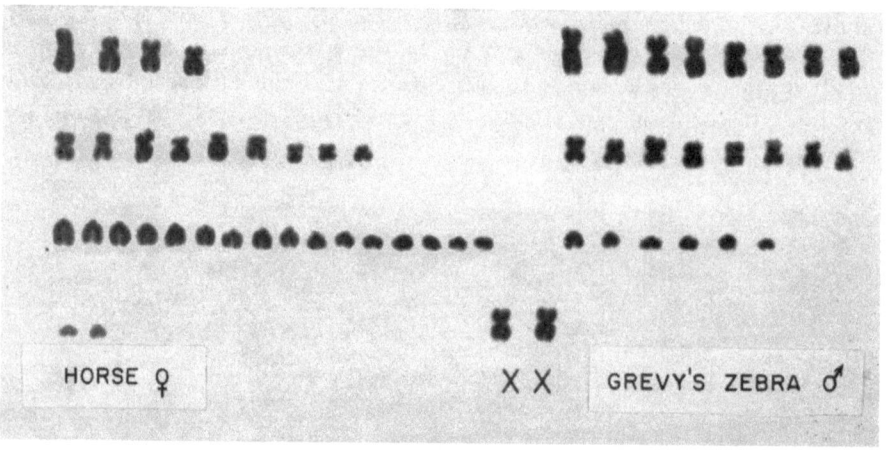

Fig. 6

Figs. 5 and 6. The chromosomes of the hybrids separated into the haploid sets of the parent species, in so far as the morphology of these parental sets is distinctive.

steroid assays performed on testicular tissue, is quite unaffected. The Leydig cells are well developed in the interstitial tissue of the hybrid and would appear to be the main source of the testicular androgens (van Oordt, 1963; Christensen and Mason, 1965). Since the oestrogen content of the testis is normal but the seminiferous epithelium poorly developed there is no reason to implicate the latter in normal testicular steroid production. The Sertoli cells of the hybrid tubule show poor differentiation and any residual lipid, shown in the Oil red 0 and haematoxylin-stained sections, is confined to the interstitial tissue. This

Steroid Activity in Horse, Zebra and Zebroid Testis

(All steroid results are corrected for extraction losses and are expressed as µg steroid/100 g testis.)

	Horse	Grant's Zebra		Grevy's Zebra		Grevy's × Horse		Grant's × Ass
Approx. age (years)	9	Mature	6	10	15	2	4	5
Testis + epididymis Wt. (g)	163	135*	475	600	550	90	100*	60*
Mean tubular diameter (µ)	189 ± 4	208 ± 2	—	192 ± 4	—	—	127 ± 3.5	126 ± 2
Progesterone	2.1	<0.1	6.9	1.8	2.6	0.4	<0.1	1.5
17 α-Hydroxyprogesterone	33.0	11.5	9.0	2.8	3.0	2.4	7.6	2.6
Androstenedione	69.0	0.9	3.8	3.2	1.5	2.4	1.4	2.2
Testosterone	205.0	424.0	154.0	41.5	59.0	34.2	211.0	268.0
Oestrone	10.2	<0.1	<0.1	0.5	<0.1	1.9	22.6	<0.1
Oestradiol-17β	11.6	14.1	13.7	7.4	4.9	0.8	44.3	5.6

* Weight estimated from testis dimensions (King, 1966).

would suggest that most of the testicular oestrogens are secreted by the interstitial tissue (Levy *et al.*, 1959) rather than the Sertoli cells (Teilum, 1950). *In vitro* incubations of rat seminiferous tubules with labelled progesterone and androgens have demonstrated that a small amount of androgen but no oestrogen synthesis can occur (Christensen and Mason, 1965). However, there is some evidence that the Sertoli cells are capable of active sterol metabolism, including small amounts of oestrogen, in association with high spermatogenic activity (Lacy and Lofts, 1965; Lacy *et al.*, 1965).

The testicular steroid production is normal in both the hybrid XY and the mosaic. This mosaicism was not confined to testicular tissue (Brøgger and Aagenaes, 1964), and may reflect the instability of the sex chromosomes *in vivo* since the somatic chromosomes show a high degree of aneuploidy (Mutton *et al.*, 1964). On the other hand it could be an *in vitro* deviation, since Sasaki and Sasaki (1962) have shown changes of somatic chromosomes of the horse in serial *in vitro* transfers.

The failure of spermatogenesis in the hybrid occurs in the meiotic division of the primary spermatocyte at late pachytene or pachytene II to diplotene. This is probably the stage at which bivalents have been formed and may be proceeding to the formation of chiasmata (Henricson and Bäckström, 1963). The meiotic block may be related to the abnormal chromosome complement of the hybrid since the morphological differences between the maternal and paternal karyotypes, associated with an extensive rearrangement of chromosome arms (Benirschke and Malouf, 1965), would cause a mechanical impediment to synapsis.

This hypothesis must also explain two apparently equivocal findings: the recovery of a few spermatozoa from the mule after centrifugation of the semen (Brantanov *et al.*, 1964); the fertility of the interspecific hybrid between the domestic and Przewalski's horse.

An explanation for the achievement of spermatogenesis in the hybrid has been provided by Ohno and Trujillo (personal communication) on examination of a chromosomally-confirmed, male hinny. They found clones of tetraploid spermatogenia, a few tetraploid meiotic figures and some spermatozoa. They suggest that by becoming tetraploid, each horse (or donkey) chromosome can pair with its true homologue, and that the tetraploid primary spermatocyte undergoes tetrapolar division to produce for secondary spermatocyes. These findings do not really affect the issue since a sperm density of $50–200 \times 10^6$/ml of ejaculate in the normal in the horse (Day, 1940) and a relatively high sperm density is required for fertility.

The chromosomes of the domestic horse (2n = 64) and the Przewalski's horse (2n = 66) are similar. This would allow the haploid sets present in a hybrid between these two species to pair during meiosis with the additional formation of a trivalent to account for the extra Przewalski chromosome (Benirschke and Malouf, 1965). This hybrid is therefore fertile (Balachov, 1961; Treus, 1963; Veselovsky and Volf, 1965).

In the light of these findings it seems likely that Darwin was misinformed when he observed in the Zoological Gardens, London, "a curious triple hybrid, from a bay mare, by a hybrid from a male ass and a female zebra" (cited by Ewart, 1896).

Acknowledgments

The field work in Kenya was made possible through the financial support of the Wellcome Trust and the Horserace Betting Levy Board. I am most grateful to Mr. Raymond Hook, of Nanyuki, for making the zebroids available to me, and to the Kenya Game Department, the East African Veterinary Research Organization, Professor A. S. Parkes and Mr. James Roberts for their help and for the provision of equipment, and to the Zoological Society of London for a biopsy specimen.

I would also like to thank Dr. Mary Hay for her advice on the study of spermatogenesis, Mrs. Iris Bavister, Miss Mary McGuire and the technical staff of the Paediatric Research Unit, for their help in the preparation of material for this paper.

I am also indebeted to the Zoological Society of London for permission to reprint Figure 3.

References

Antonius, O.: Die Tigerpferde (die Zebras). Dr. Paul Schöps, Frankfurt Main. 105, 1951.

Balachov, N. T.: Breeding of Przewalski wild horses in Askania Nova. Equus. Ed. Z. Veselovský, p. 197. Nakladatelství Ceskoslovenské Akademie Věd. 1961.

Benirschke, K., and L. E. Brownhill: The chromosomes of the Grant zebra, *Equus quagga*. Mammalian Chromosomes Newsletter *10*:82, 1963.

Benirschke, K., J. L. Low, and H. Heck: Mitotic chromosomes of Equidae. Mammalian Chromosomes Newsletter *14*:65, 1964a.

Benirschke, K., J. L. Low, M. M. Sullivan, and R. M. Carter: Chromosome study of an alleged fertile mare mule. J. Hered. *55*:3, 1964.

Benirschke, K., and N. Malouf: Chromosome studies of Equidae. Second International Symposium for the Preservation of the Feral Horse, Berlin. January 1965.

Bonadonna, T.: Possibilità per un riconoscimento differenziale tra il bardoto ed il mulo. Zootec. Vet. *2*:42, 1957.

Bielánski, W.: Observations on ovulation processes in she-mules. Bull. Acad. Pol. Sci. Cl. II Ser. Sci. Biol. *3*:243, 1955.

Bratanov, K., V. Dikov, and V. K. Dokov: Recherches sur l'infecondité chez l'hybride male. 5th International Congress on Animal Reproduction and Artificial Insemination, Trento. *3*:560, 1964.

Brøgger, A., and O. Aagenaes: Role of Y chromosome in development of testicular structures. Lancet 2:259, 1964.

Christensen, A. K., and N. R. Mason: Comparative ability of seminiferous tubules and interstitial tissue of rat testes to synthesize androgens from progesterone-4-¹⁴C in vitro. Endocrinology 76:646, 1965.

Day, F. T.: The stallion and fertility. Vet. Rec. 52:597, 1940.

Ewart, J. C.: Telegony experiments: the birth of a hybrid between a male Burchell's zebra (E. burchelli) and a mare (E. caballus). Veterinarian, Lond., 69:755, 1966.

————: On zebra-ass hybrids; with observations on the relationships of the zebras. Veterinarian, Lond., 71:185, 1898.

Gunn, W. D.: A fertile mule. Veterinarian, Lond., 71:564, 1898.

Henricson, B., and L. Bächström: Spermatocytogenesis in the boar. Acta Anat. 53:276, 1963.

Horie, T., and Y. Nishikawa: Studies on reproductive ability of mules. II. On functions of testes. Bull. Natn. Inst. Agric. Sci., Tokyo, G. 8:143, 1954.

Iwanoff, E.: Untersuchungen über die Ursachen der Unfruchtbarkeit von Zebroiden (Hybriden von Pferden und Zebra). Biol. Zbl. 25:789, 1905.

King, J. M.: Comparative aspects of reproduction in Equidae. Ph.D. Thesis, University of Cambridge, 1966.

King, J. M., R. V. Short, D. E. Mutton, and J. L. Hamerton: The reproductive physiology of male zebra-horse and zebra-donkey hybrids. In Comparative Biology of Reproduction in Mammals. I. W. Rowlands, ed. London, Academic Press, 1966.

Kupfer, M.: The behaviour of the ovary of equines during normal sexual functions. Rep. vet. Res. Un. S. Afr., 13 and 14: 1928.

Lacy, D., and B. Lofts: Studies on the structure and function of the mammalian testis. 1. Cytological and histochemical observations after continuous treatment with oestrogenic hormone and the effects of F.S.H. and L.H. Proc. roy. Soc. B 162:188, 1965.

Lacy, D., B. Lofts, G. Kinson, D. Hopkins, and H. Dott: Sertoli cells and steroid synthesis. (Abstract). Gen. and Compar. Endocr. 5:693, 1965.

Levy, H., H. W. Deane, and B. L. Rubin: Visualization of steroid-3β-ol-dehydrogenase activity in tissues of intact and hypophysectomised rats. Endocrinology 65:932, 1959.

Mutton, D. E., J. M. King, and J. L. Hamerton: Chromosome studies in the genus Equus. Mammalian Chromosomes Newsletter 13:82, 1964.

Nishikawa, Y., T. Sugie, and N. Harada: Studies on effect of day length of reproductive functions in horses. 1. Effect of day length on functions of ovaries. Bull. Natn. Inst. Agric. Sci., Tokyo, G. 3:35, 1952.

Robert, S.: Veterinary Obstetrics and Genital Diseases. Edwards, Ann Arbor, Michigan, 1956.

Sasaki, M. S., and M. Sasaki: Changes of somatic chromosomes of the horse in serial in vitro transfers. Cytogenetics 1:291, 1962.

Teilum, G.: Oestrogen production by Sertoli cells in the etiology of benign senile hypertrophy of the human prostate. Acta Endocrin. 4:43, 1950.

Treus, V. D.: Preservation of the Przewalski horse, Equus przewalski przewalski Pol., in the U.S.S.R. Int. Zoo. Yb. 4:66, 1963.

Trujillo, J. M., C. Stenius, L. Christian, and S. Ohno: Chromosomes of the horse, the donkey and the mule. Chromosoma *13*:243, 1962.

Van Oordt, G. J.: Comparative Endocrinology. Ed. U. S. von Euler and H. Heller New York, Academic Press, *1*:154, 1963.

Veselovsky, Z., and J. Volf: Breeding and care of rare Asian equids at Prague zoo. Int. Zoo. Yb. *5*:28, 1965.

Wodsedalek, J. E.: Causes of sterility in the mule. Biol. Bull. *30*:1, 1916 (cited by Benirschke, Brownhill and Beath, 1962).

DEVELOPMENTAL MALFORMATIONS AS MANIFESTATIONS OF REPRODUCTIVE FAILURE *

V. H. FERM

Department of Anatomy/Cytology, Dartmouth Medical School, Hanover, New Hampshire

Developmental malformations represent one of the most obvious manifestations of reproductive failure. This conference, dealing with a multitude of factors concerned with the comparative aspects of reproductive failure, properly focuses some attention to the overall problem of maldevelopment. Not only are developmental anomalies of interest to the experimental embryologist in his attempt to unravel the mysteries of normal development, but they are of some considerable concern to commercial animal breeders. Furthermore, the tremendous psychic and financial impact of this problem on human society is quite obvious.

The experimental teratologist is concerned with attempts to define more accurately the multitude of factors which contribute to these disasters. He is certainly aware that two important common denominators, heredity and environment, play important roles in these problems. Figure 1 is a diagrammatic attempt to place these two factors into some etiological relationship. A certain proportion of congenital malformations are probably due to a pure environmental factor (thalidomide, hyperbaric oxygen, etc.), while another group are assuredly due to specific patterns of mendelian inheritance or related to gross chromosomal abnormalities. The vast majority of congenital malformations, however, represent the subtle operation of some environmental factors upon specific genotypes. Certainly, this has been well demonstrated by the experiments of Fraser *et al.* (1954), in which two different strains of mice (A and C) respond so differently to the teratogenic effect of cortisone.

This gray zone of unexplained etiologies of specific malformations

* These studies were supported by USPHS Grant GM-10210.

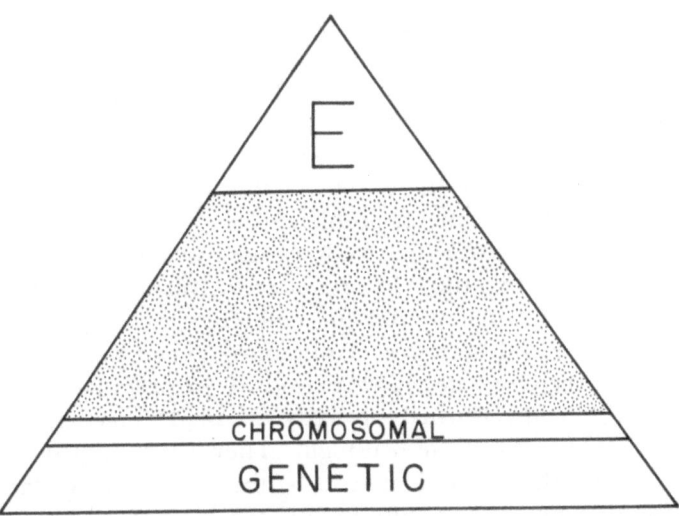

Fig. 1. A diagram to explain the relative importance of etiological mechanisms in developmental malformations. The total area of the triangle represents the total load of congenital malformations. There appear to be three primary etiological categories: environmental (E), genetic, and chromosomal. The gray stippled area represents those malformations of still unknown etiology. See text.

represents the true frontier of experimental teratology. Various other investigators (Fraser, 1963; Cohlan, 1963; Wilson, 1959) have speculated upon some so-called "principles of experimental teratology." There seems to be some common agreement that these principles can be reduced to the following general propositions:

1. The embryo is susceptible to teratogenic stimuli during the early stages of development only.
2. The genotype of the embryo influences the response to the teratogen.
3. A teratogen increases embryonic mortality.
4. The maternal system may not be noticeably affected by the teratogen.

It is not my purpose to add further proof for these statements. Rather, I would like to discuss three other important "observations" which I think are directly related to the problems of experimental mammalian teratology. These three observations deserve our close attention in the future:

1. The placental membrane plays a complex role in the etiology of mammalian malformations.
2. There is a specific teratogen:organ effect.
3. Teratogenic agents may act synergistically with other teratogens or non-teratogens.

This, then, is the basic problem. Experimental mammalian teratology is concerned with a study of a variety of animal forms, varying in phenotype and genotype, and which possess strikingly dissimilar placental membranes. It is not possible to draw accurate generalizations with all these variables but the best that we can do is to develop an experimental model, realizing all the while the limitations of this model and work toward a better definition of our general problem.

The experimental model that we have used in our laboratory utilizes the pregnant golden hamster, *Cricetus auratus,* Waterhouse. This animal is an ideal experimental animal for embryological research for these reasons: The ease and accuracy of obtaining timed matings, the large litter size, and the short gestation period of 16 days. Female hamsters are placed with males during the early evening hours, and if mating ensues, are left with the male overnight. They are separated the following morning and this is considered to be the first day of gestation (Day 1). Implantation normally occurs in this species during the fourth day of gestation (Graves, 1945), and the average number of implantation sites is about twelve (Ferm, 1966a).

The development of the blastocyst proceeds rather rapidly and on the eighth day of gestation, embryonic differentiation passes through a most critical phase (Ferm, 1965). By careful dissection, the embryonic disc can be retrieved from the gestation sac as early as 8:00 A.M. on the eighth day, and at this time the embryos are represented by a disc measuring about 1.0 mm in its long axis. By 2:00 P.M., the axis becomes slightly more obvious and at 8:00 P.M. on the eighth day, differentiation has proceeded to that of an embryo possessing five pairs of somites with a total length of about 1.5 mm.

Twelve hours later, at 9:00 A.M. on the ninth day of gestation, the neural tube is completely closed and the heart is beating. Thus, development in this species during this 24 hour period between the eighth and ninth days of gestation, compares to the same total development in the human embryo from the 18th to 28th day of gestation. From many observations, including the use of known teratogens, we have found that this particular period, the eighth day of gestation, is the critical period for the study of suspected teratogenic agents.

This is the simple experimental model which we have used and it is now our purpose to apply this model to the three observations previously mentioned.

I. THE PLACENTAL MEMBRANE PLAYS A COMPLEX ROLE IN THE ETIOLOGY OF MAMMALIAN MALFORMATIONS

The interaction of maternal and embryonic tissues during development is characteristic of all mammals. This continuous interaction during pregnancy takes place first at the surface of the blastocyst and later at that complex interfacial membrane, the placenta. If this placental membrane were similar in all mammalian species, we could easily standardize its importance in experimental teratology. However, the tremendous variety in shapes, thicknesses, and in physiological exchange of this organ in mammals makes this an extremely important additional factor to keep in mind in experimental work.

It is, therefore, worthwhile to at least mention some of the variability in placental forms and to allude to some of the obvious differences in placental permeability among species which our later discussion will emphasize. This is most important if we are to attempt to extrapolate data from one species to another. In addition to such gross morphological differences such as shape (e.g., zonary, discoid, cotyledonary), there are pronounced histological differences characterized by the number of tissue layers which separate the maternal and fetal blood streams. Thus, in some species, six different histological layers of maternal and fetal origin separate the blood streams, while in some other species, only the endothelium and perhaps mesenchyme of the fetal capillaries separate the maternal from the fetal blood streams.

In addition to these differences described for the so-called "chorioallantoic" placenta of mammals, certain other groups of mammals possess a highly developed and extremely important second placental membrane, the yolk sac placenta. This is particularly true of the rodents in which so much of the basic work in mammalian teratology has been done. This yolk sac placenta has been shown to have some interesting relationships to embryonic development and to experimental teratology. For example, Brambell *et al.* (1951) have shown that homologous antibodies are transferred from mother to embryo via this membrane and that this yolk sac is capable of determining to some degree the species of origin of certain of these antibodies. The tremendous phagocytic properties of this membrane have been shown for the teratogenic azo dyes (Ferm and Beaudoin, 1960). More recently, the ultrastructural mechanisms involved in the uptake of particulate matter, Thorotrast, have been studied (Carpenter and Ferm, 1966). All of these results show that this membrane has an important protective function for the developing embryo, but that this protection probably does not exist during the critical early stages of development when teratogens are most active (Ferm, 1956).

There are two points which are of importance with respect to the complex role of the placenta in teratogenesis. First, the permeability of the placenta to a specific teratogen apparently depends on the species of animal under study. The placenta of the rodent, for example, is relatively impermeable to the potent teratogenic azo dye, trypan blue, at least after early stages of pregnancy (Ferm, 1956; 1958). On the other hand, the placenta of the armadillo is relatively permeable to this compound, staining the placenta, fetal liver and kidneys to a prominent degree

(Ferm and Beaudoin, 1965). The mechanisms of placental storage and/ or transfer of teratogenic agents have yet to be determined.

Secondly, the placenta of a given species may behave quite differently to substances of a similar nature. Nowhere is this more evident than in viral infections in the pregnant hamster. Figure 2 shows our results obtained from four different viruses which were injected into pregnant hamsters early in gestation.

Mumps Virus

This virus was isolated from the milk of a human patient who had an acute mumps parotitis at the time of delivery (Kilham, 1951). It proved to be virulent for suckling hamsters in first passage and causes fatal meningoencephalitis (Kilham and Overman, 1953). When injected intravenously into pregnant hamsters the virus proliferates to a marked degree in the uterus and placenta but no virus could be detected in fetal tissues (Ferm and Kilham, 1963a). There was no histopathological evidence of maternal infection. A few pregnant animals injected with this virus were allowed to litter but no young became ill following three weeks of post-natal observation.

Rat Virus

This is an extremely small, single-stranded DNA virus approximately 13 mμ in diameter. This virus, isolated from rats bearing implants of human tumors, causes a remarkable stunting phenomenon when injected into new-born hamsters. This stunting has been referred to as "mongoloid." Here again, the virus grows preferentially in the uterus and placenta and, in addition, traverses the placenta and can be recovered from fetal tissues in high titer (Ferm and Kilham, 1963b). Again, no histopathological evidence of viral infection was found in any maternal or fetal tissue.

It is of interest to add that the identical experiment carried out in pregnant rats using the laboratory adapted strain of rat virus failed to show any transplacental passage of this virus (Kilham and Ferm, 1964). Recently, however, a subsequently discovered strain of the same virus crossed the placenta of rats with ease, causing hepatitis and cerebellar hypoplasia in the young (Kilham and Margolis, 1966). This is acute testimony to the problems of both virology and experimental teratology.

Herpes Simplex Virus

This virus was obtained from the American Type Culture Association (Strain HF) and had no previous known pathological implications for hamsters. It has, however, been suspected of crossing the human placenta and causing human new-born infections. Under the identical experi-

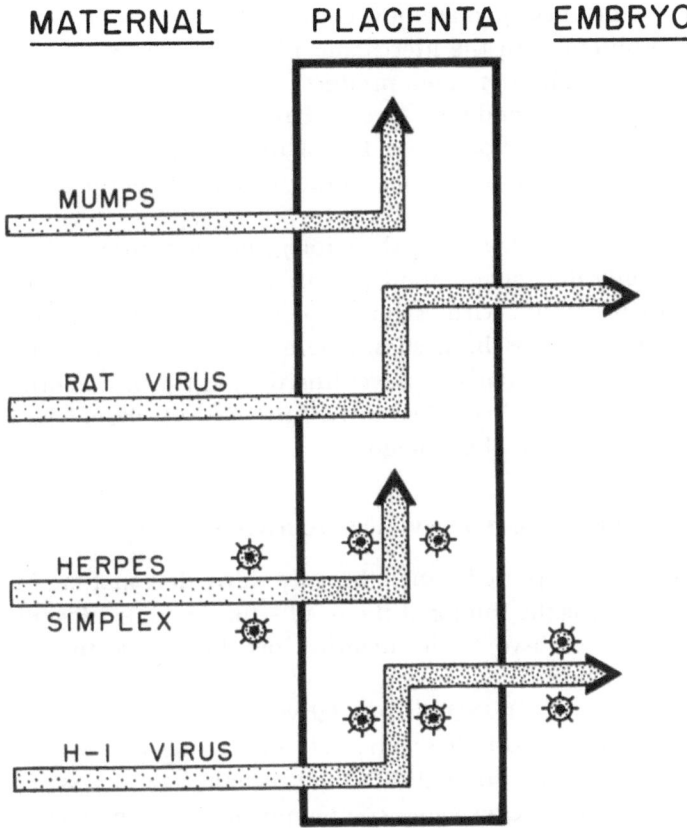

Fig. 2. A diagrammatic summary of the effect of certain viruses upon the permeability of the hamster placenta, the histopathology (✵) of the maternal, placental, and fetal tissues, and the relative titers of the specific viruses within these issues. See text.

mental conditions described for the two previous viruses, the herpes simplex agent induces marked necrotic changes in the maternal adrenal cortex with the production of the typical herpes inclusion bodies. Occasional similar foci of infection in the maternal liver were seen and frequent involvement of placental tissues were noted. Of the many specimens of fetal tissues tested for the presence of virus, only two showed positive titers and these both were extremely low. No histopathological evidence of fetal infection could be found following the intravenous injection of the mother (Ferm and Low, 1965).

H-1 Virus

This virus was reported by Toolan (1961) to have been recovered from human placentas and abortuses. It has many of the characteristics of

the rat virus and these two are now considered to be closely related. The
H-1 virus proliferates in the uterus and placenta, can be recovered from
fetal tissues in high titer, and produces a spectrum of developmental
malformations in the embryos. Areas of tissue necrosis as well as many
intranuclear inclusion bodies are found in the embryonic tissues especially those derived from the embryonic mesoderm (Ferm and Kilham,
1965).

The role of viral infections in the etiology of congenital malformations
then is not only dependent upon certain specific susceptibilities of embryonic cells to specific viruses but also to the role of the placenta in
these infections. As has been shown here, the placenta may be a very
efficient protective mechanism, preventing the passage of potentially damaging viruses to the embryo. Certainly, further work should be directed
toward this interesting phenomenon.

II. There is a Specific Teratogen: Organ Response

It has been said repeatedly that the only thing that matters in experimental teratology is the timing of the insult—that those teratogenic agents
which insult the embryo at the identical developmental time will produce the same defect.

Dimethylsulfoxide (DMSO) is a relatively simple compound with a
molecular weight of 78. It has extremely interesting and versatile solu-
bility properties. When injected into the pregnant hamster on the eighth
day of gestation, this remarkable compound produces malformations in
a great number of embryos (Ferm, 1966b). The most striking thing about
these embryos is a characteristic and strikingly specific exencephalic
lesion. In three separate litters, almost all of the surviving embryos exhibited the identical lesion. This, then, is the effect of DMSO in the
cranial portion of the embryo. Marin-Padilla (1966) has shown that
DMSO produces a marked disturbance on the head mesenchyme of these
hamster embryos. On the other hand, the intravenous injection of lead
salts under identical conditions results in a completely different anomaly.
In these animals the tail end of the embryo is affected. With this teratogen, almost all of the recovered embryos exhibited the same defect—
a complete absence of the tail (Ferm and Carpenter, *unpublished data*).
The cross defects, DMSO-tail versus lead-exencephaly, were never encountered. These differences cannot be accounted for on the basis of
the rate of placental permeability since we have injected both compounds
at earlier and later stages of gestation and did not find the opposite
anomalies. Other examples of the specificity of the teratogen: organ
effect would include the well-known thalidomide: phocomelia and the
rubella: ductus arteriosus relationships.

The conclusion that we can draw from this is that there is a specific teratogen: organ effect and that this effect is dependent upon a specific action of the teratogen at some subcellular level in a particular developing cell.

III. SYNTERATOGENS AND PROTERATOGENS

These represent the most confusing but conceivably the most rewarding areas of teratological research. Simply, two teratogenic agents which potentiate each other are called *synteratogens*. Examples of these have been described by Landauer (1962) utilizing insulin and other agents in the chick, and by Kalter (1960) using cortisone and hypocaloric diets in mice. The teratogenic responses are more varied and more severe than with one teratogen alone.

A *proteratogenic* effect can be defined as the potentiation of a single teratogen by the simultaneous treatment with a non-teratogenic stimulus. This particular phenomenon can be demonstrated most vividly in our experimental model by the effect of Vitaimn A and low temperature. Vitamin A is probably one of the most potent experimental teratogens known. Twenty thousand units of this vitamin administered orally on the eighth day of gestation to the pregnant hamster produces a wide range of congenital malformations including exencephaly, spina bifida, rib and ocular defects, and many others (Marin-Padilla and Ferm, 1965).

Pregnant golden hamsters placed in a cold environment at 1°C for 48 hours from the 7th–9th days of gestation revealed no increase in fetal resorptions and no congenital malformations. No effect on the maternal system was noted. However, when pregnant hamsters are kept in the same cold environment for the same period and fed 20,000 units of Vitamin A, on the eighth day of gestation, a much more marked teratogenic effect is noted, than with Vitamin A alone. Severe and extensive cases of spina bifida, frequently associated with symmelia, more frequent and severe cleft lips and palates, and a higher embryonic resorption rate were noted (Ferm, *unpublished data*).

Here, the potentiation effect is quite striking and it is important that further consideration be given to this phenomenon. It is probable that a considerable variety of subtle environmental influences, interacting with teratogenic agents or perhaps even potentiating sub-teratogenic stimuli, may be responsible for a significant part of the overall problem of congenital malformations.

In summary, developmental malformations represent one important chapter in the total problem of reproductive failure.

An experimental model for mammalian teratology, utilizing the pregnant golden hamster, has pointed to certain observations which should lead to further investigations in this field. These are: (1) The complex role of the placental membrane in embryonic development, (2) a specific teratogen:organ effect, and (3) the interaction of synteratogenic and proteratogenic stimuli.

References

Brambell, F. W. R., W. A. Hemmings, and M. Henderson: Antibodies and Embryos. University of London: The Athlone Press, 1951.

Carpenter, S. J., and V. H. Ferm: Electron microscopic observations on the uptake and storage of Thorotrast by rodent yolk sac epithelial cells. Anat. Rec. *154*:327, 1966.

Cohlan, S. Q.: Teratogenic agents and congenital malformations. J. Ped. *63*:650, 1963.

Ferm, V. H.: Permeability of the rabbit blastocyst to trypan blue. Anat. Rec. *125*:745, 1956.

————: Teratogenic effect of trypan blue on the golden hamster. J. Embryol. Exp. Morph. *6*:284, 1958.

————: The rapid detection of teratogenic activity. Lab. Invest. *14*:1500, 1965.

————: Observations on litter size in interrupted hamster pregnancies. Anat. Rec. *154*:460, 1966a.

————: Teratogenic effect of dimethyl sulfoxide. Lancet *1*:208, 1966b.

Ferm, V. H., and A. R. Beaudoin: Absorptive phenomena in the explanted yolk sac placenta of the rat. Anat. Rec. *137*:87, 1960.

————, ————: Studies on the effect of trypan blue in the pregnant armadillo, *Dasypus novemcinctus*. Anat. Rec. *151*:571, 1965.

Ferm, V. H., and L. Kilham: Mumps virus infection of the pregnant hamster. J. Embryol. Exp. Morph. *11*:659, 1963a.

————, ————: Rat virus (RV) infection in fetal and pregnant hamsters. Proc. Soc. Exp. Biol. and Med. *112*:623, 1963b.

————, ————: Histopathologic basis of the teratogenic effects of the H-1 virus on hamster embryos. J. Embryol. Exp. Morph. *13*:151, 1965.

Ferm, V. H., and R. Low: Herpes simplex virus infection in the pregnant hamster. J. Path. and Bact. *89*:295, 1965.

Fraser, F. C.: Experimental teratogenesis in relation to congenital malformations in man. New York: The International Medical Congress, Ltd., 1963.

Fraser, F. C., H. Kalter, B. E. Walker, and T. D. Fainstat: Experimental production of cleft palate with cortisone and other hormones. J. Cell. and Comp. Physiol. *43*:237, 1954.

Graves, A. P.: Development of the golden hamster, *Cricetus auratus* Waterhouse, during the first nine days. Am. J. Anat. 77:219, 1945.

Kalter, H.: Teratogenic action of a hypocaloric diet and small doses of cortisone. Proc. Soc. Exp. Biol. and Med. *104*:518, 1960.

Kilham, L.: Mumps virus in human milk and in milk of infected monkey. J. Amer. Med. Assn. *146*:1231, 1951.

Kilham, L., and J. R. Overman: Natural pathogenicity of mumps virus for suckling hamsters on intracerebral inoculation. J. Immunol. 70:147, 1953.

Kilham, L., and V. H. Ferm: Rat virus (RV) infections of pregnant, fetal, and new-born rats. Proc. Soc. Exp. Biol. and Med. *117*:847, 1964.

Kilham, L., and G. Margolis: Spontaneous hepatitis and cerebellar "hypoplasia" in suckling rats due to congenital infections with rat virus. Am. J. Path. *49*:457, 1966.

Landauer, W., and E. M. Clark: Teratogenic interaction of insulin and 2-deoxy-D-glucose in chick development. J. Exp. Zool. *151*:245, 1962.

Marin-Padilla, M.: Mesodermal alterations induced by dimethylsulfoxide. Proc. Soc. Exp. Biol. and Med. *122*:717, 1966.

Marin-Padilla, M., and V. H. Ferm: Somite necrosis and developmental malformations induced by vitamin A in the golden hamster. J. Embryol. Exp. Morph. *13*:1, 1965.

Toolan, H. W.: Isolation of the H-1 and H-3 viruses directly from human embryos. Proc. Amer. Assn. Cancer Res. *3*:368, 1962.

Wilson, J. G.: Experimental studies on congenital malformations. J. Chron. Dis. *10*:111, 1959.

MALFORMATIONS AND DEFECTS OF GENETIC ORIGIN IN DOMESTIC ANIMALS

F. B. Hutt

Department of Poultry Science, New York State College of Agriculture, Cornell University, Ithaca, New York

To clarify matters at the beginning, I consider that reproduction has failed whenever a fertilized egg does not produce eventually a viable animal, with normal anatomy and physiology, and capable of normal development. Because a malformation like polydactyly can be harmless, while an invisible "inborn error of metabolism" like galactosemia can be lethal, I have extended the title orginally assigned to me to permit inclusion of defects other than malformations.

This discussion deals chiefly with inherited defects in domestic animals. As time and space do not permit presentation now of a complete catalogue of such abnormalities, it is proposed to select examples to show (1) the diversity of genetic defects, and (2) the fact that similar genetic abnormalities occur in different species, even in animals as distantly related taxonomically as mammals and birds. Extensive lists of lethal genes and hereditary defects in domestic animals have been provided by Stormont (1958), Hutt (1964), and Koch *et al.* (1957).

Before proceeding to discuss abnormalities, we must recall that, in dealing with domestic animals, the word *normal* is a very elastic one. Expressed otherwise, animal breeders have always been intrigued by the unusual, the bizarre, and many a variation that is abnormal for the wild type of a species has been adopted as the normal, distinguishing feature of some new breed of dog, fowl, or cattle. For evidence on that score one need only go to a dog show to see the extremely brachycephalic English Bulldog that wins first prize in its class, and to hear the stertorous breathing which proclaims its pathological state. The tailless Manx cat and the achondroplastic Dexter cattle both owe their distinction to single dominant genes that are lethal in the homozygous state. The beautiful crest of Houdan and Polish fowls results from a gene which causes internal hydrocephaly.

Not all abnormalities are injurious. Polled cattle are abnormal by definition, but, under domestication, are much better off without their horns than with them. Fortunately, a dominant gene will remove the horns and thus make unnecessary the gory operations formerly commonly done for that purpose. Goat breeders also prefer polled animals, but they are less fortunate than cattle breeders because goats homozygous for the polling gene are intersexes if originally females, and apparently subnormal in fertility if males.

Now to examples of defects. For convenience, and to emphasize their diversity, they are arranged by structures and systems.

THE SKELETON

Entire Skeleton

A good example of a hereditary abnormality affecting the entire skeleton is *fragilitas ossium,* or brittle bones, in man. It is transmitted as a simple dominant trait. Victims have been reported with over 20 fractures in five years. A recent harrowing report in the press tells of an 18-month-old boy who had already suffered 15 fractures of the legs, five of his arms, and 10 broken ribs. Associated effects include blue sclerotics, anomalies of dentition and hypermobility of the joints.

Chondrodystrophy, or achondroplasia, affects primarily the fetal bones that are formed in cartilage, while those laid down in membrane are unaffected, but the visible effect is an apparent dwarfing of the entire skeleton. It affords a good example of mutations with similar effects in different species. Human achondroplastic dwarfs have been known for centuries. Dexter cattle are heterozygous for an incompletely dominant gene which causes characteristic achondroplasia, the animals being short-legged and of small size. The same mutation has occurred in other breeds, but is usually discarded. When Dexters are mated together, about a quarter of their progeny show extreme shortening of the legs, rounded skull, oedema, high tail-head, and other symptoms of extreme achondroplasia. All of them die at birth, or before full term. So far as is known, all surviving human chondrodystrophic dwarfs are heterozygous for the causative gene, and it is assumed that the homozygous state must be lethal as in cattle.

Similarly, poultry breeders have adopted a gene for dominant achondroplasia as a distinguishing feature of two breeds: the Creeper and Japanese Bantams. All of them are heterozygous; most homozygotes die early in incubation but a few survive to the third week.

Recessive chondrodystrophy, caused by an autosomal recessive gene, and usually lethal to homozygotes before birth, is known in cattle, in the rabbit and in the fowl. Animals carrying such a gene are phenotypically normal, and are recognizable as carriers only when they produce defective offspring.

A less extreme form of chondrodystrophy is found in the so-called "snorter" dwarfs currently of great concern to breeders of Hereford cattle. These are viable and some even survive to reproduce, but they are stunted and unthrifty, subnormal in viability and characterized by labored breathing. The condition was earlier thought to be a simple recessive defect, but Julian *et al.* (1959) now consider that there are at least four kinds of dwarfism. Attempts to identify carriers of the causative gene (none satisfactory) were reviewed by Bovard (1960).

Not all dwarfing is deleterious. The sex-linked ateleiotic dwarfism found in the fowl reduces weights of females by about 30 per cent and of homozygous males by 42 per cent (Fig. 1). Viability is normal, and egg

Fig. 1. Sex-linked recessive ateleiotic dwarfing. For these two brothers hatched on the same day, weights at 209 days of age were 3030 gm. for the heterozygous male on the left, but only 1540 gm. for his homozygous brother on the right. (From F. B. Hutt in *Genetics of the Fowl.*)

production slightly less than in the dwarfs' big sisters, but fertility and hatchability are normal (Hutt, 1959).

The Head

Genes with more localized effects cause various abnormalities of the head. Among these the most conspicuous, but, fortunately, not the most common, are probably recessive genes which cause varying degrees of reduction of the lower jaw. In Grant's (1956) 15 cases in Shorthorns, the mandible was far shorter than the maxilla, and molars of both jaws were impacted and fused. All the calves died from inability to suckle.

Elsewhere (Hutt, 1949) I have reviewed four mutations in the fowl (all recessive) which cause missing maxillae, missing mandible, short mandible and short upper break. They provide good examples of localized gene action.

The Axial Skeleton

The most common genetic modifications of the axial skeleton in domestic animals are probably those which reduce or eliminate the caudal vertebrae to give us Manx (tailless) cats, Wilson's Notail sheep and Rumpless Bantam fowls. Not all such amputations are hereditary. Nongenetic taillessness is known in the rat, and the rare persistence of an embryonic tail in our own species is probably not hereditary either.

A good example of similar genetic malformations in different species is afforded by the "short-spine" recessive mutation which occurs in the turkey, dog, pig and cattle (Fig. 2). In each case, the spine is shortened, not by elimination of any vertebrae, but, rather, by their being crowded together. In the affected mammals, the neck is so shortened that the head appears to come out from the shoulders, and the tail is shortened. In all four species the appendicular skeleton is normal.

Asmundson (1942) found that none of his short-spined turkeys hatched, and most died in the later stages of incubation. Affected calves die during birth, or soon after (Mohr and Wriedt, 1930). Most of the short-spined pigs die soon after birth, but Dabczewski (1949) reported that, with special care, some were kept alive for 8 to 12 days, and one was even raised to six months of age. By contrast, de Boom's "baboon dogs" survived, but with reduced viability, and short-spined females reproduced normally (see Hutt, 1964).

The fact that mutations causing the same remarkable abnormality have occurred in these four widely-separated species suggests that investigators studying heredity in any one species should also keep a watchful eye on what can happen in others.

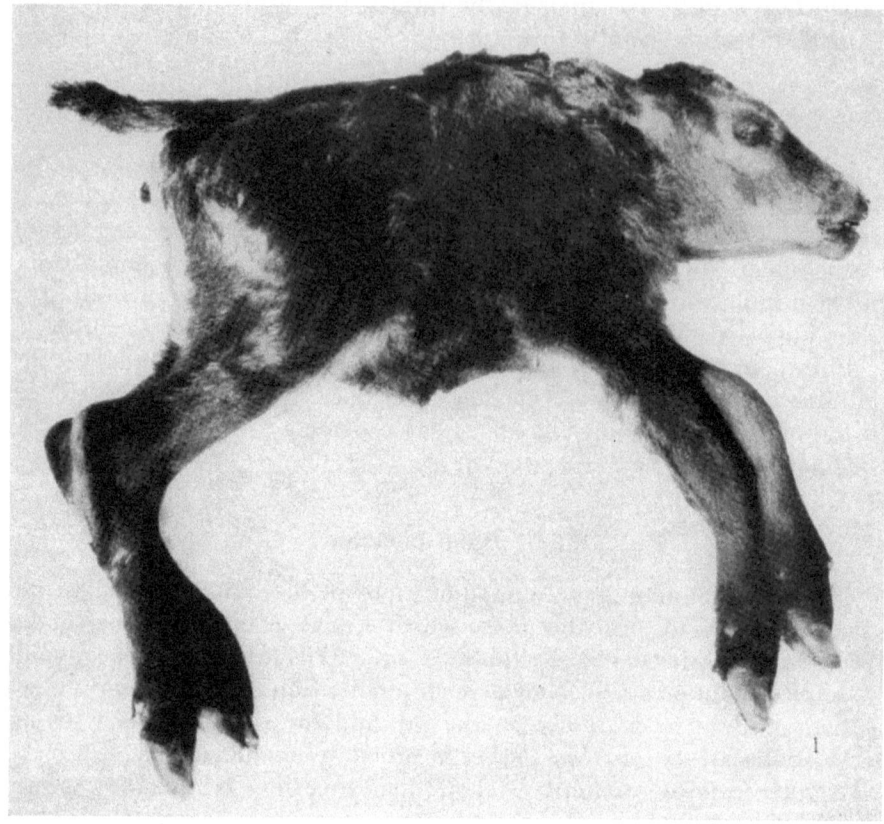

Fig. 2. Lethal short-spine in the calf. (From O. L. Mohr and C. Wriedt in J. Genetics, as cited.)

The Appendicular Skeleton

Genetic malformations of the limbs range all the way from amputations that completely remove all four legs in pigs (Johnson, 1940) (Fig. 3) down to the ungual osteodystrophy of the fowl, which eliminates or distorts the claw of the middle toe, usually on one foot only. In the latter case, it is the ungual phalanx that is defective, not the overlying keratinous nail.

In between these extremes, there are hereditary amputations in cattle (*acroteriasis congenita*) which terminate the limbs near the hock and elbow points, and are accompanied by cleft palate, short jaw and reduction in the number of teeth. Similar amputations, but without the associated defects, are also known in man, and six such cases occurred in a Brazilian family of 12. Less severe shortening of the limbs has been adopted as a breed characteristic in Dachshunds and in Seth Wight's

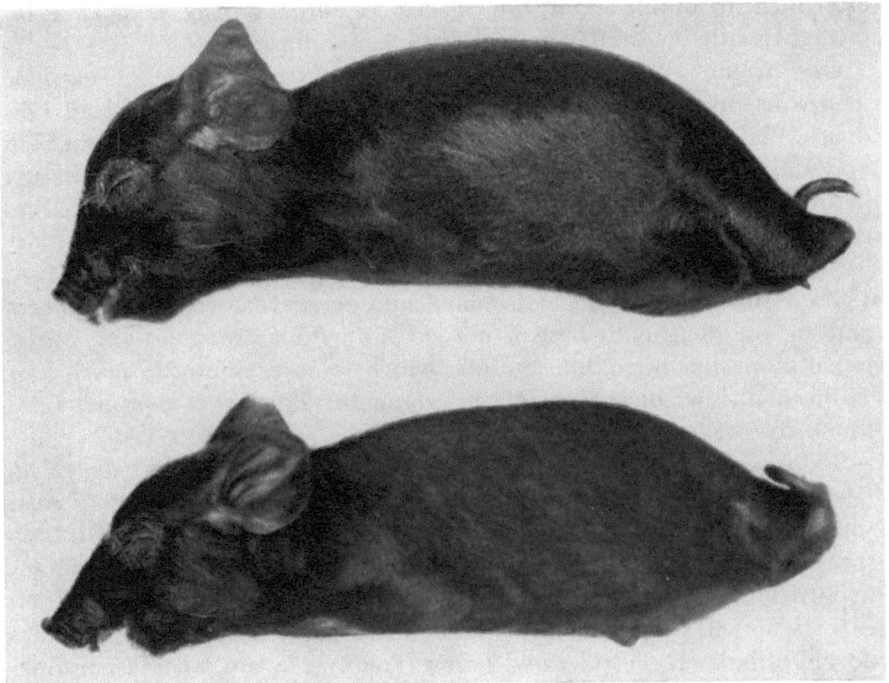

Fig. 3. Complete amputation of the limbs in piglets. (From L. Johnson in J. Heredity, as cited.)

Ancon sheep of long ago. Polydactyly is not uncommon in cats, and occurs also in other mammals, including man. Poultry fanciers have adopted it as an indispensable breed characteristic in Houdans, Dorkings, and other breeds. In most species it is transmitted as a dominant trait, but with incomplete penetrance and varying degrees of expression.

The Integument

As there are extensive monographs on genetic abnormalities of the skin and its outgrowths, all I propose to do here is to point out the wide range of such things in domestic animals.

One of them, once given the resounding name of *epitheliogenesis imperfecta neonatorum bovis*, but now generally called epithelial defects, causes (in Jerseys) large areas of the body to be devoid of normal skin. In Holstein Friesians, the lesions are smaller, and in Ayrshires they are confined to small areas above the hooves, at the knees, on the muzzle and in the ears. It is not yet known if a mutation at one locus in the chromosomes is responsible for the varying expressions in the three breeds. The defect is lethal before full term in Jerseys, and a few weeks after birth in the Friesians, but affected Ayrshires can be kept alive for several months. A similar defect occurs in the horse.

One of the worst genetic abnormalities of the skin is the extreme ichthyosis that occurred in Red Poll cattle in Norway (Tuff and Gleditsch, 1949). In affected calves, all of which died soon after birth, the skin appeared to consist of horny plates, like big scales. Hairs were present in the grooves between the plates. The abnormality was attributed to excessive production of keratin.

At least six different kinds of hypotrichosis are known in cattle (Hutt, 1963). Three of them are autosomal and recessive, two are sex-linked, and for one the genetic basis is not clear. Of the autosomal types, one is lethal soon after birth, but another, which removes almost as much hair as the first, is viable in a good environment. The lethal type has been found in Sweden, Germany and Japan.

Sex-linked recessive hypotrichosis of cattle, found by Drieux *et al.* (1950) in France appears to be homologous with the condition in man sometimes called anhidrosis with anodontia. It, too, is sex-linked. In both species there is almost complete absence of teeth. Sex-linked dominant hypotrichosis is apparently lethal to homozygous bull fetuses at an early stage of gestation. The hemizygous cows show vertical streaks devoid of hair on the sides and flanks (Fig. 4), but are otherwise normal (Eldridge and Atkeson, 1953).

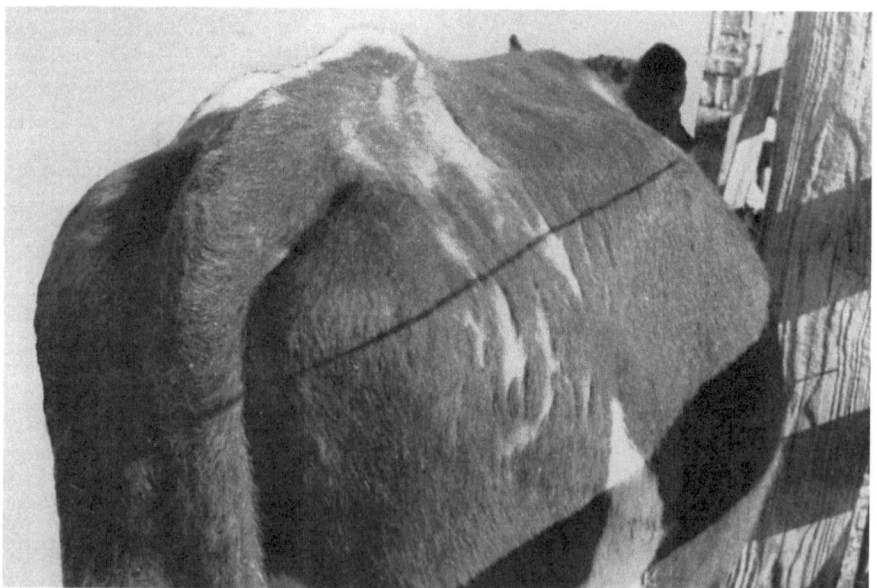

Fig. 4. Sex-linked streaked hypotrichosis in a heterozygous Holstein-Friesian cow. (From F. E. Eldridge and F. W. Atkeson in J. Heredity, as cited.)

Fig. 5. A female chick showing at three weeks of age the naked condition caused by a sex-linked recessive gene, which is lethal late in incubation to about half the naked chicks. (From F. B. Hutt in *Genetics of the Fowl.*)

Several kinds of complete or partial hairlessness are known in the mouse and Mexican Hairless dogs and Bare Neck fowls are well known to fanciers. The gene responsible for hairlessness in the dog is apparently lethal to the homozygote. This does not apply to the sex-linked gene which makes female chicks almost naked at hatching (Fig. 5), but allows more feathers at maturity. It is lethal only to some of the chicks.

DIGESTIVE TRACT

Two hereditary abnormalities of the digestive tract are noteworthy. One is *atresia coli,* an autosomal recessive malformation in the horse. In affected foals, the ascending colon is closed in the region of the pelvic flexure, and death ensues within two to four days of birth.

Another is the peculiar lethal indigestion of Karakul sheep homozygous for a certain type of grey fleece. Most of them die when four to nine months old and show at post-mortem examination abnormal contents in the rumen and abomasum (Nel and Louw, 1953).

REPRODUCTION

Information about genetic abnormalities that interfere directly with reproduction is accumulating slowly, but several distinctly different types of hereditary defects have been elucidated.

One of these is prolongation of the gestation period, with resulting gigantism and death of the calf. Birth is difficult, and in some cases is impossible without slaughter of the cow. Reports of this condition in several breeds, as summarized by Wilson and Young (1958), show the average number of days overdue to have been in Holstein Friesians 51 (21 calves), in Ayrshires 80 (9 calves), and in Swedish Red and White 125 (8 calves). Gregory and his associates recorded 30 such cases in one herd of Holstein Friesians in five years. In all three breeds a single recessive gene in the homozygous state was considered responsible for the prolongation. It is clearly the genotype of the calf that causes the trouble, and not that of the heterozygous dam.

A different gene must be responsible for the somewhat different kind of prolongation of gestation studied by Kennedy *et al.* (1957) in Guernseys. Although nine affected calves were overdue 121 days (average), they were smaller than normal, and showed varying degrees of hypotrichosis. In this case, the abnormality was attributed to aplasia of the anterior pituitary gland of the calf.

In the Netherlands, impotency of Friesian bulls was found to result from abnormality of the *retractor penis* muscles in animals homozygous for the causative recessive gene. Another sex-limited defect, and probably one affecting a smaller part of the anatomy than most other genetic malformations, is the eccentric thickening of the acrosome of sperm heads to produce what Donald and Hancock (1953) described as "knobbed" spermatozoa. They concluded that the defect, which was responsible for the sterility of 17 related bulls, is a simple recessive trait.

Atresia of the oviduct in the fowl, apparently caused by a dominant gene, does not affect the males that carry the gene, but, when transmitted to (half) their daughters, not only prevents those birds from laying, but also causes their eventual death from the peritonitis that follows return of the unlaid eggs to the body cavity.

Inability to reproduce also characterizes several kinds of intersexes found in domestic animals. The freemartin, a female born co-twin to a bull calf, is an old and familiar example. So is the sterile white shorthorn heifer, in which the reproductive tract is not fully developed, a condition estimated to occur in at least 15 per cent of the white females and found to be as high as 39 per cent in one herd. The causes of the malformation and of its association with the white coat are still unknown. Other genetic disorders of reproduction in cattle were reviewed by Rollinson (1955).

It seems probable that in the future more cases of sterility and reduced fertility in domestic animals will be found to be associated with chromosomal aberrations other than simple gene mutations. Already we know that the tortoise-shell tom-cat, nearly always sterile, is an XXY male, and hence is the feline counterpart of the Klinefelter syndrome in man. At the Royal Veterinary College in Stockholm, Knudsen (1958) found that four healthy bulls with low fertility were heterozygous for a chromosomal inversion, and three others carried translocations. More recently, Henricson and Bäckström (1964) were able to show that a boar with subnormal fertility carried a translocation from Chromosome 4 (or 3) to Chromosome 14.

A remarkable disorder of reproduction, and one of a type not likely to be found in mammals, is fatal to all chick embryos in eggs laid by hens homozygous for a recessive gene which prevents them from transferring riboflavin from the feed to their eggs. As the gene is transmitted from one generation to the next by the heterozygous carriers, which reproduce normally, it is not eliminated by natural selection. Buss and his associates found that eggs from the abnormal hens (which are themselves healthy and lay well) can be made to hatch by injecting into them enough riboflavin to permit normal development of the embryo.

OTHERS

One could continue with a catalogue of malformations and disorders affecting the nervous system, the joints, muscles, blood and metabolism, but many of these have recently been reviewed elsewhere (Hutt, 1964) and our time does not permit further enumeration of such cases at this conference.

DISCUSSION

Before concluding, however, it seems desirable to point out that not all abnormalities of form and function can be charged to the genes. Many of them result from accidents during embryonic development, and are not transmitted to later generations. In some cases, such accidents are precipitated by adverse environments, as was the case with the unfortunate victims of thalidomide. By treatment with various teratogenic compounds, Landauer (1957) has been able to induce phenocopies of various genetic mutations in the fowl, and others have done similarly with shock treatments of various kinds.

Some defects, particularly those of the skeleton, are expressed regardless of the environment. Others may be manifested in one environment, but not in others. For example, the turkeys genetically susceptible to pendulous crop, a condition associated with high intake of water during hot weather, do not develop that defect in a cooler climate (Hinshaw and Asmundson, 1936). Many similar cases could be cited.

It would be nice to know the frequency of genetic disorders in domestic animals, but statistics on that point are not available. We might remember, however, that in wild flies of *Drosophila pseudo-obscura* deliberately inbred to reveal their bad genes, Dobzhansky (1957) found that 85 per cent carried a chromosome that would be lethal in double dose, and more than half had at least one chromosome that would cause sterility in homozygotes. It seems unlikely that either we or our domestic animals are carrying less "genetic junk" than Dobzhansky's flies.

SUMMARY

Examples are cited of genetic abnormalities that affect the skeleton, integument, digestive tract or reproduction of domestic animals. The examples are selected to show the diversity of defects caused by mutant genes, and also to illustrate the fact that similar genetic defects occur in widely different species ranging from birds, through different mammals, to man.

References

Asmundson, V. S.: Skeletal abnormalities of short-spined turkeys. Proc. Soc. Exper. Biol. and Med. *50:*120, 1942.

Bovard, K. P.: Hereditary dwarfism in beef cattle. Animal Breed. Abstr. *28:*223, 1960.

Dabczewski. Z.: Studies on a teratological form of the newly born Pulawska (Golebska) swine. Bull. Acad. Polon. Sci. Lettres. B (II): 241 + Pl. 13–19. 1949.

Dobzhansky, T.: Genetic loads in natural populations. Science *126:*191, 1957.

Donald, H. P., and J. L. Hancock: Evidence of gene-controlled sterility in bulls. J. Agric. Sci. *43:*178, 1953.

Drieux, H., M. Priouzeau, G. Thiéry, and M.-L. Priouzeau: Hypotrichose congénitale avec anodontie, acérie et macroglossie chez le veau. Recueil de Méd. Vét. *126:*385, 1950.

Eldridge, F. E., and F. W. Atkeson: Streaked hairlessness in Holstein-Friesian cattle. J. Hered. *44:*265, 1953.

Grant, H. T.: Underdeveloped mandible in a herd of Dairy Shorthorn cattle. J. Hered. *47:*165, 1956.

Henricson, B., and L. Bäckström: Translocation heterozygosity in a boar. Hereditas *52:*166, 1964.

Hinshaw, W. R., and V. S. Asmundson: Observations on pendulous crop in turkeys. J. Am. Vet. Med. Assn. *88* (N.S.41): 154, 1936.

Hutt, F. B.: Genetics of the fowl. New York: McGraw-Hill Book Company, Inc., 1949.

——: Sex-linked dwarfism in the fowl. J. Hered. *50:*209, 1959.

——: A note on six kinds of genetic hypotrichosis in cattle. J. Hered. *54:*186, 1963.

——: Animal Genetics. New York: The Ronald Press Company, 1964.

Johnson, Leslie: "Streamlined" pigs. J. Hered. *31:*239, 1940.

Julian, L. M., W. S. Tyler, and P. W. Gregory: The current status of bovine dwarfism. J. Am. Vet. Med. Assn. *135*:104, 1959.

Kennedy, P. C., J. W. Kendrick, and C. Stormont: Adenohypophyseal aplasia, an inherited defect associated with abnormal gestation in Guernsey cattle. Cornell Vet. *47*:160, 1957.

Knudsen, O.: Studies on spermiocytogenesis in the bull. Internat. J. Fertil. *3*:389, 1958.

Koch, P., H. Fischer, and H. Schumann: Erbpathologie der landwirtschaftlichen Haustiere. Berlin and Hamburg: Paul Parey, 1957.

Landauer, W.: Phenocopies and genotype, with special reference to sporadically-occurring developmental variants. Amer. Naturalist *91*:79, 1957.

Mohr, O. L., and C. Wriedt: Short spine, a new recessive lethal in cattle; with a comparison of the skeletal deformities in short spine and in amputated calves. J. Genet. *22*:279, 1930.

Nel, J. A., and D. J. Louw: The lethal factor in grey Karakul sheep. Farming in South Africa *28*:169, 1953.

Rollinson, D. H. L.: Hereditary factors affecting reproductive efficiency in cattle. Animal Breed. Abstr. *23*:215, 1955.

Stormont, C.: Genetics and disease. Advances in Vet. Sci. *4*:137, 1958.

Tuff, P., and L. A. Gleditsch: Ichthyosis congenita hos kalver—en arvelig letal defekt. Nord. Vet. Med. *1*:619, 1949.

Wilson, A. L., and G. B. Young: Prolonged gestation in an Ayrshire herd. Vet. Record *70*:73, 1958.

BACTERIAL INFERTILITY IN
DOMESTIC ANIMALS

A. B. HOERLEIN

*Professor of Microbiology, College of Veterinary Medicine,
Colorado State University, Fort Collins, Colorado*

Bacterial infertilities of domestic animals have been economically serious enough to warrant extensive and intensive investigation of their cause, pathogenesis, and epizootiology so that effective control procedures could be developed. In many cases detailed investigation of the pathologic features of these diseases has been an afterthought and delayed until after a considerable body of knowledge had been developed in the other aspects of the disease. While generalized disease and localized nonspecific pyogenic infections may cause temporary or permanent infertility, the specific epizootic bacterial infections are of greatest interest. The extent of the literature precludes a thorough discussion of these diseases, but a superficial comparison of pathogenesis of some of these infections can be made within this brief summary.

Several bacterial species cause infertility and they are generally highly oriented to their specific animal hosts. In the case of brucellosis, vibriosis, and leptospirosis the several species are identified by subtle biochemic and antigenic differences, but in each case there is marked biological specificity. The pathogenesis of infections caused by similar bacteria is often unique to each of the animal species affected. While some of these infections are serious in man, they are rarely the direct cause of infertility.

The most studied infection causing infertility in domestic animals is brucellosis. *Brucella abortus, Br. melitensis,* and *Br. suis* are similar in most biochemic and antigenic aspects. These brucellae share some pathogenic characteristics in the various animal species affected, but each has its own host range and a specific pathogenesis related to each host.

Br. abortus causes a serious disease of cattle throughout the world characterized by abortion during the latter half of gestation. The infection is almost invariably transmitted to a susceptible female by direct contact with an aborted fetus or uterine exudates following abortion or by the ingestion of feed and water contaminated during or following abortion. The infection may also be spread through the milk of infected cows.

Ingestion of the organisms is followed by local lymphadenitis which develops into a transient bacteremia. The organism has a great predilection for the pregnant uterus. The bacteria are found in tremendous numbers in the chorionic epithelial cells. Intracellular multiplication appears to be an essential phase of the infection and provides large numbers of the organisms which are released before necrosis of the villi is noted. After the fetal fluids are invaded large numbers of brucellae are found in the stomach and lungs of the fetus. Other fetal tissues are not invaded. At the time of abortion great numbers of organisms are released into the environment. The gross placental changes described in bovine brucellosis are edema of the placenta and a brownish-yellow exudate adherent to the membranes and cotyledons (Roberts, 1956). The placenta may become leathery from the exudate and obvious necrosis is found in some or many of the cotyledons. Retention of the placenta is common especially if abortion occurs after the 5th month of gestation (Roberts, 1956). Inflammatory enlargement of the terminal tips of the maternal septa may result in the failure of the cotyledons to separate from the uterine caruncle (Jubb and Kennedy, 1963). Actually, the maternal tissues are not seriously affected, brucellae are rarely found in the uterine exudates after a few weeks, and most of the cattle are fertile after abortion. The lower than normal conception rates found in convalescent cattle are usually associated with complications arising from secondary infections following the retained placenta.

The infection tends to persist indefinitely in cows. While the brucellae rarely persist in the uterus, they localize in the supra-mammary lymph nodes and in the udder. It is from this remote localization that the bacteria again find their way to the uterus during subsequent pregnancies. Abortion rarely occurs during these subsequent pregnancies and a live healthy calf is usually born after a normal gestation. The calf and its placenta and fluids are, however, grossly contaminated with organisms as is the lochia following parturition. The calf is not infected with brucellosis and is not susceptible to infection until puberty.

Bulls rarely become infected with *Br. abortus*, but they may have an acute necrotic orchitis and infection of the accessory reproductive organs. Only by artificial insemination with contaminated semen is the infection spread by genital means (Manthei, 1950).

Brucella abortus infection is uncommon in sheep and they seem to have considerable resistance to infection except in France where it is an important source of brucellosis in man. Utilizing *Br. abortus* cultures of high virulence, Molello *et al. (*1963c) experienced little difficulty in infecting sheep approximately 3 months after breeding when the organisms were injected intravenously. There was a marked affinity of the bacteria for the chorionic epithelial cells and the relatively mild endometritis characteristic of all the brucella infections was present. Necrosis of the villi and the chorioallantoic membrane extending from the hilar zone of the placentome with sanguineous exudate was found in the stages preceding abortion.

The classical Malta Fever in man is contracted from goats with *Br. melitensis* infection. The disease is found throughout the Mediterranean area and in some other areas of the world including southwestern United States. It is characterized by abortions and stillbirths and especially mastitis which is severe and may be a source of organisms for years from an affected animal. *Br. melitensis* infection is not so severe in sheep and is common only in the Adriatic countries. Molello *et al.* (1963b) injected *Br. melitensis* into 120-day pregnant ewes and killed them 14 to 16 days later. Extensive placental edema and small amounts of brownish red exudate in the periplacentome were found. Exudate in the interplacentome was infrequent. Widespread necrosis of the septal tips was followed by extensive necrosis of the chorioallantoic membrane. It appeared, as in the other studies of brucellosis, that intracellular multiplication of the organisms is a major part of the pathogenesis of the infection. Ariel (1939) thought that membrane retention was due to connective tissue formation, but Molello *et al.* (1963b) considered the infection too acute to allow much fibrous tissue proliferation.

Porcine brucellosis is common in the United States and in Europe. The causative agent, *Br. suis,* differs from *Br. abortus* in some biochemical reactions, but is essentially identical in antigenic structure. While the disease may be spread by ingestion, it is more commonly spread by venereal means from an infected boar to susceptible gilts. Often the only sign of the disease is failure of gilts to settle after being bred by an infected boar (Hoerlein *et al.,* 1954). In the boar the infection often localizes in the accessory reproductive organs, the seminal vesicles being the most common. Orchitis is also common (Hutchings and Andrews, 1946). After the acute phases of the infection when little sexual interest may be shown, the boar often breeds normally and may disseminate billions of brucellae at each service. Following breeding by an infected boar, the infection causes embryonal death so early that the return to estrus is only slightly delayed. Sows in the group pregnant from previous breedings may become infected from contact with aborted feti or uterine exudates and may thus abort at any stage of gestation. A diagnosis may be made by bacteriologic examination of aborted feti large enough to be found. Recovery from the disease generally takes place in the female and is usually complete with the organisms disappearing from the animal. Little convalescent immunity has been demonstrated. A few gilts have been studied in which localized bladder infection became the source of tremendous numbers of bacteria shed in each urination. The boar generally remains infected for years. Detection of these carriers is difficult since most have no diagnostic serologic titer.

A bacteremia of weeks' or months' duration follows introduction of *Br. suis* by any means (Hoerlein *et al.*, 1954). Lymphadenitis and localization in many tissues of the body may occur. Localized granulomatous lesions may be found in the uterus and Fallopian tubes. The endometritis is more severe than that found in other brucella infections. The placenta is often covered with a slimy exudate (Jubb and Kennedy, 1963). The organisms have a definite tendency to colonize in the chorionic epithelial cells. The aborted feti or full term dead piglets have no distinctive lesions.

Cattle are occasionally infected with *Br. suis*. The resultant mastitis constitutes a serious public health hazard (Washko *et al.*, 1948). In Iowa, three epidemics of human brucellosis resulted from *Br. suis* contamination of raw milk (Jordan, 1950).

Porcine and human infections described in Iowa (Borts *et al.*, 1946) and in other corn belt states were caused by an organism considered to be *Br. melitensis* because it had the then accepted differential cultural reactions for that species. It is now thought that that organism is a variety of *Br. suis* which has been called type 3 (Meyer, 1964). There is no difference in the disease produced and control of the disease is accomplished by the same means as in the more common variety of *Br. suis* infection (Hoerlein, 1952).

While there is a small degree of cross infection with the different species of brucella it is actually so infrequent that species specificity is the rule. In each of the different animal species a unique pathogenesis is found. In cattle the disease is almost never venereal, whereas in swine venereal transmission is the most common means of spread. Bulls are rarely infected, whereas boars are commonly infected. Cows seldom recover from the infection, but sows usually rid themselves of the organisms in a relatively short time. The agglutination test can be used effectively to determine infection in individual cattle; it cannot in swine. Cattle may be vaccinated prophylactically against brucellosis; swine cannot.

Because of characteristics common to the other brucellae, Simmons and Hall (1953) and Buddle and Boyes (1953) referred to the cause of ram epididymitis as *Brucella ovis*. Meyer and Cameron (1956) found the organism to have little similarity to the genus Brucella and suggested that it be called a Neisseria. Since reference to the bacterium as the "ram epididymitis organism" (R.E.O.) probably does little to resolve the conflict of nomenclature (McGowan *et al.*, 1961), we will use the original designation with the knowledge that it may be changed.

Br. ovis is an organism of low pathogenicity and produces a chronic infection of the epididymis of rams. It localizes in the tail of the epididymis causing perivascular edema and lymphocytic infiltration. There is eventual cicatrization, tubular epithelial degeneration, and epithelial hyperplasia which together obstruct the tubular lumen and produce stasis of its contents. These changes develop slowly over a period of months. Large numbers of organisms are excreted in the ejaculate. The subsequent lesions depend on rupture of the tubule immediately proximal to the stasis, allowing extravasation of sperm. Spermatic granulomas are formed which may cause the epididymis to be enlarged 4 to 5 times its normal size. Testicular degeneration results from stasis in the seminiferous tubules. Calcification, intratesticular spermatic granulomas of microscopic size, and fibrosis eventually cause sterility of the organ (Jubb and Kennedy, 1963).

Abortion is also produced in ewes by *Br. ovis*. The limited pathogenicity of the organism is again evident. The pregnant uterus is susceptible to infection only from 30 to 90 days after breeding (McGowan *et al.*, 1961). In the experiments of Osburn and Kennedy (1966) the ewes infected by the conjunctival route from 21 to 78 days after breeding produced all of the infected lambs. Those exposed earlier or later in pregnancy did not become infected. Intrauterine inoculation of the organisms was successful throughout pregnancy, but not at or before the time of breeding. The organism apparently has little ability to survive in the non-pregnant uterus. It is remarkable that the fetus should be able to survive for so long a period in the uterus in spite of the bacteria being present. Uteri inoculated with the organism at 42 days of gestation did not produce abortion of the feti until 64 days later (Osburn and Kennedy, 1966). McGowan *et al.* (1961) commented that *Br. ovis* needs a gravid uterus and a period of 3 to 4 months within the uterus to kill the fetus. The long term survival of the fetus in the infected uterus may be related to the fetal ability to produce antibodies against the infecting organism (Osburn and Kennedy, 1966).

In the experiments of Molello *et al.* (1963a), 90-day pregnant ewes when exposed by intravenous inoculation, only showed signs of impending abortion in 2 of the 16 at 45 days after inoculation. The other ewes in the group were infected when killed 45 to 48 days postinoculation, but had only moderate placental lesions with no signs suggesting abortion. It is probable that the pathogenicity of *Br. ovis* is too low to cause serious abortion problems in the field, but might well cause a small number of stillbirths and weak lambs.

It is reasonable to assume that the prolonged bacteremia of 2 to 8 weeks in rams (Biberstein *et al.*, 1964) is also present is pregnant ewes.

This may provide inoculum for the placentomal hematomas emphasized as being an important part of the pathogenesis by Jensen *et al.* (1961) and Molello *et al.* (1963a). These hematomas are formed in the ovine placenta at about 75 days (Amoroso, 1952). From the hilar zone of the placentome the organisms invade the chorionic epithelial cells and are found in large numbers in these cells. However, *Br. ovis* preferentially localizes in the interplacentomal tissue with considerable accumulation of exudate. Necrosis of the placentome was not very evident at 45 days (Molello *et al.,* 1963a). A constant finding is that of organisms in the fetal lung with pneumonic lesions. All of the feti in Osburn and Kennedy's report (1966) had antibodies against the organism and some of the aspects of arteritis of the placental vessels were interpreted by them as evidence of a local Arthus-type reaction.

Vibriosis is an important cause of infertility in sheep and cattle in all parts of the world. Except for almost absolute host specificity, *Vibrio fetus,* the cause of the disease, was thought to be identical whether isolated from sheep or cattle. It is now recognized that there are quantitative differences in antigenic structure and biochemical characteristics adequate to differentiate two varieties of the organism (Vandeplassche *et al.,* 1963).

Vibriosis in sheep is caused by *Vibrio fetus intestinalis,* so named because it may be found in the intestinal tract of sheep, cattle and swine. It is capable of causing epizootic abortion during the last trimester of pregnancy in sheep and rarely in cattle. The organism enters the body by ingestion and after a bacteremic phase (Jensen *et al.,* 1961) localizes in the pregnant uterus. Arteriolitis and thrombosis are produced in the hilar zone of the placentome with penetration into the hematomas from vessels in the degenerated septal tips. The microaerophilic environment of the hematomas allows proliferation of the bacteria. By active penetration or by phagocytic activity the organisms are carried to the fetal placenta where intracellular multiplication is so abundant that swelling of the chorionic epithelial cells is evident. The bacteria invade the chorionic capillaries in such numbers that they become occluded with colonies of *V. fetus.* The fetus probably dies from hypoxia. Extensive edema and pale discoloration of cotyledons readily distinguish infected from normal placentas. Some exudate mixed with blood was found between the endometrium and the chorion in intercotyledonary areas (Jensen *et al.,* 1961).

Bovine vibriosis is caused by *V. fetus venerealis* and is always a venereal infection spread by infected bulls to susceptible females at the time of breeding. The bulls acquire the infection on a transient or permanent basis from breeding infected cows. The organisms merely contaminate the external genitalia and produce no evidence of pathological change. The bacteria apparently find the crypts of the prepuce a favorable environment for multiplication. Permanently infected carrier bulls may be capable of spreading the infection for years.

The susceptible female is exposed during breeding when the infected bull deposits semen and vibrios into the genital tract. Normal fertilization takes place, but the resulting embryo is killed 2 to 3 weeks later by direct action of the bacteria (Adler, 1959). The only clinical sign is return to estrus. Large numbers of *V. fetus* are found in the cervico-vaginal mucus for 3 to 6 months after exposure. Recovery from the infection generally occurs in 2 to 9 months after the first exposure and birth of a normal healthy calf results even if rebred by infected bulls. The problem is thus one of temporary infertility which in herds bred seasonally may mean the loss of a calf crop. After recovery from the initial infection with *V. fetus* cows are fertile even if subsequently bred by infected bulls (Frank, 1958). Convalescent resistance is only partial with *V. fetus* being recoverable from the cervical mucus for a week or two instead of the 3 to 6 month period following initial infection. Some cows retain the infection through a normal gestation and become a source of reinfection of bulls. Prophylactic vaccination with *V. fetus* bacterins has been shown to provide a high degree of immunity to vibriosis (Hoerlein *et al.,* 1965).

A minor inflammatory reaction of the endometrium with lymphocytic infiltration has been found 3 or more weeks after breeding and infection with *V. fetus* (Vandeplassche *et al.,* 1963). Death of the embryo usually takes place before this time which suggests that the endometritis is not the cause but a result of the embryonal death. This mild lymphocytic infiltration is also known to persist beyond the period of infertility.

Infected cows occasionally abort from the 4th month of gestation until full term when dead or weak calves may be born. The placental changes are imperfectly described, but consist of necrosis of cotyledons, edema, and areas of opacity of the membranes. No satisfactory explanation exists for the pathogenesis of these late abortions.

Thus, in the case of vibriosis of sheep and cattle, very similar organisms have almost absolute host specificity, a unique pathogenesis, and an entirely different herd infertility problem. In sheep, the infection is apparently always contracted by ingestion of contaminated feed and water while in cattle the disease is always venereal. The pathogenesis of the ovine infection is well understood and the immunity resulting from infection or vaccination logically explainable. In cattle the disease is really an "external infection"; even in the cow the effects are in the lumen of the uterus. The mechanisms responsible for convalescent and vaccinal resistance to infection cannot be explained by present immunologic knowledge.

With the exception of infections caused by *Leptospira pomona* and *L. canicola,* leptospirosis is primarily a disease of rodents with man and domestic animals being accidental hosts. More than 60 serotypes have been described and given species designation (Galton, 1962). All infections with leptospires are characterized by leptospiremia, pyrexia, hemolytic signs, and kidney localization. Most infections are inapparent and recognized only by the subsequent presence of antibodies.

L. pomona is the only serotype of reproductive disease importance in domestic animals although other serotypes may cause infections (Turner et al., 1958; Roth and Galton, 1960). Abortions may occur as a sequel to infection in cattle, swine, and sheep. Transmission of the infection results from contact with urine-contaminated feed and water. Following ingestion of the organisms leptospiremia develops with pyrexia in about 4 to 8 days. Anemia, icterus, and hemoglobinuria may be noted 7 to 14 days postexposure. Lactating cows may have agalactia and abnormal milk. Leptospires are found in the urine from about the 11th day and may persist for months. Abortion in the last trimester of gestation occurring 3 to 4 weeks after infection may be the only sign of the infection.

Numerous investigators have endeavored to determine the pathogenesis of leptospiral abortion in cattle. Inability to cultivate leptospires from the aborted feti proved puzzling and the explanation of placental damage interfering with fetal nutrition seemed most plausible. Morter et al. (1958) supported this theory by their description of placental changes in experimentally infected cattle which attributed the basic mechanism of abortion to the increased connective tissue observed in the maternal caruncles. However, the investigations of Bjorkman (1954) demonstrated that these placental changes were the normal aging process of the placenta. Of the many attempts by researchers to isolate leptospires from aborted feti, only two were successful. Te Punga and Bishop (1953), Ringen et al. (1955), Fennestad and Borg-Petersen (1958), and Dacres and Kiesel (1958) had, however, visualized leptospires in aborted feti by means of silver impregnation. They attributed failure to isolate the organisms to the fact that they died soon after the fetus was killed.

Murphy (1966) appears to have cleared much of the controversy by killing pregnant experimental heifers 10 to 20 days after intravenous inoculation. Positive cultures were obtained from 5 of 17 feti from heifers killed 10 and 15 days postinoculation. Cultures from 5 feti found dead in utero from heifers killed 17 and 20 days postinoculation and one fetus aborted on the 18th day were negative for leptospires. Histopathologic studies of placentas from heifers killed at 10 and 15 days did not reveal changes differing from negative control heifers killed at the same stage of pregnancy. Placentas containing dead feti were edematous; the villi of the cotyledons were undergoing degeneration and necrosis. The epithelium of the septa was beginning to slough and undergo degeneration and necrosis. It was concluded therefore that fetal death had resulted from acute fetal leptospirosis and that the placental changes were concomitant with the fetal death.

Listeriosis is an infectious disease of man and animals most often associated with encephalitis and meningoencephalitis. It is mentioned here because of the increasing awareness of its role in human disease. In man a rather specific neonatal infection has been commonly reported from Europe according to the extensive review of this disease by Gray and Killinger (1966). Abortion, stillbirths, and neonatal infections have been reported in many animals. There have been several epizootics of abortion in sheep.

The epidemiology and pathogenesis of infections by *Listeria mono-cytogenes* is most intriguing, but not at all understood in either man or animals. Experimental infections of pregnant ewes produced by intravenous inoculation of the organism have provided a description of the pathological changes associated with abortion in sheep (Molello and Jensen, 1964). Abortions occurred on the 6th and 7th day postinoculation. Severe placentitis was noted in ewes killed on the 6th and 9th day postinoculation with marked mesodermal infiltrations adjacent to the areas of chorionic epithelial necrosis. Placentomal infarcts were common. Fetal livers had numerous, small areas of focal necrosis similar to the lesions noted in human stillbirths. Experimental infection using more natural methods of exposure has not been successful in producing reproductive disease of sheep.

SUMMARY

While many of the bacterial infertilities of domestic animals are caused by closely related organisms, most have a very definite host range and a unique pathogenesis in each species.

The infections of the female can be placed in three general groups:

1. Early embryonal death results from venereally transmitted infections: bovine vibriosis, porcine brucellosis, and occasionally *Br. ovis* infection in sheep.

2. Placental disease results from ingestion of the organism: brucellosis of cattle, sheep, goats, and swine, and vibriosis of sheep.

3. Fetal death is the cause of abortion in leptospirosis in cattle.

The male is important in the spread of only two of these infections: bovine vibriosis which in the bull produces no pathological change, and porcine brucellosis which in the boar produces lesions of the reproductive organs. In ram epididymitis, pathological lesions are a prominent part of the infection in the male, but he spreads the disease by venereal means only rarely.

References

Adler, H. C.: Genital vibriosis in the bovine: an experimental study on the influence on early embryonic mortality. Acta Vet. Scand. *1*:1, 1959.

Amoroso, E. C.: Placentation. In: Marshall's Physiology of Reproduction, A. S. Parkes, ed. Vol. II, 3rd ed. New York: Longmans, Green and Co., 1952.

Ariel, M.: The morbid anatomy of abortion of sheep caused by *Brucella melitensis*. Vet. Bull. *9*:455, 1939.

Biberstein, E. L., B. McGowan, H. Olander, and P. C. Kennedy: Epididymitis in rams. Studies on pathogenesis. Cornell Vet. *54*:27, 1964.

Bjorkman, N.: Morphological and histochemical studies on the bovine placenta. Acta Anat. Suppl. *22*:91, 1954.

Borts, I. H., S. H. McNutt, and C. F. Jordan: Occurrence of *Brucella melitensis* in Iowa. J. Am. Med. Assn. *130*:966, 1946.

Buddle, M. B., and B. W. Boyes: A brucella mutant causing genital disease of sheep in New Zealand. Austr. Vet. J. *29*:145, 1953.

Fennestad, K. L., and C. Borg-Peterson. Studies on bovine leptospirosis and abortion. IV. Demonstration of leptospira in foetuses from field cases of abortion in cattle. Nord. Vet. *10*:302, 1958.

Frank, A. H.: Vibriosis (A situation report). Proc. 62nd Ann. Meeting of U.S. Livestock San. Assn. p. 162, 1958.

Galton, M. M.: Methods in the laboratory diagnosis of leptospirosis. Ann. N.Y. Acad. of Sci. *98*:675, 1962.

Gray, M. L., and A. H. Killinger: *Listeria monocytogenes* and listeric infections. Bact. Rev. *30*:309, 1966.

Hoerlein, A. B.: Studies in swine brucellosis. I. The pathogenesis of *Brucella melitensis* infection. Am. J. Vet. Res. *13*:67, 1952.

Hoerlein, A. B., E. D. Hubbard, T. S. Leith, and H. E. Biester: Swine Brucellosis. Bulletin. Iowa St. College, Ames, Iowa. 1954.

Hoerlein, A. B., E. J. Carroll, T. Kramer, and W. H. Beckenhauer: Bovine vibriosis immunization. J. Am. Vet. Med. Assn. *146*:828, 1965.

Hutchings, L. M., and F. N. Andrews: Studies on brucellosis in swine. III. Brucella infection in the boar. Am. J. Vet. Res. *7*:379, 1946.

Jensen, R., V. A. Miller, and J. A. Molello: Placental pathology of ovine vibriosis. Am. J. Vet. Res. *22*:169, 1961.

Jordan, C. F.: in Brucellosis: A Symposium. Am. Assn. Advancement of Science, 1950.

Jubb, K. V. F., and P. C. Kennedy: Pathology of Domestic Animals. New York: Academic Press, 1963.

McGowan, B., E. L. Biberstein, D. R. Harrold, and R. A. Robinson: Epididymitis in rams: The effect of the ram epididymitis organism (REO) on the pregnant ewe. Proc. 65th Ann. Meet. U.S. Livestock San. Assn. 291, 1961.

Manthei, C. A.: Brucellosis in cattle. In Brucellosis: A Symposium. Am. Assn. Advancement of Science, 1950.

Meyer, M. E.: The epizootiology of brucellosis and its relationship to the identification of brucella organisms. Am. J. Vet. Res. *25*:553, 1964.

Meyer, M. E., and H. S. Cameron: Studies of the etiological agent of epididymitis in rams. Am. J. Vet. Res. *17*:495, 1956.

Molello, J. A., and R. Jensen: Placental pathology. IV. Placental lesions of sheep experimentally infected with *Listeria monocytogenes*. Am. J. Vet. Res. *25*:441, 1964.

Molello, J. A., R. Jensen, J. C. Flint, and J. R. Collier: Placental pathology. I. Placental lesions of sheep experimentally infected with *Brucella ovis*. Am. J. Vet. Res. *24*:897, 1963a.

Molello, J. A., J. C. Flint, J. R. Collier, and R. Jensen: Placental pathology. II. Placental lesions of sheep experimentally infected with *Brucella melitensis*. Am. J. Vet. Res. *24*:905, 1963b.

Molello, J. A., R. Jensen, J. R. Collier, and J. C. Flint: Placental pathology. III. Placental lesions of sheep experimentally infected with *Brucella abortus*. Am. J. Vet. Res. *24*:915, 1963c.

Morter, R. L., R. F. Langham, and E. V. Morse: Experimental leptospirosis. VI. Histopathology of the bovine placenta in *Leptospira pomona* infections. Am. J. Vet. Res. *19*:785, 1958.

Murphy, J. C.: Pathogenesis of leptospiral abortion in cattle. Thesis, Colorado State University, 1966.

Osburn, B. I., and P. C. Kennedy: Pathologic and immunologic responses of the fetal lamb to *Brucella ovis*. Path. Vet. *3*:110, 1966.

Roberts, S. J.: Veterinary Obstetrics and Genital Diseases. Ithaca: Roberts, 1956.

Simmons, G. C., and W. T. K. Hall: Epididymitis in rams. Preliminary studies on the occurrence and pathogenicity of a brucella-like organism. Austr. Vet. J. *29*:33, 1953.

Te Punga, W. A., and W. H. Bishop: Bovine abortion caused by infection with *Leptospira pomona*. N. Zealand Vet. J. *1*:143, 1953.

Turner, L. W., C. S. Roberts, A. M. Wiggins, A. D. Alexander, and L. C. Murphy: *Leptospira canicola* infection in a newborn calf. Am. J. Vet. Res. *19*:780, 1958.

Vandeplassche, M., A. Florent, R. Bouters, A. Huysman, E. Brone, and P. Dekeyser: The pathogenesis, epidemiology and treatment of Vibrio fetus infection in cattle. Compt. Rend. Recherches, Brussels, No. 29, 1963.

Washko, F. V., L. M. Hutchings, and C. R. Donham: Studies on the pathogenicity of *Brucella suis* for cattle. Am. J. Vet. Res. *9*:342, 1948.

FETAL INFECTIONS IN MAN

SHIRLEY G. DRISCOLL *

*Departments of Pathology, Boston Hospital for Women and
Harvard Medical School, Boston, Massachusetts*

Despite great progress in the control of infectious diseases, infections still rank among the commonest causes of perinatal death. It is difficult to estimate the relative frequency with which infections are responsible for wastage of human lives prior to so-called viability. The morphological characteristics of the abortus are often unrevealing, although decidual necrosis and inflammation are almost invariably found. Appropriate bacteriological studies are rarely made. The majority of *septic abortions* follow instrumentation of the uterus, although the victims of these infections claim to have aborted spontaneously. However, similar phenomena occasionally evolve without external intervention, especially when the abortion takes place during the second trimester of pregnancy. Both the clinical course and the laboratory data in these *bona fide* "natural" intrauterine infections recapitulate those associated with many premature births, occurring weeks later in gestation. Review of the case histories and pertinent laboratory data indicates that intrauterine infections, notably bacterial infections, are important causes of recurrent obstetrical catastrophes, such as habitual abortions, repetitive premature labors, and consecutive perinatal deaths (Driscoll, 1965; Benirschke and Driscoll, 1967).

The teratogenicity of viral infections in the human, so clearly exemplified by rubella embryopathy, has not been matched by any other infection of the gravida. Indeed the sublethal infections of antenatal life are associated with *fetopathy* and *placental disease,* rather than influencing the development of the *embryo.* How often the residua of sublethal fetal infections are the bases of such handicaps as mental retardation or neurological defects among survivors is unknown. The obvious examples

* Supported by Grants BP 2372, National Institutes of Neurological Diseases and Blindness, and HD 00598-04, U.S. Public Health Service, Department of Health, Education, and Welfare.

of infections which may have these results are congenital syphilis, toxo-plasmosis, cytomegalovirus disease, and rubella. Since bacterial infections are much commoner, but often more difficult to define and to diagnose than these, the important question is still unanswered: how often do the commonplace infections of fetal life cause permanent damage among the survivors?

PATHOGENESIS OF FETAL INFECTIONS

During the forty weeks of gestation, the gravida may experience a number of infectious diseases, the causative agents of which may then attack the conceptus. The products of conception are also exposed to the microbial flora of the lower genital tract of the gravida, another potential source of intrauterine infection. During *embryogenesis*, infections may be teratogenic or abortogenic. Later infections, which gain access during *fetal life*, may be lethal before birth, or may be responsible for disturbed growth and development, damage to the brain or other organs, or con-genital infectious disease among those who survive (Flamm, 1959; Benir-schke, 1960; Blanc, 1961; Töndury, 1964; Weller *et al.*, 1965).

Micro-organisms reach the fetus by any of these three routes: (1) through his circulation, in continuity with that of the chorionic villi; (2) through his body ostia, which are surrounded by amniotic fluid, per-mitting inhalation, swallowing, or ingress into the middle ear; and (3) by direct contact with the skin or mucous membranes, such as the con-junctiva (Fig. 1). When maternal infections are accompanied by hematog-enous dissemination of pathogens, such organisms may be deposited in the intervillous region or in the decidua, thence to damage the placental "barriers," and penetrate chorionic tissues and vessels. If, during implanta-tion and subsequent growth, the chorionic vesicle were to encounter focal infectious lesions, such as chronic myometrial toxoplasmosis, chronic salpingitis, or endometritis, spread to the conceptus might ensue by the same mechanisms. However, the commonest fetal infections are those which are transmitted by the amniotic fluid.

Nearly all fetal infections are accompanied by placental lesions, whose characteristics aid in interpreting pathogenesis, as well as in making the diagnosis under clinical circumstances. Most *overt* maternal infections which attack the fetus are spread by the maternal circulation, destroy villi, and enter the chorionic circulation. During maternal bacteremias and septicemias, spirochetemia (syphilis), parasitemia (acute toxoplas-mosis), and viremias, villous placentitis is produced, followed by cen-trifugal spread of organisms throughout the fetus.

When maternal septicemia leads to fetal infection, septic intervillous thrombosis is common, and bacteria may be demonstrated in smears, cul-

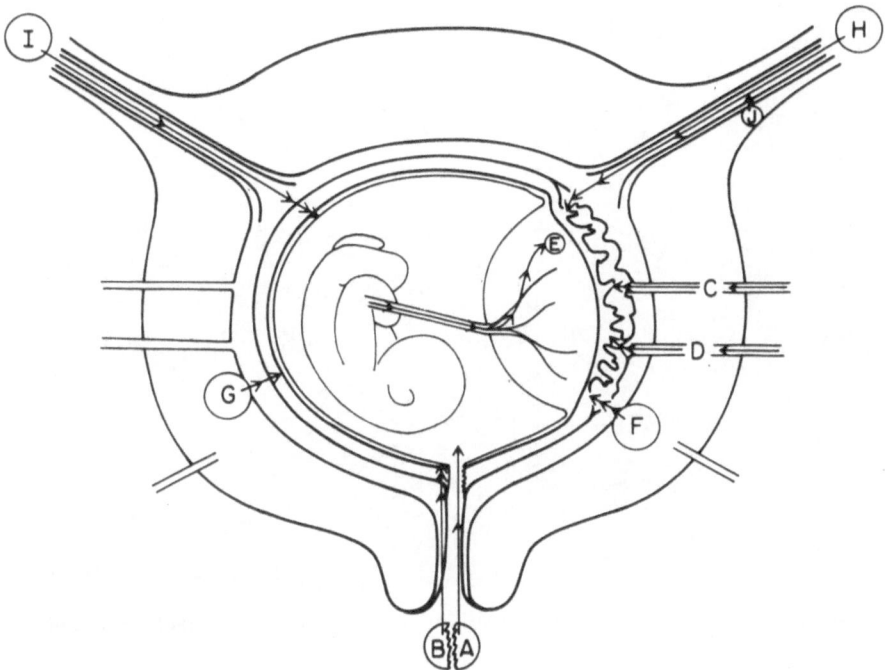

Fig. 1. Possible routes of access of micro-organisms to placenta and thence, to fetus. (A) After membranes rupture, organisms may ascend from lower genital tract. (B) With membranes intact, organisms may also ascend from lower genital tract to traverse decidua, chorion laeve, and amnion. (C) Maternal bacteremia may deposit organisms directly in intervillous space. (D) Maternal bacteremia may deposit organisms in decidua basalis, whence they may then invade contiguous placental villi. (E) During fetal bacteremia, organisms may be deposited in villous tissues. (F and G) Organisms lodged in a myometrial lesion, such as an abscess, invade contiguous decidua basalis (in F), or decidua vera (in G), thence to invade the intervillous space, the villi themselves, or the membranes. (H and I) From the peritoneal cavity, organisms traverse oviduct and lodge in intervillous space (in H), or decidua vera, thence through membranes to amniotic fluid (in I). (J) Organisms originally lodged within a tubal lesion traverse the tube to invade the intervillous space, as in H, or the decidua, as in F, G, or I.

tures, or histological sections of the thrombi (Fig. 2). The typical lesions of all blood-borne infections of the placenta comprise villous inflammation and necrosis, often affecting single villi or a few adjacent villi, which may appear to be agglutinated and are often surrounded by inflammatory cells. Such villositis is produced by toxoplasmosis, cytomegalovirus disease, a variety of bacteria, and, perhaps, by *Herpes simplex* (Witzleben and Driscoll, 1965). The extent of necrosis and the character and intensity of the inflammatory reaction vary, being most florid and acute with bacterial diseases, such as listeriosis and staphylococcemia, and relatively

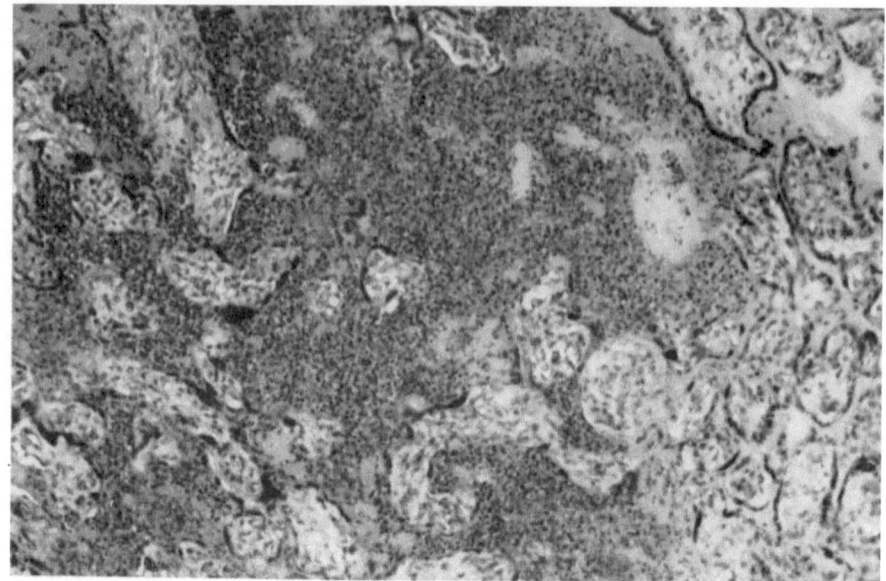

Fig. 2. Septic intervillous thrombosis, complicating maternal staphylococcemia.

indolent and chronic in toxoplasmosis. Fetal bacteremias, consequent to septic, induced abortions, are associated with severe villous placentitis, often characterized by masses of proliferating bacteria within villous and fetal vessels, obviously a source of bacterial toxins (Studdiford and Douglas, 1956). Vaccinia, variola, and varicella attack the chorionic villi, producing massive necrotizing inflammation, which involves large portions of the placenta (Lynch, 1932; MacArthur, 1952; Wielenga *et al.*, 1961; Entwistle *et al.*, 1962; Tucker and Sibson, 1962; Garcia, 1963; Killpack, 1963; Naidoo and Hirsch, 1963). Rubella placentitis is unusual in that the villous lesions are characterized by progressive sclerosing angiitis (Fig. 3). The histopathologic picture of such a placenta comprises lesions at all stages of evolution, from active, necrotizing villous inflammation, through various stages of sclerosis, to complete avascularity and atrophy. These changes suggest that rubella placentitis may be one specific cause of progressive reduction of the chorionic villous capillary surface, which may ultimately impair the exchange of gases and nutriments so necessary for fetal health and survival (Driscoll, 1966).

Infections which reach the fetus through his blood stream and produce this "parenchymatous" placentitis usually do not affect the umbilical cord and membranes. However, bacterial diseases sometimes involve these structures secondarily, as in fetal listeriosis. In toxoplasmosis, groups of parasites are seen within the connective tissues of cord and membranes, without an associated frank inflammatory reaction.

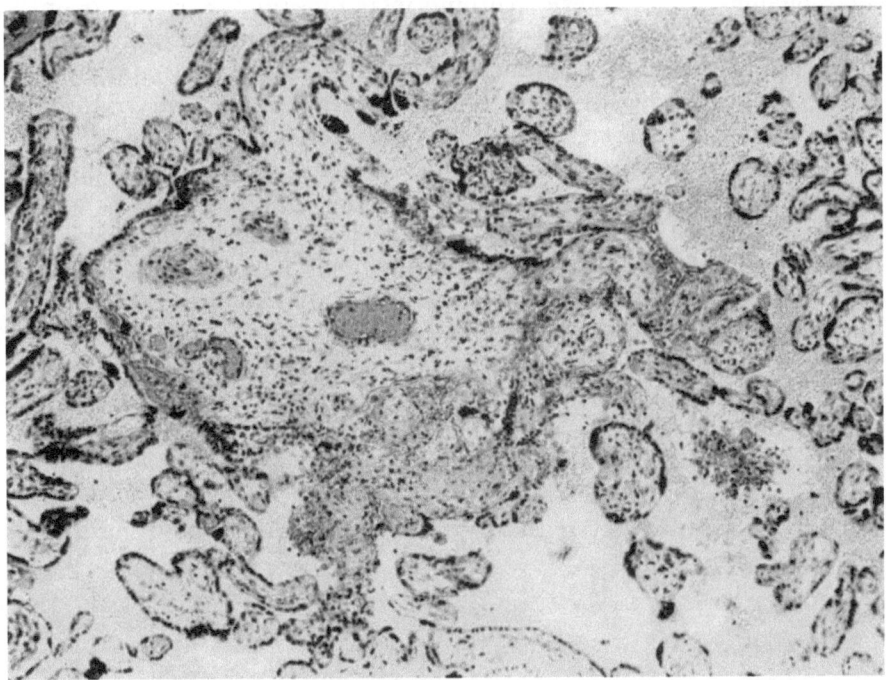

Fig. 3. Rubella placentitis, associated with persistent infection in the newborn infant. Both anomalies and "intrauterine growth retardation" were obvious. Progressive sclerosing angiitis of chorionic vessels produces a mixed picture of atrophic villi, active inflammation, and residual normal tissue.

Invasion of the chorionic circulation by micro-organisms is promptly followed by dissemination of the organisms throughout the fetus, and, perhaps, by afferent, *i.e.*, arterial, vessels, secondary spread to other portions of the placenta itself. Typical lesions may be found in viscera, brain, and bone marrow, their nature and extent being determined by the properties of the pathogen and the defenses of the new host. Listeriosis exemplifies fetal septicemia: minute foci of necrosis and inflammation are found at random in all tissues, so-called granulomatosis infantiseptica, and organisms can usually be easiliy identified (Driscoll *et al.*, 1962). Involvement of the intestine and lungs allows dissemination of bacilli to the amniotic fluid, with resulting chorioamnionitis and funisitis. Toxoplasmosis is a generalized infection, destructive lesions and typical organisms being found at autopsy in many fetal tissues, although meningoencephalitis is its most devastating result.

The clinical course of most infants with disseminated congenital infections is typical of septicemia. Depending on the sites of major involvement, icterus, hydrocephalus, neurological disturbances, pneumonia, etc., may predominate. However, chronic fetal infections, such as toxoplasmosis and syphilis, sometimes have unusual manifestations. For example, erythroblastosis fetalis is mimicked by the associated placental enlargement, neonatal edema, jaundice, hepatosplenomegaly, and erythro- or normoblastemia of some infants with these chronic infections (Fig. 4).

More infections are carried to the fetus by contaminated amniotic fluid than reach him hematogenously. The usual sequence is one of ascent of pathogens from the lower genital tract of the gravida, to invade adjacent fetal membranes, producing chorionic and amniotic infection, thence to gain access to the amniotic fluid (Fig. 5). Usually, the responsible bacteria comprise such agents as *Escherichia coli* and other coliforms, enterococci, or even *Staphylococcus albus,* common residents of the vagina in pregnancy. However, more virulent organisms, such as hemolytic *Staphylococcus aureus,* beta hemolytic Streptococcus, Pneumococcus, and Pleuropneumonia-like organisms, have followed the same pathways through the palcenta to the fetus. Except for *Candida albicans,* nonbacterial agents do not seem to produce amnionitis and transorificial fetal infections.

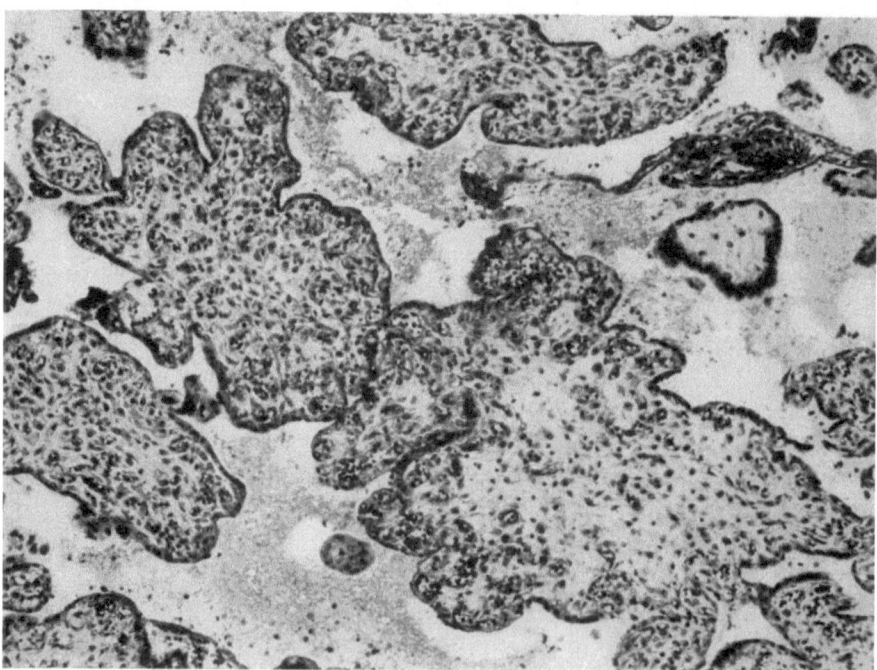

Fig. 4. Chronic fetoplacental toxoplasmosis, mimicking erythroblastosis fetalis. The villi are large, cellular and edematous, the trophoblast is well preserved and relatively thick, and there are many nucleated erythrocyte precursors in the circulating fetal blood.

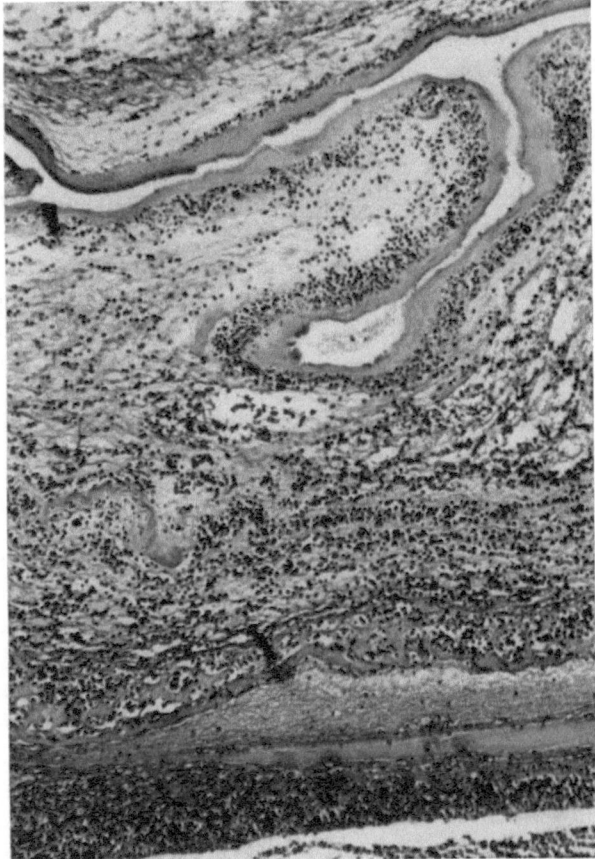

Fig. 5. Severe deciduochorioamnionitis. Massive polymorphonuclear leukocytic infiltration of the decidua vera, contiguous chorion laeve, and amnion. These granulocytes are of maternal origin.

With the flow of amniotic fluid into the fetal respiratory passages, the commonest variant of fetal and congenital infections, aspiration pneumonia, is produced. Swallowed material, similar to that aspirated, may fill the stomach, and acute mucosal inflammation sometimes ensues. Rarely, the contaminated amniotic fluid carries infectious material to the middle ear or the conjunctiva. In several instances, congenital cutaneous infections with *Candida albicans* have been produced by direct contact of fungus-bearing fluid with the fetal skin (Sonnenschein *et al.*, 1960, 1964). Of course, secondary hematogenous dissemination of any of the initially amniotic infections may occur at any time, superimposed on the initial picture.

Histologically, congenital aspiration pneumonia, secondary to bacterial amnionitis, is characterized by the presence of aspirated particulate matter in the lungs, including polymorphonuclear leukocytes, bacteria, squames, and debris, accompanied by local evidence of inflammation in response to this aspirate. Leukocytes emigrate from pulmonary vessels into the adjacent pulmonary tissues, and may become mixed with the aspirated material within the air spaces. Rarely, the tissues of the fetal lung may contain mononuclear inflammatory cells, including plasma cells.

The placental membranitis which precedes transamniotic fetal infections is commonly associated with premature and with prolonged rupture of the membranes, and prolonged labor, and is sometimes accompanied by maternal fever and tachycardia. However, unlike the blood-borne infections which spread from gravida to fetus, transorificial fetal infections may give rise to few, if any, clinical signs in the mother.

CLINICAL IMPLICATIONS

During the three years, 1963 through 1965, the perinatal death rate at the Boston Hospital for Women was 26 per 1000 births at gestational ages of 20 weeks or greater. Necropsies were performed in 90 per cent of these cases and perinatal death was attributed to infection of the fetus or neonate in 16 per cent, a rate not appreciably different from that recorded some 15 years ago. Among *fetal deaths,* so many of which are not satisfactorily explained even on study of all available clinical data as well as postmortem examination, 6 per cent were ascribed to fetal infections. Only erythroblastosis fetalis and fetal anomalies outnumbered the infections. Infections placed second among the causes of *neonatal death,* being exceeded only by "hyaline membrane disease," the prime killer of premature infants. Many infections lethal to the neonate began prior to his birth, and many of the others were acquired postnatally by infants otherwise handicapped from birth.

An accurate estimate of the frequency of neonatal infections among infants who survive is very difficult to obtain. Vague respiratory disorders are sometimes attributed to aspiration pneumonia, which cannot be proven to be of microbial etiology. Among all instances of known or suspected neonatal infections, especially those manifested within the first 48 hours after birth, an associated placentitis, usually chorioamnionitis, can often be demonstrated.

Spontaneous abortions during the second trimester are also commonly associated with placental membranitis, demonstrably caused by bacterial infection. These small fetuses and placentas show the same features of inflammation of membranes, and aspiration and deglutition of contaminated amniotic fluid as are seen with births later in gestation, and the same bacteria are implicated in their pathogenesis (Fig. 6).

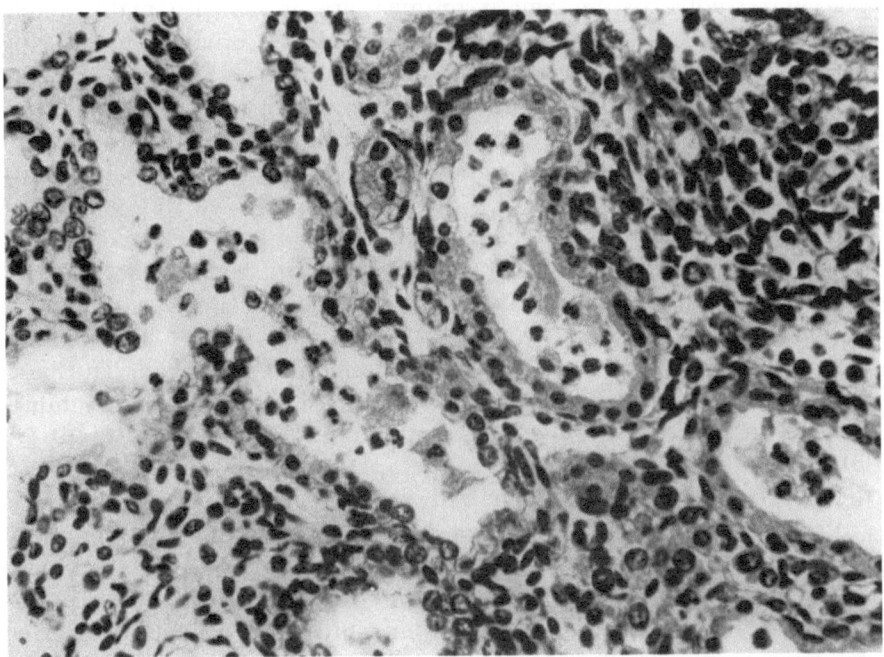

Fig. 6. Aspiration of particulate matter from the amniotic sac in the presence of severe amnionitis, chorionitis and deciduitis. Spontaneous abortion in the middle trimester.

Rarely, such organisms as *Toxoplasma gondii, Listeria monocytogenes,* and *Vibrio fetus* have been observed to produce intrauterine infections and abortions (Benirschke and Driscoll, 1967). In general, these infections of early gestation seem to have the same pathogenesis as those occurring later. Reports that chronic toxoplasmosis and listeriosis may be responsible for recurrent pregnancy wastage have been challenged (Seeliger, 1957; Rost *et al.,* 1958; Rappaport *et al.,* 1960; Weinmann, 1960; Dungal, 1961; MacNaughton, 1962; Ruffolo *et al.,* 1962; Gray *et al.,* 1963; Rabau and David, 1963; Langer, 1963; Werner *et al.,* 1963; Remington *et al.,* 1964). Further study will be required to clarify these particular issues.

CORRELATIVE STUDIES OF PLACENTITIS

Although the correlations of perinatal infections with placental inflammation are striking, examination of consecutive, unselected placentas indicates that many healthy infants have also been exposed to "amnionitis" (Driscoll, 1965). Bacteriological cultures of placentas associated with fetal and congenital bacterial infections demonstrate pathogens within and upon the placental membranes. However, many inflamed placentas, studied by the same techniques, are sterile. Also, many amnia, which have been sterile while *in situ,* become contaminated during passage through the birth canal.

In an effort to evaluate the rôle of chorioamnionitis and umbilical angiitis in the development of fetal infections and in relation to the outcome of pregnancy, we have studied 3623 unselected pregnancies, delivered at gestational ages of 20 weeks or greater. This work was carried out within the framework of the Collaborative Study on Cerebral Palsy, Mental Retardation, and Other Neurological and Sensory Disorders of Infancy and Childhood, and with collaboration of Doctors John Mac Laren, William Cochran, Luke Gillespie, and Stewart H. Clifford. Standard clinical data were coded from prenatal, intrapartum, and postnatal records of each case, and a few data were obtained from examinations of the progeny at four months and at one year of age. Cultures were taken from the amniotic surface of each placenta and from the throat of the offspring immediately after delivery. All placentas were examined grossly and histologically. Necropsies of perinatal deaths were also reviewed in detail. There were no maternal deaths.

Twenty-six and six tenths per cent of the study group were primigravid, 73.4 per cent, multigravid women. In 6 per cent of the series, delivery occurred at or prior to 36 weeks' gestation. Six and nine tenths per cent of the offspring weighed less than 2500 grams at birth. Thirty-four sets of twins were observed. The perinatal mortality of the entire series was 2.1 per cent, comprising 38 stillbirths and 39 neonatal deaths. Sixty per cent of the neonatal deaths occurred among prematures.

The placentas of twelve of the 39 infants who expired neonatally showed inflammation of the membranes, and 13, umbilical vascular inflammation. Among 12 neonatal deaths associated with chorioamnionitis, six infants weighed less than 1000 gm., and nine, less than 1500 gm. at birth. Of the 13 with umbilical angiitis, five weighed less than 1000 gm., and eight less than 1500 gm. Although some of these lesions were associated with positive cultures taken at birth, many were not so associated, and so-called positive cultures were frequent among healthy births with normal placentas and survival. Three of the stillbirths were attributed to intrauterine bacterial infections. Thirteen neonates died of bacterial infections, most commonly pneumonia. Infections were diagnosed in another 27 infants who recovered and survived the neonatal period. The perinatal incidence of infections was 1.2 per cent, or one in 80 births. However, infections again accounted for a large fraction of all perinatal deaths: 21 per cent of the total, 7 per cent of stillbirths, and 33 per cent of neonatal wastage. The three fetal deaths due to infection were associated with chorioamnionitis, two also showing umbilical angiitis. The placentas of six of the 13 infants who died of infection manifested membranitis, and six, umbilical vasculitis. Since the neonatal deaths ascribed to infection occurred at ages ranging from two hours to 26 days, the absence of placentitis in some cases is not surprising, although all 13 infants had been ill from birth. It was interesting that a total of 20 of the 38 stillbirths were associated with membranitis, probably often a sequel to fetal death, and eight had umbilical angiitis. Of the 13 lethal infections, three were in infants weighing less than one kilogram, four, less than 1500 gm., at birth.

The mothers of infants in whom perinatal infections were diagnosed, either in life or at autopsy, reported a higher frequency of previous reproductive failures than was true of the total population studied. Of 89 previous pregnancies, only 58 per cent delivered at term, compared to 87 per cent of previous pregnancies of the controls. This difference is highly significant statistically.

Maternal fever and tachycardia were common during labors of women whose babies were found to be infected. Again, prolonged rupture of the membranes was relatively common, as were longer labors. Fetal tachycardia was recorded during labor in 11 per cent of the total series; consecutive readings exceeding 160 per minute were recorded in 4 per cent of cases. Among those found to be infected, 31 per cent had exhibited some fetal tachycardia, and in 13 per cent, the increased fetal heart rate was sustained.

The poorest correlations between laboratory data, *i.e.*, results of histological and bacteriological examinations of placentas, and clinical diagnoses of infections were observed among cases of so-called aspiration pneumonia in term infants who survived. Some of these may have been instances of aspiration of sterile particulate material, rather than true bacterial infections.

"Positive" cultures were numerous, which is not surprising in view of the fact that most of these placentas had passed through the cervix and vagina, and some had also been handled, prior to culturing. Among organisms recovered, bacteria resident in the vagina and rectum predominated.

There was no clear evidence that antimicrobials given to the gravida influenced the *clinical* outcome. However, since the management of cases in the series was not intended to test the efficacy of such therapy, these findings may merely reflect the criteria upon which the clinical decision to treat had been based.

Postpartum fever and foul lochia were significantly increased in frequency when the placentas had been inflamed, indicating the possible diagnostic value of placental examination in cases of residual intrauterine infection.

What information is useful in determining which infant is actually infected? Our combined correlative review indicates certain characteristics of the "high risk pregnancy," with respect to fetal infections: previous pregnancy failures of the gravida, prolonged rupture of membranes, premature labor, prolonged labor, increased frequency of rectal or vaginal examinations during labor, and elevated maternal temperature and/or pulse rate. These factors seem to be related to increased risks of ascent of bacteria from the lower genital tract. Fetal tachycardia also suggests intrauterine infection. Bacteriological cultures provide supplemental information concerning the organisms which may be implicated in bacterial placentitis.

Longitudinal studies of obstetrical histories of some women with *recurrent* perinatal wastage have shown the syndrome of recurrent chorioamnionitis and early premature birth. In such instances, bacterial infections have been documented by cultures of the placenta and from tissues of the fetus or premature infants, the same organisms being recovered repeatedly, from successive pregnancy failures.

Obviously, a prospective series yielding a larger number of frank infections will be required for an adequate study of the many factors which may be relevant to congenital bacterial infections in man.

This study demonstrated a surprisingly high frequency of the *histological lesions,* placental membranitis and umbilical angiitis, among unselected births. The overall incidence of *chorionitis* was 16 per cent, that of *amnionitis,* 11 per cent, and umbilical *vasculitis,* 19 per cent, most of the latter being associated with some inflammation of the membranes. Obviously, these figures exceed the perinatal and maternal morbidity from all causes, and are much in excess of the clinical evidences of intrauterine infection. Cultures of these placentas sometimes yielded growth of significant bacteria, but the incidence of histological lesions was not clearly correlated with an increased frequency of such results, as compared to the non-inflamed placentas. Like the frank infections, these lesions occurred with increasing frequency following prolonged rupture of membranes and long labors. However, they were also found when membranes had been intact until moments before birth, and following labors of less than two hours' duration. Other obstetrical factors positively associated with inflammation of the fetal membranes and umbilical vessels comprised intrapartum fever, even though slight, maternal tachycardia, total number of rectal and vaginal examinations during labor, and, again, postpartum morbidity, indicated by fever, and/or foul lochia. There was no evidence of increased risks of placentitis in association with primigravidity, asymptomatic bacteriuria, anemia, third trimester vaginal bleeding, clinical vaginitis or cervicitis, or recent history of upper respiratory infection. Administration of antimicrobial agents to the gravida prior to delivery, usually begun within six hours of delivery, was associated with a decreased incidence of umbilical vasculitis and chorionitis at birth. The latter observation may be an effect of the criteria by which treatment was selected among clinic patients, rather than an indication of the efficacy of such therapy. Available data do not permit a confident interpretation of the influence of antimicrobials in comparable case material.

The inflammatory lesions were most strikingly correlated with premature birth, their incidence, especially that of chorioamnionitis, varying inversely with the gestational age at which delivery occurred. For example, among infants weighing over 2500 gm. at birth, the incidence of chorioamnionitis was 10 per cent; among those weighing less than 2500 gm., 20 per cent showed chorioamnionitis. At birth weights between 1000 gm. and 2500 gm., 18 per cent had membranitis; 36 per cent of those under 1800 gm. at birth had this lesion, as did 50 per cent of those born at weights of less than one kilogram. Similarly, umbilical angiitis was seen in 42 per cent of cases delivered between 20 weeks and 28 weeks of gestation, 26 per cent of those of gestational ages from 28 through 32 weeks, 15 per cent of cases at 32 through 36 weeks, 17 per cent of those terminating at 36 through 40 weeks, and 19 per cent of cases in which delivery took place after 40 weeks. Among middle trimester abortions, the subject of another study, chorioamnionitis was still more frequent, being found in 66 per cent of cases (Driscoll, 1965). Umbilical angiitis is less common at the early gestational ages of less than 20 weeks, a finding which seems to reflect the fetal incapacity to respond, since chorioamnionitis is often intense in the same specimen.

Correlations of pediatric interest included the obstetrical observations cited previously, especially the strong association with prematurity. Hyperbilirubinemia, defined as a maximum serum bilirubin of more than 16 mg. per cent, was significantly associated with inflammation of the umbilical vessels and fetal membranes. Abnormal Apgar scores were not more frequent in cases of placentitis.

Its striking predilection to involve the placentas of premature births may be the most important aspect of chorioamnionitis. Perhaps the onset of premature labor and the pathogenesis of the placental inflammation are sometimes linked to a common cause. The impact of chorioamnionitis on premature infants seems to be related to the circumstances of their births, rather than to the specific nature of their neonatal diseases. This became evident when two groups of liveborn premature infants of similar gestational ages and birth weights were compared, one group being associated with chorioamnionitis, and the other, without such placental inflammation (Driscoll, 1965). Prognosis for survival was unrelated to the presence of membranitis, but was strongly correlated with the degree of maturity.

Evidence that chorioamnionitis may subside, or "heal" prior to birth is rarely obtained. This fact suggests that the conditions which lead to the infiltration of the fetal membranes are parts of an irreversible sequence, which can only end with the evacuation of the uterus. On rare occasions, a frank intrauterine bacterial infection has been treated intensively, and successfully, the gestation proceeding to termination many weeks later. At the time of delivery, the connective tissue of the fetal membranes is densely hyalinized, and slightly more cellular than normal, but free of frank inflammation.

CLOSING COMMENTS AND HYPOTHESIS

Inflammation of the fetal membranes, *i.e.,* chorioamnionitis, is an abnormal, albeit a common, condition. Its incidence varies inversely with the gestational age at which spontaneous delivery takes place (Driscoll, 1965). The leukocytic infiltrate is composed exclusively of granulocytes, which first collect in necrotic decidua, then invade contiguous chorionic membrane, thence to infiltrate the apposed amnion. Similar infiltration of the film of fibrin beneath the chorionic plate, with progression to involve the latter tissue and its adjacent amnion, begins later than that of the extraplacental membranes, and is usually less intense. The migrating leukocytes which are found in these areas are *maternal* cells. *Fetal* granulocytes also emigrate, into the media of the umbilical vessels and the vessels of the chorionic plate. An unusual feature of all these reactions is the "amniotropism" of the leukocytic migration. The advancing zone of leukocytic infiltration is not uniform and concentric within the vascular wall. On the contrary, the infiltrated sector occupies that portion of the vessel which is nearest to the amniotic fluid. Both the membranitis and the umbilical and chorionic angiitis are peculiarly limited to those tissues which are bathed in amniotic fluid, or in immediate contiguity with such tissues. In spite of being in close proximity to severely inflamed membranes and in physical continuity with infiltrated connective tissues and vessels of the chorionic plate, the stroma and blood vessels of even the major stem villi are rarely involved in the reaction. Similarly, the umbilical angiitis, even when it is massive in degree, usually stops at the abdominal wall of the fetus. Even in monochorionic twin gestations, with patent anastomotic channels joining the circulations of the twins, maternal and fetal leukocytes are often drawn toward one amniotic sac, while the other remains unaffected. When chorioamnionitis and umbilical angiitis occur in twin gestations, the first-born placenta, *i.e.,* that which lay nearer the lower pole of the ovular spheroid and contiguous to the cervix and vagina, is always the one affected. When such inflammation of the fetal adnexa is found in the second placenta, its predecessor was also inflamed, and usually more intensely so (Benirschke and Driscoll, 1967).

These morphologic characteristics of so-called chorioamnionitis, umbilical vasculitis, and funisitis imply that such lesions are initiated by a leukocytotactic stimulus from the amnion, probably within the amniotic fluid. Sometimes bacterial contamination of the amniotic fluid can be demonstrated, and infection is believed to have produced the sequence of placental membranitis and fetal vasculitis. Indeed, bacterial amnionitis and transorificial fetal infections, notably aspiration pneumonia and gastroenteritis, are invariably associated with deciduochorionitis, and often with umbilical and chorionic angiitis. However, most efforts to implicate microbial agents in the etiology of membranitis, as this lesion is observed in placentas selected at random rather than those chosen because of perinatal illness, have been disappointing; smears, cultures, and stained sections commonly fail to demonstrate organisms in the affected tissues. Moreover, clinical evidences of overt infection are often absent in both the gravida and the fetus or neonate, in spite of florid chorioamnionitis.

On the other hand, meconium, or its products, mixed with the amniotic fluid and incubated for hours *in utero,* provokes acute inflammation of the membranes and fetal vessels, without the participation of microbial agents. Also, chorioamnionitis develops following fetal death, if the conceptus is retained, even in the absence of secondary bacterial invasion of the uterine contents.

These observations are compatible with the hypothesis that the leukocytic infiltration of membranes and fetal vessels is commonly initiated by factors other than microbial infections. Because of the decidual necrosis which is invariably associated with these infiltrates, exceeding in degree and extent that seen in normal gestation, a product of degenerating decidua is suggested as the leukocytotactic substance. The etiology of the postulated initial necrosis of the decidua vera might be ischemic, hormonoprival, a result of mechanical pressure, or even, in some instances, bacterial damage. However, the usual bacterial infection of the membranes is relegated to secondary importance, as it is regarded as a complication of chorioamnionitis. In such instances, the bacterial flora of the cervix and vagina participate as opportunists, invading the already devitalized contiguous decidua and membranes, thence to contaminate the amniotic fluid and reach the fetus. While the damaged tissue may thus act as a *locus minoris resistentiae,* permitting fetal infection, more stress should be placed on the possible relationships of deciduochorionitis to the causes of premature spontaneous labor. The impact of premature birth on the individual and the community demand that any clue to pathogenetic mechanisms be investigated both epidemiologically and experimentally.

It would also be important to determine whether fetal membranes and decidua which are sufficiently damaged to permit egress of leukocytes from vessels and their migration to the amnion and the bathing fluids still act as effective barriers, or become more permeable, permitting leakage of other materials to and from the ovisac. Data on the functional effects of placental membranitis may have some additional bearing on fetal welfare and the success of pregnancy.

References

Benirschke, K.: Routes and types of infection in the fetus and the newborn. A.M.A. J. Dis. Child. *99:*714, 1960.

Benirschke, K., and S. G. Driscoll: The pathology of the human placenta, in: Placenta, Band VII/5, Handbuch der speziellen pathologischen Anatomie und Histologie. Berlin: Springer-Verlag, 1967.

Blanc, W. A.: Pathways of fetal and early neonatal infection. Viral placentitis, bacterial and fungal chorioamnionitis. J. Pediat. *59:*473, 1961.

Driscoll, S. G.: Histopathological observations in gestational rubella. (*In preparation*) 1966.

——: Pathology and the developing fetus. Pediat. Clin. N. Amer. *12:*493, 1965.

Driscoll, S. G., A. Gorbach, and D. Feldman: Congenital listeriosis: diagnosis from placental studies. Obstet. Gynec. *20:*215, 1962.

Dungal, N.: Listeriosis in four siblings. Lancet *ii:*513, 1961.

Entwistle, D. M., P. T. Bray, and K. M. Laurence: Prenatal infection with vaccinia virus: report of a case. Brit. Med. J. *ii:*238, 1962.

Flamm, H.: Die pränatalen Infektionen des Menschen. Stuttgart: Georg Thieme Verlag, 1959.

Garcia, A. G. P.: Fetal infection in chicken pox and alastrim, with histopathologic study of the placenta. Pediatrics *32:*895, 1963.

Gray, M. L., H. P. R. Seeliger, and J. Potel: Perinatal infections due to Listeria monocytogenes. Do these affect subsequent pregnancies? Clin. Pediatrics *2:*614, 1963.

Killpack, W. S.: Prenatal vaccinia. Lancet *i:*388, 1963.

Langer, H.: Repeated congenital infection with *Toxoplasma gondii.* Obstet. Gynec. *21:*318, 1963.

Langer, H., and H. Geissler: Nachweis von Toxoplasmosen bei Aborten und Frühgeburten. Arch. Gynäk. *192:*304, 1960.

Lynch, F. W.: Dermatologic conditions of the fetus. Arch. Derm. Syph. *26:*997, 1932.

MacArthur, P.: Congenital vaccinia and vaccinia gravidarum. Lancet *ii:*1104, 1952.

MacNaughton, M. C.: *Listeria monocytogenes* in abortion. Lancet *ii:*484, 1962.

Naidoo, P., and H. Hirsch: Prenatal vaccinia. Lancet *i:*196, 1963.

Rabau, E., and A. David: *Listeria monocytogenes* in abortion. Lancet *i:*228, 1963.

Rappaport, F., M. Rabinovitz, R. Toaff, and N. Krochik: Genital listeriosis as a cause of repeated abortion. Lancet *i:*1273, 1960.

Remington, J. S., J. W. Newell, and E. Cavanaugh: Spontaneous abortion and chronic toxoplasmosis. Obstet. Gynec. *24:*25, 1964.

Rost, H. F., H. Paul, and H. P. R. Seeliger: Habitueller Abort und Listeriose. Dtsch. Med. Wschr. *83:*1893, 1958.

Ruffolo, E. H., R. B. Wilson, and L. A. Weed: *Listeria monocytogenes* as a cause of pregnancy wastage. Obstet. Gynec. *19*:533, 1962.

Seeliger, H. P. R.: Some new aspects of human listeriosis. In: Human Listeriosis. Its Nature and Diagnosis. United States Dept. of Health, Education, and Welfare. Public Health Service Communicable Disease Center, Atlanta, Ga., 1957.

Sonnenschein, H. H., H. L. Clark, and C. L. Taschdjian: Congenital cutaneous candidiasis in a premature infant. A.M.A. J. Dis. Child. *99*:81, 1960.

Sonnenschein, H., C. L. Taschdjian, and H. L. Clark.: Congenital cutaneous candidiasis. A.M.A. J. Dis. Child. *107*:260, 1964.

Studdiford, W. E., and G. W. Douglas: Placental bacteremia: a significant finding in septic abortion accompanied by vascular collapse. Am. J. Obstet. Gynec. *71*:842, 1956.

Töndury, G.: Über den Infektionsweg und die Pathogenese von Virusschädigungen beim menschlichen Keimling. Bull. Schweiz. Akad. Med. Wissensch. *20*:379, 1964.

Tucker, S. M., and D. E. Sibson: Foetal complication of vaccination in pregnancy. Brit. Med. J. *ii*:237, 1962.

Weinmann, D.: Toxoplasma and abortion. Fertil. Steril. *11*:525, 1960.

Weller, T. H., C. A. Alford, and F. A. Neva: Changing epidemiologic concepts of rubella, with particular reference to unique characteristics of the congenital infection. Yale J. Biol. Med. *37*:455, 1965.

Werner, H., L. Schmidtke, and G. Thomascheck: Toxoplasma-Infektion und Schwangerschaft. Der histologische Nachweis des intrauterinen Infektionsweges. Klin. Wschr. *41*:96, 1963.

Wielenga, G., H. A. E. Van Tongeren, A. H. Ferguson, and T. G. Van Rijssel: Prenatal infection with vaccinia virus. Lancet *i*:258, 1961.

Witzleben, C. L., and S. G. Driscoll: Possible transplacental transmission of Herpes simplex infection. Pediatrics *36*:192, 1965.

TOXOPLASMOSIS *

J. K. Frenkel

University of Kansas Medical Center, Kansas City, Kansas

The problem of fetal infection has been associated with toxoplasmosis ever since the discovery by Wolf and Cowen (1939) of the occurrence of congenital toxoplasmosis in man. Whether transmission occurs only during the acute or also the chronic stage of maternal infection, and whether only single or also multiple pregnancies can be infected is still now being debated. Fetal infection has been observed spontaneously also in sheep, pigs, dogs and mice, and produced experimentally in a variety of animals (Siim *et al.*, 1963). Additional modes of transmission include carnivorous habits (Desmonts, Couvreur, Alison *et al.*, 1965) and the ingestion of Toxoplasma-infected nematode eggs (Hutchison, 1965).

To gain sufficient perspective for evaluation of the various claims and hypotheses, we will first survey toxoplasmosis in general, and discuss the problem of fetal infection, possible dystrophic effects on the placenta and habitual abortion. Finally, suggestions concerning diagnosis, prevention and treatment will be made.

MICROBE

Toxoplasma gondii is an obligate intracellular protozoan which appears related to the Sporozoa. Qualities contributing to its pathogenicity include the capacity to utilize substrates offered in a wide variety of host cells, its rate of multiplication and invasiveness, *i.e.*, its ability to spread, and finally, its ability to resist intracellular digestion, especially after immunity has developed. Dissemination throughout the body is mainly by blood stream and lymphatics, both free as well as within leucocytes. Contiguous extension from cell to cell plays an important role since contact with extracellular antibody is avoided.

* Supported in part by Grants AI-7489 and AI-3344 from the National Institutes of Allergy and Infectious Diseases, United States Public Health Service.

Primary infection is accompanied by an active period of multiplication, resulting in "proliferative forms" or "trophozoites" (Fig. 1); the stage of intracellular proliferation just prior to rupture of the cell has been called a "terminal colony." During chronic infection, in the presence of pre-munition, "cysts" of Toxoplasma are found in many organs, such as brain, eye, muscle, and depending on the host, also in lung, uterus, adrenals and elsewhere (Fig. 1). The cysts are characterized by an argy-rophilic, PAS-positive (diastase resistant) cyst wall and by large glycogen granules in each organism. Trophozoites also persist during the chronic infection. The term "pseudocyst" has been originally applied to the form now called "cyst," and subsequently to "terminal colony," with the result that the term has lost all precision. Attempts to homologize stages in tissue culture with those found in intact hosts remain debatable in view of the absence of the prime cyst characteristics of argyrophilia and glycogen content. Instead, the characteristic of resistance to peptic diges-tion is used, which has first been described for intracerebral cysts, and later for forms in tissue culture. This subject has been discussed exhaus-tively elsewhere (Survey of Ophthalmology, 6: (6, part II) 724–758, 1961). It is summarized mainly to clarify the terminology used here.

INFECTION

The biologic fact of *"infection,"* and clinical *"disease"* are distin-guished, although they are not necessarily different immunologically. *Immunity* (acquired, total) is used here to describe the altered host responses developing one to several weeks after infection which tend to contain, diminish and terminate infection. In toxoplasmosis this includes cellular immunity, reflected by enhanced phagocytosis and digestion of microorganisms (Frenkel and Lunde, 1966; Frenkel, 1967). Likewise, there is antibody-related immunity as measured by the neutralization test, dye test, hemagglutination test, fluorescent antibody test, precipitin test, and the direct agglutination test (Jacobs, 1963). Seven-S and 19-S antibodies have been separated (Remintgon and Miller, 1966). Lysis of extracellular Toxoplasma has been shown by Desmonts (1955) to take place during the *in vitro* dye test. If, as is probable, this phenomenon occurred *in vivo* it would curb hematogenous and intracellular dissemi-nation of infection. However, as shown vividly by Sabin and Feldman (1948), antibody does not affect intracellular Toxoplasma.

Hypersensitivity is used to characterize the exaggerated inflammatory response, and frequently necrosis, following prior sensitization, with harmful effect on the host. It appears as early as a few days after infection (in hamsters infected with the related Besnoitia organism), although immunity is not acquired until the end of the third week (Frenkel, 1967). While frequently occurring together in a host, phenomena accompanying

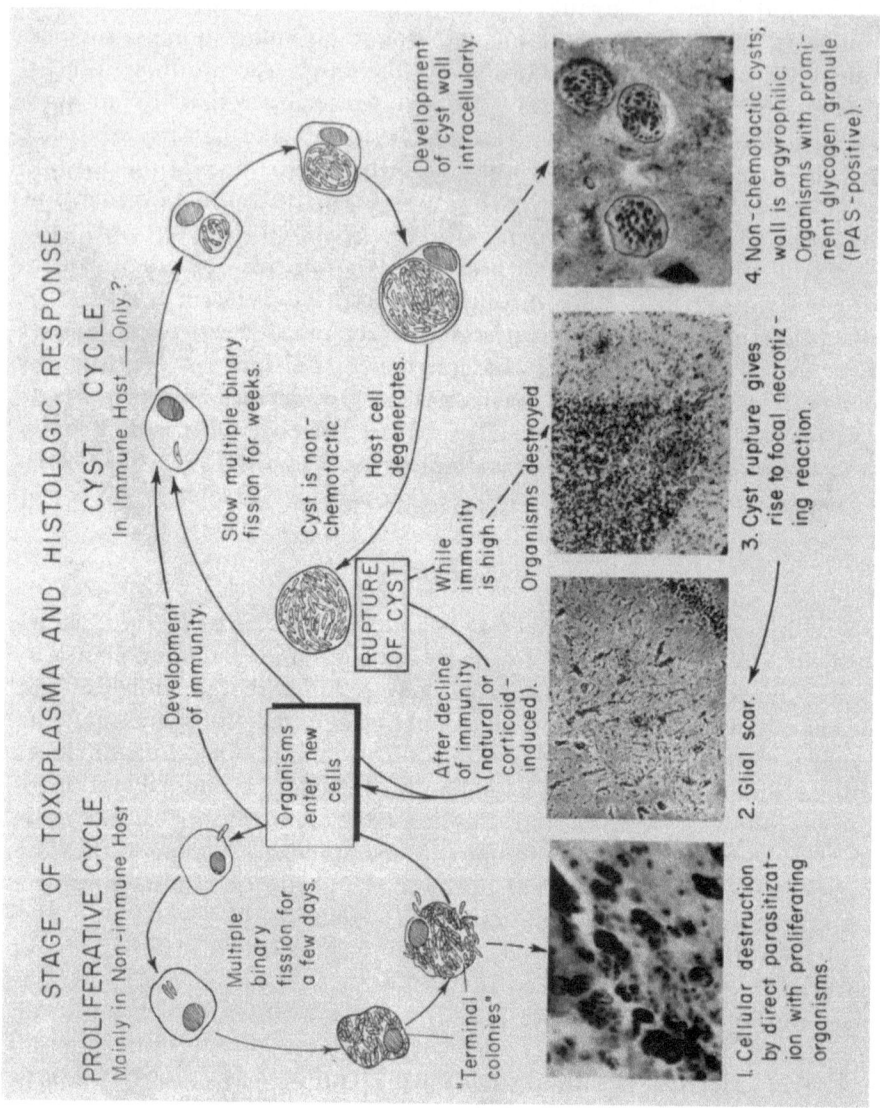

Fig. 1. Toxoplasma cycle.

infection are assigned to either immunity or hypersensitivity, according to the criteria stated. Pharmacologic amounts of corticosteroids interfere both with immunity and hypersensitivity, initially diminishing the latter. Frenkel and Lunde (1966) showed that established antibody levels, as well as the interaction of antibody with the Toxoplasma-related Besnoitia organisms are not affected by corticosteroids, and that the immune failure brought about by hypercorticism appears directly related to changes in the intracellular milieu, probably the digestive capacity of immune macrophages. Since it is possible to separate hypersensitivity from certain aspects of immunity, not only pharmacologically, but immunologically by desensitization, wherever possible these two concepts will be kept clearly separate.

Acute Toxoplasmosis

This is a generalized infection of many tissues. Trophozoites of Toxoplasma proliferate actively, and destroy their host cells. Nevertheless, most spontaneous infections appear to be asymptomatic. Clinical disease is accompained by fever, rash, signs of pneumonia, sometimes with myocarditis and encephalitis and it may result in death such as has been observed in humans, dogs, cats, and many other animals (Siim *et al.*, 1963). In babies, jaundice, purpura, spleno- and hepatomegaly may be prominent, clinically suggesting erythroblastosis fetalis. Neutralizing type antibodies measurable with a dye test make their appearances within 1 to 2 weeks of infection (Frenkel, Weber and Lunde, 1960). Lymphadenopathy with reticular cell hyperplasia of the medullary cords and often accompanied by peripheral lymphocytosis, is sometimes observed in the recovery stage from asymptomatic infection (Siim, 1960).

Subacute Toxoplasmosis

When infected in utero, progress of fetal infection is moderated by the almost simultaneously transferred antibody from the mother (Frenkel and Friedlander, 1951). Active immune response may be deficient and a protracted central nervous system infection may follow. Although largely cleared from the extraneural viscera, Toxoplasma continues to proliferate in the central nervous system, giving rise to (1) scattered small foci of necrosis with glial nodules and (2) to confluent focal lesions in the cortex secondary to vascular involvement. In human infants a third, rather unique lesion has been observed: multiple ependymal ulcers progressing to necrotic zones surrounding lateral and third ventricles, and the aqueduct. The ventricular fluid serves to disseminate Toxoplasma; on account of aqueductal blockage, it also contains much toxoplasmic antigen. This antigen seeps through the ependymal ulcers and reacts with antibody carried in the blood stream, resulting in vascular occlusion and necrosis described in detail by Frenkel and Fried-lander (1951). The fourth ventricle ependyma shows ulceration without necrosis if the foramina of Luschka and Magendi remain open. Retino-choroiditis is common and particularly destructive in the newborn. About a third of the children infected *in utero* show stigma of severe cerebral and retinal damage with mental retardation, and some die from infection (Eichenwald, 1960; Thalhammer, 1962; Couvreur and Desmonts, 1962). Two-thirds of the neonatally infected babies appear to remain asympto-matic for varying periods of time (Desmonts and others, 1965).

Chronic Toxoplasmosis

Immunity, as indicated by subsidence of toxoplasmic proliferation, is acquired after infections of varying duration, whether symptomatic or not. In many animals and probably man, immunity is not sterilizing and cysts of Toxoplasma persist for long periods of time. A time correlation for numbers of microbes in different tissues has been made for the Toxoplasma-related organism Besnoitia (Frenkel and Lunde, 1966). The cysts are most common in the brain where they are spherical, and in the heart and the skeletal muscle where they are elongate; in both locations they are surrounded by an argyrophilic, PAS-positive, diastase-resistant cyst wall (Frenkel, 1956). Such cysts appear to be long-lived, the intra-cystic organisms multiply slowly, and each organism contains a large glycogen granule giving the cyst a PAS-positive appearance. During chronic infection antibody titers usually drop, probably due to decreased antigenic stimulation, suggesting that cysts rupture only rarely (Fig. 1). Even when they do, hypersensitivity with necrosis and inflammatory reaction locally fixes antigen and minimizes the antigenic stimulus. While the intact cysts appear to be innocuous, cyst rupture gives rise to clinical lesions especially in the retina with its highly concentrated func-tion. A similarly sized lesion resulting from cyst rupture in heart, skeletal muscle, and even the brain might go unnoticed. However, multiple lesions in small areas of the brain have been diagnosed in man (Bobow-ski and Reed, 1958), in golden hamsters (Frenkel, 1956) and in other animals (Siim and others, 1963).

Relapse

The presence of chronic toxoplasmic infection carries an additional risk from the possibility of resurgent proliferative activity, following corticosteroid therapy and possibly for unknown reasons (Frenkel, 1956, 1961). Corticoid-treated patients can die with disseminated encephalitis

(Cheever *et al.*, 1965; Lemaire *et al.*, 1962). Relapse in hamsters results in diffuse encephalitis or pneumonia (Frenkel, 1956). Chemotherapy remains effective.

Comparative Clinical Aspects

Infections vary considerably in severity. In part, this is due to strains of Toxoplasma with different rates of multiplication. However, after repeated passages in mice and hamsters, strains may adapt to the new host, so that the asymptomatic infections observed originally give way to fatal ones. The nature of this adaptation has not been studied in detail. It may be a true adaptation, with more efficient use of the available substrates. Alternately, rapid passage may lead to a build-up of organisms, so that finally 5 to 7 $logs_{10}$ of Toxoplasma are injected, producing fatal infections, whereas previously, immunity cut short an infection initiated with a small inoculum. The occasionally reported increase of virulence of organisms by passage through multi-mammate mice or through canaries (Lainson, 1955) appears to illustrate the latter phenomenon, where, after passage through very susceptible animals, higher numbers of organisms are available to produce fatal infections in mice and other conventional laboratory hosts.

Of special interest are differences in disease patterns observed in various hosts. Toxoplasma encephalitis has been observed occasionally in dogs and cats, sheep, pigs, cattle, mice, hamsters, rabbits and man (Siim *et al.*, 1963). Naturally infected rabbits, dogs and cats frequently died from acute generalized infection, with mesenteric adenitis and enteric ulceration (Siim, 1963). In sheep, abortion with fetal encephalitis occurs if maternal infection is contracted during pregnancy (Hartley and Marshall, 1957; Jacobs, 1961). Amongst birds, encephalitis has been observed in young chickens and in pigeons; however, when ocular involvement exists, retinitis is observed in chickens, and choroiditis in pigeons (Frenkel, 1953).

Possible differences in pathogenesis are better observed in laboratory animals where strains of similar virulence can be compared quantitatively. Differences in organs involved are less marked in acute than in chronic infection. With sulfadiazine-treated RH strain infection, for example, cysts developed prominently in cardiac muscle of guinea pigs, whereas they are formed primarily in the brains of mice and hamsters (Frenkel, 1956). Mice develop immunity with difficulty, requiring repeated courses of chemotherapy. Once they have acquired immunity, cysts give rise to only insignificant lesions. Hamsters, who acquire immunity more easily, develop a progressive encephalitis with irritability, choreiform movements and paralysis from lesions in the neuraxis, from the cumulative effects of cyst rupture, probably due to the greater degree of hypersensitivity present (Frenkel, 1961). Retinochoroiditis develops during such chronic infections both with trophozoites and cyst rupture (Frenkel, 1961). Ordinarily cysts are uncommonly found in the uterus since their numbers are low; however, peptic digestion has shown that cysts can be present in the uterus of many animals and man (Remington *et al.*, 1960).

Epidemiologic studies in animals and man are necessary in each geographic-climatic area to establish the frequency of asymptomatic infection, especially in order to evaluate new disease syndromes in animals and man, such as the relation of abortion and retinochoroiditis to toxoplasmosis.

DIAGNOSIS

Toxoplasma can be isolated by the inoculation of susceptible laboratory animals like mice, hamsters or rabbits free from spontaneous toxoplasmosis. Freedom from infection is shown by "blind passages," as well as by serologic examination of the animals prior to inoculation (conveniently done by obtaining a small amount of blood from the retroorbital sinus). Any organism isolated should be compared serologically and immunologically with known strains of Toxoplasma.

The presence of stable antibody titers, whether high or low, suggests the persistence of chronic infection, generally associated with toxoplasmin sensitivity (Frenkel and Jacobs, 1958). The presence of high antibody does not necessarily imply clinical activity, however. A few well verified instances have been observed in rats where antibody has disappeared although cysts persisted. However, this does not sanction a diagnosis of toxoplasmosis from supposed isolations in the absence of serologic evidence. Quite on the contrary, infection and antibody are associated as a rule, and it would take extensive evidence to document infection without antibody.

To prove that a syndrome is associated with toxoplasmosis, it is important to establish the incidence of latent infection, which needs to be determined simultaneously from the exact area where the survey is made using comparable age groups, and similar selection factors. The antibody titers must be consistent with the diagnosis to be related. If in a certain region 20 per cent of the adults had antibodies, and positive dye tests for toxoplasmosis were found in 70 per cent of patients with retinochoroiditis, then approximately 70 minus 20 = 50 per cent of the patients with retinochoroiditis can be concluded to have lesions as a result of toxoplasmosis. The same type of proof would have to be brought forth for patients with habitual abortions if this is to be established as a frequent event. Subgroups of patients and control groups differing in age should show a consistent pattern.

Morphologic documentation of Toxoplasma is a useful adjunct to isolation and serologic evidence. However, in view of the simple organization of Toxoplasma, morphologic evidence alone is insufficient. One cannot over-emphasize the need for permanently stained preparations to recheck visual impressions gained in a moment of enthusiasm. This is as important for the identification of trophozoites of Toxoplasma as for cysts. Spherical structures resembling cysts in size are often encountered in histologic specimens, representing pollen or fungal spores; on Giemsa stained smears, ovoid yeasts resembling Toxoplasma have been found, requiring PAS-staining to show the individual cell walls. Cysts cannot be identified from the size and structure alone, but have to be squashed to show the crescentic organisms within. Organisms resembling Toxoplasma have been reviewed by Frenkel (1956) and this author has seen a number of other organisms, as yet not described, which probably are not Toxoplasma but have been impossible to identify morphologically. In the case of doubtful organisms, smears should be stained with Giemsa and with PASH. Sections should be stained with hematoxylin and eosin, Giemsa, with PASH, with an ammoniacal silver stain, such as Wilder's, and with Goodpasture's, or Wright and Craighead's stain, to differentiate Encephalitozoon (Nosema). Fresh preparations are useful to recognize the motility of trophozoites and cyst organisms; however, the intact cyst structure is not sufficiently characteristic.

FETAL TOXOPLASMOSIS AS SEQUEL TO ACUTE INFECTION OF THE MOTHER

In all congenital human infections so far studied, the mothers had high antibody titers suggesting recent primary infections such as during pregnancy (Sabin *et al.,* 1952; de Roever Bonnet, 1961; Desmonts, Couvreur and Ben-Rachid, 1965; Sever *et al.,* 1967). Although the high titers might be interpreted as indicating reactivation of chronic infection, this has never been observed in a study of 216 mothers through a total of 380 pregnancies subsequent to their giving birth to a toxoplasmic baby (Eichenwald, 1961). Furthermore, in several studies women have been followed serologically throughout pregnancy, in whom negative tests were found early and positive ones at delivery. Desmonts *et al.* (1965), updated by personal communication to the author (1966), had so far studied 47 women who acquired toxoplasmosis in pregnancy, of whom 43 per cent had infected their babies, and 57 per cent had not. Kräubig (1966) found 44 per cent transmission in 18 mothers infected during pregnancy, but only 7 per cent when it was not certain that the high titers resulted from acute infection during pregnancy. Sever *et al.*'s percentages are in a same order of magnitude (1967). It is most likely that Toxoplasma reaches the placenta hematogenously since groups of trophozoites can be found widely dispersed in the chorionic plate, decidua

and the amnion (Glasser and Delta, 1965; Driscoll, 1966). Although
necrosis of individual cells or larger lesions are not usually prominent,
Mellgren *et al.* (1952) reported Toxoplasma in necrotic placental foci
and followed this by isolation in animals. Neghme *et al.* (1952), Beckett
and Flynn (1953), Farber and Craig (1956), and Cardoso *et al.* (1960) also
reported Toxoplasma in placental villi and umbilical cord without associated significant lesions. It is notable that the organisms are generally
found more frequently on the fetal side of the placenta, perhaps reflect-
ing the lesser immunity present here when compared to the maternal
tissues. Once in the placenta there is no difficulty explaining infection of
the fetus proper, either hematogenously, or by ingestion of amniotic
fluid (Figs. 2–4).

An excellent retrospective statistical study by Robertson (1960) has
linked Toxoplasma to the increased perinatal mortality observed over a
3-year period in Lincolnshire, England. In the mothers who had experienced stillbirths, Toxoplasma antibodies were increased in frequency and
higher in titer than in matched controls. In support of a recent acute
infection of the mother was that the medians of the maternal dye test
titers were the higher the less the interval between stillbirth and testing.

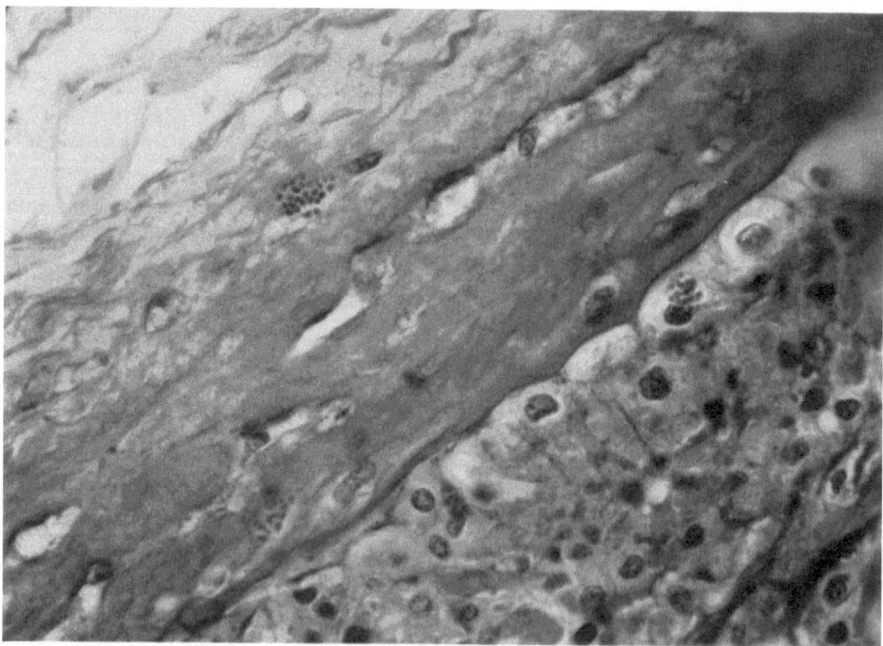

Fig. 2. Groups of Toxoplasma trophozoites proliferating intracellularly in chorion
(left above) and decidua (right). Hematoxylin and eosin, ×500. (Section S 62-3327
courtesy of Dr. S. G. Driscoll.)

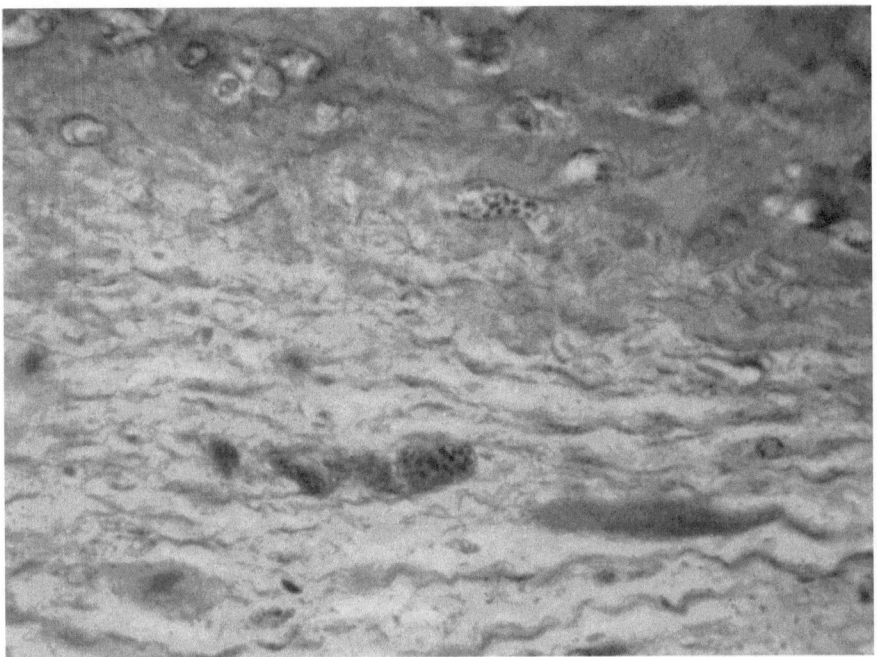

Fig. 3. Two groups of Toxoplasma in chorion, the lower having the appearance of a cyst. H & E. ×500. (Section S 62-3327 from Dr. S. G. Driscoll.)

In sheep, all evidence indicates that transmission took place during the pregnancy in which primary infection occurred (Jacobs, 1961; Beverley and Watson, 1962). Organisms were found in small necrotic foci in the fetal cotyledons (Hartley and Marshall, 1957).

Cowen and Wolf (1950, 1951) in a series of 5 papers described the pathology, spread and intrauterine transmission of Toxoplasma after vaginal installation. Their strains which were of intermediate virulence produced some acutely fatal and some chronic infection in mothers and offspring, providing a great spectrum of lesions. Placental involvement was by hematogenous dissemination rather than by direct extension from the vagina (Cowen and Wolf, 1950, 1951). Infection during early pregnancy resulted in fetal death; transmission to mouse fetuses followed intravaginal infection between days 7 and 9 of pregnancy.

Werner and Seidlitz (1960) studied various means of infection in mice and hamsters to see how Toxoplasma reaches the uterus; they found no infection after intraperitoneal, rare infections after intravenous, and heavy acute infection after intrauterine inoculation. Virulent Toxoplasma were used which caused death in 4–6 days. The organisms observed were trophozoites or "terminal colonies," although the term "echte Pseudocysten" (real pseudocysts = in the sense of cysts) was used

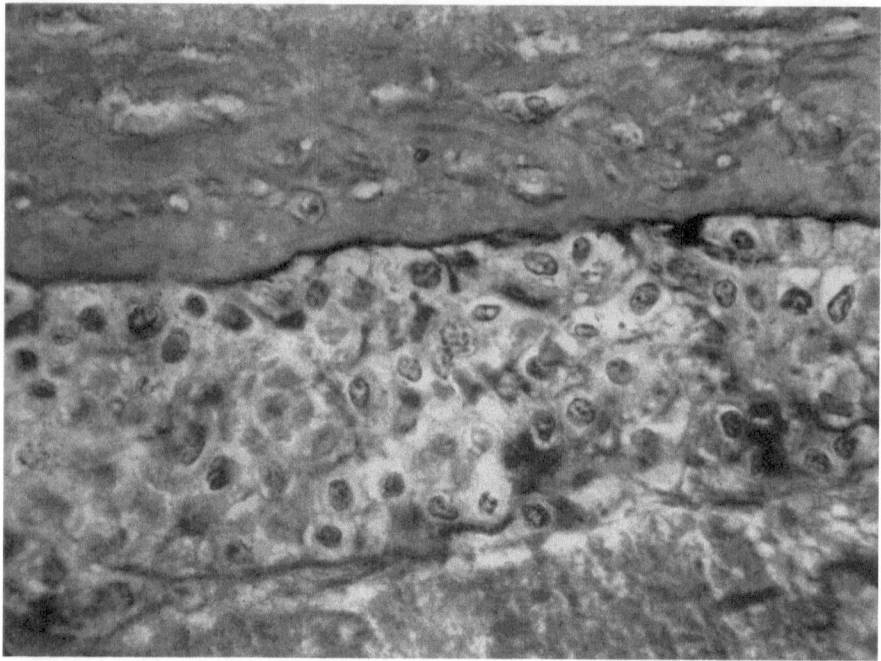

Fig. 4. Toxoplasma in decidua, resembling a young cyst. H & E. ×500. (Section S 62-3327 from Dr. S. G. Driscoll.)

erroneously; based on supposed cysts from the acute infection studied in various stages of murine pregnancy, speculations were offered that might somehow be applicable to chronic infection of man (Werner, 1966).

Occasional observations suggest persistent or recurrent parasitemia also in humans which could result in placental infection some time after the acute infection. A recent report by Miller *et al.* (1966) illustrates parasitemia 1, 4 and 5 months after delivery of a toxoplasmic infant, proven at autopsy, with the isolation of Toxoplasma from the endometrium 3 weeks after delivery, and dye test titers ranging between 1:65,000 and 1:1,000,000.

TOXOPLASMOSIS TRANSMITTED DURING CHRONIC MATERNAL INFECTION

Eichenwald (1948) first observed the transmission of Toxoplasma *in utero* and by lactation, however, only when mice were reinfected during pregnancy and not during unchallenged chronic infection. Repeated transmission of Toxoplasma in several pregnancies and through 5 successive generations has been observed in mice by Beverley (1959) using a strain of low virulence. The infection reduced the life span of the mice, the number of litters and the number weaned, but it permitted sufficient infected mice to survive for vertical transmission to be accomplished. These findings were confirmed by van der Waaij (1960) and by Remington, Jacobs and Melton (1961).

Similar findings have been made in rats, by Hellbrügge (1954), Wildführ (1954), Thierman (1957) and Remington *et al.* (1958, 1961, 1963). The latter authors observed only strains of lower virulence to be transmitted congenitally. However, in guinea pigs, congenital infection occurred even with the virulent RH strain (Huldt, 1960). Transmission in 2 successive pregnancies of dogs has been observed by Koestner and Cole (1960).

Transmission during chronic infection of animals could be correlated to some degree with the size of the inoculum (Hellbrügge, 1954), the extent of tissue parasitization (Remington, Jacobs and Kaufman, 1958) and the occurrence of recurrent parasitemia (Hellbrügge, Remington).

The existence of chronic toxoplasmosis in the non-pregnant human uterus has been shown by Remington, Melton and Jacobs (1960), who isolated Toxoplasma from 4 out of 32 peptic-digested uteri coming from patients with low-stabile dye test titers. This finding led to an intensive search for toxoplasmosis in liveborn infants and placentas, in perinatal deaths, aborted fetuses, and products of conception of mothers with low-stabile dye test titers by Remington (1963, 1966). However, Toxoplasma was isolated from the products of conception of only 2 women with low-stabile dye test titers (Remington, Newell and Cavanaugh, 1964) out of a group of 155 women with 29 per cent dye test positivity, indicating positive results in 4 per cent of those at risk, and no isolation from mothers with negative serology. Another group of 50 women with habitual abortions and matched controls showed similar incidence of positive tests for toxoplasmosis. It was not possible to incriminate Toxoplasma as a significant cause of perinatal mortality in a sample of 146 babies, from San Salvador, Central America, 65 per cent of whose mothers had toxoplasmic antibody (Remington, 1966).

Langer (1963, 1966) summarized his studies reporting positive isolation in 23 of 70 cases (33 per cent) of one series, and in 15 of 50 cases (30 per cent) of a second series, from material including curettings, placenta, fetal brain, amniotic fluid, menstrual fluid and milk. The high incidence appears remarkable, especially since 7 of the 23 mothers had negative dye tests in the first series (1963) and 3 out of 7 from the second series whose detailed history was given (1966). In a number of instances where Toxoplasma was identified visually or by isolation, subsequent histologic examination failed to show either Toxoplasma or inflammation, although the fetus was sufficiently mature to react. However, in his most recent publication, Langer (1966) mentions having identified pollen grains which were indistinguishable from Toxoplasma cysts, and stated that it was impossible in retrospect to state in which cases real cysts or pollen grains had been seen.

Werner, Schmidtke and Thomascheck (1963) examined material from 47 women with a suspicious history, and found 10 instances of Toxoplasma-like parasites in endometrium and fetal material with isolation from some of these. Surprisingly, 71 isolation attempts with fetal material from an unknown number of cases remained negative. The number of identifications from the placenta was not stated although 4 are illustrated. Thomascheck (1965) mentions 24 isolations, 12 of which came from placental material. Although the total number of patients is not mentioned in this paper, he appears to refer to his earlier study, and if so, had obtained positive results in 24 out of 47 cases. According to Langer (1966) Thomascheck had examined 175 patients with 60 (34 per cent) positive isolations. However, on examination of Werner, Schmidtke and Thomascheck's illustrations (1963), Figures 6, 7, 9, and 10 do not resemble Toxoplasma morphologically. The authors themselves raise the issue obliquely, by referring to observations supposedly showing that the morphology of Toxoplasma was changed in animals with antibody (Werner, 1963), but that by use of special staining techniques and filters (Werner and Seidlitz, 1960) it was comparatively easy to distinguish Toxoplasma from cell debris.

It should be commented here, that the morphologic identification of Toxoplasma in necrotic tissue is difficult or impossible, and that its identification should be based on intracellular cysts or trophozoites which are not exposed to antibody (Figs. 2–4) and which have an unimpaired staining intensity. Indeed if cysts were present, their vivid glycogen-dependent PASH staining would afford an additional recognition characteristic.

It is conceivable that in certain areas Toxoplasma strains exist that do infect the fetus during chronic maternal infection, such as the Beverley strain in mice. However, the observations by Langer and Thomascheck referred to above, derived from patients from Giessen, Goslar and Berlin, must be considered together with the negative isolation attempts by two other groups of German investigators. Thalhammer and Kräubig examined the products of conception and placentas from 89 women with repeated abortions from Göttingen (Thalhammer, 1966). Vorherr and Piekarski examined the products of conception, lochia and menstrual blood of 57 women with 112 spontaneous abortions from the area of Frankfurt on the Main (Vorherr, 1965). In summary, it appears fair to conclude that although occasional fetal infections during chronicity are probable, procedural defects and inconsistencies in the evidence make it unlikely that their frequency has been statistically evaluated.

HABITUAL ABORTIONS AND EMBRYOPATHIES

In 1957 Jirovec, Jira, Fuchs and Peter reported finding an increased incidence of toxoplasmin skin test positivity in groups of women with habitual abortion, and in those who repeatedly delivered premature, stillborn or malformed children. These findings were detailed by Cech and Jirovec (1960). Fifty of these women were treated with Daraprim after which the incidence of reproductive failure decreased from 96 per cent to 16 per cent. In 12 toxoplasmin negative women, Daraprim treatment reduced reproductive failure only from 66 per cent to 50 per cent. As the authors report that their earlier skin test antigens were less potent than later, they identify the years in which several of their studies were performed. However, age-matched controls tested with the same strength antigen are not easily identifiable. In 1965, Jira summarized their findings, "We examined 1331 normal women by means of the toxoplasmin test. The obstetric histories were analyzed in relation to the results of this test. Women with a positive toxoplasmin test had a 2.2-fold increase of abortions, 6.6-fold increase of premature deliveries, and a 3-fold increase of dead newborns. Furthermore, positive women had a 3.6-fold increase in abnormal pregnancies and a 2.6-fold increase in operative deliveries. The newborns in this group showed a 1.8-fold increase in prematurity and a 1.7-fold increase in abnormalities." Jira believes, along with Langer, Thomascheck, Werner and others, that chronic toxoplasmosis can affect embryogenesis and fetogenesis, resulting in fetal malformations, infections and repeated abortions. The previous observation that toxoplasmosis is transmitted only in one pregnancy (Sabin *et al.*, 1952) is believed to apply only to women infected for the first time during pregnancy, and it is postulated that repeated transmission could be the consequence of infection acquired before pregnancy, perhaps in the long distant past (Langer, 1963). Although the cogency of this view cannot be denied, the evidence submitted for it is insufficient. Espinosa *et al.* (1964) from the Hospital de Gineco-Obstetricia No. 1, Mexico City, found a greater frequency of abortion in dye test positive mothers; however, their figures were not statistically significant.

The necessary requisites are congruent findings serologically, histopathologically and by isolation, and exclusion of other infections with rigorous age and area controls if statistical evaluation is sought. An exemplary case report, although apparently of an acute infection, has recently been published by Miller, Aaronson and Remington (1966) of a mother with intermittent parasitemia after delivery of a toxoplasmic infant, with isolation of Toxoplasma from the endometrium and high dye test titers.

Toxoplasmic infants commonly are born prematurely, which has been discussed monographically by Paul (1962).

It is difficult to visualize at present how Toxoplasma affects the embryo causing only developmental defects such as cleft palate and harelip, in which the incidence of toxoplasmin positivity was reported to be about twice normal by Jirovec et al. (1957). Rubella, the only well known infection affecting the embryo is a relatively non-destructive infection compared to toxoplasmosis. As tolerance is acquired to rubella after infection of the embryo and the virus persists until long after delivery, so should Toxoplasma be easily identifiable, both histologically and by isolation. With the occurrence of tolerance, antibodies should be absent, at least early after delivery. The time of appearance of antibody and of skin sensitivity would be hard to predict in the case of embryonal infections. If maternal toxoplasmosis gave rise to significant vascular changes in the placenta leading to embryonal damage, such has not been demonstrated.

Still more difficult to explain would be a causal relationship between toxoplasmosis and mongolism. Jirovec et al. (1957) reported positive tests in 42 per cent of mongoloid children in 84 per cent of their mothers but only 28 per cent of their fathers. In view of the association of chromosomal aberrations, it would have to be shown that toxoplasmosis were related to the non-hereditary form, the non-disjunction 21-trisomy type of this syndrome, and it would postulate an influence of Toxoplasma on the ovum which so far has not been explained. Another explanation would be that mongoloids developed delayed hypersensitivity more readily to small antigenic stimuli. This might be correlated with the observation that mongoloid children and their mothers, but not fathers, have a significantly increased frequency of thyroid autoantibodies at an early age (Fialkow, 1966). Thalhammer (1966) examined 71 mongoloid children and also found an increased incidence of toxoplasmic antibody; however, he attributed it to their mental deficiency, perhaps the institutional environment, which favored a higher infection rate, as the incidence was normal in the 0–5 year group and rose beyond the control levels only in later years. In other studies mongoloids were found to be poor antibody formers (Siegel, 1948).

STERILITY

Toxoplasmosis has been suspected as a cause of orchitis resulting in sterility of the human male as claimed by Labady et al. (1962). This was based on positive skin and complement-fixation tests; however, no age-matched controls for the same area are mentioned. Experimental toxoplasmosis in hamsters is sometimes associated with orchitis and sterility (Frenkel, *unpublished*).

Prevention of Toxoplasmosis

In many areas of the world more than half of the population contracts toxoplasmosis during its lifetime. In view of the potentially serious effects on the fetus it is reasonable to consider measures that can be used to prevent primary maternal infection. Kimball and others (1958) observed a conversion rate of 1 per cent per year in obstetrical patients from Minneapolis, Minnesota. According to the calculations of Thalhammer (1966) there are 6–7 infants with congenital toxoplasmosis per 1000 liveborn in Vienna. Kräubig (1966) observed an incidence of 5 per 1000 in Göttingen, Germany. Desmonts *et al.* (1965) computes the frequency of primary toxoplasmosis in women of child bearing age as 8 per 1000 per 9 months' observation. If one recalls that children become infected in 43 per cent of such pregnancies, and disease manifestations appear in a third to a half of these infants, neonatal toxoplasmosis is not too rare. The rate of acquisition of infection varies in different localities and by age, as summarized by Thalhammer (1966) for some European cities, by Couvreur and Desmonts (1962) for Paris, and by Feldman for 10 populations (1956) and for the United States as a whole (1965).

The three main preventive measures are avoidance of raw meat, avoidance of contact with cats (and dogs), and avoidance of farm animals including chickens. Ingestion of raw and undercooked meat has been linked to the acquisition of toxoplasmosis in a carefully conducted observational and experimental study of Desmonts, Couvreur, Alison and others (1965). The experimental transmission of Toxoplasma from cats to mice with the eggs of the nematode Toxocara cati by Hutchison (1965) has been confirmed by Jacobs and Melton (personal communication) and suggests another transmission route which may be operative in man. In an epidemiologic study, Kimball *et al.* (1960) found a significantly high correlation of dye test positivity in humans in contact with farm animals, including chickens and cattle, but excluding pigs, horses or pets at home.

In animals, the prevention of toxoplasmosis would also include the avoidance of raw meat, if this is part of the diet, and avoidance of contact with nematodes. However, the actual transmission of toxoplasmosis requires much further investigation. The epidemiology of toxoplasmosis has been reviewed recently by Piekarski (1966).

Chemotherapy of Toxoplasmosis

The treatment of toxoplasmosis during pregnancy requires special consideration of the drug hazards for the fetus especially if treatment were necessary in the first two trimesters. Nevertheless, treatment is feasible if intelligently considered and carefully supervised. Thalhammer (1957, 1966) developed a useful diagnostic outline which is reproduced in slightly modified form as Table 1. This presupposes the classic view that toxoplasmosis is principally a problem of primary infection acquired during pregnancy. Clinically useful signs in humans include febrile illness, especially with rash, pneumonia, myocarditis, meningoencephalitis, lymphadenopathy and lymphomonocytosis of blood with a negative heterophile test.

As pyrimethamine (Daraprim) and the sulfonamides inhibit the endogenous synthesis of folinic acid (Leucovorin), fetal damage would be most likely if these drugs were used in the first trimester. A deficiency would give rise to embryopathies such as have been observed with methotrexate by Thiersch (1952), Nelson (1955) and Murphy *et al.* (1954) and as reviewed earlier in this volume by Warkany. Krahe (1965) described fetal resorption in rats treated with 3–10-fold "therapeutic doses" mathematically derived from a human dose, which are illustrative although not suggestive of the dangerous dose level in man.

Chemoprophylaxis against fetal infection was described by Kräubig (1966). In 59 of a series of 143 pregnant women with serologic indications of acute or recent toxoplasmosis, serologically diagnosed infection in the neonate was reduced from 17 per cent to 5 per cent. The drugs used were Supronal, 4 gm/day for 10–12 days, generally repeated after 1 month, and Daraprim 25–50 mg for 28 to 30 days, although apparently administered in variable doses.

Table 1

TESTING FOR THE ACQUISITION OF TOXOPLASMOSIS DURING PREGNANCY
(According to Thalhammer, 1966, slightly modified)

Tests	Possible Results		
1. Toxoplasmin, 3rd Month	positive	negative	negative
2. Dye Test 8th Month	unnecessary	positive (high titer)	negative
INTERPRETATION:			
Mother Infected	before pregnancy	during pregnancy	uninfected
Risk for Fetus	none	40–50% infection, 20% signif. lesions	none
Treatment	none	chemotherapy	none
Follow-up on Infant	none	for rise, or disappearance of antibody	none

Combined therapy with sulfadiazine and Daraprim has been developed by Eyles and Coleman (1953). That folinic acid (Leucovorin) and fresh yeast improves the therapeutic index has been shown by Frenkel and Hitchings (1957). This depends on the fact that man and mouse can utilize pre-formed Leucovorin, whereas Toxoplasma must start its biosynthesis with para-aminobenzoic acid. The sulfadiazine-Daraprim combination blocks the conversion of para-aminobenzoic acid to the intracellular folic and folinic acid equivalents necessary for parasite and host. The effect on the host can be counteracted by Leucovorin administered in 3-5-10 mg amounts daily, as needed, together with 5–10 gm of fresh, refrigerated baker's yeast (Fig. 5).

The indications and conduct of treatment are summarized below (Frenkel, 1964):

1. Positive tests for toxoplasmosis in a mother without symptoms of disease indicate immunity. Although chronic infection persists, there is no routine indication for treatment. In this connection it is important to recall that no cases of toxoplasmosis have been found in 380 children whose 260 mothers had previously given birth to a toxoplasmic baby (Eichenwald, 1961).

2. Mothers without antibody run the risk of a primary infection, which

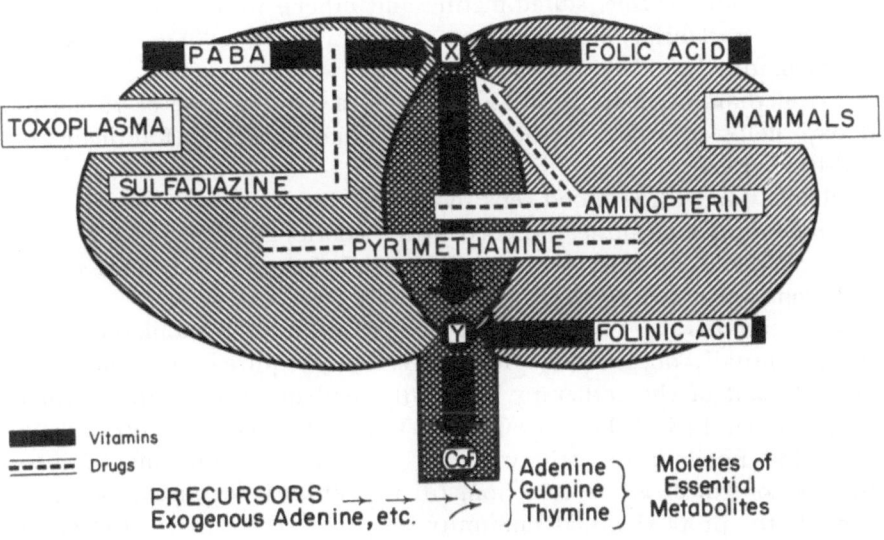

PATHWAYS to BASIC MOIETIES of NUCLEIC ACIDS

Fig. 5. Biosynthetic pathway from para-aminobenzoic acid to folinic acid, and its blockage by sulfadiazine and pyrimethamine (Daraprim). The mammalian host can utilize pre-formed folinic acid, but Toxoplasma cannot.

usually is asymptomatic, and during which about half of the fetuses become infected.

3. After primary infection of a mother, one should try to avoid fetal infection but also consider the risk of specific therapy for the fetus.

4. With infection during the first trimester of pregnancy, the probability of survival of an infected fetus is small, whereas the danger of a congenital malformation due to treatment is great. For this reason, treatment appears contraindicated. During the third trimester of pregnancy, treatment would appear less dangerous to the fetus and as it might prevent or control fetal infection, treatment is recommended. During the second trimester, treatment should be conducted only together with folinic acid (Leucovorin) and yeast.

5. Suspected habitual abortion due to toxoplasmosis must be proven in each patient by isolation, histologic identification, and by positive serologic tests for toxoplasmosis. In proven cases, treatment is suggested for 4–8 weeks, followed by the use of the antagonists for 2 weeks, prior to attempting another pregnancy. As a matter of principle, chemotherapy should be avoided if possible during pregnancy. If another abortion due to Toxoplasma should be proven, treatment should be guided by the previous paragraph.

6. *Chemotherapy:* Sulfadiazine 0.5 gm 4–6 times daily and alkalinization of the urine. Sulfamerazine, sulfamethazine or the triple combination can also be used, and other sulfonamides which reach high concentrations in the intracellular fluid where Toxoplasma multiplies. However, sulfisoxazole, sulfadimetine and others, which are distributed in the extracellular fluid are not effective (Eyles, 1963). Long-acting sulfonamides should be avoided.

Pyrimethamine (Daraprim) 25 mg daily; a loading dose of 75 mg for 3 days helps to attain therapeutic levels more quickly (Kaufman and Caldwell, 1959).

Platelets and white blood cells should be counted twice weekly.

Antagonists: Folinic acid (Leucovorin, Lederle) 2-5-10 mg once or twice daily. (Note: folic acid is not useful.) Fresh, refrigerated baker's yeast 5–10 gm once or twice daily.

7. Chemotherapy suppresses the proliferation of Toxoplasma, however, it usually does not eradicate infection. Acquired immunity, even with the aid of chemotherapy, is usually inadequate to eradicate infection. Similar as in other chronic infections, the mere presence of antibody does not indicate activity or need of treatment. Treatment is used for the control of active infection, till immunity is acquired, and occasionally for prophylaxis, if immunity-depressing doses of corticosteroids are used during chronic infection.

8. An infant resulting from a pregnancy in which toxoplasmosis was acquired should be followed clinically and serologically. In the patients

of Desmonts (1965) and Kräubig (1966) where antibody titers rose from zero to a high level during pregnancy, 43–44 per cent of infants became infected, a surprising agreement. Infection in these infants is demonstrated by a rise in complement-fixing antibody titers, or by a rise of 19-S and 7-S dye test antibodies (Remington and Miller, 1966). In noninfected infants the passively transferred 7-S antibodies generally disappear in 4–6 months (Brownlee, Miller, Remington, 1966).

Even in asymptomatic infected children chemotherapy is indicated, since immunity in the neonate may be marginal. Delayed damage has been found in children who were believed to be asymptomatic previously (Glasser and Delta, 1965), or who showed signs of subsiding infection (Neumann *et al.*, 1960).

Recommended drug doses for infants are:

> Sulfadiazine 100 mg/kg/day in divided doses.
> Daraprim 1 mg/kg/day (double dose for first 3 days).
> Leucovorin 1 mg/day.
> Baker's yeast (fresh and refrigerated) 100 mg/day.

Platelet and white blood counts to be performed twice weekly. The optimal duration of such a treatment has not been determined. Theoretically, it should keep the number of Toxoplasma organisms minimal until immunity is acquired. A rise in antibody would provide some indication of immunity. Thalhammer (1966) and Kräubig (1966) recommend the treatment of such infants. Although Eichenwald (1960) observed what was probably a shorter duration of acute illness, he concluded that the effects of treatment could not be regarded as strikingly beneficial for the prevention of sequelae, although a small favorable effect could not be ruled out. It is suggested here, that higher and more effective doses of sulfadiazine and pyrimethamine can be employed if Leucovorin is added to counteract the limiting factor of drug toxicity.

SUMMARY

The effects of toxoplasmosis are reviewed in normal hosts, in pregnancy, fetus and infant. It is concluded that toxoplasmosis is a danger to the fetus principally when maternal infection results during pregnancy. Toxoplasmic babies have not been observed in subsequent pregnancies, and the assertion that a mother will have only one toxoplasmic baby is sanctioned by experience. Although it is likely that abortions result occasionally from chronic maternal toxoplasmosis their frequency has not been evaluated statistically. The manner of transmission and prevention of toxoplasmosis has been reviewed. The indications, strategy and conduct of chemotherapy in mother and infant are discussed.

References

Beckett, R. S., and F. J. Flynn: Toxoplasmosis; report of two cases with a classification and with a demonstration of the organisms in the human placenta. N. Eng. J. Med. *249*:345, 1953.

Beverley, J. K. A.: Congenital transmission of toxoplasmosis through successive generations of mice. Nature *183*:1348, 1959.

Beverley, J. K. A., and W. A. Watson: Further studies on toxoplasmosis and ovine abortion in Yorkshire. Vet. Record *74*:548, 1962.

Bobowski, S. J., and W. G. Reed: Toxoplasmosis in an adult, presenting as a space-occupying cerebral lesion. AMA Arch. Path. *65*:460, 1958.

Brownlee, I. E., M. J. Miller, and J. S. Remington: Techniques for demonstration of 19S anti-Toxoplasma antibodies. Am. Soc. Parsitologists, 1966.

Cardoso, R. A. de A., F. Nery-Guimaraes, and A. P. Garcia: Congenital toxoplasmosis. In: Siim, J. C. (ed.), Human Toxoplasmosis. Copenhagen: Munksgaard, 20–33, 1960.

Cech, J. A., and O. Jirovec: The importance of latent maternal infection with Toxoplasma in obstetrics. Fortsch. d. Geburtsh. und Gynäkol. (Bibliotheca Gynaecologica, Fasc. 22) *11*:41, 1961.

Cheever, A. W., M. P. Valsamis, and A. S. Rabson: Necrotizing toxoplasmic encephalitis and herpetic pneumonia complicating treated Hodgkin's disease. N. Eng. J. Med. *272*:26, 1965.

Couvreur, J., and G. Desmonts: Congenital and maternal toxoplasmosis. A review of 300 congenital cases. Develop. Med. and Child Neurol. *4*:519, 1962.

Cowen, D., and A. Wolf: Experimental congenital toxoplasmosis. I. The vagina as a portal of entry of Toxoplasma in the mouse. J. Exp. Med. *92*:393, 1950.

———, ———: Experimental congenital toxoplasmosis. II. Transmission of toxoplasmosis to the placenta and fetus following vaginal infection in the pregnant mouse. J. Exp. Med. *92*:403, 1950.

———, ———: Experimental congenital toxoplasmosis. III. Toxoplasmosis in the offspring of mice infected by the vaginal route. Incidence and manifestations of the disease. J. Exp. Med. *92*:417, 1950.

———, ———: Experimental congenital toxoplasmosis. IV. Genital and secondary lesions in the mouse infected with Toxoplasma by the vaginal route. J. Neuropath. and Exp. Neurol. *10*:1, 1951.

———, ———: Experimental congenital toxoplasmosis. V. Lesions in the offspring of mice infected with Toxoplasma by the vaginal route. Observations on an associated hepatic injury. J. Neuropath. and Exp. Neurol. *10*:142, 1951.

Desmonts, G.: Sur la technique de l'épreuve de lyse des toxoplasmes. La Semaine des Hôpitaux de Paris. *31*:1, 1955.

Desmonts, G., J. Couvreur, F. Alison, J. Baudelot, J. Gerbeaux, and M. Lelong: Étude épidémiologique sur la Toxoplasmose: de l'influence de la cuisson des viandes de boucherie sur la fréquence de l'infection humaine. Rev. Franc. Études Clin. et Biol. *10*:952, 1965.

Desmonts, G., J. Couvreur, and M. S. Ben-Rachid: Le toxoplasme, la mere, et l'enfant. Arch. Franc. pediatrie *22*:1183, 1965.

Driscoll S. G.: Fetal infections in man. This volume, p. 279, 1967.

Eichenwald, H.: Personal communication, 1961.

————: Experimental toxoplasmosis. I. Transmission of the infection in utero and through the milk of lactating female mice. Am. J. Dis. Child. 76:307, 1948.

Eichenwald, H. F.: A study of congenital toxoplasmosis. In: Human Toxoplasmosis. (ed.) Siim, 42, Munksgaard, Copenhagen, 1960.

Espinosa de los Reyes, V. M., A. E. Machain, A. Estrada-Viesca, and P. Garcia-Medrano: Toxoplasmosis humana. Memoria de la Primera Jornada Medica Bienal, Hospital de Gineco-Obstetricia Numero Uno, Mexico, D.F. 2:491, 1964.

Eyles, D. E.: Chemotherapy of toxoplasmosis. In: Experimental Chemotherapy, 1:641, 1963. Schnitzer and Hawking (eds.). New York–London: Academic Press.

Eyles, D. E., and N. Coleman: Synergistic effect of sulfadiazine and Daraprim against experimental toxoplasmosis in the mouse. Antibiot. and Chemoth. 3:483, 1953.

Farber, S., and J. Craig: Clinical Pathological Conference. J. Pediat. 49:752, 1956.

Feldman, H. A.: The clinical manifestations and laboratory diagnosis of toxoplasmosis. Am. J. Trop. Med. and Hyg. 2:420, 1953.

————: A nationwide serum survey of United States military recruits, 1962. VI. Toxoplasma antibodies. Am. J. Epidemiol. 81:385, 1965.

Feldman, H. A., and L. T. Miller: Serological study of toxoplasmosis prevalence. Am. J. Hyg. 64:320, 1956.

Fialkow, P. J.: Autoimmunity and chromosomal aberrations. Am. J. Human Genet. 18:93, 1966.

Frenkel, J. K.: Host, strain and treatment variation as factors in the pathogenesis of toxoplasmosis. Am. J. Trop. Med. Hyg. 2:390, 1953.

————: Pathogenesis of toxoplasmosis and of infections with organisms resembling Toxoplasma. Ann. N.Y. Acad. Sci. 64:215, 1956.

————: Pathogenesis of toxoplasmosis with a consideration of cyst rupture in Besnoitia infection. Surv. Ophthal. 6:799, 1961.

————: La toxoplasmosis y la infeccion citomegalica durante el embarazo y en el recien nacido. Hosp. Gineco-Obstetricia No. 1, Mexico, Memoria 1. Jornada Medica Bienal 2:675, 1964.

————: Adoptive immunity to intracellular infection. J. Immunol. (in press), 1967.

Frenkel, J. K., and S. Friedlander: Toxoplasmosis. Pathology of Neonatal Disease. Pathogenesis, Diagnosis, and Treatment. Public Health Service Publi. No. 141, U.S. Government Printing Office. Washington, D.C., 1951.

Frenkel, J. K., and G. H. Hitchings: Relative reversal by vitamins (p-amino-benzoic, folic and folinic acids) of the effects of sulfadiazine and pyrimethamine on Toxoplasma, mouse and man. Antibiot. and Chemoth. 7:630, 1957.

Frenkel, J. K., and L. Jacobs: Ocular toxoplasmosis. A.M.A. Arch. Ophth. 59:260, 1958.

Frenkel, J. K., and M. N. Lunde: Effects of corticosteroids on antibody and immunity in Besnoitia infection of hamsters. J. Infect. Dis. 116:414, 1966.

Frenkel, J. K., R. W. Weber, and M. N. Lunde: Acute toxoplasmosis. Effective treatment with pyrimethamine, sulfadiazine, leucovorin calcium and yeast. JAMA *173*:1471, 1960.

Glasser, L., and B. G. Delta: Congenital toxoplasmosis with placental infection in monozygotic twins. Pediatrics *35*:276, 1965.

Hartley, W. J., and S. C. Marshall: Toxoplasmosis as a cause of ovine perinatal mortality. New Zealand Vet. J. *5*:119, 1957.

Hellbrügge, T.: Die fetale Infektion im Verlauf der akuten und chronischen Phase bei der latenten Rattentoxoplasmose. Arch. Gynäk. *186*:384, 1954.

Huldt, G.: Experimental toxoplasmosis. Transplacental transmission in guinea pigs. Acta Path. et micro. Scand. *49*:176, 1960.

Hutchison, W. M.: Experimental transmission of *Toxoplasma gondii*. Nature *206*:961, 1965.

Jacobs, L.: Toxoplasma and toxoplasmosis. Ann. Rev. Microbiol. *17*:429, 1963.

————: Toxoplasmosis in man and animals. N. Zealand Vet. J. *9*:85, 1961.

Jira, J.: Some aspects of the diagnosis, pathogenesis and treatment of toxoplasmosis. Wiadomoschi Parazytologiczne *11*:139, 1965.

Jirovec, O., J. Jira, V. Fuchs, and R. Peter: Studien mit dem Toxoplasmintest. I. Bereitung des Toxoplasmins. Technik des intradermalen Testes. Frequenz der Positivität bei normaler Bevölkerung und bei einigen Krankengruppen. Zbl. Bakt. (I. Orig.) *169*:129, 1957.

Kaufman, H. E., and L. A. Caldwell: Pharmacological studies of pyrimethamine (Daraprim) in man. AMA Arch. Ophth. *61*:885, 1959.

Kimball, A. C., H. Bauer, C. G. Sheppard, J. R. Held, and L. M. Schuman: Studies on toxoplasmosis. III. Toxoplasma antibodies in obstetrical patients correlated with residence, animal contact and consumption of selected foods. Am. J. Hyg. *71*:93, 1960.

Kimball, A. C., M. K. Cooney, H. Bauer, and C. G. Sheppard: Studies on toxoplasmosis. II. Toxoplasma antibodies in obstetrical patients with extensive serological follow-up. J. Immunol. *81*:187, 1958.

Koestner, A., and C. R. Cole: Neuropathology of canine toxoplasmosis. Am. J. Vet. Res. *21*:831, 1960.

Krahe, M.: Untersuchungen über die teratogene Wirkung von Medikamenten zur Behandlung der Toxoplasmose während der Schwangerschaft. Arch. Gynäk. *202*:104, 1965.

Kräubig, H.: Präventive Behandlung der konnatalen Toxoplasmose. In: H. Kirchhoff and Kräubig (eds.): Toxoplasmose. Stuttgart: Georg Thieme Verlag, 1966.

Labady, F., V. Vrsansky, G. Catar, S. Repas, and J. Podoba: Beitrag zur Frage der möglichen Beziehungen der Toxoplasmose-Infektion zur männlichen Sterilität, Z. f. Urologie *55*:75, 1962.

Lainson, R.: Toxoplasmosis in England. I. The rabbit, *Oryctolagus cuniculus* as a host of *Toxoplasma gondii*. II. Variation factors in the pathogenesis of Toxoplasma infections: the sudden increase in virulence of a strain after passage in multimammate rats and canaries. Ann. Trop. Med. & Parasit. *49*: 384, 1955.

Langer, H.: Die Bedeutung der latenten mütterlichen Toxoplasma-Infektion für die Gestation. In: H. Kirchhoff and H. Kräubig (eds.): Toxoplasmose. Stuttgart: Georg Thieme Verlag, 1966.

Langer, H.: Intrauterine Toxoplasma-Infektion. Stuttgart: G. Thieme, 1963.

———: Repeated congenital infection with Toxoplasma gondii. Obstet. Gynec. *21*:318, 1963.

Lemaire, A., G. Boudin, A. Lauras, J. Debray, and G. Lyon: Toxoplasmose cerebrale decouverte a l'examen anatomique d'une leucemie myeloide compliquee de troubles psychiques. In: Bull. et Mem. de la Soc. med. des Hop. de Paris *113*:1102, 1962.

Mellgren, J., L. Alm and Å. Kjessler: The isolation of Toxoplasma from the human placenta and uterus. Acta path. et microbiol. Scand. *30*:59, 1952.

Miller, M. J., W. J. Arronson, and J. S. Remington: Persistent parasitemia in human toxoplasmosis. Clin. Res. *14*:145, 1966.

Murphy, M. L., R. R. Ellison, D. A. Karnofsky, and J. H. Burchenal: Clinical effects of the dichloro- and monochlorophenyl analogues of diamino pyrimidine antagonists of folic acid. J. Clin. Inves. *33*:1388, 1954.

Neghme, A., E. Thiermann, F. Pino, R. Christen, and M. Agosin: Toxoplasma humana en Chile. Bol. inform. parasit. Chile 7:6, 1952.

Nelson, M. M.: Mammalian fetal development and antimetabolites. In: R. F. Sognnaes (ed.), Antimetabolites and Cancer. Amer. Assn. for the Advanc. Sci., Washington, D.C., 1955.

Neumann, C. G., C. Hilton, and A. Barreda: Acquired toxoplasmosis in a child. Am. J. Dis. Child. *100*:117, 1960.

Paul, J.: Frühgeburt und Toxoplasmose. München, Berlin: Urban & Schwarzenberg, p. 253, 1962.

Pierkarski, G.: Zur Epidemiologie der Toxoplasmose. In: H. Kirchhoff and H. Kräubig (eds.). Toxoplasmose. Stuttgart: Georg Thieme Verlag 1966.

Remington, J. S.: Toxoplasmosis and human abortion. Progr. in Gyn. *4*:303, 1963.

———: Personal Communication, 1966.

———: Congenital transmission of Toxoplasma during chronic infection of the mother. Proc. 1. Int. Congr. of Parasitol. A. Corradetti, editor, *1*:184, 1966.

Remington, J. S., L. Jacobs, and H. E. Kaufman: Studies on chronic toxoplasmosis. The relation of infective dose to residual infection and to the possibility of congenital transmission. Am. J. Ophth. *46*:261, 1958.

Remington, J. S., L. Jacobs, and M. Melton: Congenital transmission of toxoplasmosis from mother animals with acute and chronic infections. J. Inf. Dis. *108*:163, 1961.

Remington, J. S., M. L. Melton, and L. Jacobs: Chronic Toxoplasma infection in the uterus. J. Lab. Clin. Med. *56*:879, 1960.

Remington, J. S., and M. J. Miller: 19S and 7S anti-Toxoplasma antibodies in diagnosis of acute congenital and acquired toxoplasmosis. Proc. Soc. Exp. Biol. Med. *121*:357, 1966.

Remington, J. S., J. W. Newell, and E. Cavanaugh: Spontaneous abortion and chronic toxoplasmosis. Obstet. Gynec. *24*:25, 1964.

Robertson, J. S.: Excessive perinatal mortality in a small town associated with evidence of toxoplasmosis. Brit. Med. J. *ii*:91, 1960.

Roever-Bonnet, H. de: Congenital toxoplasmosis. Trop. Geogr. Med. *13*:27, 1961.

Sabin, A. B., H. Eichenwald, H. A. Feldman, and L. Jacobs: Present status of clinical manifestations of toxoplasmosis in man. Indications and provisions for routine serologic diagnosis. JAMA *150*:1063, 1952.

Sabin, A. B., and H. A. Feldman: Dyes as microchemical indicators of a new immunity phenomenon affecting a protozoon parasite (Toxoplasma). Science *108*:660, 1948.

Sever, J. L., H. Berendes, W. Weiss, J. S. Drage, J. Hardy, M. R. Gilkeson, and J. M. Roberts: Toxoplasmosis: serological and clinical studies of 23,000 pregnant women. In: H. F. Eichenwald (ed.), Mental Retardation and Infectious Diseases. Washington, D.C., U.S. Government Printing Office, 1967.

Siegel, M.: Susceptibility of mongoloids to infection. II. Antibody response to tetanus toxoid and typhoid vaccine. Am. J. Hyg. *48*:73, 1948.

Siim, J. C.: Clinical and diagnostic aspects of human acquired toxoplasmosis. In: Siim (ed.) : Human Toxoplasmosis, p. 54, 1960.

Siim, J. C., U. Biering-Sørensen and T. Møller: Toxoplasmosis in domestic animals. Adv. Vet. Sci. *8*:335, 1963.

Thalhammer, O.: Die angeborene Toxoplasmose. In: H. Kirchhoff and H. Kräubig (eds.): Toxoplasmose. Stuttgart: Georg Thieme Verlag, 1966.

———: Congenital toxoplasmosis. Lancet *i*:23, 1962.

———: Die Toxoplasmose. Wien/Bonn: Verlag Wilhelm Maudrich, 1957.

Thiermann, E.: Transmission congenita del *Toxoplasma gondii* en ratas con infeccion leve. Biologica *23*:59, 1957.

Thiersch, J. B.: Therapeutic abortions with a folic acid antagonist, 4-aminopteroylglutamic acid (4-amino PGA) administered by the oral route. Amer. J. Obstet. Gynec. *63*:1298, 1952.

Thomascheck, G.: Toxoplasmainfektion und Schwangerschaft. Arch. Gynäk. *202*:93, 1965.

van der Waaij, D.: Congenital transmission of avirulent *Toxoplasma gondii* after experimental infection in mice prior to gestaton. Trop. Geogr. Med. *12*:251, 1960.

Vorherr, H.: Discussion of intrauterine transmission of toxoplasmosis. Arch. Gynäk. *202*:116, 1965.

Weinman, D.: Chronic toxoplasmosis. J. Inf. Dis. *73*:85, 1943.

Werner, H.: Experimentelle Untersuchungen über den intrauterinen Infektionsweg von Toxoplasmen. In: H. Kirchhoff and H. Kräubig (eds.): Toxoplasmose. Stuttgart: Georg Thieme Verlag, 1966.

———: Über die Formvariabilität von *Toxoplasma gondii* unter dem Einfluss von Antikörpern. Zbl. Bakt. Parasit. Infekt. und Hyg. (I. Orig.) *189*:497, 1963.

Werner, H., L. Schmidtke, and G. Thomascheck: Toxoplasma Infektion und Schwangerschaft. Klin. Wschr. *41*:96, 1963.

Werner, H., and P. Seidlitz: Experimenteller Beitrag zur connatalen Toxoplasmose. I. Mitteilung: Über den Befall der weiblichen Genitalorgane von Maus und Goldhamster durch *Toxoplasma gondii* nach intravenöser, intraperitonealer und intrauteriner Injektion. Zbl. Bakt. Parasit. Infekt. und Hyg. (I. Orig.) *178*:250, 1960.

Wildführ, G.: Tierexperimentelle Untersuchungen zur Frage der diaplazentaren Übertragung der Toxoplasmen beim vor der Gravidität infizierten Muttertier. Z. Immunitätsforsch. *111*:110, 1954.

Wolf, A., D. Cowen, and B. Paige: Toxoplasmic encephalomyelitis. III. A new case of granulomatous encephalomyelitis due to a protozoon. Am. J. Path. *15*:657, 1939.

MYCOTIC DISEASES IN MAMMALIAN REPRODUCTION

CH. H. BRIDGES

Department of Veterinary Pathology, College of Veterinary Medicine and Texas Agricultural Experiment Station, Texas A&M University, College Station, Texas

The role of fungi in influencing reproduction is gradually being defined through the endeavors of many investigators with a variety of interests, some contributions being made by those who are interested in the sire or dam as an individual or as an experimental model and others being made by those who are interested primarily in reproductive performance and thus devoting most of their efforts to the study of the reproductive organs. A study of these contributions reveals, as might be expected, many gaps in our knowledge of details of the problems. However, as with mycoses in general, it is becoming more apparent that mycotic diseases of reproductive organs are of considerable importance in some species under certain conditions of environment and, particularly, altered host susceptibility. This is especially true in cows and horses which are exposed to heavily contaminated and spoiled forages or to birds whose embryonating ova are exposed to a contaminated environment.

One finds in studying mycoses that on one hand the reproductive organs are, indeed, just another part of the whole individual and that, as expected, systemic infections with pathogenic fungi will in some cases include the testes or the uterus and the appropriate accessory sex organs, while the survival of the animal is in question. Similarly one finds that the reproductive organs are occasionally affected by localized infection with fungi. However, in the pregnant viviparous female a relationship of two individuals exists which creates as unique a situation in the pathogenesis of mycoses as it does with viral or bacterial disease. One should suspect that many other unnamed fungi will be found to be capable of attacking the reproductive system, particularly that of the pregnant female under appropriate host-fungus relationship, but the important ones appear to have been identified.

Mycosis of the Uterus, Placenta and Fetus

Among the various forms of mycotic infections of the reproductive organs, the involvement of the pregnant uterus and fetus of cattle and horses stands out most prominently. The earliest report of mycotic infection was that of Theobald Smith (1920) in which he described his observation of the uterus of a cow at an abattoir. The absence of the carcass, which had been processed for food, no doubt robbed us of greater contribution by him, his precise description of the diseased uterus and fetus being a classic. Practically all later contributions have been made from study of aborted dead fetuses and/or their placentae. The observations of abortion in cattle due to mycotic infection have been reported and extended, especially by Bendixen and Plum (1929), Austwick and Venn (1961), Cordes *et al.* (1964), Weikl (1964) and others since, the condition being recognized as the most frequently diagnosed cause of abortion in some studies (realizing that 50 to 75 per cent of such cases have no etiologic diagnosis) (King, *et al.,* 1965) (O'Hara, 1966). One finds reports of mycotic placentitis and abortion involving 1.4 per cent (Cordes *et al.,* 1964), 6.2 per cent (Austwick and Venn, 1957), 8.7 per cent (Weikl, 1964) and 8.4 per cent (King *et al.,* 1965) of those abortions in cattle studied in series which involved up to 20,000 specimens in two reports. Most frequently there are only one to a few cases recognized in each herd (Figs. 1–3). Mycotic placentitis and abortion have been reported to have occurred also in horses (Weikl, 1964) (Hensel *et al.,* 1961) (Mahaffey and Adam, 1964). A report of probable abortion in a dog caused by a fungus has been made by McGaughey *et al.* (1961). Sonnenschein *et al.* (1960) described a case of congenital cutaneous candidiasis in a newborn infant and Benirschke and Raphael (1958) described a mycotic placentitis associated with an anencephalic human fetus. The validity of interpretations in reports of congenital or neonatal cryptococcosis of newborn children (Heath, 1950) (Nassau and Weinberg-Heiruti, 1958) (Neuhauser and Tucker, 1948) has been questioned by Littman and Zimmerman (1956). Secondary involvement of the pregnant uterus and ovary of women and suspected primary (ascending) endometrial infection with *Coccidioides immitis* has been reported, but the number of cases is very limited (Page and Boyers, 1945) (Small and Birshner, 1949) (Conan and Hyman, 1950). The last named authors found no involvement of the fetuses or placentae of five women with fatal systemic coccidioidomycosis but report having reviewed a case with placental involvement. Blastomycosis was found affecting the ovary of one dog with systemic infection (Robbins, 1954).

The fungi which have been found to infect the pregnant uterus are various species of *Aspergillus* (*A. flavus, A. fumigatus, A. niger, A. terreus, A. versicolor*), *Absidia* (*A. corymbifer, A. ramosa*), *Rhizopus* (*R. arrhizus, R. cohnii, R. microsporus*), *Allescheria boydii, Monotospora lanuginosa, Mortierella polycephala, Nocardia asteroides, Polystictus versicolor, Candida tropicalis* (reviewed by Ainsworth and Austwick, 1959), *Scopulariopsis bovicaulis* (McGaughey *et al.*, 1961), *Syncephalastrum racemosum* (Turner, 1964), *Candida albicans* (Benirschke and Raphael, 1958), *Candida parapsilosis* (Bisping *et al.*, 1964), *Candida krusei* (Aldasy and Hajdu, 1961), *Trichosporon capitum* (Hellman and Ruethel, 1964) and *Coccidioides immitis* (Conan and Hyman, 1940). One finds that species of *Aspergillus, Absidia, Rhizopus* and *Mucor* appear to be the major offenders in domestic animals with the other organisms being of considerably less significance. *Aspergilli* are involved in greater than two-thirds of the cases (Plum, 1932). Intrauterine infection of infants appears to be quite uncommon and apparently only by *Candida* species.

The pathologic change associated with mycotic infection of the bovine placenta and fetus is quite remarkable in comparison with those changes associated with bacterial or viral infection of these systems. The classical report of Theobald Smith (1920) is the only one of spontaneous mycotic infection reviewed which described lesions present prior to abortion or normal parturition. He found a turbid fluid containing many small flakes filling the uterochorionic space and five cotyledons had separated from the maternal caruncles, taking with them a good portion of the caruncles, the remaining portions of the caruncles being short, blunt grayish projections from the uterine wall which contained some hemorrhagic spots.

The affected cotyledons were enlarged, having thickened margins which rolled cup-shaped over a central necrotic portion because of the retention of necrotic maternal tissue from the caruncles in their periphery. The allantoic fluid was clear but that of the amnion contained fecal material. The subchorionic tissue was edematous, measuring up to 2 cm. in thickness. The fetus, approximately 25 cm. long (approximately 120 days old) had slight perirenal capsular edema and interlobular edema of the lungs. Non-septate fungal hyphae were found in material scraped from the cotyledons. Histologically there was necrosis, masses of leucocytes, and hemorrhage in the cotyledons in which sloughed portions of maternal caruncle were embedded. Fungal hyphae were found penetrating the necrotic and purulent foci. The fungus which was identified as *Mucor rhizopodiformis* (*R. cohnii*) was isolated from the placenta, fetal lungs and digestive tract. Similar pathologic change was produced experimentally by Gilman and Birch (1925).

Subsequent studies have added observations on the occurrence of cutaneous and pulmonary lesions in occasional infected fetuses, the occurrence of a thickening of the chorion which becomes somewhat leathery (Gilman and Birch, 1925), and data on the incidence and pathogenesis. The degree of pathologic change may vary from involvement of a few to many placentomes and little to much intercotyledonary tissue. The uterus may be involved with an acute purulent and necrotizing endometritis in areas between caruncles (Cordes et al., 1964). Reports of pathologic changes in bovine tissues associated with *Candida* or *Trichosporon* species have been inadequate to suggest different pathologic changes; however, the morphology of the infecting organism in the tissue is different, as would be expected.

The lesions of the fetal skin consist of irregular grayish elevated plaques, usually over the shoulders, back, sides, occiput and about the eyes. The dermis and hair follicles may be invaded by the fungus with the resultant edema, cellular exudation and parakeratosis. In the lungs there occasionally is a bronchopneumonia with hyphae appearing in clumps in the alveoli.

The aborted fetuses usually are 6 to 7 months old although some have been only 3 months old (Austwick and Venn, 1957). The fungus is found in the stomach contents and in amniotic fluid of the amniotic sac as well as in that in the lungs in about 25 per cent of the cases. Histologic examination of sections of placenta and smears of affected

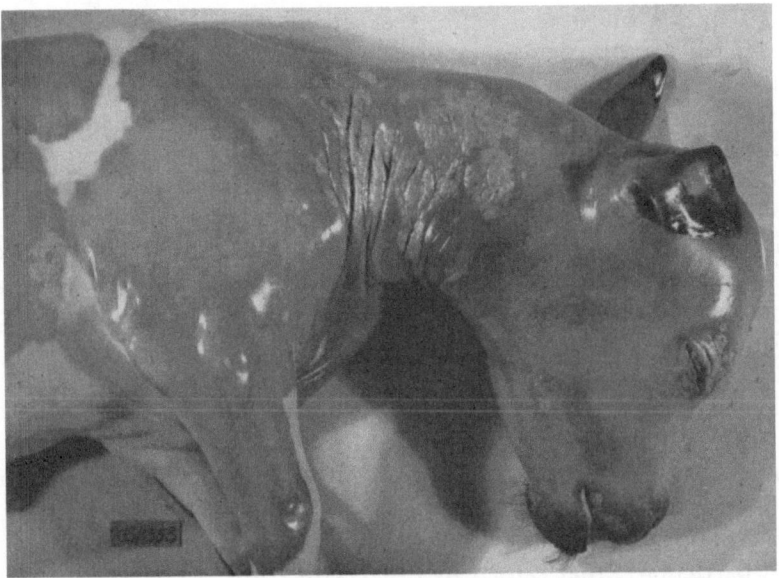

Fig. 1. Cutaneous mycosis in an aborted bovine fetus. (Courtesy of D. O. Cordes.)

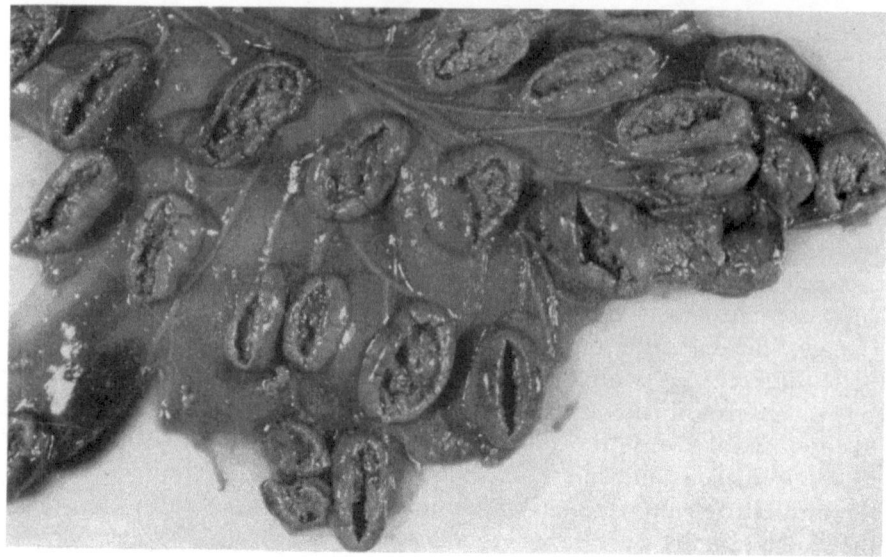

Fig. 2. Characteristic peripheral thickening of cotyledons of a bovine placenta with mycotic infection. (Courtesy of D. O. Cordes.)

tissues and stomach contents appear to be the most reliable approach to diagnosis.

The data on mycotic infection of the reproductive organs of mares and of their fetuses, of which there are only a few reports, have been based upon clinical observations of aborting dams, and pathological examination of the fetus and its placenta. Mahaffey and Adam (1964) identified mycotic placentitis in 9 mares. There was extensive necrosis of the maternal surface of the chorion in which were many fungal hyphae. In some cases the affected areas were pale, grayish yellow, thickened and fissured. Whereas the lesions in the tissues of other species have been characterized by necrosis and purulent inflammation, these had in addition a granulomatous response which was present also in the lungs and liver of one foal. Granulomatous inflammation about fungal hyphae was found also by Hensel *et al.* (1961) in aborted foals and their placentae.

The *Candida albicans* infection of the amniotic sac of a human placenta reported by Benirschke and Raphael (1958) was characterized by acute inflammation in the walls of umbilical vessels and substance. *Candida* was plentiful on the surface of the umbilical cord, but none was found in the placenta.

The pathogenesis of most fungal infections of the female reproductive tract, particularly that caused by fungi with fruiting bodies (*Aspergillus, Rhizopus,* etc.) presently remains in a state of deductive reasoning with the major evidence pointing to hematogenous spread of the fungus from the lung which itself obviously is exposed to inhaled spores. However, spread from initial focal lesions in the digestive tract, which are not uncommon in animals, is possible. Instillation of fungal spores into the uterus at time of artificial insemination or feeding of fungi has not caused abortion to occur. Although Cordes *et al.* (1964) found eight cases of fatal mycotic pneumonia associated with characteristic mycotic abortion in 39 cows, others have not found a similar association. Mycotic abortion has been reproduced experimentally by intravenous injection of spores of fungi (Gilman and Birch, 1925) (Bendixen and Plum, 1929). Attempts to do so by forced inhalation of fungal spores has not been successful. Austwick (1962) has called attention to the occurrence of minute lesions containing *A. fumigatus* in the lungs of two-thirds of otherwise normal cattle which he examined.

Cows and horses, animals in which the greater incidence of mycotic abortion is recognized, are frequently exposed to massive numbers of spores of *Aspergillus, Rhizopus* and *Mucor*. Baruah (1961) found the atmospheric concentration of fungal spores in a cowshed to range between 95,000 and 16,000,000 spores per cubic meter of air. Certainly this concentration is much lower than what an animal encounters when it buries its muzzle into a shock of very moldy hay. The incidence of mycotic abortion in cattle is known to be higher in the winter months in which hay is fed (Weikl, 1964). Emmons (1960) found that *A. fumigatus* composed 90 to 95 per cent of the colonies on a culture plate inoc-

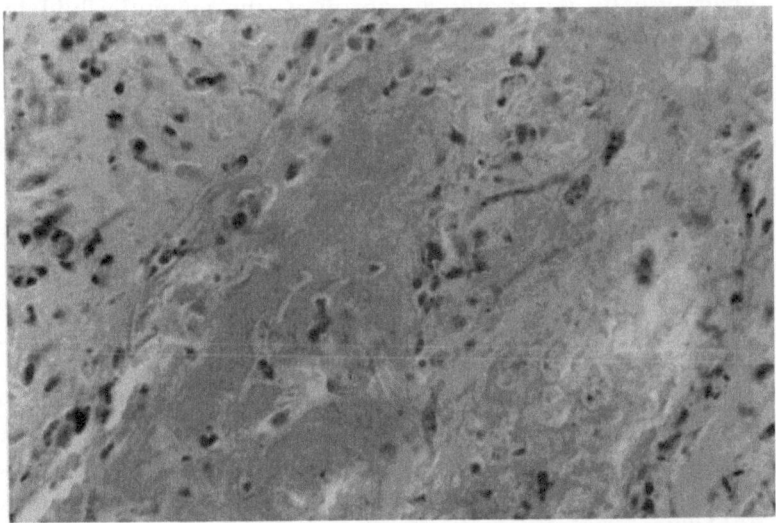

Fig. 3. Thrombosed blood vessel in placenta of cow containing fungal hyphae. ×450.

ulated with material from a coarse mulch in an early stage of decomposition. Such predisposing factors as diabetes, malignancy, Cushing's syndrome and other conditions which have been shown to predispose to mycotic infection of man have not accompanied most cases of mycotic infection of the reproductive tract of animals. However, pregnancy in itself has been accused of enhancing the manifestations of systemic mycotic disease, even to the point of bringing about relapses of systemic coccidioidomycosis of women (Utz, 1962). The endometrium definitely is more susceptible to infection while under the influence of progesterone. Exogenous estrogens among other things have been shown to favor mycotic infection. The basic immunologic status of animals with reference to the fungi has not been investigated and the metabolic derangement of acidosis or ketosis is common in pregnant cows in the winter. Acidosis is a common occurrence in diabetics which are very susceptible to systemic mycotic infections. Other factors in addition to the hyperglycemia of diabetes appear to influence the susceptibility to mycotic infection. Consideration should be given, in future investigations, to the determination whether or not the pregnant animals which are used in experimental exposures by inhalation of spores of fungi have immunological resistance such as occurs with viral disease because of a prior exposure when not pregnant. The report of Austwick (1962), who found two-thirds of otherwise normal cattle to have many small pulmonary lesions containing fungus, suggests that prior local infection of the lung may be common.

The pathogenetic mechanisms within the pregnant uterus deserve consideration. Although the placenta of cattle is infected frequently, the fungus usually is recognized only in it, the amniotic fluid in the amniotic sac and in the lungs as well as in the skin with which it has primary contact. The fungus floats freely in the alveolar fluids of many of these fetuses causing no inflammatory reaction or less than would be expected if the fetus were infected by hematogenous spread through the umbilical blood vessels. Consideration then may be focused on the anatomical relationships within the pregnant uterus. The maternal and fetal vascular systems are generally autonomous. The basic pathologic change in the maternal tissue appears to be a vascular thrombosis and infarction which, in turn, brings about anoxic death of the adjacent fetal placenta and effectively eliminates fetal blood flow in the area. The suggestion is that the fungus grows into the fetal portion of the placentome, through the membranes and into the amniotic sac to seed the amniotic fluid. The retention of the maternal caruncular tissue by the fetal cotyledons is indicative of the infarctive process and associated loss of tensile strength. The finding of hepatic lesions in the foal (Mahaffey and Adam, 1964) does not deny the validity of the observations because the placenta of the horse has a diffuse contact with the uterus, thus permitting the spread of the fungus in the fetal placenta and into viable placental tissue and blood vessels. Even so, the distribution of the fungus in the aborted equine fetus is quite similar to that of the bovine fetus. Experimental support of this concept is offered by Flamm *et al.* (1958) who found that in pregnant mice infected intravenously with *Candida albicans* the fungus gained access to the amniotic fluid and then the embryo after multiplying within the membranes. The fetal blood did not contain organisms though the maternal blood did. A similar behaviour has been found in experimental infection of the pregnant uterus of rats by certain bacteria (Payne, 1958).

Although Benirschke and Raphael (1958) did not find the placenta of the human fetus affected, there was a healing lesion which they suspected of having occurred as a part of a temporary rupture of the membranes during a prior time when the mother had clinical signs of vaginal candidiasis.

Mycosis of the Male Reproductive System

Fungal infections of the male reproductive system are quite uncommon except as a part of systemic mycoses such as those of coccidioidomycosis, blastomycosis, and histoplasmosis. Robinson and McVickar (1952) reported phagocytosed *Histoplasma capsulatum* in the testes of two dogs with spontaneous systemic infection; however, these were the only testes available in the twenty-one cases which they studied. No mention of

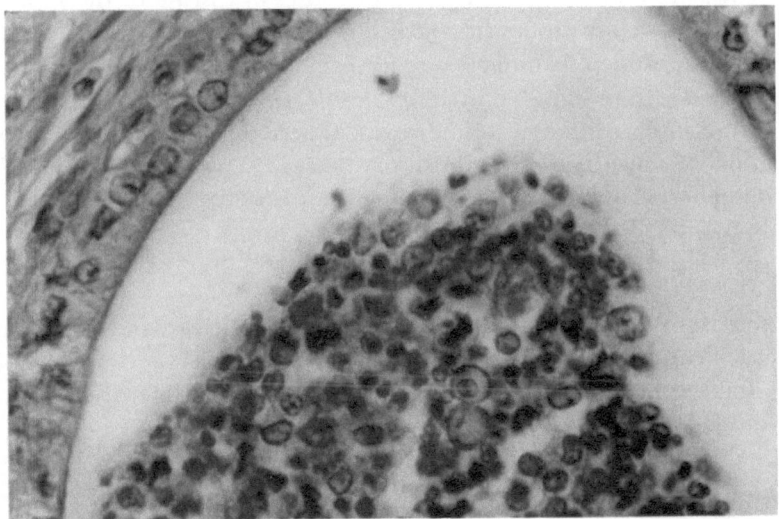

Fig. 4. *Blastomyces dermatidis* in tubular lumen of epididymis of dog with systemic blastomycosis. ×450.

the testes had been made in previous reports of twenty-five other cases which they reviewed. Phagocytes containing the organism were found in the tubular portions of the testes but the possibility of artificial translocation from interstitial granulomas was considered. Lesions have been found in the testes in experimental histoplasmosis also (Kipkie and Howell, 1951). *Blastomyces dermatidis* does get into the epididymal lumens of dogs with metastatic infection (Fig. 4). Menges (1960) found lesions in 5 of 56 dogs with systemic infection with the latter organism.

Coccidioidomycosis has been found to affect the human prostate in 6 per cent of 95 cases in one series (Forbus and Bestebreurtje, 1946) and coccidioidal epididymitis occurs in some cases as the only extrapulmonary lesion. Seminal ejaculate has been found contaminated (Weyrauch *et al.*, 1950). Nocardiosis occurs rarely in the testes of bulls, causing purulent to granulomatous inflammations (Awad, 1960). A special propensity for localization of *C. albicans* in the seminal vesicles of rats, but not other experimental animals, has been recognized (Fuentes *et al.*, 1952). Other organisms may cause their characteristic pathologic change on rare occasions. The pathologic appearance of the lesions in the genital tract are not different from that seen elsewhere in the body. That the testis is a good site for fungous growth is attested to by the frequent injection of laboratory materials into the testes of small animals in order to isolate the organisms in pure culture.

Conclusions

Mycoses are now recognized as major causes of disease of the reproductive tract of certain domestic animals and, less so, that of man. Much basic data on the epidemiology and behaviour of the infections are available. However, considerable investigation is needed to define more clearly the pathogenesis of the mycotic diseases. These should include controlled experiments concerned with the immunologic status of the animal and the influence of more common metabolic disorders.

Acknowledgments

The author wishes to thank Donald Cordes, New Zealand Department of Agriculture, Ruakura Animal Research Station, Hamilton, New Zealand, for the privilege of reviewing pathological specimens and the use of Figures 1 and 2.

References

Ainsworth, G. C., and P. K. C. Austwick: Fungal Diseases of Animals. Farnham Royal England Commonwealth Agriculture Bureau, 1959.

Aldasy, P., and G. Hajdu: Peneszes sarjadzogombak okozta elvetelesek szarvas-marha-allomanyokban. Mag. allator. Lapja. *16*:448, 1961.

Austwick, P. K. C.: The presence of *Aspergillus fumigatus* in the lungs of dairy cows. Lab. Invest. *11*:1065, 1962.

Austwick, P. K. C., and J. A. J. Venn: Mycotic abortion in England and Wales. Proc. IVth Int. Congr. Anim. Reprod., The Hague. *3*:562, 1961.

———, ———: Routine investigations into mycotic abortion. Vet. Rec. *69*:488, 1957.

Awad, F. I.: Nocardiosis of the bovine udder and testis. Vet. Rec. *72*:341, 1960.

Baruah, H. K.: The air spora of a cowshed. J. Gen. Microbiol. *25*:483, 1961.

Bendixen, H. C., and N. Plum: Schimmelpilze (*Aspergillus fumigatus und Absidia ramosa*) als Abortusursache beim Rinde. Acta pathol. et microbiol. Scand. *6*:252, 1929.

Benirschke, K., and S. I. Raphael: *Candida albicans* infection of the amniotic sac. Am. J. Obstet. Gynec. *75*:200, 1958.

Bisping, W., M. Refai, and G. Trautwein: *Candida parapsilosis* as the cause of abortion in the cow. Berl. Münch. tierärztl. Wschr. *77*:260, 1964.

Conan, N. J., and G. A. Hyman: Disseminated coccidioidomycosis: treatment with protoanemonin. Am. J. Med. *9*:408, 1950.

Cordes, D. O., D. C. Dodd, and P. J. O'Hara: I. Bovine mycotic abortion. N.Z. Vet. J. *12*:95, 1964.

———, ———, ———: Acute mycotic pneumonia of cattle. N.Z. Vet. J. *12*:101, 1964.

Emmons, C. W.: The Jekyll-Hydes of mycology. Mycologia. *52*:669, 1960.

Flamm, H., W. Kovac, and C. Kunz: Experimentelle Pilzinfektion der weissen Maus während der Gravidität. Zbl. Bakt. (I. Orig.) *172*:449, 1958.

Forbus, W. D., and A. M. Bestebreurtje: Coccidioidomycosis: a study of 95 cases of the disseminated type with special reference to the pathogenesis of the disease. Mil. Surgeon. *99*:653, 1946.

Fuentes, C. A., J. Schwarz, and R. Abdoulatia: Some aspects of the pathogenicity of *Candida albicans* in laboratory animals. Mycopathologia. *6*:176, 1952.

Gilman, H. L., and R. R. Birch: A mould associated with abortion in cattle. Cornell Vet. *26*:81, 1925.

Heath, P.: Massive separation of the retina in full term infants and juveniles. J. Amer. Vet. Med. Assn. *144*:1148, 1950.

Hellman, E., and S. Ruethel: *Trichosporon capitum* als Ursache eines Abortes beim Rind. Berl. Münch. tierärztl. Wschr. *77*:380, 1964.

Hensel, L., W. Bisping, and H. Schemmelpfennig: Aspergillusabort beim Pferde. Berl. Münch. tierärztl. Wschr. *74*:290, 1961.

King, S. J., B. L. Munday, and W. J. Hartley: Bovine mycotic abortion and pneumonia. N.Z. Vet. J. *13*:76, 1965.

Kipkie, G. F., and A. Howell, Jr.: Histopathology of experimental histoplasmosis. AMA Arch. Path. *51*:312, 1951.

Littman, M. L., and Lorenz E. Zimmerman: *Cryptococcosis, torulosis or European blastomycosis*. New York: Grune & Stratton, 1956.

Mahaffey, L. W., and N. M. Adam: Abortions associated with mycotic lesions of the placenta in mares. J. Amer. Vet. Med. Assn. *144*:24, 1964.

McGaughey, C. A., G. H. Corray, P. Seneviratna, and C. St. George: Scopulariopsosis in dogs in Ceylon. *Scopulariopsis brevicaulis* infection causing systemic mycosis and death of foetuses. Ceylon vet. J. *9:*119, 1961.

Menges, R. W.: Blastomycosis in animals. A review of an analysis of 116 canine cases. Vet. Med. *55:*45, 1960.

Nassau, E., and C. Weinberg-Heiruti: Torulosis of the newborn. Harefuah. *35:* 50, 1948.

Neuhauser, E. B. D., and A. Tucker: The roentgen changes produced by diffuse torulosis in the newborn. Am. J. Roentgenol. *59:*805, 1948.

O'Hara, P. J.: Personal communication, 1966.

Page, E. W., and L. M. Boyers: Coccidioidal pelvic inflammatory disease. Am. J. Obstet. Gynec. *50:*212, 1945.

Payne, J. M.: Changes in the rat placenta and foetus following experimental infection with various species of bacteria. J. Path. Bact. *75:*367–385, 1958.

Plum, N.: Verschiedene Hyphomyceten-Arten als Ursache sporadischer Fälle von Abortus beim Rinde. Acta path. et microbiol. Scand. *9:*150, 1932.

Robbins, E. S.: North American blastomycosis in the Dog. J. Amer. Vet. Med. Assn. *125:*391, 1954.

Robinson, V. B., and D. L. McVickar: Pathology of spontaneous canine histoplasmosis. A study of twenty-one cases. Am. J. Vet. Res. *13:*214, 1952.

Small, L. E., and J. W. Birshner: Maternal deaths from coccidioidomycosis. J. Amer. Vet. Med. Assn. *140:*1152, 1949.

Smith, Theobald: Mycosis of the bovine fetal membrane due to a mold of the genus *Mucor.* J. Exptl. Med. *31:*115, 1920.

Sonnenschein, H., H. L. Clar, and C. L. Taschdjian: Congenital cutaneous candidiasis in premature infant. Am. J. Dis. Child. *99:*81, 1960.

Turner, P. D.: Syncephalastrum associated with bovine mycotic abortion. Nature *204:*399, 1964.

Utz, J. P.: The spectrum of opportunistic fungus infections. Lab. Invest. *11:* 1018, 1962.

Weikl, A.: Beitrag zum Schimmelpilzabort. Berl. Münch. tierärztl. Wschr. *77:* 293, 1964.

Weyrauch, H. M., F. W. Norman, and J. B. Bassett: Coccidioidomycosis of the genital tract. California Med. *72:*465, 1950.

COMPARATIVE ASPECTS OF REPRODUCTIVE FAILURE INDUCED IN MAMMALS BY VIRUSES

D. N. Medearis, Jr.

Department of Pediatrics, University of Pittsburgh School of Medicine, and The Children's Hospital of Pittsburgh, Pittsburgh, Pennsylvania

If certain broad areas within the field of virus induced reproductive failure had not been reviewed recently, the scope of this discussion would have perforce been much greater. Fortunately, pregnancy wastage induced in man was recently discussed by Hardy (1965); in addition, some of the basic biologic aspects of the interactions of animal viruses and embryos were reviewed by Ebert in 1960. In contrast, reproductive failure induced in mammals other than man by viral infections natural to the respective species has been very infrequently considered. Koprowski reviewed this matter in 1958. He listed attenuated blue-tongue virus vaccination as a cause of abortion and malformation in sheep, equine rhinopneumonitis (equine abortion) virus as an inducer of abortion and stillbirth in mares, and modified hog cholera virus vaccine as a cause of abortion, stillbirth, and malformations in swine. It should be pointed out that of those, the viruses which induce alterations of development are those altered by laboratory attenuation, and in a recent text (Blood and Henderson, 1963a), it was stated that there were no congenital anomalies induced by viruses natural to the host known to veterinary medicine. Two recent reviews of comparative medicine did not discuss the problem (International Symposium on Comparative Medicine, 1962; Schwabe, 1964).

This lack of attention is not entirely deserved. Pregnancy interruption, fetal and neonatal infections, and congenital anomalies can be induced in subprimate species by viruses. In this review a context will be provided by presenting certain aspects of the reproductive failure induced by rubella and cytomegaloviruses in man; then equine abortion virus (rhinopneumonitis virus), equine infectious anemia in horses, infectious bovine rhinotracheitis, bovine virus diarrhea in cattle, attenuated blue-tongue virus vaccine and scrapie in sheep, and modified hog cholera virus vaccine will be reviewed in more detail. In so doing special attention will be given to whatever data is available concerning the pathogenesis of these diseases in order to determine areas which might be profitably explored in the future.

It should be stressed that this review is not meant to be comprehensive. Reproductive failure induced in mammals by viruses foreign to the animal will not be discussed, and examples have been chosen, either on the basis of the extent of information available, or on relevance to certain aspects of the problem. The superb work of Ferm, Kilham and Margolis with Rat virus, H-1 virus and the virus of spontaneous ataxia in cats (Kilham, 1966; Kilham and Margolis, 1966) will not be discussed since their discussion appears elsewhere in this conference.

Rubella Virus Infections of Man

Rubella virus has many properties in common with myxoviruses. It is about 200 mμ. Electronmicroscopically it has a dense core and a wide envelope and resembles mouse leukemia virus. Total loss of infectivity occurs after one hour at 56°C; it is ether sensitive and its replication is unaffected by 5-iodo 2' deoxyuridine. A subtle cytopathic effect is induced in cell cultures of primary human amnion and no cytopathic effect is seen in monkey kidney in which its presence can be detected by interference. The virus can be recovered from the blood, the pharynx, and the feces from 3 to 7 days before the exanthem appears, and lymphadenopathy appears one week before the rash. The incubation period varies from 13 to 16 days. Histopathologic studies reveal edema and reticular hyperplasia of involved lymph nodes. Thrombocytopenia and occasionally a unique encephalopathy occur.

Spontaneous abortion and stillbirth are two to four times as frequent in pregnancies complicated by rubella in the first trimester as in pregnancies not complicated by rubella. The placenta is involved, the fetus is infected and remains so for prolonged periods of time after the infant is born. Approximately 20 per cent of live born infants of mothers who had rubella in the first trimester of pregnancy will have malformations. Despite this, rubella is the cause of only 3 to 4 per cent of all congenital heart disease. Some of the effects of rubella induced embryopathy (for example, deafness) may not be noted for many years following birth. Prenatal infections acquired *in utero* may be inapparent for the first few weeks of life. It is important to emphasize that the rubella syndrome can occur in infants born of mothers who, although they were infected, had no apparent disease. Although pregnancy wastage in rubella is much more frequently induced when the infection occurs in the first trimester of pregnancy, there is some evidence that reproductive failure may be induced when rubella occurs at other times. The fetal effects of maternal rubella include cataract, glaucoma, patent ductus arteriosus, peripheral pulmonic stenosis, microcephaly, deafness, and mental retardation. The neonatal rubella syndrome is characterized by small size, thrombocytopenic purpura, jaundice, hepatosplenomegaly, cataracts, congenital heart disease, encephalopathy, diarrhea, failure to thrive, and hearing deficit. The reader is referred to excellent discussions by Weller and Neva (1965) and Hardy (1965) for more detailed information.

Cytomegalovirus Infection of Man

Human cytomegalovirus (CMV) is highly species specific; it is about 100 mμ in diameter and contains DNA. It is ether sensitive, heat sensitive, and has been classified as a herpesvirus. Most cytomegalovirus infections are inapparent. Notable exceptions are the pneumonia induced in patients with malignancies of the reticuloendothelial or hematopoietic systems and neonatal cytomegalic inclusion disease (CID). The virus induces distinctive, almost pathognomonic cytomegalia. Viremia is induced in mice by mouse cytomegaloviruses but viremia has not been demonstrated in man even though pathologic observations indicate it occurs. The placenta is involved and infants born with cytomegalic inclusion disease are infected. Thus, if the infant recovers from generalized neonatal cytomegalic inclusion disease he may be left with the residual of a necrotizing meningoencephalitis: mental retardation and microcephaly. A very interesting and significant observation is that of Hanshaw (1966a): the incidence of cytomegalovirus antibody in the serum of patients with microcephaly is significantly higher than that of a control population. (For detailed discussion of cytomegalovirus infection the reader is referred to Weller, 1965; Hanshaw, 1966; Medearis, 1964b.)

Other viral infections of man may induce a degree of reproductive failure, but these have not been so clearly defined as rubella and CID. One should note especially the discussion by Hardy (1965) of influenza.

In summary, both rubella and cytomegalovirus induce reproductive failure as a result of inducing fetal infection, which probably occurs secondary to viremia and diaplacental passage of the virus. In rubella, there is a critical gestation time (first trimester); for cytomegalovirus fetal infection may follow more frequently if the maternal infection occurs after the 4th month of gestation. In both, fetal infection persists a long time after birth. Rubella induces its effect as a result of having a much less severe destructive effect (one without inflammatory change) on organs at a time that is critical to their development. In contradistinction, cytomegalovirus induces a generalized destructive process at a less critical time (and one which is accompanied by inflammatory response); in survivors of CID all the damaged organs can be repaired except the brain. In addition, Hanshaw's observation indicates that microcephaly can be induced without accompanying CID.

Equine Virus Abortion

In 1933, Dimock and Edwards first reported a form of epizootic abortion in mares which was shown to be due to a virus (see Dimock, 1940). Their work was confirmed, and subsequently a relationship between equine abortion in mares and equine influenza was considered. Viruses recovered from equine abortion and from horses incurring respiratory infection were subsequently shown to be identical. It is not completely clear, however, whether the virus of equine rhinopneumonitis and the virus of equine abortion as first recovered are identical. Sufficient evidence exists, however, to justify considering these agents as very closely related, if not identical. They are not related to what is now known to be equine influenza. The virus induces characteristic large eosinophilic intranuclear inclusions with margination of nuclear chromatin. The virus contains double stranded DNA and is 180–250 millimicrons in size. It is ether sensitive. Accordingly, its proposed grouping would be with herpesvirus.

The clinical disease is characterized by mild upper respiratory infection manifested by fever, conjunctivitis, and coughing. It follows an incubation period of two to three days. The febrile reaction is diaphasic in character and persists for about 10 days. Often the infection is inapparent. Diseases are more often seen in weanling animals than in older animals. Mild leukopenia has been observed and mild mucopurulent nasal discharges may follow inoculation of viruses obtained from aborting mares. Hagan and Bruner state that the disease in mares consists only in abortions and that this clearly differentiates it from abortion induced by bacterial infection. If fetuses are inoculated *in utero,* they are aborted in a few days. If mares are inoculated by other routes, the abortion occurs three to four weeks later. Pathologic findings are limited to rhinitis and pneumonitis. In the fetus there are greyish-white foci of degeneration in the liver, and focal myocardial, splenic and pulmonary hemorrhages. Pleural fluid accumulates in many of the aborted animals. In the liver and in the lungs intranuclear inclusions are seen. In mares abortions occur accompanied by few or only mild clinical symptoms. The involution of the uterus occurs in a normal manner following abortion.

The pathogenesis of the disease (according to Blood and Henderson, 1963) is characterized by an initial viremia followed by localization in the uterus and respiratory tract. Abortion is most frequently induced during the latter third of pregnancy. The placenta is apparently involved even though specific statements as to the recovery of virus from the placenta and the nature of the pathology induced in it are not available. The fetus is infected; it can be used to induce abortions and respiratory disease in susceptible recipients. The question has been raised as to whether the fetuses are decreased in size for the gestational time. Foals born alive frequently die within the neonatal period. It is not known whether the foals have a persistent infection. Congenital anomalies have not been described. Abortions seldom if ever have been observed in the same mare in succesive seasons and this suggests that immunity is long lasting. Evidence does not indicate that the virus is carried for a long period after abortion by pregnant mares. (For additional information see Blood and Henderson, 1963; Hagan and Bruner, 1957; Doll, 1961.)

Equine Infectious Anemia (Swamp Fever)

This is a disease of horses, seen throughout the world, and characterized by a protracted anemia. Horses, mules, and donkeys of all ages are susceptible to the infection and the infection can be transmitted by inoculation of blood from a diseased or healthy carrier. Its natural mode of transmission is probably through an insect vector. The stable fly can transmit the virus from infected to normal horses, as can horse flies; mosquitos can transmit the anemia in an inapparent subclinical form. The disease is most frequently seen where horses are pastured in swampy land.

The disease is due to a virus whose properties have not been determined. The outstanding pathogenic characteristic of this disease is the protracted viremia which may persist for a period of 18 years during which the blood of the infected horse can be shown to be constantly infective for susceptible animals. Virus can be recovered from the blood, milk, urine, saliva, feces, and all of the parenchymatous organs. The virus seems to be very resistant to heat, chemicals and putrefaction, but other properties have not been defined.

The disease is characterized by an incubation period of 2 to 4 weeks, weakness, depression, fever, incoordination and edema. The anemia is characterized by marrow suppression, anemia, proliferation of reticulo-endothelial cells, erythrophagocytosis, decreased red cell survival time, and decreased incorporation of iron into erythrocytes. Inapparent infections occur also. Whether or not the infection induces abortion with any degree of frequency is debated.

Pathologic changes reveal hemorrhages in most of the viscera. In the marrow there is marked erythroid hyperplasia. There is hemosiderosis in various reticuloendothelial organs, and marked erythrophagia.

This discussion of equine infectious anemia is included because of the reports by several authors (Ishii, 1963; Hagan and Bruner, 1957) that infected mares gave birth to foals which immediately after parturition were found to be infected. These infections were inapparent and their duration is not known. Since infected mare's milk occasionally contains virus, it was noteworthy that Stein and Mott (as quoted by Hagan and Bruner, 1957) were unable to obtain clear evidence that foals could contract the disease by suckling infected dams or associating with them. The nature of the disease, if any, in foals born of infected dams, and the immune response in these foals is not known. Certainly, however, this is an area which deserves further investigation. (For detailed reviews the reader is referred to the two veterinary texts previously cited and to Ishii, 1963, and to the comprehensive monograph by Dreguss and Lombard, 1954.)

Infectious Bovine Rhinotracheitis (IBR, Rednose)

This disease is characterized by fever, rhinitis, and conjunctivitis; and the syndromes of vulvovaginitis, conjunctivitis, balanoposthitis and encephalomyelitis have also been reported. Recently it has been one of the most carefully studied virus diseases of domestic animals. IBR virus contains double-stranded DNA and is about 100 millimicrons in size. It is ether sensitive. Accordingly, it can be tentatively grouped as a herpesvirus (Armstrong *et al.*, 1961). The agent has been grown in cell cultures (Gillespie *et al.*, 1957).

The clinical disease is characterized by an incubation period of three to seven days in experimental infections, and 10–20 days after the introduction of susceptible cattle into feedlots containing infected cattle. The sudden onset of anorexia and fever, accompanied by marked hyperemia of the nasal mucosa, rhinitis and conjunctivitis and an increased degree of salivation characterize the disease. There is tachypnea and respiratory distress, as well as a short explosive cough on occasion. The temperature may return to normal in a day or two and recovery may be complete in two weeks. Deaths are infrequent; when they occur, a secondary bronchopneumonia is frequently responsible. Australian investigators revealed that encephalitis was an infrequent but clearly demonstrable manifestation of IBR infection. In doing so they also demonstrated viremia by recovering the virus from circulating leukocytes. This, of course, is an important aspect of the pathogenesis of this disease. Young calves are very much more susceptible to the infection than are adults and after experimental inoculation they incur a fatal illness characterized by generalized lesions and virus can be recovered from many organs. Experimental infections have been induced by inoculating virus intramuscularly (IM). Fever begins 2 to 7 days later and is followed by vulvar edema and papulovulvovaginitis (Owen *et al.*,

1964). In these experiments there was a 60 per cent fetal mortality but the fetuses revealed no gross lesions and there were no congenital anomalies. There were diffuse microscopic changes characterized by a coagulative necrosis in the liver and spleen. Interestingly enough, there were no inclusions despite the fact that this virus induces distinct intranuclear inclusions. Focal renal necrosis with a mononuclear infiltrate is frequent. Virus was recovered from the placenta, amniotic fluid, liver, spleen, and lungs. Similar corroborative data was reported by McKercher and Wada (1964). Six isolates were recovered in cell cultures from the organs and fluids of fetuses aborted by heifers between the sixth and seventh month of gestation. The animals were not sick and they had not been previously vaccinated with IBR. Subsequently seven heifers were inoculated IV, IM, or IN. All became febrile and most developed conjunctivitis. Six of these animals aborted and virus was recovered from each of the fetuses. These investigators recovered IBR from 6 to 12 fetuses from a herd suffering from epizootic abortion. Fetal tissues which contained virus included the placenta, liver, spleen, kidney, lungs, blood and pleural fluid. One isolation came from one fetus aborted after vaccination with modified IBR virus.

These authors discuss the possibility that a mutation has occurred endowing the virus with a capacity to induce abortion. They point out that the disease was known for some time before abortions, encephalitis, or viremia were thought to be associated with it. Others differ, noting that abortions were mentioned early, and that the viremia may not have been detected because of technical difficulties. No serologic differences between strains have been detected.

Chow *et al.* (1964) demonstrated that abortions could be induced during the first and during the third trimester in experimental infections. In most experiments in which abortion has been induced, heifers have had disease. Serologic studies of herds in which clusters of abortions are occurring also indicate an association of IBR antibody with abortion. No gross anomalies have been demonstrated and persistent neonatal infections have not been detected.

Bovine Viral Diarrhea Virus (BVD)

The virus is apparently similar to the virus of hog cholera and current information indicates that it possesses many of the properties of myxoviruses. It is 55 millimicrons in size, contains RNA and is ether and acid sensitive. There are both cytopathic and non-cytopathic strains. The virus can be propagated in cattle, sheep, and goats, but not in dogs, cats, mice, suckling mice, hamsters, parrots, pigeons, one day old chicks, or guinea pigs. It can be cultivated in cell cultures of bovine skin and muscle and bovine embryonic kidney. Neutralization tests indicate that the BVD virus and mucosal disease complex virus (MDV) are immunologically related. The clinical and pathologic manifestations of BVD and MDV resemble rinderpest, but no immunological relationship has been found between them. Antibody against the following viruses does not neutralize BVD: bluetongue virus of sheep, hog cholera, infectious bovine rhinotracheitis, infectious ulcerative stomatitis, sporadic bovine encephalitis, winter dysentery and mycotic stomatitis. However, Kniazeff and Pritchard (Pritchard, 1963) have shown by gel diffusion studies that several strains of BVD-mucosal disease complex virus are related to hog cholera virus.

The clinical illness can be reproduced experimentally. The principal clinical signs include fever, nasal discharge, ulcerative erosions of the oral mucosa, lameness, and diarrhea. Characteristically, there is a diphasic fever (Pritchard, 1963) curve, with fever occuring 2–4 days and 7–10 days after inoculation. Temperature elevation varies from 1°–6°F. Concurrently there is a leukopenia which lasts for 1–6 days.

Diarrhea occurs about a week after the febrile period and persists for days to weeks; feces contain mucus and occasionally considerable quantities of blood. The diarrhea can cause severe dehydration; a 1000 pound steer can lose 150–250 pounds. Morbidity and mortality are significant. Clinically apparent infection induces a long-lasting immunity.

Principal pathologic changes induced by the infection are hemorrhage, edema and ulceration of the intestine. The clinical and pathologic changes of bovine virus diarrhea are less extensive than those of the mucosal disease complex virus infections. Mononuclear infiltrates, vacuolation, and marked enlargement of the cytoplasm of cells, and pyknotic nuclei are characteristic of the histopathology. No significant changes have been reported in the central nervous system.

The disease can be transmitted by inoculating animals with blood or spleen emulsions obtained during the viremic stage of the disease. Abortions have occurred from 10 days to several months following the acute illness. Abortions have not been consistently induced in animals experimentally infected. Recent evidence suggests, however, that antibodies to bovine virus diarrhea are more frequent in herds incurring an epizootic form of abortion than in herds not so affected (Kniazeff, 1966). Available evidence suggests that abortion occurs during the third trimester, that neonatal deaths have occurred following maternal BVD, that the fetuses are normal on examination. Virus has not been recovered from fetuses.

Hog Cholera (Swine Fever)

The virus which causes hog cholera resembles the myxoviruses. Its size is about 55 mμ. Its nucleic acid is RNA (as determined by a failure of 5-bromo, 2′ deoxyuridine, and 5-iodo 2′ deoxyuridine to inhibit its multiplication in cell cultures) and it is ether and acid sensitive (Loan, 1964).

The disease is a highly infectious one characterized by a very rapid spread and a high mortality rate. The incubation period is about 5 to 10 days, although periods as long as 35 days have been recorded. The first clinical manifestation is the onset of a high temperature. Other manifestations include depression, anorexia, and a drooped attitude with tail hanging and an unsteady swaying gait. A purplish discoloration of the skin occurs and is followed by the appearance of necrotic areas on the ears, tail, and vulva. Signs of central nervous system involvement are common and are characterized by circling gait, incoordination, muscle tremor and convulsions. Blindness has also been observed. When the virus is administered with immune serum as a vaccine a chronic form of the disease may be induced in which the incubation period is longer and signs of illness include emaciation and characteristic skin lesions including alopecia and blotching of the ears. Terminally, there is a deep purple discoloration of the abdominal skin. A significant leukopenia accompanies the disease.

Pathologic changes include submucosal and subserosal petechial hemorrhages in the kidney, the ileocecal valve, cortical sinuses of the lymph nodes, the bladder, and the larynx. The liver, the bone marrow, and the lungs are congested. The encephalitis is non-suppurative.

Young *et al.* (1955) were the first to demonstrate that edema of the newborn swine could be induced by modified hog cholera viruses. Fifty-six per cent of 61 fetuses born in six litters of sows not previously immunized and given modified virus between the 10th and the 16th days of gestation were abnormal, whereas only four per cent of 83 fetuses in 8 litters of sows not previously immunized (and not injected with virus) were abnormal. It was interesting that seven per cent of 20 fetuses born of two litters previously given modified virus during gestation were abnormal. However, this compared with the four per cent in a control group. It was clear that prior administration of inactivated vaccine or serum and virulent virus did not prevent the subsequent birth of abnormal fetuses when modified virus vaccine was given at that particular time in pregnancy (the 10th to the 16th day). The anomalies noted were deformed noses and kidneys, edema and ascites. The virus was recovered from the spleens of the abnormal fetuses. The virus strain which has induced reproductive failure is one attenuated by passage in rabbits. Inactivated vaccines have not induced reproductive failure. Additional experimental infections reveal that reproductive failure is most consistently induced when sows are inoculated between the 20th and 100th days of gestation. Mild febrile reactions do occur in the inoculated animals and there is a viremia. Recent observations by Emerson and Delez (1965) reveal that

the lesion induced in piglets by prenatal vaccination of sows with modified hog cholera vaccine is a cerebellar hypoplasia and hypomyelinogenesis manifested by congenital tremors. Myelination was deficient in the
cerebrum, pons, medulla, cerebellum, and spinal cord. Anatomic structures were poorly developed and the molecular layer of the cerebellum
was decreased. Clinical signs included muscular tremors, head nodding,
inability to stand or nurse. Liver, kidney, and spleen homogenates from
trembling pigs inoculated into susceptible pigs induced fever and
leukopenia and these pigs were resistant to challenge with virulent hog
cholera virus four weeks later. Another interesting observation has been
made by Baker and Sheffy (1960) who found that a persistent viremia
can be induced in sucklings using modified virus. If piglets are inoculated
when 6 weeks old the viremia is prolonged, whereas when the animal is
3 months old only a transient viremia occurs. Viremia was accompanied
by a significant retardation in weight gain.

Bluetongue

This is a viral infection of sheep transmitted by insect vectors and
characterized by stomatitis, rhinitis, enteritis, and lameness. Presently
available information suggests that it is a small (20 millimicron) RNA
containing a virus that is neither ether nor desoxycholate sensitive
(Studdert, 1965). There are a large number of antigenic strains which
vary considerably in their pathogenicity.

There is an incubation period of 2 to 4 days in experimental infections.
This is followed by a severe febrile reaction which continues for the
next 5 or 6 days. After 2 days of elevated temperature and nasal discharge, salivation and hyperemia of the buccal and nasal mucosa became
apparent. Subsequently the rhinitis becomes mucopurulent and blood-
stained. There is edema and swelling of the oral mucosa and organs.
These lead to ulcers. Swallowing is difficult and subsequently respiration
becomes stertorous and there is a tachypnea. The syndrome has two
varieties. In the abortive type the febrile period is not followed by
localized lesions. In the less acute variant, local lesions are minimal but
emaciation and weakness are continued over a protracted period. The
disease is easily confused with foot and mouth disease although its spread
is much slower. The infection is thought to be transmitted by flies
(Foster *et al.*, 1963). It is not spread by contact and there is a marked seasonal incidence. Viremia occurs in the early stages of the infection and
blood of diseased animals can be used to transmit the infection to susceptible sheep.

Schultz and Delay (1955) noted that vaccination of pregnant ewes with
modified bluetongue virus was attended by a risk of deformity in the

lambs. They did not recover virus from the abnormal fetuses, however. There is a critical gestational time for fetal damage; the most critical period is between the fifth and the sixth weeks though it may extend from the fourth to the eighth week. Recently Young and Cordy (1964) described the encephalopathy induced by bluetongue vaccine virus. When vaccine was given subcutaneously on the fortieth day of gestation six of twenty-nine fetuses had an encephalopathy characterized by acute, diffuse and focal necrotizing meningoencephalitis. There was a plasmacytic invasion in the pia around blood vessels. There were intraneural hemorrhages and early mineralization and microcavitation. Hydranencephaly was found. Most importantly, virus was recovered from the lesions. Ewes were sick; they had fever, but there was no correlation of the fever with encephalopathy. These authors emphasize that villae first appear on the surfaces of the chorionic sac on the thirty-first day of gestation and question whether this is related to the capacity of the virus to pass diaplacentally.

Scrapie (Visna)

This non-febrile chronic disease ends fatally and is characterized by a very long incubation period. The most outstanding change is a degenerative vacuolization in the neurons of the medulla oblongata, the pons, and the cervical portion of the spinal cord. Affected animals weave as they walk, the head is carried higher than normal and the animal stares. There is marked incoordination and falling when the animal tries to run. There is paralysis of the hind quarters and there may be convulsions. The disease is included for discussion here because there is evidence that congenital infection may derive either from the ram or the ewe at a time when neither parent manifests disease. Scrapie was introduced into previously clean flocks by means of a vaccine for louping-ill made from the brain, cord, and spleen of sheep infected with louping-ill, but the sheep, however, were the progeny of animals known to have had scrapie (Hagan and Bruner, 1957). The tissues had been treated with 0.35 per cent formalin. This is indicative of the marked resistance of this virus to physical agents. It is destroyed neither by boiling, or by rapid freezing and thawing. There is recent clear evidence that the virus can be spread by contact of inoculated and uninoculated mice (Morris *et al.*, 1965). Sheep infected with large quantities of tissue culture infective virus develop antibodies and despite this the virus persists in the blood (Gudnádottir and Pálsson, 1965). The virus can be grown from the spleen, the choroid plexus, the salivary gland, the lung and the kidney. (The reader is referred to the monograph on "Slow Viruses" by Gibbs *et al.*, 1966.)

Abortion

It is interesting to compare the results of two recent attempts to recover viruses from human and cattle abortuses (see Table 1). The reasons for this difference are unknown. If, despite this evidence, viruses (other

Table 1

VIRAL CAUSES OF ABORTION IN MAN AND IN CATTLE
(Two Recent Studies)

Man—Boue *et al.*[1] recovered no viruses from multiple organ explants from 74 human embryos.

Cattle—Sattar *et al.*[2] recovered IBR from 6 fetuses (from 3 herds), parainfluenza type 3 from 1 fetus (28 aborted fetuses from 13 herds tested)

---

[1] Proc. Soc. Exp. Biol. Med. *122*:11, 1966.
[2] J.A.V.M.A. *147*:1207, 1965.

than rubella, CMV and influenza) do induce some of the abortions in human beings, the failure to recover them may reflect a very different pathogenesis in man than in domestic animals.

SUMMARY

Here we have considered those instances of reproductive failure induced in man and domestic animals by viral infections which cause only mild disease or none at all in the pregnant animal.

A precise definition of those changes in the pregnant animal which accompany and are responsible for abnormal termination of pregnancy when the pregnant animal is very ill would be worth while also. The greater problem however would seem to be those illnesses which are mild when a specific and direct effect of the virus induces either an interruption of pregnancy or an altered development of the embryo-fetus.

It is very interesting that of the six instances in which viruses clearly induce reproductive failure three have the characteristics of herpesvirus. Two are myxovirus like, and one (bluetongue virus) cannot be easily grouped (see Table 2). So far there are not sufficient data to determine whether or not the stage in pregnancy is a clear determinant of whether or not reproductive failure will be induced if an infection occurs, even though with each virus-host system reproductive failure is most frequently induced when maternal infection occurs at a particular time. Certain aspects of the pathogenesis are quite clear. In every case a viremia is induced (with the possible exception of cytomegalovirus). In each instance

Table 2

VIRUSES WHICH INDUCE REPRODUCTIVE FAILURE IN ANIMALS

Virus	Proposed Grouping
Cytomegalovirus	Herpesvirus
Rubella	Myxovirus
Equine rhinopneumonitis (Equine abortion virus)	Herpesvirus
Equine infectious anemia virus	Unknown
Infectious bovine rhinotracheitis virus	Herpesvirus
Bovine virus diarrhea	Myxovirus
Modified bluetongue virus	Unknown
Scrapie	Unknown
Modified hog cholera virus	Myxovirus

the placenta is infected and altered (although it should be emphasized that whether this alteration of the placenta is important in allowing diaplacental passage of the virus or inducing an alteration of the physiologic functioning of the placenta is not known). In every case fetal infection is induced and in most every instance of abortion (or stillbirths or neonatal deaths) a significant histopathologic alteration is induced in the fetus and usually it is generalized. It is extremely interesting that significant central nervous system pathology is induced by four of the agents: Cytomegalovirus and rubella in man and modified bluetongue virus and modified hog cholera virus in the sheep and the pig respectively (see Table 3).

More is now known about cytomegalovirus and rubella in man than is known about any of the corresponding diseases in domestic animals. This is paradoxical in view of the fact that one would have thought that the domestic animal could be more precisely studied. The explanation undoubtedly lies in the fact that the diseases of man provide a very much greater stimulus to investigation by man. There are certain aspects of the problem of reproductive failure induced by viruses which should be very much more easily approachable in domestic animals than in man. Briefly expressed, precise pathogenetic studies should be carried out. One needs to know exactly where the virus multiplies at what time and to what extent. One needs to know precisely what terminates infection or what allows for the persistence of viral replication. The exact characteristics of the host response must be determined. Similarly, the exact pathology induced in various tissues in fetal animals needs to be determined. The need is very clear that we must define placental morphology and its relation to placental physiology and how these may be altered by viral infections. There is much to do for it is clear from many examples

Table 3

COMPARATIVE ASPECTS OF REPRODUCTIVE FAILURE IN MAN AND DOMESTIC ANIMALS

Species	Virus	Disease in Pregnancy	Trimester of Pregnancy	Viremia	Infection		Abortion	Fetal Pathology	
					Placental	Fetal		CNS	Other
Man	CMV	0	second and third	+?	+	+	+	++	++
Man	Rubella	±, 0	first	++	+++	+++	+++	+	++
Horse	EVA	±, +, 0	third	++	+++	+++	+++	0	++
Cow	IBR	±	first and third	++	++	++	++	0(?)	++
Sheep	MBTV	±	first	++	++	++	?	++	+·+
Pig	MHCV	±	first	++	++	++	+	++	

that the gross effects we have considered today may be but the protruding tip of an iceberg. Especially is this so when we consider the late-appearing and very subtle effects of some virus infections: deafness in rubella, microcephaly in cytomegalovirus infections, the previously un-suspected viral etiology of ataxia in cats, the physiologic effect of sigma factor in Drosophila, and the long-term effects of lymphocytic chorio-meningitis virus infections in mice.

Acknowledgment

The author wishes to express his gratitude to the authors of the general references cited, Doctors Weller, Neva, Blood, Henderson, Hagan, Bruner, Ishii, and Pritchard, for this review has drawn heavily on them. Doctor Alex Kniazeff of the Naval Biological Laboratory, Oakland, aided this author greatly in a personal communication.

References

Armstrong, J. A., H. G. Pereira, and C. H. Andrewes: Observations on the virus of infectious bovine rhinotracheitis and its affinity with the herpesvirus group. Virology *14*:276, 1961.

Baker, J. A., and B. E. Sheffy: A persistent hog cholera viremia in young pigs. Proc. Soc. Exp. Biol. Med. *105*:675, 1960.

Blood, D. C., and J. A. Henderson: "Diseases of the Newborn." Chap. 3 in: Veterinary Medicine, 2nd ed. Baltimore: Williams & Wilkins, 1963.

——, ——: "Diseases Caused by Viruses, I and II." Chap. 20 in Veterinary Medicine, 2nd ed. Baltimore: Williams & Wilkins, 1963.

Boue, A., C. Hannoun, J. B. Boue, and S. A. Plotkin: Cytological, virological and chromosomal studies of cell strains from aborted human fetuses. Proc. Soc. Exp. Biol. Med. *122*:11, 1966.

Chow, T. L., J. A. Molello, and N. V. Owen: Abortion experimentally induced in cattle by infectious bovine rhinotracheitis virus. J. Amer. Vet. Med. Assn. *144*:1005, 1964.

Dimock, W. W.: The diagnosis of virus abortion in mares. J. Amer. Vet. Med. Assn. *96*:665, 1940.

Doll, E. R.: Immunization against viral rhinopneumonitis of horses with live virus propagated in hamsters. J. Amer. Vet. Med. Assn. *139*:1324, 1961.

Dreguss, M. N., and L. S. Lombard: Experimental Studies in Equine Infectious Anemia. Philadelphia: University of Pennsylvania Press, 1954.

Dunne, H. W., J. L. Gobble, J. F. Hokanson, D. C. Kradel, and G. R. Bubash: Porcine reproductive failure associated with a newly identified "SMEDI" group of picorna viruses. Am. J. Vet. Res. *26*:1284, 1965.

Emerson, J. L., and A. L. Delez: Cerebellar hypoplasia, hypomyelinogenesis, and congenital tremors of pigs associated with prenatal hog cholera vaccination of sows. J. Amer. Vet. Med. Assn. *147*:47, 1965.

Foster, N. M., R. H. Jones, and B. R. McCrory: Preliminary investigations in insect transmission of bluetongue virus in sheep. Am. J. Vet. Res. *24*:1195, 1963.

Gibbs, C. J., D. C. Gajdusek, and M. Alpers (eds.) : Slow, Latent and Temperate Virus Infections. NINDB monograph 2. Bethesda: National Institutes of Health, 1966.

Gillespie, J. H., B. E. Sheffy, and J. E. Baker: Propagation of hog cholera virus in tissue culture. Proc. Soc. Exp. Biol. Med. *105*:679, 1960.

Gudnádottir, M., and P. A. Pálsson: Host-virus infection in Visna infected sheep. J. Immun. *95*:1116, 1965.

Hagan, W. A., and D. W. Bruner: Infectious Diseases of Domestic Animals, 3rd edition. New York: Comstock Pub. Assoc. 1957.

Hanshaw, J. B.: Congenital and acquired cytomegalovirus infection. Ped. Clin. of North Amer. *13*:279, 1966.

————: Cytomegalovirus complement fixing antibody in microcephaly. New Eng. J. Med. *275*:476, 1966.

Hardy, J. B.: Viral infection in pregnancy: A review. Am. J. Obstet. Gynec. *93*:1052, 1965.

Ishii, S.: Equine infectious anemia or swamp fever. Adv. Vet. Sci. *8*:263, 1963.

Kilham, L.: "Viruses of Laboratory and Wild Rats," in: Viruses of Laboratory Rodents. National Cancer Institute Monograph 20, U.S. Dept. Health, Education and Welfare, Bethesda, 1966.

Kilham, L., and G. Margolis: Viral etiology of spontaneous ataxia of cats. Am. J. Path. *48*:991, 1966.

Kniazeff, A.: Personal communication, 1966.

Koprowski, H.: Counterparts of human viral disease in animals. Ann. New York Acad. Sci. *70*:369, 1958.

Loan, R. W.: Studies of the nucleic acid type and essential lipid content of hog cholera virus. Am. J. Vet. Res. *25*:1366, 1964.

McKercher, D. G., and E. M. Wada: The virus of infectious bovine rhinotracheitis as a cause of abortion in cattle. J. Am. Med. Assn. *144*:136, 1964.

McKercher, D. G., E. M. Wada, E. A. Robinson, and J. A. Howarth: Immunologic and epizootiological studies of epizootic bovine abortion. J. Amer. Vet. Med. Assn. *147*:1666 (abstr.) , 1965.

Medearis, D. N.: Viral infections during pregnancy and abnormal human development. Am. J. Obstet. Gynec. *90*:1140, 1964.

————: Observations concerning human cytomegalovirus infection and disease. Bull. Johns Hopkins Hospital *114*:181, 1964.

Moll, T.: Abortion and stillbirth of guinea pigs resulting from experimental exposure to bovine enteric virus. Am. J. Vet. Res. *25*:1727, 1964.

Morris, J. A., D. C. Gajdusek, and C. J. Gibbs: Spread of scrapie from inoculated to uninoculated mice. Proc. Soc. Exp. Biol. Med. *120*:108, 1965.

Moulton, J. E., and L. M. Frazier: Cytopathic changes in tissue cultures inoculated with infectious hepatitis virus and treated with 5-fluoro deoxyuridine. Am. J. Vet. Res. *25*:41, 1964.

Owen, N. V., T. L. Chow, and J. A. Molello: Bovine fetal lesions experimentally produced by infectious bovine rhinotracheitis virus. Am. J. Vet. Res. *25*: 1617, 1964.

Pritchard, W. R.: The bovine viral diarrhea mucosal disease complex. Adv. Vet. Sci. *8*:1, 1963.

Sattar, S. A., E. H. Bohl, and M. Senturk: Viral causes of bovine abortion in Ohio. J. Amer. Vet. Med. Assn. *147*:1207, 1965.

Schwabe, C. W.: Veterinary Medicine and Human Health. Baltimore: Williams & Wilkins, 1964.

Schultz, G., and P. D. Delay: Losses in newborn lambs associated with Blue-tongue vaccination of pregnant ewes. J. Amer. Vet. Med. Assn. *127*:224, 1955.

Studdert, M. J.: Sensitivity of Bluetongue virus to ether and sodium desoxy-cholate. Proc. Soc. Exp. Biol. Med. *118*:1006, 1965.

Studdert, M. J., and P. C. Kennedy: Enzootic abortion of ewes. Nature *203*:1088, 1964.

Taylor, D. O. N., D. P. Gustafson, and R. M. Claflin: Properties of some viruses of the mucosal disease virus diarrhea complex. Am. J. Vet. Res. *23*:143, 1963.

Watson, W. A.: Abortion and stillbirth in sheep. II. Viral and rickettsial infections. Vet. Bull. *32*:335, 1962.

Weller, T. H.: "Cytomegaloviruses." Chap. 43 in: Viral and Rickettsial Infections of Man. Philadelphia: J. B. Lippincott Company, 1965.

Weller, T. H., C. A. Alford, and F. A. Neva: Changing epidemiologic concepts of rubella, with particular reference to unique characteristics of the congenital infection. Yale J. Biol. Med. *37*:455, 1965.

Weller, T. H., and F. A. Neva: "Rubella Virus." Chap. 36 in: Viral and Rickettsial Infections of Man. Philadelphia: J. B. Lippincott Company, 1965.

Young, G. A., R. L. Kitchell, A. J. Luedke, and J. H. Sautter: The effect of viral and other infections of the dam on fetal development in swine. Modified live hog cholera viruses—Immunological, virological and gross pathological studies. J. Amer. Vet. Med. Assn. *126*:165, 1955.

Young, S., and O. R. Cordy: An ovine fetal encephalopathy caused by Blue-tongue vaccine virus. J. Neuropath. Exp. Neurol. *23*:635, 1964.

A MODEL FOR VIRUS INDUCED REPRODUCTIVE FAILURE: THEORY, OBSERVATIONS AND SPECULATIONS [1]

G. Margolis, L. Kilham and J. Davenport [2]

Departments of Pathology and Microbiology,
Dartmouth Medical School, Hanover, New Hampshire

If a virologist-pathologist team were to undertake to construct a virus designed to induce reproductive failure the most important attribute sought would likely be the ability to enter and to destroy mitotic cells. Such an agent would, theoretically, be infectious for the fetus throughout gestation. An attack upon sites of cell proliferation at critical stages of organogenesis could result in malformations incompatible with life, or induce severe developmental defects. Considering that critical phases of growth and differentiation of various organs occur sequentially throughout fetal development, infections even in late gestation or in the perinatal period would still be capable of producing disabling or fatal sequelae. In a relatively stable adult cell population only mild infection would be expected. Hence, a pregnant host could harbor an unapparent or latent infection with an agent which would be highly virulent or lethal for her progeny.

Significantly, this hypothetical agent is not merely a fantasy. A "family" of viruses with the prerequisite propensities is already at hand. This group includes the Kilham rat virus (RV) (Kilham, 1965; Margolis and Kilham, 1965; Margolis and Kilham, *in press*), the closely related H-1 virus of Toolan (Toolan, 1961; Chandra and Toolan, 1961), and the recently isolated feline ataxia virus (FAV) (Kilham and Margolis, 1966). Abundant information is available regarding the physical and chemical properties of RV and H-1, the spectrum of pathologic states produced by these viruses (Kilham, 1965; Margolis and Kilham, *in press*), and the

[1] Supported by U.S.P.H.S. Grants NB-05545 and CA-06010, and by National Cancer Institute Research Career Award CA-22,652.
[2] Second year medical student.

range of host susceptibility among laboratory animals. In nature, anti-bodies to these agents are widely prevalent in rat populations, possibly indicating a latent infection. The range of host susceptibility to one or both of these viruses includes the hamster, mouse (Margolis and Kilham, *unpublished observations*), mastomys (Rabson *et al.*, 1961), and cat (Kilham and Margolis, 1965), as well as the rat. RV and H-1 are DNA viruses, heat and ether resistent. They attack proliferating cell populations and their early cytopathic effects are featured by the formation of intranuclear inclusion bodies. Both embryocidal and teratogenic effects have been induced by infections in early gestation (Ferm and Kilham, 1964; Ferm and Kilham, 1965; Kilham and Ferm, 1964; Kilham and Margolis, 1966). Late *in utero* infections have led to profound destructive effects upon the cerebellum, and severe neonatal hepatitis (Kilham and Margolis, 1966). Infection of neonates and sucklings has led to four different disease states, including a fulminant fatal disease (Kilham, 1961, 1965), a "mongoloid" state (Kilham, 1961) characterized by stunting of growth and tooth dysplasia (Baer and Kilham, 1962), and the cerebellar hypoplasia and hepatic involvement noted above. These disease states vary with virulence of the inoculum and species of host, as well as being influenced by age of animal upon inoculation. Adult differentiated tissue appears receptive only when responding to a strong stimulus for proliferation, as exemplified by the attack of RV upon bone at the site of healing fractures (Engler *et al.*, 1966) and by the action of H-1 upon the regenerating liver following subtotal hepatectomy (Ruffolo *et al.*, 1966) and administration of hepatotoxic doses of carbon tetrachloride (Margolis and Kilham, *unpublished data*). The action of these agents upon the placenta remains to be explored. It appears, however, that virulent strains of RV pass readily through the placenta of its natural host, the rat, to produce the spectrum of diseases described above. Although H-1 and RV have been isolated from human embryos (Toolan, 1962), Monif *et al.* (1965) have found no causal relationship between these agents and abnormal outcome of pregnancy in man. The teratogenic action of these viruses has been discussed by Ferm (Ferm, this volume) at this conference.

The properties and pathogenic effects of the newly discovered agent FAV are less well known. Although serologically distinct from RV and H-1 it possibly belongs to the same class of virus. Like these agents it is resistant to heat and ether. Again, its attack is directed against mitotic cells, and its cytopathic effects include a prominent intranuclear inclusion body phase. Further, it is the only virus other than RV and H-1 found to selectively attack the external germinal layer of the cerebellum during the active phase of the histogenesis of this structure. Interestingly, this latter action of FAV, resulting in a congenital cerebellar "hypoplasia" (Herringham and Andrewes, 1888) was masked for over 75 years, being generally interpreted as a genetically determined developmental defect. The relationship of FAV to other feline viruses is yet undetermined.

At first consideration it may seem strange that the attack of a group of viruses upon the cerebellum could lead to the recognition of the affinity of these agents for the mitotic cell. The cerebellum, however, is the last portion of the central nervous system to complete its cytoarchitectonic development, undergoing its major growth and differentiation postnatally (Uzman, 1960; Miale and Sidman, 1961; Altman and Das, 1966). In contrast to the early development of the Purkinje cells and neurones of the deep cerebellar nuclei, which undergo their final divisions midway in gestation, the external germinal layer initiates a period of remarkable proliferative activity late in gestation or at birth (Fig. 1), and continues to divide rapidly and continuously postnatally, until it becomes richest in DNA of all brain tissues (Zamenhof et al., 1964; Gaito, 1966).

Further data may be cited in support of the postulate that these viruses attack the mitotic cell. One channel of evidence is obtained from studies of laboratory infections with these agents. In the face of widespread and

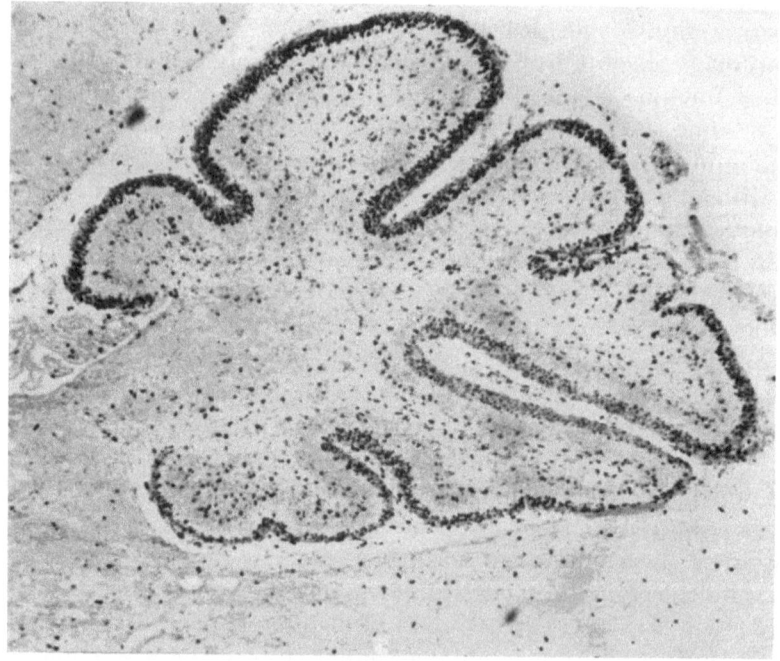

Fig. 1. Perinatal development of the cerebellum, demonstrated by H3-thymidine autoradiography. In this 4-day-old hamster, sacrificed 24 hours after intracerebral inoculation of H3-TdR, the external germinal layer is heavily labelled, as a consequence of a high mitotic activity of this layer. ×60.

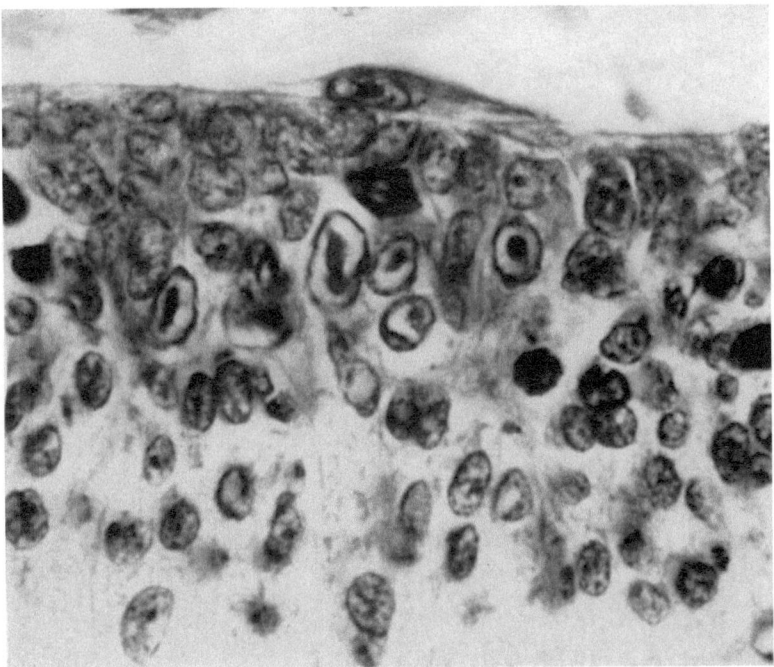

Fig. 2. Selective attack of a virus with affinity for mitotic cells upon the external germinal layer of the developing cerebellum. This term caesarean-delivered kitten was inoculated intraperitoneally *in utero* 6 days previously with feline ataxia virus (FAV). The presence and localization of the virus is indicated by the many intra-nuclear inclusion bodies. Hematoxylin-eosin. ×1500.

varied manifestations of viral action a single common feature prevails. Wherever viral cytopathic effects are observed, the involved tissue shows a high level of mitotic activity (Figs. 1–5). Another channel of evidence is the similarity, if not identity, of the viral cerebellar lesion with that induced by varied antimitotic agents. This resemblance is manifest in investigations of the effects of ionizing radiation upon the developing cerebellum (Schmidt, 1962; Hicks and D'Amato, 1965). More recent corroborative data are derived from studies of cerebellopathic effects (Fig. 6) of the antimetabolite cytosine arabinoside (Chu and Fischer, 1962; Kimball *et al.*, 1966) when administered to suckling hamsters (Fischer and Jonas, 1965) and to pregnant and suckling rats (Margolis and Kilham, *unpublished observations*). The external germinal layer of the developing cerebellum, therefore, represents a singularly strategic area for the study of the fetocidal propensities of antimitotic agents, viral, chemical and physical. Not only are cytopathic effects recognizable in the active phases of a destructive process, but the lack of a reparative potential in this cell population makes it possible to identify readily sites of destruc-

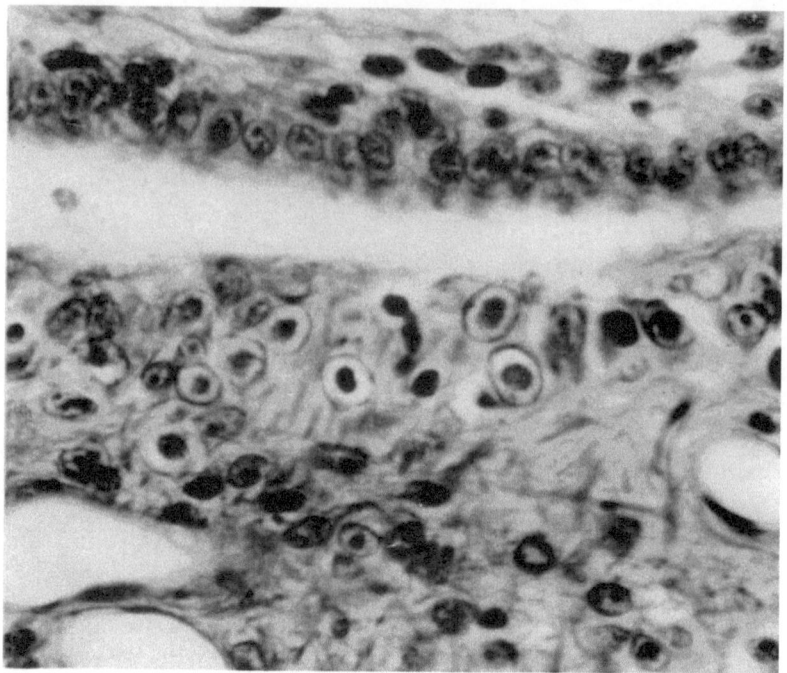

Fig. 3. Relationship of viral susceptibility to mitotic activity. This photograph is from the brain of a newborn kitten inoculated with FAV *in utero* 20 days prior to delivery. Below is the proliferating cell population of the subependymal cell plate (Smart, 1961; Smart & LeBlond, 1961) which contains several intranuclear inclusion bodies. Above is the apposing, non-proliferating ependyma without evidence of viral cytopathic effects. Hematoxylin-eosin. ×600.

tive action and to reconstruct a timetable of these events even at periods remote from the episode of injury.

It is pertinent to compare the action of rubella and the RV-FAV viruses in the production of birth defects. In human embryo tissue cultures arrest of mitotic activity dominates other cytopathic effects of rubella (Plotkin *et al.*, 1965). Naeye (1965) has postulated that the disturbances of growth and the specific organ malformations associated with congenital rubella are the result of the inhibition of cell multiplication by this agent. He presents evidence, both from his own studies and from the literature that, "By arresting mitoses, viral multiplication at one point in the embryogenesis of an organ might lead to an overt malformation, while viral propagation at a later period might merely limit organ growth." To date, our studies suggest that RV and FAV, because they destroy dividing cells rather than simply arrest mitosis, may be more embryocidal and less teratogenic than rubella, as well as remaining virulent for the fetus throughout gestation. This embryocidal effect has been well illustrated by Ferm and Kilham (1965) and by Kilham and Margolis (1966).

The significance of virus-embryo interactions for an understanding of fetal development has been emphasized by Ebert and Wilt (1960). The effects of the RV-FAV agents during early stages of embryogenesis and upon the ovum have not been explored, nor have placental-viral relationships been adequately investigated. No ovapathological effects have been observed in perinatal infections, a finding consistent with the suspended metabolic status of the ovum (Blandau, this volume; Ebert, 1965). Only a few studies have been made of the action of viruses on the ovum. Transovarial transmission of avian lymphomatosis virus with heavy infection of chick embryos has been demonstrated by Rubin *et al.* (1961). That a picorna virus can penetrate the zona pellucida and arrest development of the mammalian ovum has been observed by Gwatkin (1963, 1966), using Mengo virus, and the mouse ovum cultivated *in vitro* at both the unfertilized and the two-cell stage. A faulty and inefficient synthesis of new virus units by these infected ova has also been shown (Gwatkin and Auerbach, 1966). Mims (1966) and Mims and Subrahmanyan (1966), using

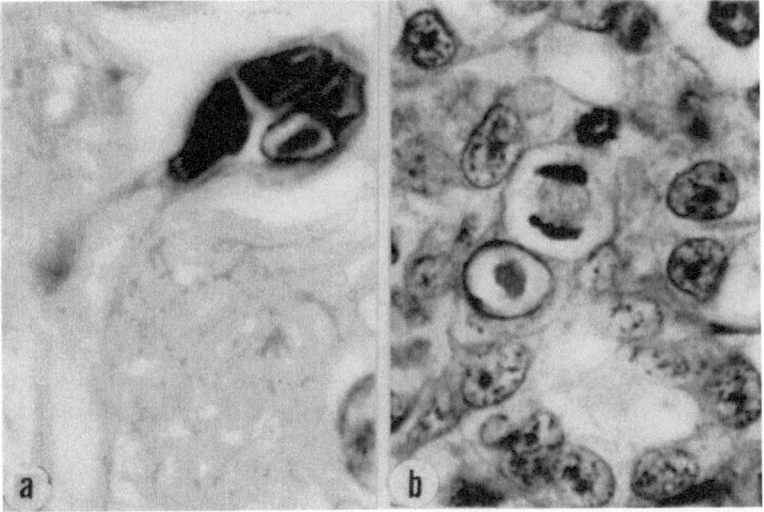

Fig. 4. Relationship of viral susceptibility to mitotic activity. (A) This photograph shows a cerebral venule of a 6-day-old kitten inoculated at birth with rat virus (RV). The four parasitized endothelial cells in a single cluster suggest viral invasion of proliferating endothelium and subsequent parallel replication of virus and host cells. Hematoxylin-eosin. ×600. (B) Kidney. This photograph shows detail of the hypoplastic neonatal zone of the kidney from the animal described in Fig. 3. The height of the viral attack has passed, resulting in premature arrest in neogenesis of nephrons. That there is some residual growth potential and viral activity is demonstrated by the mitotic figure and the intranuclear inclusion body. Hematoxylin-eosin. ×1500.

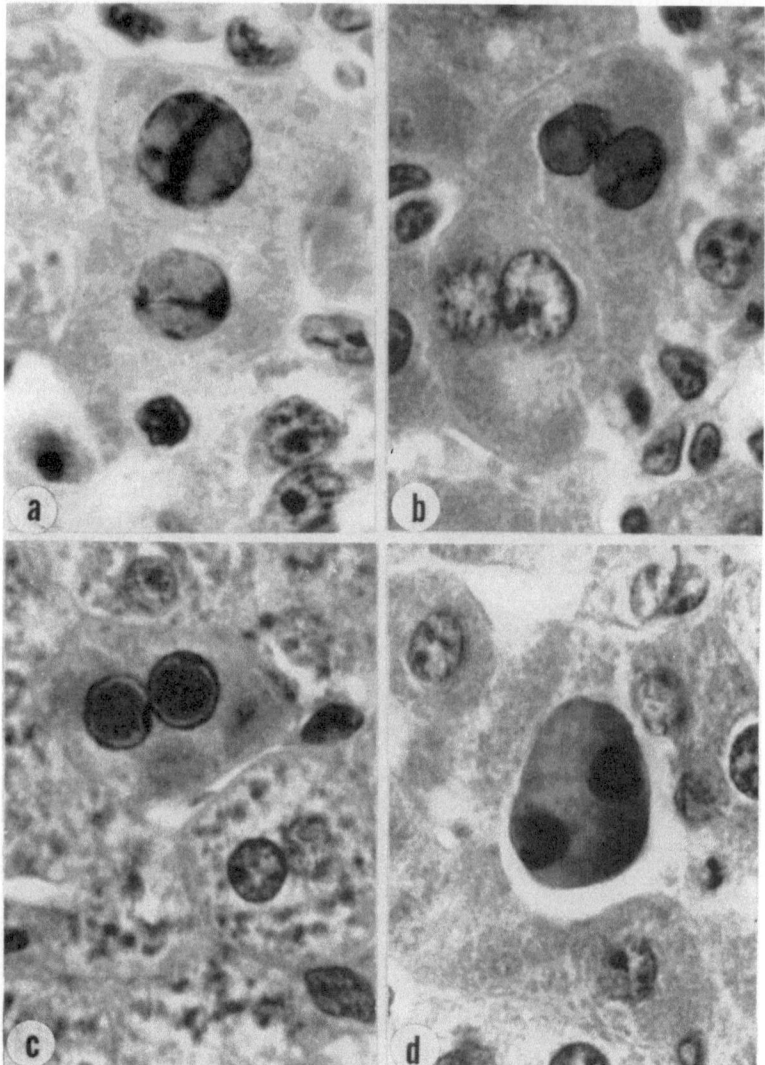

Fig. 5. Relationship of viral susceptibility to mitotic activity. The adult rat liver, refractory to the action of the H-1 virus, becomes susceptible again if exposure occurs during the active phase of repair of a deficiency of liver tissue, produced either by subtotal hepatectomy or by a hepatotoxic dose of carbon tetrachloride. Figures a to d have been selected to demonstrate twin intranuclear inclusion bodies in hepatocytes with double or partially cleaved nuclei. These relationships suggest parallel replication of host cell and virus. Recent studies of the replication of binucleate cells by Sandberg *et al.* (1966) have demonstrated an asynchrony in replication of binucleate cells, an observation supporting the possibility of viral invasion prior to cell division. Evidence of a temperate relationship does not persist long, as indicated by the progressive viral cytopathic effects on hepatocytes depicted by this series. Hematoxylin-eosin. ×750.

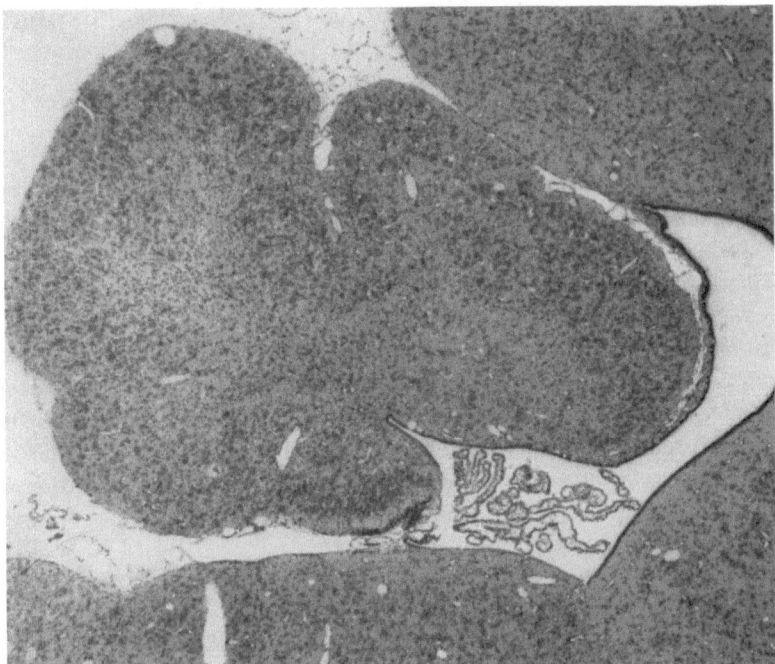

Fig. 6. Granuloprival cerebellar "hypoplasia" in rat. This is the cerebellar lesion resulting from late antenatal or early postnatal destruction of the external germinal layer of the cerebellum. It can be produced by the action of virus, X-ray, or anti-metabolite. The lesion illustrated here has been induced by cytosine arabinoside. Hematoxylin-eosin. ×40.

immunofluorescence technics to study the carrier state and the mechanism of vertical transmission in lymphocytic choriomeningitis (LCM) virus infections in mice have added startling dimensions to our concept of congenital viral infections. They have found that this temperate virus can infect the ovum and induce the formation of abundant viral antigen in the absence of recognizable cytopathic effects. Observing that viral antigen is demonstrable in virtually all cells of very early mouse embryos and of adult carriers, they present strong arguments favoring transovarial transmission as a major pathogenetic mechanism for congenital viral infections. LCM antigen may be produced by both replicating and non-replicating cells. The interaction of the mature fertilized ovum and the group of viruses whose affinity for the dividing cell has been described in our report, offers another provocative model for studies of host-virus relationships in the earliest phases of embryogenesis. Although the embryocidal effect of these agents may dominate (Ferm and Kilham, 1965; Kilham and Margolis, 1966), recent observations suggesting that they may have a slow or temperate action (Kilham and Margolis, 1966, Margolis and Kilham, *unpublished observations*) make such a study particularly inviting.

SUMMARY

A group of animal viruses is reported with properties answering an ideal theoretical design for an experimental model of virus induced reproductive failure. The major attribute of this group, which includes rat virus, the closely related H-1 virus, and a feline virus with similar properties, is the ability to destroy mitotic cells. By attacking proliferating tissues at critical stages of organogenesis, these agents, presumably infectious for the fetus throughout gestation, exert embryocidal or teratogenic effects, or induce severe or fatal neonatal disease states. Although not implicated as causative agents of reproductive failure in man, they offer promising approaches to studies of virus-ovum and virus-host interactions at all stages of embryogenesis.

References

Altman, J., and G. D. Das: Autoradiographic and histological studies of postnatal neurogenesis. J. Comp. Neur. *126*:337, 1966.

Baer, P. N., and L. Kilham: Rat virus and periodontal disease. Oral Surg., Oral Med., and Oral Path. *15*:756, 1962.

Blandau, R. J.: Oogenesis—Ovulation and Egg Transport. This volume, p. 194.

Chandra, S., and H. W. Toolan: Electron microscopy of the H-1 virus. J. Nat. Cancer Inst. *27*:1405, 1961.

Chu, M. Y., and G. A. Fischer: A proposed mechanism of action of I-p-D-arabinofuranosylocytosine as an inhibitor of the growth of leukemic cells. Biochem. Pharmacol. *11*:423, 1962.

Ebert, J. D., and F. H. Wilt: Animal viruses and embryos. Quart. Rev. Biol. *35*:261, 1960.

Ebert, J. D.: Interacting Systems in Development. New York: Holt, Rinehart & Winston, 1965.

Engler, W. O., P. N. Baer, and L. Kilham: Effects of rat virus on healing osseous wounds. Arch. Path. *82*:93, 1966.

Ferm, V. H.: Developmental malformations as manifestations of reproductive failure. This volume, p. 246.

Ferm, V. H., and L. Kilham: Congenital anomalies induced in hamster embryos with H-1 virus. Science *145*:510, 1964.

———, ———: Histopathologic basis of the teratogenic effects of H-1 virus. J. Embryol. Exp. Morph. *13*:151, 1965.

Fischer, D. S., and A. M. Jonas: Cerebellar hypoplasia resulting from cytosine arabinoside treatment in the neonatal hamster. Clinical Research *13*:540, 1965.

Gaito, J.: Macromolecules and brain function, in Macromolecules and Behavior, J. Gaito, editor. New York: Appleton-Century-Crofts, 1960, page 89.

Gwatkin, R. B. L.: Effect of viruses in early mammalian development. Proc. U.S. Nat. Acad. Sc. *50*:576, 1963.

———: Effect of viruses on early mammalian development. III. Further studies concerning the interaction of Mengo encephalitis virus with mouse ova. Fertility and Sterility *17*:411, 1966.

Gwatkin, R. B. L., and S. Auerbach: Synthesis of a ribonucleic acid virus by the mammalian ovum. Nature *209*:993, 1966.

Herringham, W. P., and F. W. Andrewes: Two cases of cerebellar disease in cats, with staggering. St. Barth. Hosp. Rep., London *24*:112, 1888.

Hicks, S. P., and C. J. D'Amato: Effects of ionizing radiations on mammalian development, in Advances in Teratology, D. H. M. Woollam, M.D., Editor. London: Logos Press, Ltd., 1966, page 195.

Kilham, L.: Rat virus (RV) infections in hamsters. Proc. Soc. Exper. Biol. Med. *106*:825, 1961.

———: Viruses of laboratory and wild rats, in Viruses of Laboratory Rodents, National Cancer Institute Monograph #20, 1965, page 117.

Kilham, L., and V. H. Ferm: Rat virus (RV) infections of pregnant, fetal and newborn rats. Proc. Soc. Exper. Biol. Med. *117*:874, 1964.

Kilham, L., and G. Margolis: Cerebellar disease in cats induced by inoculation of rat virus. Science *148*:244, 1965.

———, ———: Viral etiology of spontaneous ataxia of cats. Am. J. Path. *48*:991, 1966.

———, ———: Spontaneous hepatitis, jaundice and cerebellar disease in suckling rats due to congenital infections with rat virus. Am. J. Path. *49*:457, 1966.

Kimball, A. P., B. Bowman, P. S. Bush, J. Herriot, and G. A. LePage: Inhibitory effects of the arabinosides of 6 mercaptopurine and cytosine in purine and pyrimidine metabolism. Cancer Res. *26*:1337, 1966.

Margolis, G., and L. Kilham: Virus induced cerebellar hypoplasia, in Infections of the Nervous System, Proceedings of the 44th Annual Meeting of the Association for Research in Nervous and Mental Diseases, New York City, 1964. (*In press*)

———, ———: Rat virus, an agent with an affinity for the dividing cell, in Slow Latent and Temperate Virus Infections, National Institute of Neurological Diseases and Blindness Monograph #2, 1965, page 361.

———, ———: *Unpublished observations.*

Mims, C. A.: Immunofluorescence study of the carrier state and mechanism of vertical transmission in lymphocytic choriomeningitis virus infection in mice. J. Path. and Bacteriol. *91*:395, 1966.

Mims, C. A., and T. P. Subrahmanyan: Immunofluorescence study of the mechanism of resistance to superinfection in mice carrying the lymphocytic choriomeningitis virus. J. Path. and Bacteriol. *91*:403, 1966.

Miale, I. L., and R. L. Sidman: An autoradiographic analysis of histogenesis in the mouse cerebellum. Exper. Neurol. *4*:227, 1961.

Monif, G. R. G., J. L. Sever, and W. D. Cochran: The H-1 and the RV viruses and pregnancy: Serological study of certain groups of pregnant women. J. Pediatrics *67*:253, 1965.

Naeye, R. L., and W. Blanc: Pathogenesis of congenital rubella. J.A.M.A. *194:* 1277, 1965.

Plotkin, S. A., A. Boue, and J. G. Boue: The in vitro growth of rubella virus in human embryo cells. Am. J. Epidem. *81:*71, 1965.

Rabson, A. S., L. Kilham, and R. L. Kirschstein: Intranuclear inclusions in rattus (Mastomys) natalensis infected with rat virus. J. Nat. Cancer Inst. *27:*1217, 1961.

Ruffolo, P. R., G. Margolis, and L. Kilham: The induction of hepatitis by prior partial hepatectomy in resistant adult rats injected with H-1 virus. Am. J. Path. *49:*795, 1966.

Rubin, H., A. Cornelius, and L. Fanshier: The pattern of congenital transmission of an avian leukosis virus. Proc. Nat. Acad. Sc. *47:*1058, 1961.

Sandberg, A. A., T. Sofini, and G. E. Moore: Chronology and pattern of human chromosome replication. IV. Autoradiographic studies of binucleate cells. Proc. Nat. Acad. Sc. *56:*105, 1966.

Schmidt, R.: Die postnatale Genese der Kleinhirndefekte roentgenbestrahlter Hausmäuse. J. f. Hirnforschung *5:*163, 1962.

Smart, I.: The subependymal layer of the mouse brain and its cell production as shown by radioautography after Thymidine-H³ injection. J. Comp. Neurol. *116:*325, 1961.

Smart, I., and C. P. LeBlond: Evidence for division and transformations of neuroglia cells in the mouse brain, as derived from radioautography after injection of Thymidine-H³. J. Comp. Neurol. *116:*349, 1961.

Toolan, H. W.: A virus associated with transplantable human tumors. Bull. N.Y. Acad. Med. *37:*305, 1961.

————: Isolation of H-1 and H-3 viruses directly from embryos. Proc. Am. A. Cancer Res. *3:*368, 1962.

Uzman, L. L.: The histogenesis of the mouse cerebellum as studied by its tritiated thymidine uptake. J. Comp. Neurol. *114:*137, 1960.

Zamenhof, S., H. Bursztyn, K. Rich, and P. J. Zamenhof: The determination of deoxyribonucleic acid and of cell number in brain. J. Neurochem. *11:*505, 1964.

ROUND TABLE DISCUSSION ON
PLACENTAL PATHOLOGY

K. Benirschke, S. G. Driscoll, V. H. Ferm, P. Gruenwald,
E. S. E. Hafez, and C. D. King

Department of Pathology, Dartmouth Medical School, Hanover, New Hampshire; Departments of Pathology, Boston Hospital for Women and Harvard Medical School, Boston, Massachusetts; Department of Anatomy/ Cytology, Dartmouth Medical School, Hanover, New Hampshire; Departments of Pathology, Sinai Hospital of Baltimore and the Johns Hopkins University, Baltimore, Maryland; Department of Animal Sciences, Washington State University, Pullman, Washington; The University of Wisconsin, Wisconsin Regional Primate Research Center, Madison, Wisconsin

INTRODUCTION

In this conference on reproductive failure, the placenta has been incriminated on several occasions. Dr. Hertig, whose numerous studies in this field mark an important milestone in our understanding of the normal anatomy as well as many pathologic lesions of this organ, has dwelt some time emphasizing the need for examination of the placenta. We have heard cytogenetic results obtained from the examination of cultures of this organ; placental pathology was mentioned in hybrid gestation failure, and in teratologic studies; it participates importantly in infectious diseases, etc. None of us here doubts that the placenta is a vital organ to the fetus whose normality or dysfunction often governs reproductive success. It is surprising then, how rarely this expendable organ receives the thorough examination which is given for instance to surgical specimens or, for that matter, the embryos in our experimental studies of pregnant animals. Indeed, enormous deficiencies exist in our knowledge of this organ and, in particular, we know little of the many aspects of comparative pathology of the placenta which, potentially, is a rich source for study and might contribute importantly to the understanding of many questions posed here in the last days. Despite the excellent and comprehensive treatment of the comparative normal anatomy of the placenta by Grosser (1927), Mossman (1937) and Wimsatt (1962), many gaps exist. For instance, in gathering material for this conference I was unable to find really good descriptions of mule, zebra and donkey placentas, and only one comparative paper on the goat placenta was found (Krölling, 1929), organs which should in fact be quite readily available. When one wishes to understand possible hybrid gestational failure (as in sheep × goat hybrids) or the altered PMSG production of mule gestations (Clegg *et al.,* 1962) it is likely that the knowledge of comparative anatomy, histology and physiology of this organ will be of some importance.

Much less known still are the pathologic events occurring in this tissue. What constitutes a normal placenta in man, where more numerous studies exist, is still a difficult question to answer; and which pathologic changes are significant to the gestation? One is awed by the paucity of information available on the interaction of many diseases with the gestational product and can only hope that the availability of this organ will make it a subject for continuing inquiry.

It is for these reasons that we thought it important to have at least a brief period of this conference solely devoted to the comparative pathology of the placenta. In an attempt not to duplicate any of the material presented elsewhere in these pages and, foregoing the temptation of having a haphazard round table discussion, I have asked some of my friends who possess special knowledge to make brief presentations and hope that these will highlight the need for comparative studies and be a challenge to all who come into contact with relevant material.

THE NUMBER OF VESSELS IN THE UMBILICAL CORD

K. BENIRSCHKE

*Department of Pathology, Dartmouth Medical School,
Hanover, New Hampshire*

One of the simplest observations one can make about the afterbirth is to determine its size, shape and weight and to examine the umbilical cord. Here, *i.a.*, the number and size of vessels is of interest, usually a macroscopic observation although I grant that in rodents microscopy is needed.

The subject has interested me considerably since we observed rather frequently in human placentas that one umbilical artery was absent (Benirschke and Brown, 1955). The matter has been taken up by numerous investigators since then and we have reviewed it comprehensively recently (Benirschke and Driscoll, 1966).

Briefly, the absence of one umbilical artery is undoubtedly the commonest anomaly in man, occurring in about 1 per cent of all deliveries, Negroes having a slightly lower frequency. Also, it has been observed that it is often, although not always, associated with other congenital anomalies in the infant and that it is more frequent in twins and, when velamentous insertion of the cord exists.

Numerous questions can and have been asked about this anomaly of which, to me, the following are the more pertinent in the present context:

a) Is this absence of one artery the result of aplasia or is it the end stage of atrophy?

b) Is it etiologically related to the other malformations or the frequently smaller size of the affected infants?

c) What is the relationship to twinning and velamentous insertion of the cord?

d) What is the vasculature of the cord like in various mammals and are numerical anomalies observed?

The answers to most of these questions are unknown; however, it is perhaps worthwhile, in an audience representing many different disciplines and working with numerous species, to restate briefly findings presented

363

previously (Benirschke *et al.*, 1964). Hopefully, the study of other comparative material will lead to a better understanding of this common problem in the future.

a) There are arguments favoring both points of view. Thus, the association of the anomaly with some chromosomal anomalies make an aplasia perhaps more likely. On the other hand, the anomaly has been found in the thalidomide embryopathy (Dunn *et al.;* Russell and McKichan; Thomas, 1962; Kajii *et al.*, 1963). Further, not infrequently can one find an atrophic remnant of a vessel. It has an association with closely approximated twin placentas and nearly invariably is only one artery found in the acardiac twin, all of which favor a secondary, atrophic event.

b and c) It is not conceivable to me that this anomaly should be a primary factor causing fetal defects at a very early time in embryogenesis. Rather, when such defects occur one may perhaps assume a simultaneous damage to several systems, perhaps best exemplified by the few cases described of its association with thalidomide injury. If, as we envisage it, a vessel atrophies as a consequence of the obstructive presence of a twin placenta or because of trophotropism (Benirschke, 1965) and during the course of the development of a velamentous cord, then perhaps it is of some importance to know when such a compromise occurs. If it were to take place early in embryonic life then the effect on embryogenesis may be more severe than if it occurs later when perhaps reduced fetal weight may be a consequence.

d) Very few descriptions of the vasculature of the umbilical cord in animals are found in the literature and none concerns itself with possible anomalies. For these reasons and because polyembryony, resulting in the birth of identical quadruplets, is a regular phenomenon in the nine-banded armadillo (*Dasypus novemcinctus*), we studied the placentas and fetuses of this species. In 4 of 210 specimens (0.9 per cent) we found anomalies: one cord had only one artery; in two cases one artery was atrophic and once one artery and one vein were atrophic. The circulation of the armadillo placenta is such that there are no anastomoses between the littermates. Moreover, contrary to the relationship of vessels in the cord of man, no anastomoses exist between the vessels in the armadillo cord. From a detailed dissection of relevant armadillo placentas it appears, however, that loops interconnecting chorionic vessels govern the development of placental (and umbilical cord) vascular structure. None of the four armadillos showed any other congenital anomaly. Nevertheless, the presence of anomalous vessels in surely monozygous quadruplets is of interest and correlates with such a finding in man.

The only point I wish to make is that rather simple observations on the afterbirth in various species may contribute to a greater understanding of the pathophysiology of pregnancy.

References

Benirschke, K., and W. H. Brown: A vascular anomaly of the umbilical cord. The absence of one umbilical artery in the umbilical cords of normal and abnormal fetuses. Obstet. Gynec. *6:*300, 1955.

Benirschke, K., and S. G. Driscoll: The Pathology of the Human Placenta. Handb. Path. Anat., Henke-Lubarsch VII/5 Berlin: Springer-Verlag, 1966.

Benirschke, K., M. M. Sullivan, and M. Marin-Padilla: Size and number of umbilical vessels. A study of multiple pregnancy in man and armadillo. Obstet. Gynec. *24:*819, 1964.

Clegg, M. T., H. H. Cole, C. B. Howard, and H. Pigon: The influence of foetal genotype on equine gonadotrophin secretion. J. Endocrinol. *25:*245, 1962.

Dunn, P. M., A. M. Fisher, and H. G. Kohler: Phocomelia. Amer. J. Obstet. Gynec. *84:*348, 1962.

Grosser, O.: Frühentwicklung, Eihautbildung und Placentation des Menschen und der Säugetiere. München: Bergmann, 1927.

Kajii, T., M. Shinohara, K. Kikuchi, S. Dohmen, and M. Akichika: Thalidomide and the umbilical artery. Lancet *ii:*889, 1963.

Krölling, O.: Über den Bau der Plazentome der Ziege und Gemse (*Rupicapra rupicapra*). Baum Festschr. p. 125, 1929.

Mossman, H. W.: Comparative morphogenesis of the foetal membranes and accessory uterine structures. Contrib. Embryol. *26:*129, 1937.

Russell, C. S., and M. D. McKichan: Thalidomide and congenital abnormalities. Lancet *i:*429, 1962.

Thomas, J.: Die Entwicklung von Fetus und Placenta bei Nabelgefässanomalien. Arch. Gynäk. *198:*216, 1962.

Wimsatt, W. A.: Some aspects of the comparative anatomy of the mammalian placenta. Amer. J. Obstet. Gynec. *84:*1568, 1962.

GESTATIONAL PATHOLOGY AND MATERNAL DIABETES MELLITUS

Shirley G. Driscoll

Departments of Pathology, Boston Hospital for Women and Harvard Medical School, Boston, Massachusetts

Among diabetic women, reproductive failures occur three to 12 times as often as among the general population (White, 1949; Gellis and Hsia, 1959; Kyle, 1963). Although fetal wastage is greatest with poor clinical care of the diabetes, the risk of failure is still appreciably increased when the maternal disease has been carefully monitored and controlled throughout gestation. The gestational catastrophes experienced by diabetics comprise chiefly those conditions which are associated with perinatal death or with residual damage among surviving progeny. However, *previable wastage* is also increased when the maternal diabetes is not controlled and/or when it is complicated by angiopathy (White, 1965; Driscoll, 1965). Insight into mechanisms of reproductive failure among diabetic women requires study of the entire conceptus, *i.e.*, the embryo-fetus and its adnexa, the placenta, membranes, and umbilical cord, and as much of the environment as possible.

The three major hazards to the conceptus of a diabetic woman are these: (1) antepartum fetal death, which occurs without clinical warning, at 34 to 38 weeks' gestation, (2) prematurity and all its handicaps, the price of elective early delivery to circumvent fetal death, and (3) anomalous development. The diabetic gravida is also unusually susceptible to polyhydramnios. In addition, the progeny of diabetic mothers may exhibit these stigmata of the maternal disease: (1) fetal "gigantism," *i.e.*, increased birth weight, principally a reflection of obesity, (2) hyperplasia of the beta cells of the pancreatic islets, associated with increased insulin production (Driscoll, 1965; Steinke and Driscoll, 1965), (3) organ weights differing from those of comparable infants whose mothers do not have diabetes (Cardell, 1953; Driscoll *et al.*, 1960), (4) abnormally persistent extramedullary hematopoiesis, and (5) physiological "fragility." The fetus of the diabetic mother is especially vulnerable to bland thrombosis of the renal vein (Avery *et al.*, 1957; Takeuchi and Benirschke, 1961). When vascular disease, notably arterio- and arteriolosclerosis, complicates the maternal diabetes, the progeny may be undersized and fragile, even for the early gestational age at which elective early delivery has taken place.

That diabetes mellitus of the gravida is associated with disturbances of the placenta, cord, and membranes is not surprising. Typically, the obese, cherubic infant is accompanied by a large, moist, cellular, relatively "immature" placenta, with clear histological evidence of active trophoblastic proliferation. Smaller placentas, which often contain infarcts and may appear relatively mature or even senescent, accompany the smaller babies born to mothers with toxemia of pregnancy and/or diabetic angiopathy (Driscoll, 1965; Benirschke and Driscoll, 1967). Structural anomalies, notably the absence of one umbilical artery and diffuse angiomatosis of the chorionic villi, are relatively common. The chorionic veins, like those of the fetal kidney, are peculiarly susceptible to bland thrombosis, which may be extensive and lethal.

Within the endometrial shell which encases the ovisac, the decidua, the vascular lesions of diabetes may be found. Both arterioles and larger arterial channels of the uterus may be narrowed by such lesions as atherosis and hyalinosis, thus impairing deciduoplacental perfusion. The poor fetal growth and nutrition which has been observed to accompany advanced maternal vascular disease, with or without diabetes mellitus, may be attributed to these phenomena.

The complex pathophysiology of pregnancy complicated by diabetes mellitus involves all tissues of the conceptus, often producing similar lesions at more than one locus. Thorough study of the large "biopsy" which the availability of the placenta presents, can, therefore, be expected to clarify some of the responsible mechanisms, whether these be hereditary factors shared with the fetus, or a disturbed intra-ovular *milieu* provided by the diabetic mother. Many aspects of diabetes in pregnancy and of pregnancy in the diabetic may be approached in diabetic animals.

References

Avery, M. E., E. H. Oppenheimer, and H. H. Gordon: Renal-vein thrombosis in newborn infants of diabetic mothers. New Engl. J. Med. *256:*1134, 1957.

Benirschke, K., and S. G. Driscoll: The pathology of the human placenta, in: Placenta. Band VII/5. vom Handbuch der speziellen pathologischen Anatomie und Histologie. Berlin: Springer-Verlag, 1967.

Cardell, B. S.: The infants of diabetic mothers: a morphological study. J. Obstet. Gynaec. Brit. Emp. *60:*843, 1953.

Driscoll, S. G.: Pathology of pregnancy complicated by diabetes mellitus. Med. Clin. N. Amer. *49:*1053, 1965.

Driscoll, S. G., K. Benirschke, and G. W. Curtis: Neonatal deaths among infants of diabetic mothers. A.M.A. J. Dis. Child. *98:*818, 1960.

Gellis, S. S., and D. Y.-Y. Hsia: The infant of the diabetic mother. A.M.A. J. Dis. Child. *97:*1, 1959.

Kyle, G. C.: Diabetes and pregnancy. Ann. Int. Med. *59:* Suppl. 3, 1, 1963.

Steinke, J., and S. G. Driscoll: The extractable insulin content of pancreas from fetuses and infants of diabetic and control mothers. Diabetes *14:*573, 1965.

Takeuchi, A., and K. Benirschke: Renal vein thrombosis of the newborn and its relation to maternal diabetes. Biol. Neonat. *3:*237, 1961.

White, P.: Pregnancy and diabetes, medical aspects. Med. Clin. N. Amer. *49:* 1015, 1965.

———: Infants of diabetic mothers. Amer. J. Med. 7:609, 1949.

EXPERIMENTAL VIRAL INFECTION
OF THE PLACENTA *

V. H. FERM

Department of Anatomy/Cytology, Dartmouth Medical School, Hanover, New Hampshire

The herpes simplex virus (HSV) has been implicated in transplacental infections of human newborns leading to severe postnatal infections, including, typically, hepato-adrenal necrosis (Hass, 1935; White, 1963; Witzleben and Driscoll, 1965). In most instances of intra-uterine viral infection, the natural history of the disease as it relates to the placenta is not known.

* This study was supported by USPHS Grant GM-10210.

In controlled studies of viral infection during pregnancy in the golden hamster, the use of HSV has led to some interesting observations on placental involvement and the apparent protection of the embryo from the devastating effects of this virus (Ferm and Low, 1965).

When HSV is injected intravenously into the pregnant hamster early in gestation, it proliferates to a marked degree in uterine and placental tissues as compared to non-pregnant uterine tissues. The maternal adrenal gland is often severely involved showing foci of necrosis with typical inclusion bodies. These may be solitary or involve the entire adrenal cortex. This effect on the maternal adrenal is transitory and healing begins in about four days. There are no hormonal or other physiological manifestations of this adrenal involvement and pregnancy continues in an uninterrupted manner.

Uterine and placental tissues collected 48 hours following intravenous injection of the mother also show some interesting involvement. Sections of the chorio-allantoic placenta revealed small areas of necrosis which were in close proximity to maternal blood vessels. Inclusion bodies, typical of HSV infection, were noted in trophoblastic nuclei and in some decidual cells. Sections of fetal tissues showed no histopathological evidence of fetal infection and attempts to recover the virus from the fetal tissues by tissue culture methods revealed an occasional trace of virus in only a few specimens. In essence then, the maternal adrenal, uterine, and placental tissues were involved in the acute infection but there was no definite penetration of the placenta by the HSV virus. It is rather surprising that in view of the frequency of placental lesions, the integrity of this membrane remained intact and insufficient virus "leaked" over to the fetus to cause any histopathology.

A second experiment involved the direct inoculation of fetal tissues *in utero* (Fig. 1). This was accomplished by exposing the uterus after a mid-line abdominal incision and injecting 0.05 ml of HSV directly into the fetus through the uterine wall and amniotic cavity with a No. 30 gauge needle. Forty-eight hours following this procedure, the fetuses were recovered and examined histologically. Small foci of necrosis and inflammation were noted in the fetal adrenals and liver and HSV inclusion bodies were occasionally found in these organs. These fetuses had failed to grow in comparison to their uninjected littermates. The intra-embryonic injection of HSV also caused some histopathological involvement of the placenta but no effect whatsover on the maternal adrenal gland.

In summary, the placenta appears to be an effective barrier by protecting the fetus against the penetration of HSV experimentally introduced into the maternal system. No lesions or evidence of growth retardation were found in the fetal tissues. This protective effect of the

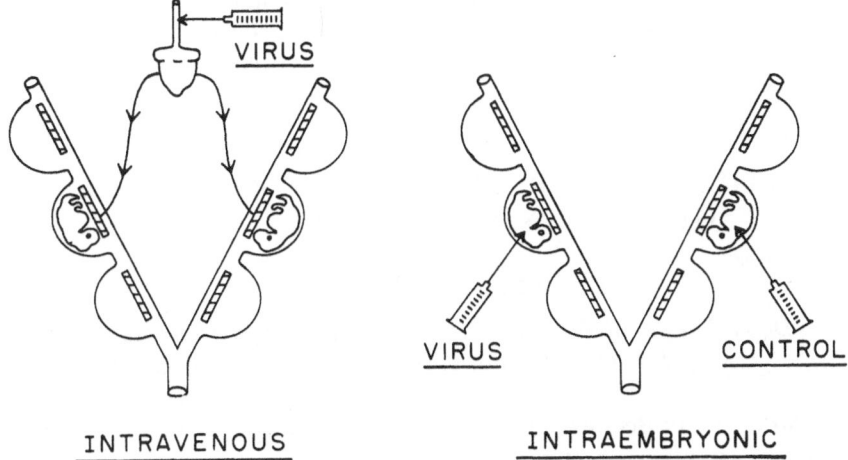

Fig. 1. A diagrammatic representation of the experimental method utilized in experiments with herpes simplex virus in the pregnant golden hamster. See text.

placenta is substantiated by the fact that direct intra-embryonic injection of HSV *in utero* causes marked fetal pathology. There is a suggested similarity of this experimental situation to human rubella infection in pregnancy. Alford *et al.* (1964) have examined the products of conception of some women who had a diagnosis of clinical rubella in early pregnancy. The overall isolation rate of rubella virus was 47 per cent; in those cases in which fetal and placental tissues could be studied separately, 61 per cent of placentas and 30 per cent of fetal tissues revealed detectable virus. The role of placental infection in human rubella disease is thus considerable and complex, and more attention should be given to the role of the placenta in intra-uterine infections. Moreover, the entire problem of viral disease in pregnancy including maternal, placental, and fetal production of interferon, the placental permeability of antibodies and the phenomenon of viral persistence in the fetus and newborn as so importantly demonstrated in rubella infection, should be studied in animals with various placental types and with other viruses.

References

Alford, C. A., F. A. Neva, and T. H. Weller: Virologic and serologic studies on human products of conception after maternal rubella. New Eng. J. Med. *271*:1275, 1964.

Ferm, V. H., and R. J. Low: Herpes simplex infection in the pregnant hamster. J. Path. Bact. *89*:295, 1965.

Hass, G. M.: Hepato-adrenal necrosis with intra-nuclear inclusion bodies. Am. J. Path. *11*:127, 1935.

White, J. G.: Fulminating infection with herpes simplex virus in premature and newborn infants. New Eng. J. Med. *269:*455, 1963.

Witzleben, C. L., and S. G. Driscoll: Possible transplacental transmission of herpes simplex infection. Ped. *36:*192, 1965.

THE LOBULAR ARCHITECTURE OF THE HUMAN AND RHESUS PLACENTA

Peter Gruenwald

Departments of Pathology, Sinai Hospital of Baltimore and the Johns Hopkins University, Baltimore, Maryland

In order to understand the variations of villous structure within a given human placenta it was desirable to study the gross architecture of the organ, with reference to the structural unit and its relationship to the maternal circulation. The same was subsequently done with the placenta of the rhesus monkey since many of the studies of maternal circulation have been done in this species on the tacit assumption that its placenta is similar to that of man. In both species, studies have been limited to the mature placenta. The most valuable information was obtained from reconstructions based on serial sections cut at right angles to the chorionic plate, including a number of specimens with vessels injected with India ink during life or after death. The author's own material was supplemented by that of the Department of Embryology of the Carnegie Institution of Washington.

THE HUMAN PLACENTA

The present study (Gruenwald, 1966) led to the characterization of the fetal placental lobule, an approximately spherical unit of densely packed villi with a loose center. The lobule is not identical with the cotyledon which is outlined more or less distinctly on the maternal surface by fissures. The literature relating to the subject has been reviewed elsewhere (Gruenwald, 1966) and only a few studies will be mentioned here. Crawford (1962) has estimated that there are more than 200 lobules (he called them cotyledons) in the human placenta. Wilkin (1954) has particularly emphasized the course of villous stems along the border of the lobule, forming what he has called the *tambour*. This aspect has been investigated repeatedly and examination of the sections suggests that the current views of the disposition of villi in relation to the lobule are correct.

As is shown diagrammatically in Figure 1 and on sections in Figure 2, the lobules are separated from one another by loose interlobular areas

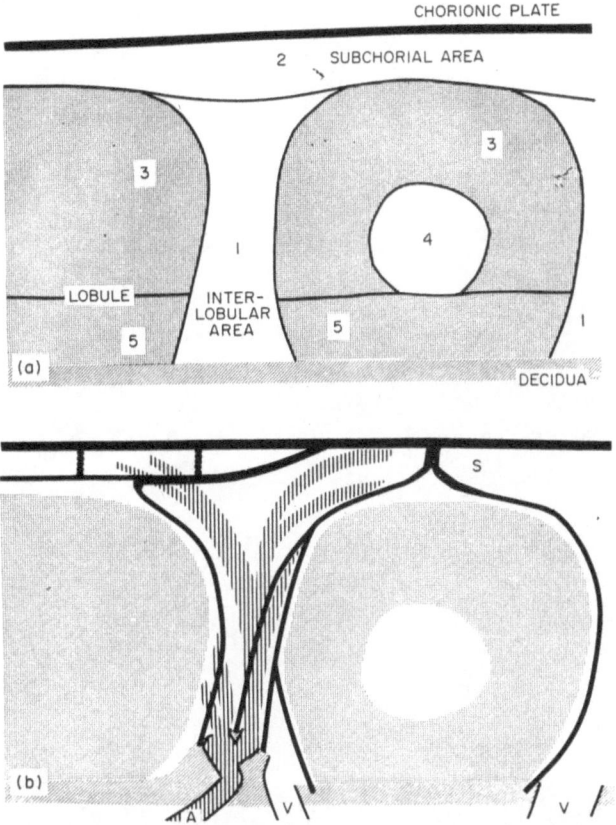

Fig. 1. Diagrams of lobules in a portion of human placenta. The dense portions are stippled, the loose portions left clear. The heavy line at the top is the chorion, the hatched strip at the bottom, the decidua. In *a* the areas to be examined separately are outlined by thin lines. In *b* a maternal artery (A) and the presumed spread of arterial blood in the intervillous space are indicated by vertical lines; veins (V) are shown as clear openings in the decidua. The heavy lines indicate villous stems (S). (From Gruenwald (1966), courtesy Bulletin of the Johns Hopkins Hospital.)

except where they are fused. A similar loose subchorial area separates the lobules from the chorionic plate. For purposes of histologic and pathologic study it has been suggested to consider separately five areas as characterized in Figure 1. Area 1, the interlobular portion, contains relatively few terminal villi, and many villous stems running along the border of the lobules toward their point of anchoring in the decidua. Area 2, the subchorial lake, often contains even fewer intact terminal villi, but the larger stems. Stacks of villi held together by fibrin are frequently seen extending across this area at right angles to the chorionic plate. They invariably connect the plate with one or several stems running nearly parallel to it, and may serve an important function in securing

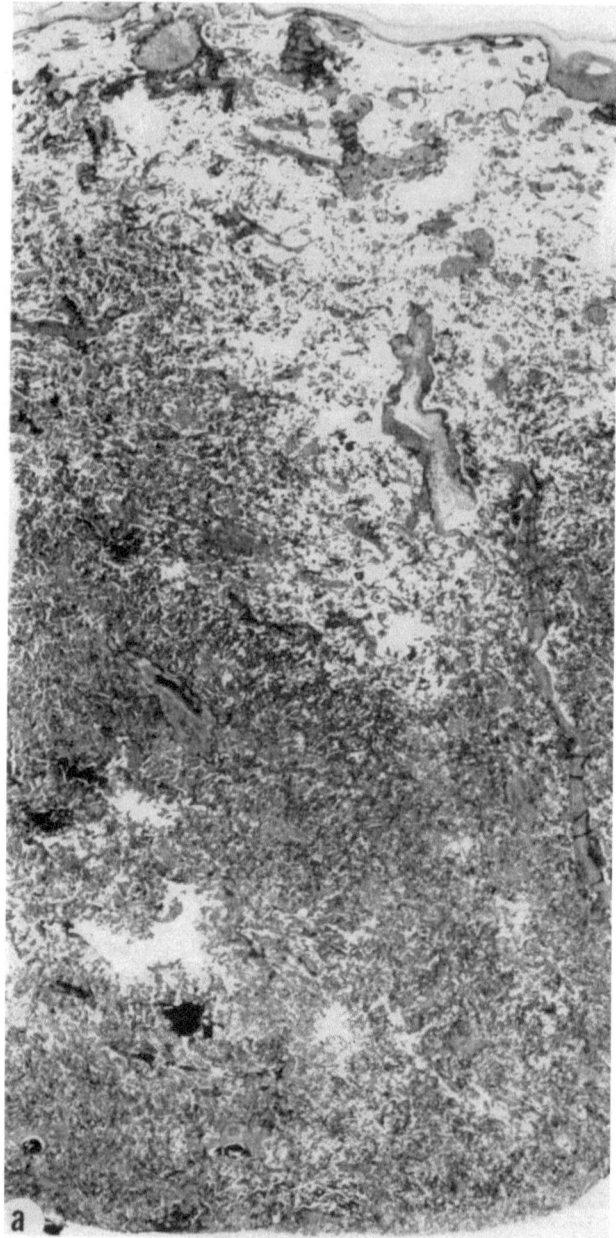

Fig. 2. Sections from full-term human placentas. *a* at right angles to the chorion (top) showing a lobule with its loose center (lower left). *b* parallel to the chorionic plate, showing several lobules, and villous stems running through the interlobular areas near the border of the lobules. *c* also parallel to the chorion but nearer the basal plate, showing lobules with their loose centers, and two septa.

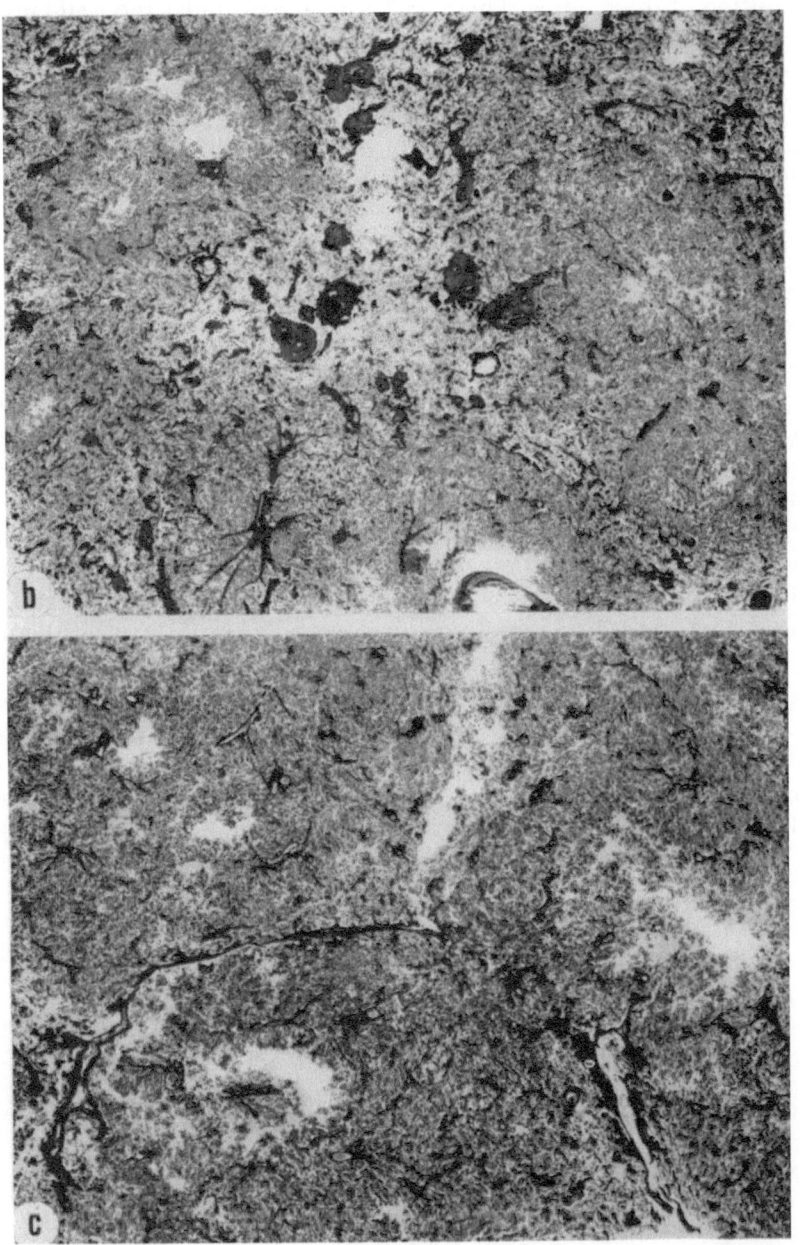

the position of these stems. These stacks are absent under the well-known thick plaques of subchorionic fibrin. The manner in which they develop in some areas, and the reason why they fail to associate themselves with the subchorionic plaques are unknown. The lobule itself shows terminal villi most densely packed near its periphery and somewhat less densely further in, with a very loose center. For purposes of description the dense portion has been divided into two areas, one adjacent to the decidual plate and the other comprising the rest. The latter portion is designated in Figure 1a as 3 and that near the decidua as 5. Area 4 is the loose central portion of the lobule. The villi in areas 3 and 5 are likely to be larger than in the other areas and approach in some placentas the appearance commonly considered to be characteristic of prematurity even when the villi in other areas have the structure usually associated with mature placentas. Clumping of syncytial nuclei as well as severe congestion of capillaries in the villi are often more pronounced in areas 3 and 5 than elsewhere in the placenta. The normal and abnormal distribution of structural changes in the villi and on their surface remains to be studied in detail with regard to the areas just outlined.

The distribution of maternal vessels was ascertained by means of reconstructions based on serial sections. One example is shown in Figure 3. It is apparent that the great majority of arterial openings into the intervillous space are located between the lobules, in area 1. They are frequently found in elevations of the decidua (which are not septa, see below), and are characteristically associated with anchoring points of several large villi (Fig. 4). It is likely that of the many arteries shown in Figure 3 some have a common origin in one spiral arteriole, with the

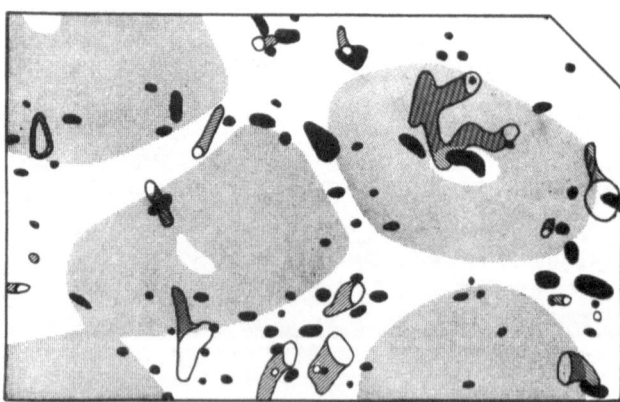

Fig. 3. Graphic reconstruction of the base of a piece of a mature human placenta. Anchoring points of villous stems: black; arteries: hatched, with openings clear; superimposed outlines of lobules: dotted. Drawn at ×10, reduced to ½ of original. (From Gruenwald (1966), courtesy Bulletin of the Johns Hopkins Hospital.)

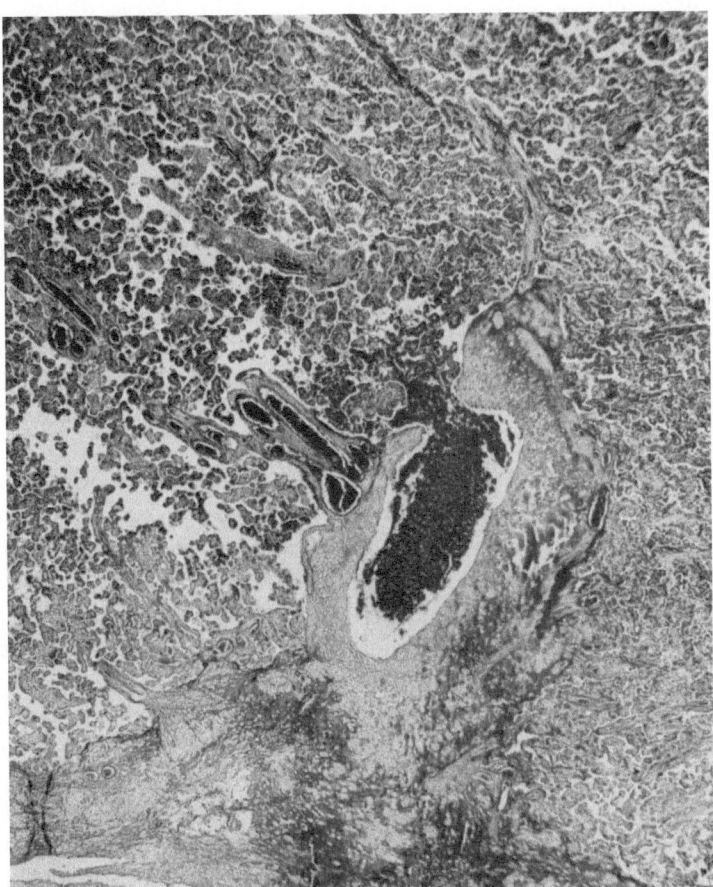

Fig. 4. Opening of a maternal artery into the intervillous space, in an elevation of the decidua on which several large villi anchor between lobules.

connection not included in that portion of the decidua which is found on the separated placenta. Veins were studied in a similar manner, but considerable difficulty was encountered because not all ostia could be reliably recognized. The portion of venous wall running through the decidua of the delivered placenta may be very short, and only present itself as an opening in the decidua through which villi protrude for a short distance. However, the findings were confirmed on sections of placentas *in situ* from the Carnegie collection. Many venous openings are close to those of arteries, others are located elsewhere in area 1 as shown diagrammatically in Figure 1b. It has been surmised that the arteries open under the center of the placental lobules and that the organization of the lobule and the existence of the loose central portion of the lobule bears a relationship to the inflow of arterial blood (Reynolds, 1966). Contrary to this, the present study has shown that the arterial blood flows through the interlobular areas toward the subchorial lake. The loose center of the lobules (which is never directly adjacent to the basal plate) does not appear to have any relationship to the pattern of blood flow. It is more likely that its loose structure is the result of a greater distance from the villous stems. A free central cavity in a nearly normal lobule is an artefact, produced by cutting it open with resulting evacuation of maternal blood and entry of air while the few terminal villi fall back against the denser part of the lobule. Arts (1961) found in his injection studies that a loose or empty center is seen only when the fetal vessels are incompletely injected, but not when injection is complete and includes the capillaries. On the other hand, pathologic changes which will be mentioned below, suggest that there is in reality a loose, though not empty center of the lobule.

The septa consist of thin folds of the decidua extending into the intervillous space for varying distances in the direction of the chorionic plate. If the two layers of the septum separate in the delivered placenta, a fissure forms. Few large villous stems anchor in these thin septa in contrast to the broad, short, cone-shaped elevations of the decidua on which many large villi anchor; only the latter elevations contain maternal arteries and their openings. It has been suggested that the septa divide the intervillous space below the subchorial lake into compartments corresponding to the lobules or cotyledons, and serve a function in directing the flow of maternal blood (Strauss, 1964). As a matter of fact, the number, distribution and height of the septa are quite variable (Becker and Jipp, 1963), and in some instances septa run through lobules rather than between them (Gruenwald, 1966). It is therefore clear that the septa do not have a significant relationship to the distribution of lobules, or to the flow pattern of maternal blood.

THE RHESUS PLACENTA

The lobular architecture as based on the density of terminal villi is not as distinct in the placenta of the rhesus as in man. Large areas which might be considered as cotyledons are outlined in the rhesus placenta by grooves which are obvious on the fetal surface if the placenta is examined *in situ,* and on both the fetal and the maternal surface in the delivered placenta (Fig. 5a). These areas have almost the size of the cotyledons outlined on the maternal surface of the human placenta and are therefore much smaller in number. Each corresponds to one lobule, and incomplete fusion of these areas is occasionally found. A looser portion may be seen in the center of some lobules. The pattern of distribution of the larger villous stems is not nearly as distinctly

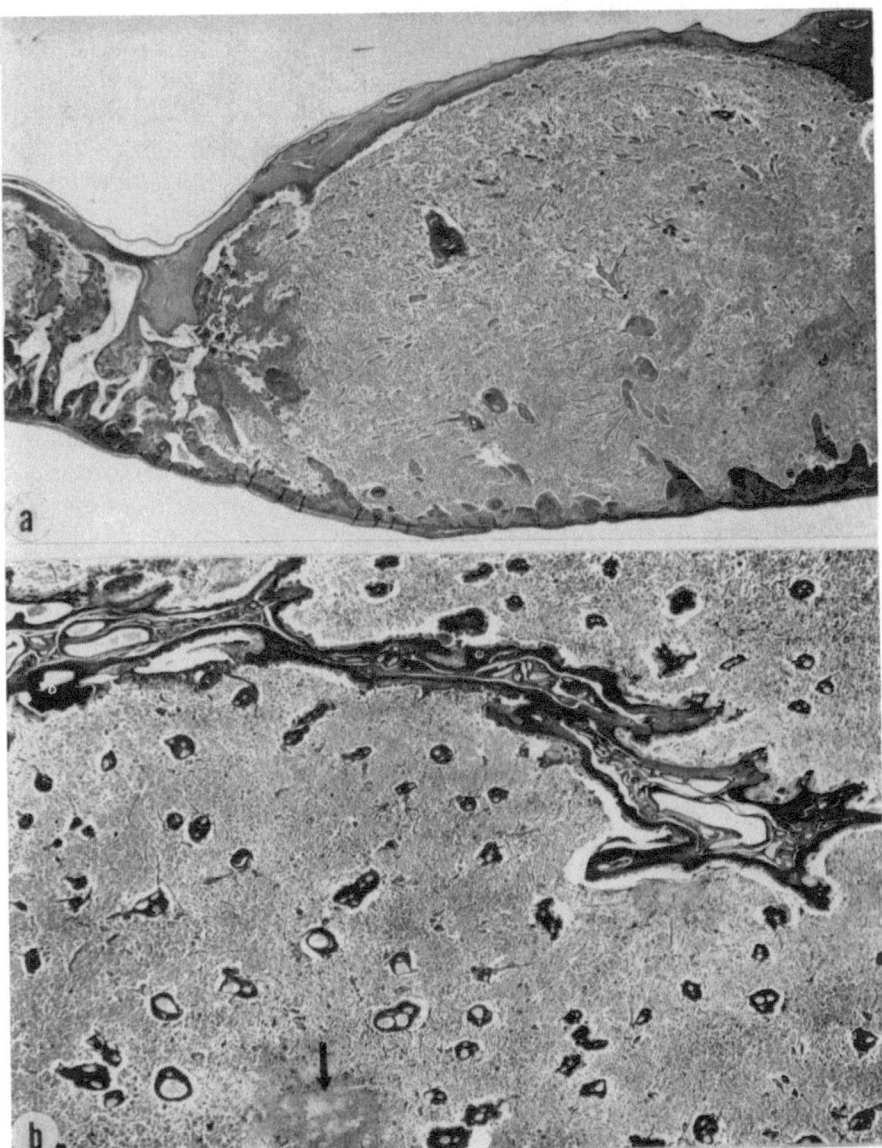

Fig. 5. Sections of mature rhesus placentas. *a* cut at right angles to the surfaces, shows a lobule on the right and a thin, interlobular area with large venous channels on the left. *b* is cut parallel to the surfaces and shows near the top an interlobular area in which the chorion is cut as it dips into the placenta. Villous stems are distributed almost at random with only a slight concentration in the interlobular area. A spurt from an artery near the center of a lobule is seen at the bottom (arrow).

associated with the border of the lobules in the rhesus as it is in man
(Fig. 5b). On the other hand, the border of the lobules corresponding
to thin portions seen grossly in the rhesus placenta is often marked
on section by large venous channels extending through the entire height
of the placenta as will be described below. Septa and fissures rarely
occur in the rhesus placenta, but in some areas the chorionic plate dips
down into the placenta forming a fissure on the fetal side at the border
of two lobules.

The arteries were studied in a similar manner as those of the human
placenta with greater use of specimens injected with India ink, including
those of the Carnegie collection. Reconstructions showed convolutes of
arteries below the thin, interlobular areas in a location corresponding
with that in man. However, contrary to the human arteries which open
at that point, the arteries in the decidua of the rhesus monkey frequently
take a straight course from this convolute towards a point below the
center of the lobule and open there (Figs. 6, 7), in a manner consistent
with the observations of Freese *et al.* (1966). Some arterial openings are
found in other locations relative to the lobules.

The maternal venous drainage of the rhesus placenta is much more
complex than its human counterpart. The veins leading from the
placenta are very large and their points of origin can be found in three
different forms as shown diagrammatically in Figure 7. One of these
forms is a simple opening at the base of the placenta as found in man,
and on rare occasions groups of villi protrude into it. A second form is
associated with large numbers of anchoring villi particularly in the

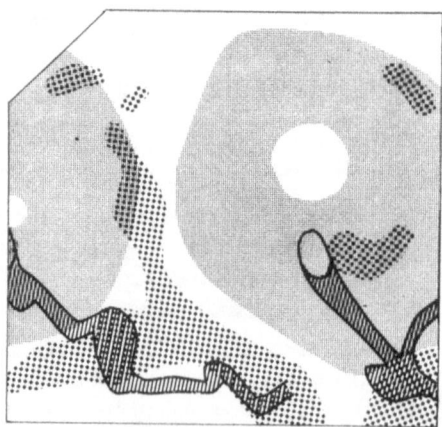

Fig. 6. Reconstruction of a portion of a mature rhesus placenta, showing lobules
(finely stippled) and arteries (cross-lined) much as in Fig. 3. The coarsely stippled
areas are those in which many villi anchor in the decidua. The wide portions of the
arteries are actually outlines of convolutes.

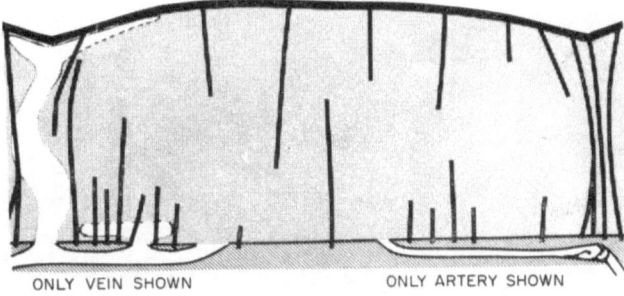

ONLY VEIN SHOWN   ONLY ARTERY SHOWN

Fig. 7. Diagram of the placental lobule of the rhesus monkey. The lobule is faintly outlined by grooves on the fetal surface which in some areas (left) are associated with wide venous channels. Other maternal venous pathways are indicated on the left; an artery is shown on the right. Villous stems are indicated by heavy lines, the decidua is lined obliquely.

vicinity of the border between lobules. Between the anchoring points of these villi, in areas containing large amounts of trophoblast, there are numerous channels which are not located in the maternal tissue of the decidua, but are entirely surrounded by fetal tissue (Figs. 8 a, b). The lining of these channels may consist either of syncytium or of delicate connective tissue which is presumably of fetal origin. These labyrinths of channels open into veins in the decidua. The third form of opening consists of continuation of large veins across the intervillous space up to the chorionic plate in the thin, interlobular areas (Fig. 8c). These channels do not have a typical venous wall from the point where they leave the decidua, but are rather formed by obliterated and conglutinated villi which apparently contribute to a smooth and sometimes quite distinct wall of connective tissue. As these channels extend along the chorionic plate (Fig. 8c) they are bounded on the side of the chorion either by cytotrophoblast or by a connective tissue wall as described; on the opposite side, facing the intervillous space, there may also be a connective tissue wall similar to that described before, but with many openings draining blood from the subchorial area. It is thus apparent that in the rhesus, arterial blood enters the lobules near their centers, and venous blood is carried away largely at the periphery. This drainage at the periphery can occur at various points at the base but also in the subchorial area through the channels just described. These channels are found at scattered points between lobules and also at the margin of the placenta and come close to the hypothetical, but actually nonexistent, marginal sinus of the human placenta (Wilkin and Picard, 1961). However, they do not completely encircle either the lobules or the entire disc. The extension of veins into channels beyond the endometrium, and the existense of "marginal sinuses" has been described by Ramsey (1954).

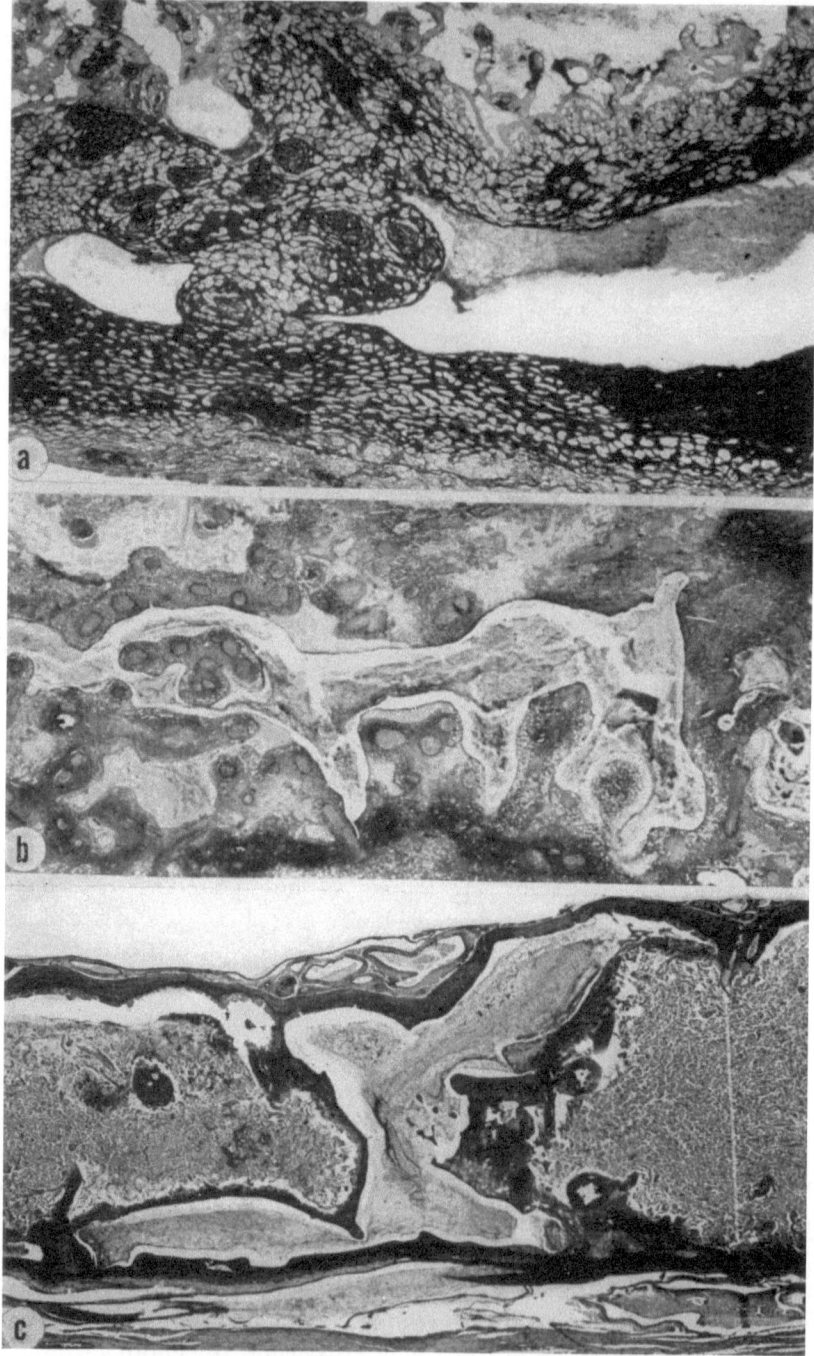

Fig. 8. Maternal venous channels within the intervillous portion of the rhesus placenta. *a* basal portion of the placenta cut at right angles to the maternal surface;

It is obvious that the intricate structure of the base of the rhesus placenta, including the decidua but particularly a layer of tissue of trophoblastic origin, is prominently concerned with the arrangement of the vessels. This area needs further study which is now in progress.

COMPARISON OF THE HUMAN AND RHESUS PLACENTA

It is interesting to note that the entry of arterial blood below the center of the placental lobules which had been postulated to exist both in man and the rhesus monkey, actually occurs in the rhesus but not in man. In both species arterial convolutes are found at the border of lobules and in man they open into the intervillous space at that point; in the rhesus on the other hand, a straight portion of entry continues towards the center of the lobule in many instances. The distribution of the veins is also different in the two species. While veins originate in the intervillous space predominantly at the base of the interlobular areas in both species, the manner of origin is quite different in the two, and the rhesus has in addition a drainage system which collects blood directly from the subchorial area.

The entry of arterial blood occurs, according to radiographic studies, in the form of jets or spurts which extend from the arterial opening far across the placenta toward the chorionic plate. It has been tempting to look for areas in which such flow would preferentially occur, and this has been done at two points, namely, the center of the lobule with its loose texture and the interlobular tissue. If one examines the path of flow of arterial blood in rhesus specimens injected with India ink, it becomes quite obvious that the injected material which in many instances was carried under maternal blood pressure, extends from the arterial opening in the form of a cone with its base near the chorionic plate, usually between densely packed villi and quite obviously without any need for preformed spaces or loosely textured areas (Fig. 9). While in both species the placenta contains denser and looser portions, the latter particularly in man, these are not necessary to conduct the flow of large amounts of blood either from the arteries or back into the veins.

---

a small portion of decidua is seen at the lower left. Venous channels are surrounded by trophoblast and lined by either syncytium or connective tissue probably derived from nearby anchoring villi. *b*, at lower magnification, is a section cut parallel to the maternal surface at the level of the channels seen in *a*. The latter are seen forming a labyrinth between anchoring villi. *c* shows an interlobular area cut at right angles to the surfaces. A large channel connecting with maternal veins has been walled off by degenerating villi and a layer of connective tissue derived from them. It reaches the chorionic plate (top) and extends along it to the right.

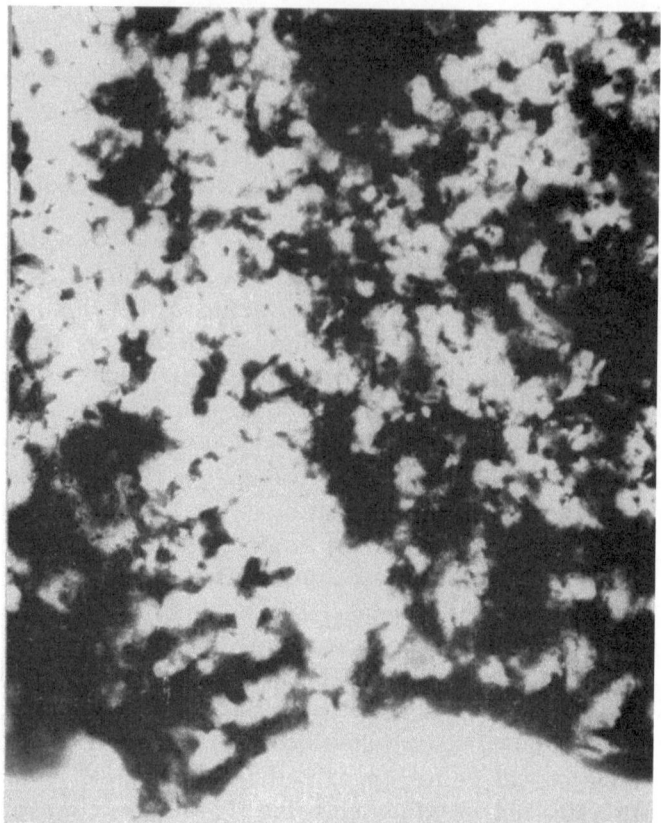

Fig. 9. India ink outlining the intervillous space of a mature rhesus placenta, after injection into the aorta with intact maternal circulation. Openings of one artery were at the bottom, one at the left and another one beyond the right margin of the figure. Blood appears to flow freely through the intervillous space without a need for preformed channels or loose portions.

RELATION OF PATHOLOGIC CHANGES TO THE LOBULAR ARCHITECTURE

Some lesions of the *human* placenta have a predilection for certain areas outlined above (Fig. 1a). Infarcts may occupy the dense portion of the lobule and thus form a hollow sphere which greatly accentuates the outline of the affected lobule (Fig. 10). The so-called red infarcts have a lobular distribution. Groups of avascular villi are most commonly seen in area 5. Intervillous thrombi occupy in some placentas selectively area 4. Fibrin in considerable amounts may surround villous stems and thus occupy much of area 1. Small infarcts at the base of area 1 occur near obliterated maternal arteries.

Little is known about pathologic changes in the *rhesus* placenta. In

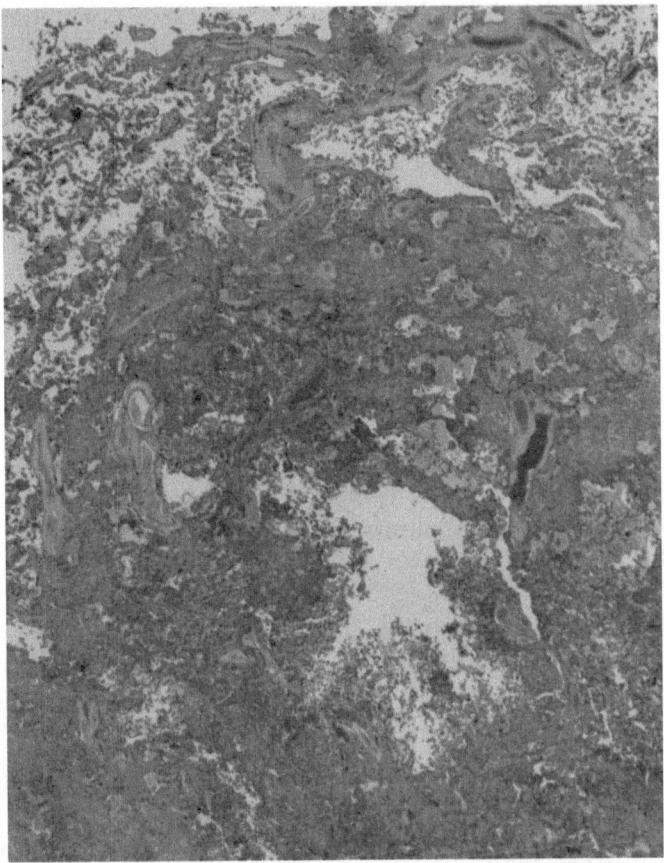

Fig. 10. An infarct developing in the dense parts of a lobule of a human placenta forms a hollow sphere. (From Gruenwald (1966), courtesy Bulletin of the Johns Hopkins Hospital.)

the few specimens in which there was either spontaneous fetal death, or a change following experimental interference, obliteration of the intervillous space by fibrin followed a lobular pattern much like that occurring in man. In addition, engorgement and thrombosis of the larger and more elaborate venous drainage system was conspicuous in rhesus specimens. Rounded, small islands of calcification in the trophoblast surrounding anchoring villi near the base, appear to be a normal and regular occurrence in the mature rhesus placenta.

SUMMARY

The placental lobule characterized as a dense mass of terminal villi with a more or less distinct loose center, is described for the human and rhesus placenta. Some pathologic changes have a characteristic distribution in relation to the lobule. Convolutes of maternal arteries occur in the decidua below the border between lobules in both species; in man the arteries open at that point, whereas in the rhesus they often take a straight course to a point below the center of the lobule, and open there. Most veins open between lobules. In the rhesus, channels form in the intervillous space which provide an elaborate drainage system connecting with the veins in the decidua, and extending to the subchorial area.

Acknowledgments

This work was supported by grant no. HD-00547-05 of the National Institute of Child Health and Human Development, Public Health Service.

The author is also greatly indebted to the Department of Embryology, Carnegie Institution of Washington, and particularly Dr. Elizabeth M. Ramsey for the use of their material and valuable advice.

References

Arts, N. F. T.: Investigations on the vascular system of the placenta. Part I. General introduction and the fetal vascular system. Am. J. Obstet. Gynec. *82*:147, 1961.

Becker, V., and P. Jipp: Ueber die Trophoblastschale der menschlichen Plazenta. Geburtsh. Frauenheilk. *23*:466, 1963.

Crawford, J. M.: Vascular anatomy of the human placenta. Am. J. Obstet. Gynec. *84*:1543, 1962.

Freese, U. E., K. Ranniger, and H. Kaplan: The fetal-maternal circulation of the placenta. II. An x-ray cinematographic study of pregnant rhesus monkeys. Am. J. Obstet. Gynec. *94*:361, 1966.

Gruenwald, P.: The lobular architecture of the human placenta. Bull. Johns Hopkins Hosp. *119*:172, 1966.

Ramsey, E. M.: Venous drainage of the placenta of the rhesus monkey (*Macaca mulatta*). Contr. Embryol. *35*:151, 1954.

Reynolds, S. R. M.: Formation of fetal cotyledons in the hemochorial placenta. Am. J. Obstet. Gynec. *94*:425, 1966.

Strauss, F.: Bau und Funktion der menschlichen Plazenta. Fortschr. Geburtsh. Gynäk. *17*:3, 1964.

Wilkin, P.: Contribution à l'étude de la circulation placentaire d'origine foetale. Gynéc. Obstet. *53*:239, 1954.

Wilkin, P., and C. Picard: La rupture du sinus marginal existe-t-elle? Étude anatomo-clinique des lésions vasculaires de la région marginale du placenta humain. Bull. Féd. Soc. Gynéc. Obstet. Franç. *13*:507, 1961.

DEGENERATION OF OVERCROWDED PLACENTAE

E. S. E. Hafez

Professor, Department of Animal Sciences,
Washington State University, Pullman, Washington

Overcrowding *in utero* can be induced experimentally by superovulation with gonadotropins or by transferring an excessive number of fertilized eggs. When rabbits are superovulated, most embryos are resorbed after implantation, and in contrast to mice, fewer living fetuses survive than after natural mating. In superovulated sheep and cattle (Hafez, 1964a) with the increasing number of implantations, the number of surviving fetuses reaches a ceiling level somewhat at or under four per pregnancy. Every species seems to have a "ceiling value" for the number of implantation sites and the number of implantations that can be maintained successfully throughout pregnancy.

Also the females, of a given species, vary in their capacity to maintain successful multiple pregnancy. The first signs of placental degeneration are the absence of vascularization (Fig. 1). In advanced stages of degeneration, the conceptus becomes a greyish mass.

When excessive numbers of embryos are transferred within appropriate pseudopregnant rabbits, as many blastocysts as physically possible are implanted (Fig. 2). However, most of these implanted blastocysts degenerate within 24 to 48 hours after implantation. The degenerating placentae could be identified at the termination of pregnancy as brownish placental remnants.

There is no uniform pattern for the distribution of degenerating placentae along the uterine horn (Hafez, 1964b). It was expected that each viable placenta would be surrounded by one or more placental remnants on each side. This did not seem to be so in most of the recipients. Factors other than limited uterine space may contribute to early embryonic death in overcrowded uteri. The local effects of overcrowding, however, are apparent.

With advancing pregnancy, the embryos become increasingly dependent upon the placenta for their survival. The degree of placental development is primarily influenced by the availability of space and vascular supply within the uterus. In polytocous species, as the number of implantations rises, the vascular supply to each site is reduced; this restricts placental development and causes high embryonic and fetal death. The physiological mechanisms involved in the complete loss of a whole litter are not fully understood.

In polytocous species overcrowding within one area of the uterus is

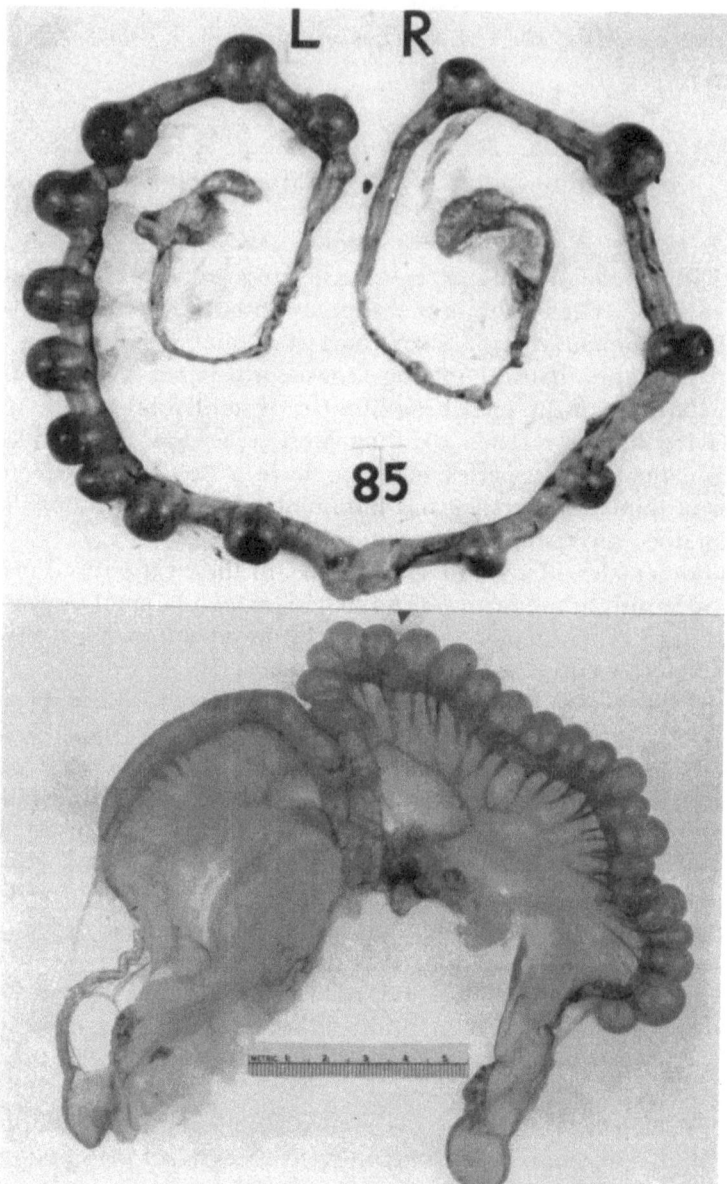

Fig. 1. Rabbit uteri 9 days post coitum. Top: Control uterus following natural breeding. Bottom: Overcrowded placentae in the right uterine horn. Some 35 embryos were transferred in the recipient at two days of pseudopregnancy. Note the uterine vasculature supplying the implantation sites; and the variability in the morphology of implantations.

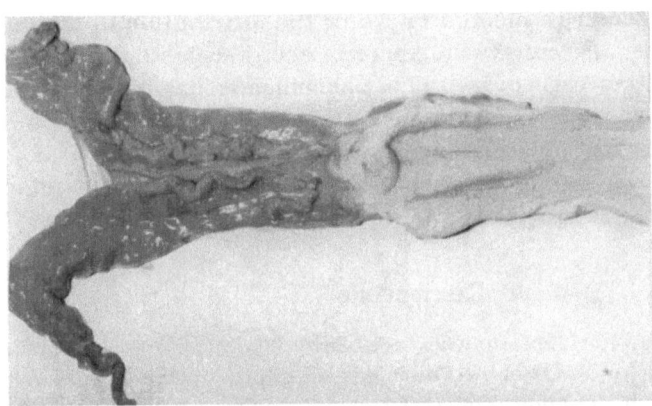

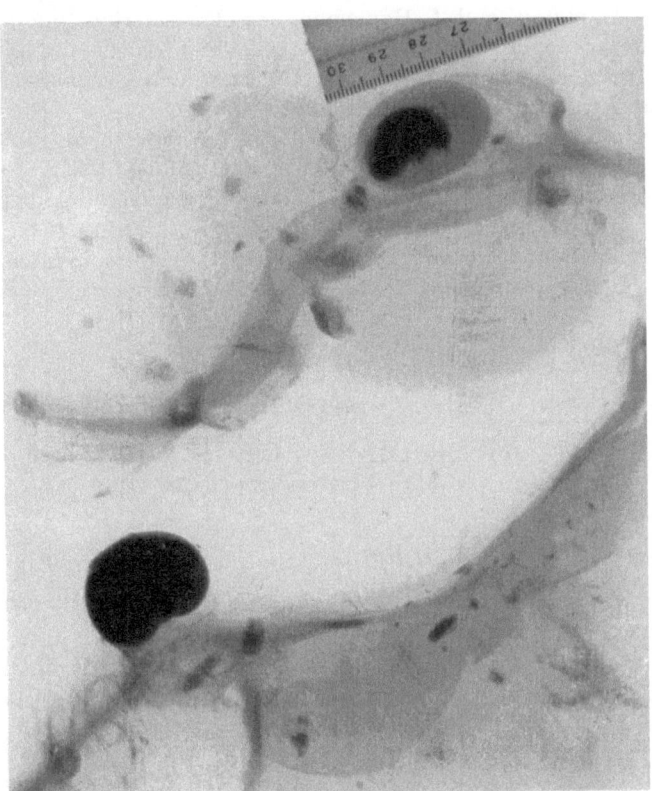

Fig. 2. Degeneration of overcrowded placentae in cattle in which multiple pregnancy was induced by an intramuscular injection cf pregnant mare serum and an intravenous injection of human chorionic gonadotropins. Right: Placenta with five degenerating embryos; note the fusion of adjacent allantochorions and the early stages of degeneration and the lack of vascularity of the placenta. Left: Reproductive tract of the cow (13 weeks of gestation) and the advanced stage of degenerating conceptus which was represented by greyish mass. (Hafez et al., 1965.)

prevented by two mechanisms: (a) a mechanism for the transport of the embryos and (b) a mechanism that prevents the implantation of one blastocyst in close proximity to another. Transuterine migration of embryos can be regarded as a method by which the distribution of embryos could be equalized in cases where there is a disparity in the number of ova ovulated from each ovary. This phenomenon has been reported for ungulates (Boyd and Hamilton 1958 and Dziuk, Polge and Rowson 1962), some carnivores and insectivores. In Chiroptera, the blastocysts of certain bats always implant in the right horn, although ovulation occurs in both ovaries; hence, migration must occur (cf. Boyd *et al.*, 1944).

References

Boyd, J. D., and W. J. Hamilton: Cleavage, early development and implantation of the egg. In: Marshall's Physiology of Reproduction. Ed. by A. S. Parkes, Longmans. London. Vol. II. Chap. 14:1, 1958.

Boyd, J. D., W. J. Hamilton, and J. Hammond, Jr.: Transuterine (internal) migration of the ovum in sheep and other mammals. J. Anat. *78*:5, 1944.

Dziuk, P. J., C. P. Polge, and L. E. A. Rowson: Migration of pig embryos following egg transfer. J. Anim. Sci. *21*:1021, 1962.

Hafez, E. S. E.: Transuterine migration and spacing of bovine embryos during gonadotropin-induced multiple pregnancy. Anat. Rec. *148*:203, 1964a.

———: Effects of overcrowding *in utero* on implantation and fetal development in the rabbit. J. Exp. Zool. *156*:269, 1964b.

Hafez, E. S. E., M. R. Jainudeen, and D. R. Lindsay: Gonadotropin-induced twinning and related phenomena in beef cattle. Acta Endocrin. Suppl. *102*:43, 1965.

A UTERINE-AMNIONIC PATHWAY OF INFECTING THE RABBIT FETUS WITH LISTERIA MONOCYTOGENES *

C. D. KING

Department of Pathology, University of Wisconsin, Madison, Wisconsin

In light of the concepts presented at this conference in regard to bacterial infection of the fetus *in utero*, I would like to propose a uterine-amnionic pathway by which *Listeria monocytogenes* might invade the 20-day-pregnant rabbit fetus. The following work represents part of a project which will be published in its entirety elsewhere.

* Published with the approval of the Director of the Wisconsin Agricultural Experiment Station and supported in part by a Public Health Service Animal Pathology Traineeship Grant 2G817, Division of General Medical Sciences, U.S. Public Health Service, under the supervision of Dr. Carl Olson, Department of Veterinary Science.

The pathology of Listeric abortion has been studied in many mammalian species (Gray and Killinger, 1966), but usually following the introduction of large numbers of Listeria into the pregnant animals or at a time when abortion was imminent or had occurred. The results of such experiments have been valuable in understanding the gross and microscopic lesions found late in the abortive period, but have accomplished little in elucidating the pathogenesis of this condition. The following experiment was undertaken to study the early changes in the rabbit uterus, placenta and fetus when *L. monocytogenes* was inoculated in relatively small numbers.

Thirty-eight 20-day-pregnant rabbits were killed at either 12, 18, 24, 48 or 72 hours after receiving an intravenous injection of either 100,000 or 50,000 viable bacterial cells. Attempts to isolate the organism from various maternal and fetal tissues were made using standard bacteriological culture techniques. Similar tissues were fixed in formalin for histological evaluation using the Brown and Brenn stain for Gram positive bacteria.

The results of the bacterial isolation data indicated that the first signs of fetal infection were dependent upon the size of the inoculum. Bacterial recovery from uterine and fetal tissues was first made at 18 hours in the 100,000 cell group while no isolates were made in the 50,000 cell group until 48 hours post-inoculation. Histologic examination of like tissues utilizing a modified Gram staining technique, indicated the *L. monocytogenes* could be seen in uterine and fetal sections at 24 hours in the 100,000 cell inoculum group and at 48 hours in the 50,000 cell group. The distribution of the microorganisms in these sections suggested a possible pathway by which Listeria might pass through the placenta of a 20-day-pregnant rabbit (Fig. 1).

Briefly, the maternal blood flow through the placentome originates at the mesometrial aspect of the uterus and flows to the surface sinuses at the top of the fetal cotyledon. From here the blood passes through the complex series of trophoblastic tubules which together with the fetal vessels comprise the labyrinth. The deoxygenated blood then collects in rather large venous sinuses in the intermediate zone of the maternal caruncle. At this point the blood flow slows considerably before being returned to the maternal venous circulation. Hence, in this portion of the decidua are pools of slow flowing, oxygen-poor blood, an ideal environment for the growth of a microorganism such as *Listeria monocytogenes.*

Listeric cells were first seen penetrating the walls of the maternal venous sinuses and invading the adjacent intermediate zone of the maternal caruncle. These organisms seemed to move freely through this tissue, but rarely passed into the zone of separation. The microorganisms were found next in relatively large colonies in the periplacentomal

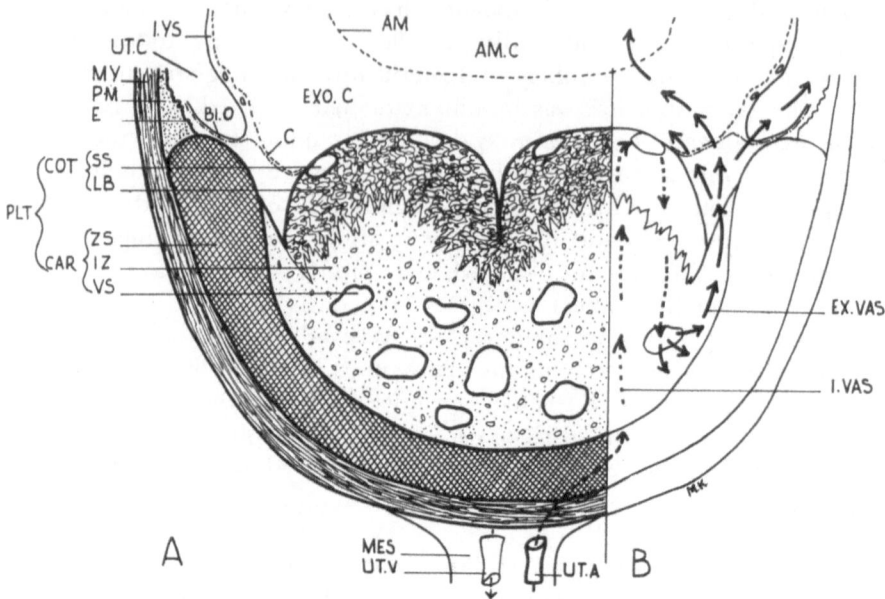

Fig. 1. (A) A diagrammatic representation of the anatomical relationships of the uterus, placentome and fetal membranes of a 22-day pregnant rabbit. AM, amnion; AM.C, amnionic cavity; BI.O, bilaminar omphalopleur; C, chorion; CAR, caruncle; COT, cotyledon; E, modified endometrium; EXO.C, exocoelomic cavity; I.YS, inverted yolk sac; IZ, intermediate zone; LB, labyrinth; MES, mesometrium; MY, myometrium; PLT, placentome; PM, propia mucosa; SS, surface sinus; UT.A, uterine artery; UT.C, uterine cavity; UT.V, uterine vein; VS, venous sinus; ZS, zone of separation. (B) A diagrammatic representation of a possible route by which *Listeria monocytogenes* may, under certain dose and time conditions, infect the rabbit fetus. The dotted arrows represent the intravascular (I.VAS) course of the organisms through the placentome to the venous sinuses in the intermediate zone of the decidua. The solid arrows represent the extravascular route (EX.VAS) by which the organisms can enter the uterine cavity, infect the chorion, bilaminar omphalopleur and inverted yolk sac. Penetration of these membranes is then followed by infection of the exocoelomic cavity, amnion and amnionic cavity. Ingestion of amnionic fluid by the fetus provides a means of fetal infection.

regions where the intermediate zone and the zone of separation become exposed to the uterine cavity. Payne (1958) has pointed out that Listeric infection in the rat placenta localizes in the junctional zone of the placental disc. In the rabbit, the invasion then spreads to the modified endometrial epithelium which is characterized by multi-nucleated, ciliated cells which are found lining the uterine cavity adjacent to the placentome. The inverted yolk sac which lies in juxtaposition to the lining of the uterus became heavily infected and the microorganisms were seen penetrating this membrane, but were never seen within the vitelline vessels. It has been shown by Brambell and Hemmings (1954) that maternal plasma proteins favor the inverted yolk sac of the rabbit placenta to reach the fetus. Fetal involvement was first noted in this work by the presence of bacteria on the skin and in the oral and nasal cavities. Generally these microorganisms superficially invaded the mucosal layers of the upper respiratory and digestive systems and did not stimulate an inflammatory response. Colonies of Listeria were found in the posterior pharynx and in the bronchioles of the lungs. Isolated, non-inflammatory, areas of mucosal invasion were seen in the esophagus, stomach and intestine. Invasion of the fetal liver was seen only after the preceding changes were noted. Likewise, bacterial invasion of the chorio-allantoic placenta in the labyrinth was noted only after the bacteria seemed to be well established in the various organs of the fetus. The possibility that the bacteria may penetrate the labyrinthine tubules more readily from the fetal side than from the maternal aspect is speculative; however, recent studies by Tillack (1966) on the rat placenta indicate that ferritin granules do diffuse more readily from fetal blood to maternal blood.

To briefly summarize the results of this work, it is believed that a non-hematogenous pathway exists for the transmission of *Listeria monocytogenes* across the rabbit placenta. This pathway is initiated when the organisms leave the venous sinuses of the maternal decidua, pass into the uterine lumen, transverse the inverted yolk sac, exocoelomic cavity and amnion to be ingested by the fetus in the amnionic fluid. Infection of the chorio-allantois originating from the fetus appears to be possible when low numbers of *L. monocytogenes* are inoculated into 20-day-pregnant rabbits.

References

Brambell, F. W. R.: Transport of proteins across the fetal membranes. Cold Spring Harbor Symposia on Quantitative Biology. *19*:71, 1954.

Gray, M. L., and A. H. Killinger: *Listeria monocytogenes* and Listeric infections. Bacteriol. Rev. *30*:309, 1966.

Payne, J. M.: Changes in the rat placenta and foetus following experimental infection with various species of bacteria. J. Path. Bact. *75*:367, 1958.

Tillack, T. W.: The transport of ferritin across the placenta of the rat. Lab. Invest. *15*:896, 1966.

ONTOGENESIS OF THE IMMUNE RESPONSE *

ARTHUR M. SILVERSTEIN

*The Wilmer Institute, The Johns Hopkins University School of Medicine,
Baltimore, Maryland*

The earliest studies of any phase of developmental immunity date back some three-quarters of a century to the investigations of Paul Ehrlich and others. They were based upon an understandable interest in one of the significant comparative aspects of reproductive success—the mechanisms by which the newborn is able to defend itself against the host of infectious disease processes that were only then being discovered by the nascent fields of bacteriology and immunology. It soon became apparent that most, if not all, of the newborn's immunity was passively acquired from the mother, via the yolk in oviparous animals, across the placenta or other fetal membranes in certain mammals, or from the colostrum and early milk in yet other mammals. The extensive body of literature which has developed in this area has been reviewed in detail by Brambell and co-workers (1951, 1958). With the increasing appreciation of the protective role of the mammalian placenta, and in view of the obvious benefits derived from passively acquired maternal antibody, there appeared to be little reason to question the reasonable assumption that the mammalian fetus and neonate are immunologically incompetent (the so-called immunologic null state) and only develop their own capabilities at some time after birth, presumably prior to complete disappearance of the protective maternal antibodies. This assumption was further reinforced by observations on the relative inability of the human newborn to respond to the standard pediatric immunizations immediately after birth, and by the repeated demonstration that the familiar small laboratory animals (primarily rodents) were unable to respond to antigenic

* Supported by Contract No. DA49-193-MD-2640 from the United States Army Research and Development Command, Washington, D.C., by Grant No. AI06713 from NIAID, the National Institutes of Health, Bethesda, Maryland, by an unrestricted grant from the Alcon Laboratories, Inc., and by an International Order of Odd Fellows Research Professorship.

stimuli during the early postnatal period and, in fact, could easily be rendered immunologically tolerant of a variety of antigens during this same time (Hašek *et al.*, 1961; Smith, 1961). Finally, these empirical observations were furnished a theoretical basis by the explanations provided by Burnet (1959) for the nature of the immunologic null period, the significance of the development of immunologic competence, and the basis for immunologic tolerance, incorporated in his clonal selection theory of immunity.

However, the past decade or so has witnessed the appearance of a number of observations that require a revision of our earlier beliefs about the lack of immunologic competence of the mammalian fetus *in utero*. It has been demonstrated that maternal antibody may actually inhibit the active immune response of the neonate (Perkins *et al.*, 1961; Uhr and Baumann, 1961), thus suggesting that many of the earlier studies did not fairly give the young neonate a chance to demonstrate its full capabilities to respond to antigenic stimuli. It has further been shown that while the subject may respond inadequately if at all to one antigen, it may simultaneously be able to furnish a very respectable antibody response to a second antigen, thus raising to question the general basis of immunologic maturation. Finally, a number of reports have demonstrated that the fetus of at least some mammalian species is capable of a very extensive repertory of immunologic responses *in utero*, and that the genesis of these responses may be somewhat more complicated than had previously been believed.

Since a number of reviews of the ontogenesis of the immune response have appeared recently (Ebert and DeLanney, 1960; Silverstein, 1964; Good and Papermaster, 1964; Miller and Davies, 1964; Šterzl and Silverstein, 1967), we shall not attempt here an exhaustive compilation of facts bearing on the subject. Rather, we shall outline some of the techniques employed in the study of immune responses in the mammalian fetus, since several of them will undoubtedly find application to the study of other aspects of reproduction and fetal development. We will then review the most significant aspects of what is known about the maturation of immune responses in the developing animal, and finally discuss the implications of immunologic competence by the fetus for the ultimate success or failure of the reproductive sequence in which it is participating.

TECHNIQUES OF INTRAUTERINE FETAL SURGERY

It would be inappropriate to present here a detailed review of the many surgical procedures that have been developed primarily during the past decade, permitting direct manipulation of the mammalian fetus *in utero*. A detailed outline of the various procedures employed for immunologic and other studies may be found in the reports of the respective investigations, and have been reviewed recently by Kraner (1966). Since, however, many of these techniques are of relatively recent vintage, and may not be widely known to workers in the many disciplines concerned, it does seem appropriate to mention a few of the more generally useful techniques. In this way investigators interested in any aspect of develop‑ mental mammalogy may gain an appreciation of the wide range of experimental approaches permitted by recent developments in intra‑ uterine fetal surgery. It will be obvious, of course, that the extension of any given technique from one species to another will depend upon a great number of variables, including the length of gestation, the size of the fetus at any given stage of gestation, the type of placentation involved, and the anatomic and physiologic variations encountered in the uteri of different species.

Immunization and Skin Grafting of the Fetus

In the fetal lamb, as in other species, a variety of immunization pro‑ cedures may be undertaken during the latter part of gestation by intro‑ ducing the needle through the relatively thin uterus directly into the fetus, which has been manipulated into appropriate position (Fig. 1). In this manner, intradermal, subcutaneous, intramuscular, and intraperito‑ neal injections may be accomplished with little difficulty. We have on occasion even been able to administer transuterine intraocular injections into the fetal lamb late in gestation. When the animal is too small to be suitably manipulated, or when it is desired to do an intravenous or intra-arterial injection, it is necessary to expose the fetus to view through an incision in the uterus and fetal membranes. This may readily be accom‑ plished in a number of species. For instance, the fetal lamb between 50 and 100 days' gestation may conveniently be removed *in toto* from the uterine cavity to permit a variety of procedures, attached only by the umbilical cord. It may then be replaced within the uterus without interruption of pregnancy, since the ovine uterus is extremely flaccid and shows little tendency to contract during this procedure. On the other hand, the uterus of the Rhesus monkey, for example, is wont to contract when an appreciable portion of the fetus is removed from the uterine cavity. In this species, therefore, care is generally taken to remove through an appropriate uterine and membrane incision only a limb, the head, or as small a portion of the fetus as will permit execution of the desired procedure.

With the fetus appropriately exposed, it is generally a simple matter to perform intravascular injections, to take blood samples from the fetus, or to do orthotopic skin grafts or biopsies of fetal skin (Bangham *et al.*, 1960; Silverstein *et al.*, 1963a; 1963b; 1964). Lymph node biopsies as well

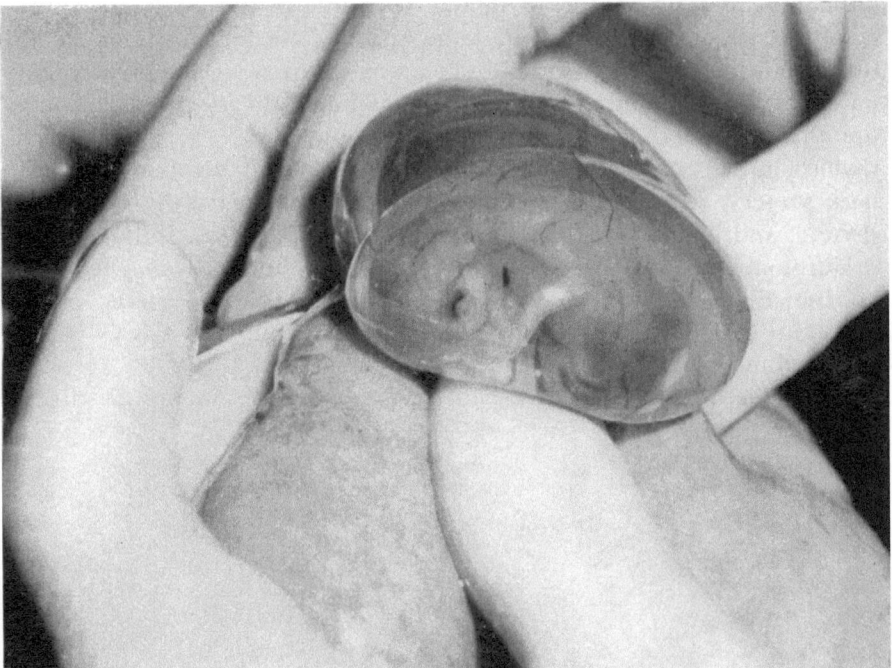

Fig. 1. Immunization of very small fetal lambs. An incision is made in the uterus, and the fetus expressed into an amniotic fluid-filled sac formed by the fetal membranes, through which the immunization is performed. Fetuses as young as 35 days' gestation have been immunized in this manner.

as the sampling of other tissues are also possible in these circumstances. These procedures are generally so well tolerated, especially in such species as the fetal lamb, that it is possible to repeat them at intervals. Thus we have not infrequently had occasion to perform as many as five to seven different surgical procedures involving the same lamb, without interruption of gestation, while four to five procedures have been performed on the same fetal Rhesus monkey. It must be confessed, however, that the success rate in any procedure on the fetal Rhesus monkey is less than that found for the fetal lamb.

Thymectomy of the Fetus *in Utero*

We may cite as perhaps the best example of a radical surgical procedure which can be performed with success on a fetal animal *in utero* that of thymectomy of the fetal lamb (Silverstein *et al.*, 1966). This is an interesting example of a situation in which even so complicated a procedure as thymectomy can more simply be performed in the fetus than in the adult

animal, since the wide-spread thymic remnants in the adult are appreciably less accessible than is the large and discrete fetal thymus. In the
ovine young the thymus is found extending down along the great vessels
on either side of the neck from the submaxillary salivary glands, through
the thoracic inlet, and into the mediastinum (Figs. 2, 3). Complete
thymectomy of these animals therefore involves what amounts to radical
neck surgery, requiring dissection of the thymus from salivary glands,
thyroid, and jugular vein and carotid artery. This is then followed by a
thoracotomy and blunt-dissection of the mediastinal thymus from its
attachments to the pericardium and pleural surfaces. The success of the
procedure is later confirmed by serial section of the soft tissues of the
neck and mediastinum.

This procedure is cited here because the extensive amount of surgery

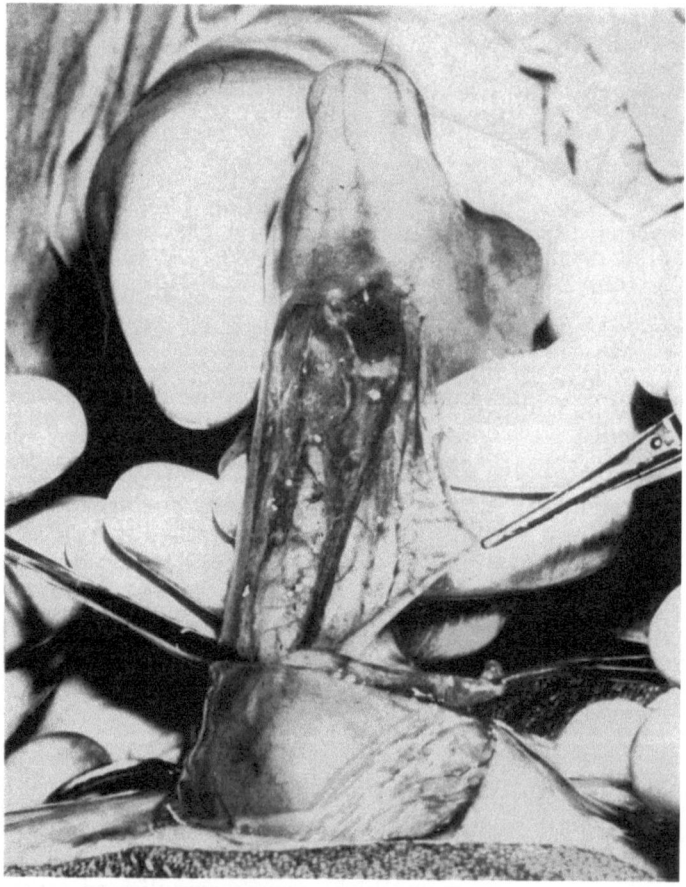

Fig. 2. Thymectomy of the fetal lamb. In this midgestation animal, the cervical
thymus is removed through an incision from the angle of the jaw to the thoracic inlet.

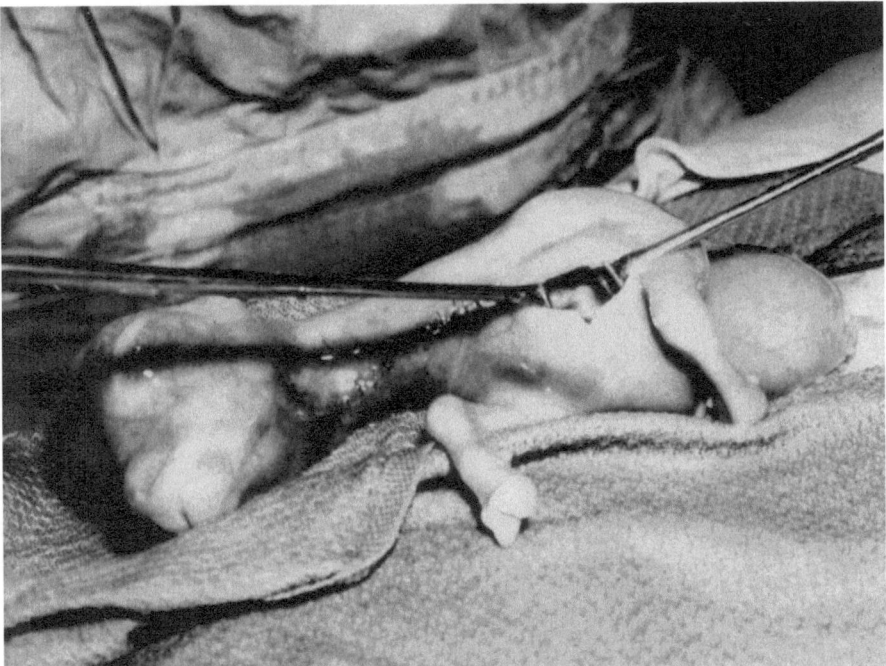

Fig. 3. Thymectomy of the fetal lamb. After closure of the neck, the mediastinal portion of the thymus is removed through a thoracotomy.

required and the duration of the procedure (the fetus lying outside the uterus for upwards of an hour) do not appear to embarrass the successful completion of an otherwise normal pregnancy. Preliminary attempts at thymectomy of the fetal Rhesus monkey have also succeeded, in this case involving splitting of the sternum to gain access to the thymus which lies entirely within the mediastinum. It is therefore our current feeling that almost any surgical procedure that might be successful in an adult animal of comparable size could be undertaken with success in the fetal animal *in utero*.

Permanent Indwelling Catheterization of the Fetus *in Utero*

The technique of fetal intravascular catheterization is not mentioned here so much because it is a more complicated surgical procedure as because of its potentially broad applicability to a great variety of approaches to the study of developmental biology (Fig. 4). To the best of our knowledge it provides for the first time a continuous access to the fetal circulation, allowing introduction of a variety of substances through the catheter, and periodic removal of samples of fetal blood for assay without necessi-

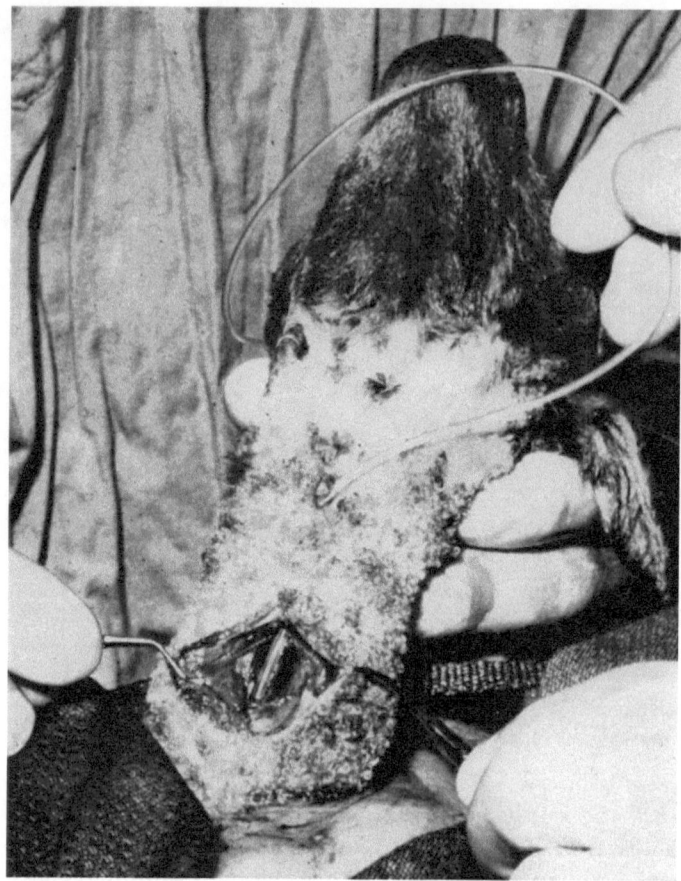

Fig. 4. Permanent indwelling catheterization of the fetal lamb. The catheter has been inserted into the proximal carotid artery of this 125-day-old fetus. After closure of the incision and replacement of the fetus, the catheter is brought in turn through the fetal membrane closure, the uterine closure, the maternal abdomen, and run subcutaneously to emerge through a skin incision on the maternal back.

tating surgical penetration of the uterus on each occasion (Silverstein *et al.*, 1966b; see also Bangham *et al.*, 1960).

Thus far we have employed these catheters only in the fetal lamb during the latter half of gestation. Initially the catheters were inserted into the proximal jugular vein, tying off the distal portion of this vessel. Despite anticoagulant therapy of the fetus, blood clots formed within the catheter or vascular lumen within a week or so of insertion. The results improved considerably with the insertion of the catheter into the proximal carotid artery, in which case patency was maintained in some animals for periods of from 2 to 4 weeks. Most recently, we have employed a modified T-tube, so that blood flow through the artery is not interrupted, and on occasion have been able to maintain such preparations patent for up to 6 weeks.

The catheter is inserted into a vessel of the fetal neck, and carried in turn through the closures of the fetal membranes, uterus, and maternal abdomen. It is then run subcutaneously to emerge through a skin incision on the back of the pregnant ewe.

THE TIMING OF IMMUNOGENESIS IN THE DEVELOPING ANIMAL

Antibody Formation

A comparative study of the developmental age at which different species attain immunologic competence discloses an extremely wide range of values. Thus the ability to form circulating antibodies in response to antigenic stimulus is demonstrated by the young of some species only after birth, while in other species this capability is already present during intrauterine life. Since there exists a wide variation in the general level of biological maturity of the various species at birth, it might have been anticipated that a correlation would be observed between the rate of immunologic maturation and the maturation of other biological functions. At the present writing, this appears not to be the case—there is even some question as to whether immunologic maturation parallels the maturation of functioning lymphoid tissue.

To cite a few pertinent examples, the newborn mouse and rat seem to develop antibody responses only some days after birth, while the newborn rabbit, thought to be immunologically incompetent prior to the third week of life, forms antibodies against certain antigens during this period even while rendered tolerant of other antigens (Smith, 1961; Hašek *et al.*, 1961). Even at earlier stages of development, the fetal calf *in utero* (Fennestad and Borg-Petersen, 1962) and the opossum embryo in the pouch (Kalmutz, 1962; LaVia *et al.*, 1963) form antibodies, while the fetal guinea pig develops delayed hypersensitivity (Uhr, 1960) before the end of gestation. The observations on the developing opossum are especially interesting, in that this animal is born "prematurely" and accomplishes the greater part of its development in the maternal pouch. Antibody formation in the opossum embryo has been found as early as the 8th day after entry into the pouch, and has been correlated with the first appearance of lymphoid elements in the developing thymus of this animal (Rowlands *et al.*, 1964).

It would appear, therefore, that the development of immunologic capabilities is not keyed to the birth process, and in fact may exhibit great variation among different species. There is also evidence that immunogenesis in the developing animal is not a single act of maturation, but may occur as a slow and perhaps stepwise process. Since there are more data on the response of the fetal lamb to antigenic stimuli than for any other species, it may be useful to outline these results in some detail to emphasize the characteristics of this maturation process and some of the problems posed by the data.

The lamb is born after a gestation period of some 150 days. At this point in its development it has hair, can walk immediately, and is not as dependent upon its mother as are the newborns of some other species. The greater part of this maturation takes place during the final third of gestation. Prior to this time the lamb is quite immature, and yet it is able to form circulating antibodies *in utero* in response to intrafetal immunization (Table 1). This response has been observed in lambs immunized as early as the 35th day of gestation and tested for a response 6 days later (Silverstein *et al.,* 1963a; Silverstein and Kraner, 1965a). It is precisely during this period that the thymus of the fetal lamb begins to show some degree of morphologic maturity, and only a little later that peripheral lymphoid tissues begin to mature. Unfortunately, the earliest age at which the fetal lamb is able to form antibody has yet to be established; it has proved technically difficult to immunize these small fetuses at an earlier age.

A study of the immunologic response of fetal lambs at different gestational ages stimulated with a variety of different antigens led to a somewhat surprising observation (Table 1). Some of the antigens employed, such as diphtheria toxoid, *Salmonella typhosa,* and Bacillus Calmette-Guerin (BCG) were found not to stimulate the formation of detectable circulating antibodies at any time during fetal life or even in the first weeks of extrauterine life. On the other hand, the bacteriophage virus

Table 1

TIMING OF DEVELOPMENT OF ACTIVE IMMUNOLOGIC RESPONSES TO DIFFERENT ANTIGENS BY THE FETAL AND NEONATAL LAMB

Antigen	Gestational Age (days)	Postpartum Age (days)
Bacteriophage ϕX 174	<41	—
Ferritin	66	—
Skin homografts	80	—
Hemocyanin (Busycon)	120	—
Ovalbumin (hen)	125	—
Salmonella typhosa O	—	>42
Diphtheria toxoid	—	>42
BCG	—	>42

$_\phi$X 174 stimulates active antibody formation in the fetus at the earliest age thus far injected—35 days' gestation. There was no question that the antibodies observed in the blood of these fetuses were of fetal rather than maternal origin, since the syndesmochorial placentation of the ovine does not permit the transfer of γ-globulins from mother to fetus, and in no instance did the mother's blood contain detectable antibody.

For almost a month after the fetal lamb is able to initiate active antibody formation against the bacteriophage virus, it still cannot respond immunologically to such protein antigens as ferritin and egg albumin. Only at about the 65–70th day of gestation does the fetal lamb begin to produce antibody to ferritin, whereas not until about 120 days' gestation for hemocyanin and 125 days' gestation for ovalbumin is the first circulating antibody detectable in the fetus.

Immunologic maturation in the fetus appears not to represent the attainment of competence by a single general mechanism by which the fetus can thenceforth respond to all antigens. There seems rather to be a slow, stepwise sequence of events permitting the fetus and newborn to respond first to one, then to another, and ultimately to all antigens. The basis for this antigenic and temporal hierarchy is at present not clear. This stepwise maturation may reflect a purely immunologic process, representing the development at different times in gestation of cells capable of manifesting each of the specific immunologic capacities. Alternatively, it is possible that cellular maturation of the ability to respond to the various antigenic stimuli appears very early in gestation, but that nonimmunologic factors necessary for the response may be delayed in their appearance. Thus, enzyme systems required to degrade each antigen into a useful form (Benacerraf *et al.,* 1963) might not all appear simultaneously. Whatever the explanation, it is clear that before a given age the immunologic mechanism of the fetal lamb does not recognize certain substances as antigenic; after that time recognition occurs, and a specific immune response ensues.

It is apparent that the stepwise maturation described above is not peculiar to the ovine species, but in all probability represents an observation of broad biologic applicability. Preliminary observations in the fetal Rhesus monkey have demonstrated a similar sequence of events, although the order of appearance of immunological reactivity to the several antigens may be different, as may also be the stage of gestation at which the fetus can first respond specifically to a given antigen.

The first antibodies to appear in the fetal circulation after immunization are the high molecular weight γ-M macroglobulins, sensitive to the action of 2-mercaptoethanol (Silverstein *et al.,* 1963a). Later γ-G globulin antibodies of lower molecular weight appear in the circulation. This sequence of events is typical of the immune response of both newborns and adults of many species, and is reviewed in detail elsewhere by Uhr (1964).

Far less is known about the maturation of immunologic competence in the human. Early observations on the immature response of the human newborn to pediatric immunizations (Osborn *et al.*, 1952), coupled with observations that human fetuses had only very immature lymphoid tissues and no plasma cells that would signal the production of antibody (Bridges *et al.*, 1959; Black and Speer, 1959), strongly implied that the human fetus was immunologically incompetent throughout gestation. However, as is true of so many other biologic responses, the given effect may only be produced by an appropriate stimulus, and the mammalian fetus is normally well protected from exogenous stimuli by the efficient placenta. If a breakdown in placental function should allow passage to the fetus of pathogenic organisms, then such diseases as congenital syphilis and toxoplasmosis may occur accompanied by the stimulation of plasma cell differentiation and undoubtedly antibody formation (Silverstein and Lukes, 1962).

Homograft Rejection

The rejection of skin homografts is widely recognized as a specific immunologic process. Here too the fetal lamb demonstrates a competence which seems to develop at a discrete time in gestation (Schinkel and Ferguson, 1953; Silverstein *et al.*, 1964). Skin homografts applied to the fetus before about midgestation are accepted as though they were not of foreign origin and survive unmolested in the recipient. After this time, however, the fetus is able to cope with the homograft in the typical manner, rejecting it specifically within 7–10 days. This is true regardless of the origin of the graft, whether from unrelated fetal or adult sheep or even from its own mother and siblings.

Specific homograft rejection *in utero* has also been observed in the fetal Rhesus monkey, occurring as early as about the middle of the normal 160-day gestation period in this species (Fig. 5).

It is of some interest that the study of certain aspects of the mechanism of homograft rejection can be pursued more effectively in the fetus than in the adult animal. Taking advantage of the agammaglobulinemic condition of the fetal lamb, and of the inability of the fetus to derive any γ-globulins passively from the mother, it was possible to show that homograft rejection could proceed without the benefit of participation in the process of circulating immunoglobulin antibodies (Silverstein and Kraner, 1965b). These observations provide strong support for the concept that homograft rejection is based upon a strictly cellular mechanism of immunologic response rather than upon the obligatory participation of humoral antibodies.

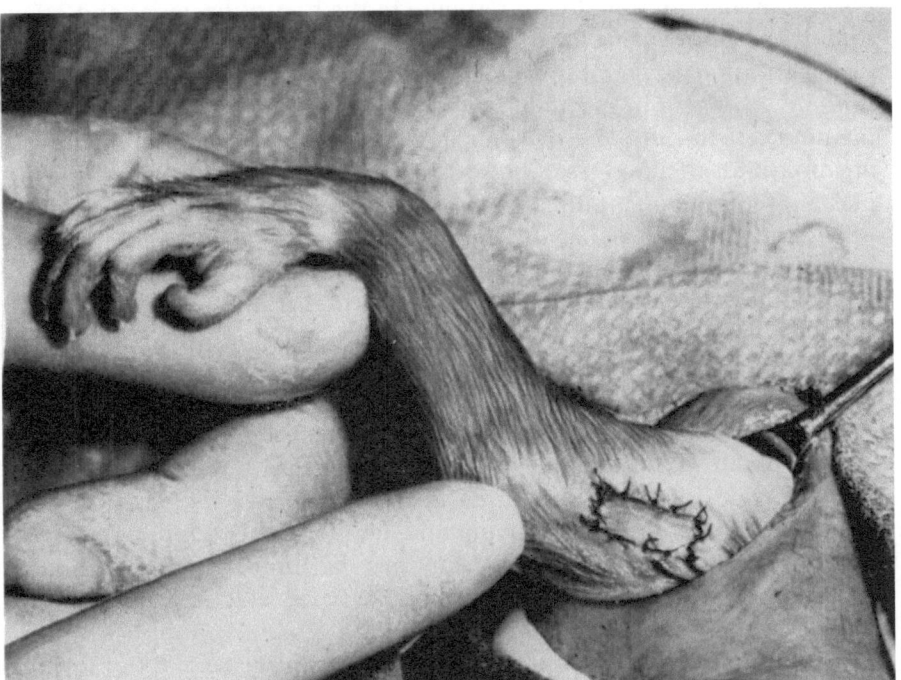

Fig. 5. Orthotopic skin grafting of the fetal Rhesus monkey. The limb of this 130-day fetus has been exposed through an incision in the uterus and fetal membranes. The fitted skin homograft is sutured in place, and may be biopsied repeatedly employing the same approach.

Fetal Lymphoid Development

The lymph nodes and spleen are recognized as constituting the primary seat of immunologic activity in the body. Most immunologically functioning cells appear first in these tissues, and the greater part of the circulating antibodies that appear in response to immunization is formed in them.

The maturation of lymphoid tissues in the normal mammalian fetus is a slow and protracted sequence (Silverstein and Lukes, 1962). The unstimulated lymph nodes of the younger fetus show primarily medullary channels lined by endothelium, an assortment of rather immature-appearing mesenchymal cells, and only minimal lymphopoiesis. There is little demarcation between the cortex and medulla, and no follicular activity. The tissue is only slowly populated by lymphocytes as gestation continues, until at birth there may be primary lymphoid nodules, generally without the formation of secondary follicles. Cells identifiable as having immunologic function are not seen in these normal immature nodes.

After birth there is an abrupt acceleration of this development in the normal animal, presumably in response to the host of bacterial and other stimuli that flood in from its new environment. Thus the final stages of the maturation of immunologic function appear to require a suitable stimulus, resulting in morphologic as well as serologic signs of the active immune response. In the development of certain types of cells and cell organizations, the mere readiness to furnish an immune response on the part of the developing animal may offer no overt morphologic indicators.

When the fetus is artificially immunized *in utero,* or the placenta fails in its protective function and permits infection of the fetus, then stimulation of the lymphoid tissue occurs. In both the infected human fetus and the immunized fetal lamb and monkey, the stimulus to antibody formation is accompanied by the differentiation of plasma cells and by a greater or lesser degree of precocious lymphoid activity. Whether there is an exact one-to-one relationship between immunologic capability and lymphoid maturation still remains to be seen. It is not unusual in fetal infection to see a pronounced plasmacytosis in a lymph node with no other signs of precocious lymphoid development. Similarly, as was mentioned above, circulating antibody may be found in the very young fetal lamb possessing little or no lymphoid tissue that seems capable of mediating the response.

Great attention has been paid in recent times to the role of the thymus in the maturation of lymphoid tissues and of their immunologic capabilities. These factors have recently been reviewed in great detail (Good and Gabrielson, 1964; Miller, 1964), and will not be dwelt upon here at any great length. It may be mentioned in passing, however, that the immunologically incompetent thymectomized female mouse may have this competence restored upon becoming pregnant, presumably due to the transplacental passage of some necessary factor provided by the developing thymus in the fetus (Osoba, 1965). It is also of some interest that thymectomy of the fetal lamb *in utero* at midgestation does not seem to impair its abilities to form circulating antibodies to reject homografts, or to continue the normal sequence of immunologic maturations expected of the fetal lamb (Silverstein *et al.,* 1966a). It is not yet clear from these studies whether at this gestational age the thymus had already performed any necessary function which it might have had, or whether the maternal thymus might contribute to the well-being of the fetus in a manner converse to that cited above in the case of the mouse.

Significance of Immunologic Immaturity

The term immunologic immaturity has generally been applied with reference to the initial attempts at immune response by the young animal which has already achieved immunologic competence. It is usually interpreted to signify the initially feeble or halting attempt at response to antigenic stimulus by the animal that has not yet realized its full immunologic capacity. In a sense it would be analogous to the requirement that an animal learn to crawl and then walk before it is able to run.

We have been interested in attempting to define more precisely the nature and characteristics of the state of immunologic immaturity. Three different immunologic assay systems have thus far been employed, using both the fetal lamb and fetal Rhesus monkey in these studies, and in each instance the earliest fetal response observed was as highly developed as that exhibited by the mature adult animal.

As was mentioned above, the application of orthotopic skin homografts at any time after about the 80th day of gestation results in specific immunologic rejection of these grafts by the fetus (Silverstein *et al.*, 1964). A careful study of the temporal and histopathologic aspects of this fetal response demonstrated that fetal skin graft rejection in no way differed from that observed in the adult. The intrauterine rejection process intervenes just as rapidly as was found in the adult animal, and with the same cytologic sequence of events. The fetus therefore shows no hesitancy or immaturity with respect to this form of immunologic response.

Again, employing the technique of Jerne and co-workers (1963) for the study of the cellular kinetics of the primary antibody response by counting single antibody-producing cells, the response to sheep erythrocytes was studied in the fetal Rhesus monkey near the end of the second third of gestation (Silverstein *et al.*, 1966c). Counting the number of antibody-forming cells within a given time after immunization in both the Rhesus fetus and adult, it was observed that the expansion of the population of antibody-forming cells was identical within the limits of experimental error. Since this approach permits an estimate of the *proportion* of the total number of lymphoid cells in the spleen which are involved at any given time in the active antibody response, thus correcting for the difference in size and relative cellularity of the tissues of the fetus and adult, it is apparent that on this basis also the fetus shows no signs of immunologic immaturity.

Finally, the primary response of the fetal lamb *in utero* to bacteriophage ϕX 174 antigen was studied, using the catheterization technique described above (Silverstein *et al.*, 1966b). In this study, antigen was injected into the catheterized fetus, and the animal bled as often as 6 to 8 times a day through the catheter in order to assess the initial rate of clearance of antigen from the circulation, the subsequent immune elimination of antigen which accompanies the earliest active antibody formation, and the rate at which free circulating antibody ultimately appears in the fetal circulation. With this assay system also, the fetal response was in no way different from that usually observed in the adult. Thus, immune elimination of circulating antigens started as early as 41 to 48 hours after immunization, and the earliest antibody appearing in the circulation increased in concentration at the same rate and to approximately the same levels as found in the adult.

As judged by the results obtained with each of these different approaches, it would appear that the earliest immunological responses by the fetus *in utero* are in no way immature, but are generally characteristic of those found in the normal adult animal. Within the limits imposed by the experimental approaches thus far employed, the classical notion of immunologic immaturity in the developing animal would appear to be without significance. Prior to a critical age, different for each antigen, the fetus seems truly incompetent and unable to respond to the antigenic stimulus. After this age, the response appears to be mature in all respects. The transition from immunologic incompetence to an adult competence seems to occur with impressive rapidity.

THE CONSEQUENCES OF ACTIVE IMMUNE RESPONSES BY THE FETUS

The study of immunity and immunopathology over the past 80 years has disclosed that, like so many other biological systems, the same immunologic mechanisms may have either beneficial or deleterious consequences for the host. On the one hand, the production of bactericidal, bacteriostatic, antitoxic, or opsonizing antibodies may contribute significantly to the defense of the body against infection by bacterial or viral pathogens. On the other hand, however, the inability of immunologic mechanisms to distinguish between benign and pathogenic antigens sometimes results in the development of the familiar gamut of allergic disorders, autoimmune diseases, and other immunopathologic conditions. In view of the persistence and apparently continuing elaboration of these mechanisms in evolution, we may assume that their benefits for the continuing survival of the species outweigh their drawbacks. While the same general conclusion may be applicable to the mammalian fetus able to develop its own active immunologic responses, certain factors peculiar to the intrauterine existence of the fetus render a consideration of these problems of great interest in the present context.

Beneficial Consequences of Active Fetal Immunity

Antibody Formation and Hypersensitivity

Embarrassingly little comparative data is available on the species in which the developing fetus may become immunologically competent to any of the pathogens which are of significance for that species, or on the gestational ages at which this competence develops. It is nevertheless quite clear from a study of the histopathology of those congenital infections accompanied by chronic inflammatory responses that the fetus is frequently able to call upon its own immunologic resources to help protect itself from the persistence or spread of an infectious process. This seems certainly to be the case in such congenital infections in the human as syphilis, toxoplasmosis, or cytomegalic inclusion body disease. It is unfortunate that in any retrospective examination of an individual process in an affected fetus, it is difficult to assess the benefits derived by the fetus from its own efforts at self-protection.

We must also be aware that some pathogens may cross the placenta from mother to fetus *prior* to acquisition by the fetus of immunologic capabilities. This seems certainly to be the case in congenital rubella infections of the human fetus during the first trimester, which are characterized by anomalies apparently resulting from a direct interference by the pathogen with normal organogenesis, rather than by the chronic inflammatory events generally associated with such diseases as syphilis and toxoplasmosis. Only later in gestation, presumably too late to do appreciable good, is the fetus able to form circulating antibodies against the rubella virus. Even here, the antibody is not always effective, since continued coexistence in the host of virus and circulating antibody suggests that the former may not always be accessible to the protective benefits of the latter.

Homograft Rejection

It is quite clear that the developing fetus of a number of species achieves *in utero* the immunologic capacity to reject tissue homografts. At first sight this would seem to be a mechanism superfluous to the needs of the developing fetus, since it has been amply demonstrated that the trophoblast provides a very effective immunologic barrier between mother and fetus (see Galton, this volume). Thus it has been demonstrated that sensitization of a mother with fetal skin and of a fetus with maternal skin, resulting in mutual immunologic incompatibility of the two, does not embarrass the normal continuation of pregnancy. Nevertheless we have become aware in recent years that there may be an appreciable exchange of cellular elements between mother and fetus during the course of a presumably normal gestation. Not only may erythrocytes cross the placenta from fetus to mother in the human and other species (leading to the familiar consequences of erythroblastosis fetalis), but apparently leukocytes may also cross the placenta in either direction. Such a situation may occur more frequently than one might currently anticipate. Should sufficient numbers of maternal circulating lymphocytes gain entry into the circulation of an immunologically incompetent fetus, then one would expect to see signs of a graft-versus-host reaction acting detrimentally on the fetus (*i.e.,* the attempt by the immunologically competent maternal lymphocytes to reject the fetus in whose tissues they have established residence). As Galton points out elsewhere in this volume, scattered reports of just this situation have appeared in the literature. It may be suggested that the rarity of this finding is due to the fact that at the time when maternal leukocytes are able to penetrate to the fetus, the fetus has already become immunologically competent to reject such homografts. Therefore one would expect that the intrauterine development of the ability to perform this immunologic function may have developed in certain species as a means by which the fetus may protect itself from this form of maternal threat to its continued well-being.

The recent popularity of the technique of intrafetal blood transfusion leads to a similar line of speculation. In the course of administration of upwards of 100 ml of whole maternal or other blood intraperitoneally into a fetus requiring such transfusion because of the development of erythroblastosis fetalis, great numbers of foreign lymphocytes may be introduced. By analogy with experiments in laboratory animals, it might be expected that given an immunologically incompetent fetus, these lymphocytes would be more than sufficient to initiate a graft-versus-host reaction in the recipient. Sufficient numbers of such transfusions have been done to suggest that this reaction in fact does not take place with any great frequency, a result which may probably be best ascribed to the capability of the human fetus at these gestational ages to reject the foreign lymphocytes before they can reject the fetus.

Deleterious Consequences of Active Fetal Immunity

Many disease processes cannot be explained adequately solely on the basis of the direct effects of the pathogenic agent on infected host tissues. A number of pathologic changes also reflect the active response of the host in its contest with the pathogen. Two instances may be cited suggesting that when the host for some reason cannot respond to the pathogenic organism, disease may not ensue despite the continued presence of the pathogen within host tissues. The response of the host to infection may in some instances be kindled by purely immunologic mechanisms contributing to the development of the pathologic process.

Lymphocytic choriomeningitis (LCM) is a naturally occurring viral disease of mice, leading generally to a severe and often fatal illness in the infected animal. It has been demonstrated, however, that under certain conditions this virus may persist in the blood and tissues of mice without causing signs of disease (Hotchin, 1962). Mice thus infected but not diseased have been shown to have received the virus *in utero* from their nonlethally infected mothers; thus infections with virus during early developmental stages seem to render the mouse thenceforth resistant to the disease. This curious phenomenon was interpreted by Burnet and

Fenner (1949) with the suggestion that immunologic tolerance of the virus had occurred owing to the presence of the antigen during maturation. With the establishment of such a tolerance, and with persistence of the virus in the host to maintain the tolerant state, the virus would never thereafter be able to induce an immune response in the host.

These findings in lymphocytic choriomeningitis present a paradox. We are accustomed to consider that only a well functioning immunologic system will offer protection against extensive disease induced by viral pathogens. Here, on the contrary, is a situation where the very inability to develop an immune response appears to render the pathogen innocuous, while the "proper" function of the immunologic apparatus spells disease for the host. In the main, therefore, lymphocytic choriomenigitis disease in mice would appear to be not so much a "viral" disease as an immunologic disturbance triggered by the virus.

In the instance cited above, absence of the expected disease resulted from the specific suppression of the immune mechanisms, due to an immunologic tolerant state in the host. In another type of situation, where the developing young has not yet matured its immunologic capabilities, the immune response to an antigenic pathogen may be equally lacking. We have discussed elsewhere (Silverstein, 1962; 1964) the interesting implications of a study of the natural history of the transuterine infection of the human fetus with *Treponema pallidum*, resulting in the typical picture of congenital syphilis of the fetal host accompanied by extensive inflammatory reactions. In this instance, immunologic factors appear to participate, since the fetus forms plasma cells as an integral part of its response to the highly antigenic organism. Judging from a histopathologic study, the lesions of congenital syphilis would seem to include at best only a minor immunologic component. Typically, congenital syphilis is never seen much before the 5th to 6th month of gestation, the age at which the placenta was supposed first capable of transmitting treponemes from mother to fetus. However, it is of no little interest that treponemes have been reported in rare instances in younger fetuses, in whom the typical pathologic picture of the disease, congenital syphilis, was absent (Stowens, 1959). The possibility exists, therefore, that the organism might infest and grow in the fetus without eliciting a typical host response, because the fetus at that young age may not yet be capable of responding.

Here then is another instance of a "pathogen" appearing to be innocuous and failing to incite a disease process. The disease proper, with its attendant lesions and embarrassment of the fetus, might only appear when the fetal response mechanisms become mature, presumably some time around the 5th to 6th month of gestation.

Thus the possibility exists that the development of active immunologic function by the mammalian fetus *in utero* may not be an unmixed blessing, and may ultimately prove to have its drawbacks as well as its advantages.

SUMMARY

Several points emerge from this brief examination of immunogenesis in the mammalian fetus. These are:

1. It has been demonstrated that the fetus of at least certain species may engage in active immunologic responses *in utero,* if appropriately stimulated with antigen. The relative paucity of comparative data in this area points up the pressing need for further studies along these lines.

2. The attainment of immunologic competence during development appears to be a complex sequence of events, depending not only upon the species in question, but also upon the antigen involved in the stimulus.

3. While the acquisition of immunologic capabilities by the fetus *in utero* may provide it with the means to participate in its own self-defence against infection, the same immunologic mechanisms may also function in the initiation of or contribution to disease processes.

References

Bangham, D. R., K. R. Hobbs, and D. E. H. Tee: Transmissions of serum proteins from foetus to mother in the Rhesus monkey. Indwelling cannulation of foetus without interruption of pregnancy. Lancet *2:*1173, 1960.

Benacerraf, B., A. Ojeda, and P. H. Maurer: Studies on artificial antigens. J. Exp. Med. *118:*945, 1963.

Black, M. M., and F. D. Speer: Lymph node reactivity. II. Fetal lymph nodes. Blood *14:*848, 1959.

Brambell, F. W. R.: Passive immunity in young mammals. Biol. Rev. *33:*488, 1958.

Brambell, F. W. R., W. A. Hemmings, and M. Henderson: Antibodies and Embryos. London: Athlone Press, 1951.

Bridges, R. A., R. M. Condie, S. J. Zak, and R. A. Good: The morphologic basis of antibody formation development during the neonatal period. J. Lab. Clin. Med. *53:*331, 1959.

Burnet, F. M.: The Clonal Selection Theory of Acquired Immunity. Nashville: Vanderbilt Univ. Press, 1959.

Burnet, F. M., and F. Fenner: The Production of Antibodies. Melbourne: Macmillan, 1949.

Ebert, J. D., and L. E. DeLanney: Ontogenesis of the immune response. Natl. Cancer Inst. Monogr. 2:73, 1960.

Fennestad, K. L., and C. Borg-Petersen: Antibodies and plasma cells in bovine fetuses infected with *Leptospira saxkoebing.* J. Infect. Diseases *110:*63, 1962.

Good, R. A., and A. E. Gabrielson (eds.): The Thymus in Immunobiology. New York: Hoeber-Harper, 1964.

Good, R. A., and B. W. Papermaster: Ontogeny and phylogeny of adaptive immunity. Advan. Immunol. *4*:1, 1964.

Hašek, M., A. Lengerová, and T. Hraba: Transplantation immunity and tolerance. Advan. Immunol. *i*:1, 1961.

Hotchin, J.: The biology of lymphocytic choriomeningitis infection: virus-induced immune disease, in Cold Spring Harbor Symp. Quant. Biol. *27*:475, 1962.

Jerne, N. K., A. A. Nordin, and C. Henry: The agar plaque technic for recognizing antibody-producing cells, in Cell-Bound Antibodies (B. Amos, ed.). Philadelphia: Wistar Inst. Press, p. 109, 1963.

Kalmutz, S. E.: Antibody production in the opossum embryo. Nature *193*:851, 1962.

Kraner, K. L.: Intrauterine fetal surgery. Advances Vet. Sci. *10*:1, 1966.

LaVia, M. F., D. T. Rowlands, and M. Block: Antibody formation in embryos. Science *140*:1219, 1963.

Miller, J. F. A. P.: The thymus and the development of immunologic responsiveness. Science *144*:1544, 1964.

Miller, J. F. A. P., and J. S. Davies: Embryological development of the immune mechanism. Ann. Rev. Med. *15*:23, 1964.

Osborn, J. J., J. Dancis, and J. F. Julia: Studies on the immunology of the newborn infant. I. Age and antibody production. Pediatrics *9*:736, 1952.

Osoba, D.: Immune reactivity in mice thymectomized soon after birth: normal response after pregnancy. Science *147*:298, 1965.

Perkins, F. T., R. Yetts, and W. Gaisford: Poliomyelitis immunization in infants in the presence of maternally transmitted antibody. Brit. Med. J. *5223*:404, 1961.

Rowlands, D. T., M. F. LaVia, and M. H. Block: Studies of the blood and blood-forming tissues of the newborn opossum. II. J. Immunol. *93*:157, 1964.

Schinckel, P. G., and K. A. Ferguson: Skin transplantation in the fetal lamb. Austral. J. Exp. Biol. Med.-Sci. *6*:533, 1953.

Silverstein, A. M.: Congenital syphilis and the timing of immunogensis in the human fetus. Nature *194*:196, 1962.

——: Ontogeny of the immune response. Science *144*:1423, 1964.

Silverstein, A. M., and K. L. Kraner: Studies on the ontogenesis of the immune response, in Mechanisms of Antibody Formation (J. Šterzl, ed.). Prague: Czechoslovak Academy of Sciences, 1965a.

——, ——: On the role of circulating antibody in the rejection of homografts. Transplantation *3*:535, 1965b.

Silverstein, A. M., and R. J. Lukes: Fetal response to antigenic stimulus I. Plasmacellular and lymphoid reactions in the human fetus to intrauterine infection. Lab. Invest. *11*:918, 1962.

Silverstein, A. M., J. W. Uhr, K. L. Kraner, and R. J. Lukes: Fetal response to antigenic stimulus. II. Antibody production by the fetal lamb. J. Exp. Med. *117*:799, 1963a.

Silverstein, A. M., G. J. Thorbecke, K. L. Kraner, and R. J. Lukes: Fetal response to antigenic stimulus. III. Gamma globulin production by the fetal lamb. J. Immunol. *91*:384, 1963b.

Silverstein, A. M., R. A. Prendergast, and K. L. Kraner: Fetal response to antigenic stimulus. IV. The rejection of skin homografts by the fetal lamb. J. Exp. Med. *119*:955, 1964.

Silverstein, A. M., C. J. Parshall, Jr., and R. A. Prendergast: The effect of intrauterine thymectomy on immunologic responses of the fetal lamb. 1966a (*in preparation*).

Silverstein, A. M., C. J. Parshall, Jr., and J. W. Uhr: Immunologic maturation *in utero:* kinetics of the primary antibody response in the fetal lamb. 1966b (*in preparation*).

Silverstein, A. M., R. A. Prendergast, and C. J. Parshall, Jr.: Cellular kinetics of the primary antibody response in the fetal Rhesus monkey. 1966c (*in preparation*).

Smith, R. T.: Immunological tolerance of nonliving antigens. Advan. Immunol. *1*:67, 1961.

Šterzl, J., and A. M. Silverstein: Developmental aspects of immunity. Advan. Immunol. *6*:1967 (*in press*).

Stowens, D.: Pediatric Pathology. Baltimore: Williams and Wilkins, 1959.

Uhr, J. W.: Development of delayed hypersensitivity in guinea pig embryos. Nature *187*:957, 1960.

————: The heterogeneity of the immune response. Science *145*:457, 1964.

Uhr, J. W., and J. B. Baumann: Antibody formation. I. The suppression of antibody formation by passively administered antibody. J. Exp. Med. *113*:935, 1961.

IMMUNOLOGICAL INTERACTIONS
BETWEEN MOTHER AND FETUS [1]

M. GALTON [2]

*Department of Pathology, Dartmouth Medical School,
Hanover, New Hampshire*

Development of the capacity to bear live young constitutes one of the
most significant events in the evolution of mammals, for it permits the
fetus to acquire complex organization, particularly of the brain, in a
more favorable environment than is afforded by other forms of reproduction. The successful introduction of viviparity necessitated its integration
with previously existing specializations, the most important of which for
the present discussion are immunological competence of the individual
and the diversity of antigenic protein variants within species. Since the
fetus inherits paternal allotypes (*see glossary on page 431 for definitions
of certain terms*), and fails to acquire all maternal genes, both fetus and
mother possess antigens foreign to the other. Although immunological
conflict between mother and fetus is commonly avoided, pregnancy
may be complicated by immunological interactions between mother and
fetus, with consequences ranging in severity from relatively trivial manifestations revealed only by sensitive laboratory tests, to catastrophic
results for one or even both partners.

Immunological phenomena, other than those attributable to direct
interaction between mother and fetus, also occur in pregnancy and include
changes in the overall maternal immunological economy and the effects
on the developing organism of pre-existing maternal immunity initiated
by factors unrelated to gestation.

In this review I propose to consider the factors responsible for the
immunologically privileged position of the fetus and then to outline the
types of altered immunological reactivity peculiar to pregnancy. Certain
aspects, such as hemolytic disease of the newborn infant due to maternal

---

[1] Supported by USPHS research grant HD 00544.
[2] John and Mary R. Markle Scholar in Academic Medicine.

Rh or ABO isoimmunity, are so well known that they will receive only passing mention. The immunological implications of viviparity, especially when the conceptus is considered in the capacity of a foreign graft, have been the subject of much recent discussion (Medawar, 1953; Woodruff, 1958; Boyd, 1959; Hašek *et al.*, 1962; Simmons and Russell, 1963; Billingham, 1964; Breyere, 1965; Dancis, 1965; Douglas, 1965; Lanman, 1965; Brent, 1966; Lloyd and Weisz, 1966; Papiernik-Berkhauer, 1966).

PREIMPLANTATION IMMUNOLOGICAL PHENOMENA

The results of extensive attempts to influence fertility by immunizing the female against spermatozoa have been summarized by Tyler (1961), who concluded that there is no reliable method for the immunological control of fertility. Incompatibility of the ABO(H) blood groups has been implicated in a relatively small proportion of cases of partial infertility in man (see Tyler, 1961). Since spermatozoa carry a variety of antigens (Barth and Russell, 1964; Edwards *et al.*, 1964) it is perhaps surprising that sensitization of the female is not more common. The nature of the antibody and its site of action remain to be elucidated. The suggestion that maternal immunity to male antigen may be harmful to subsequent male fetuses and thus influence the human sex ratio (Renkonen *et al.*, 1962; Renkonen, 1963) is open to serious question (Edwards, 1963). It is impossible to alter the sex ratio in mice by immunizing the female against the transplantation antigen determined by the Y-chromosome (McLaren, 1962; Eichwald *et al.*, 1964). In cattle, serum β-globulin polymorphism affects fertility, matings between homozygotes being significantly more fertile than matings involving heterozygotes (Ashton and Fallon, 1962). There is no evidence, however, that immunological mechanisms are involved in this phenomenon, and the fact that protein polymorphisms affect fertility in the domestic fowl (Morton *et al.*, 1965) suggests that immunological interactions might not be operative.

The fertilized egg in many species is amazingly robust and withstands surgical transplantation, even to xenogenic hosts. The failure of $1\frac{1}{2}$–$2\frac{1}{2}$ day fertilized ova from C3H/HeJ strain mice to undergo trophoblastic proliferation in specifically presensitized C57BL/6 strain recipients provides evidence that transplantation antigens are present in early trophoblast or even in the pre-trophoblastic tissues of ectopically implanted mouse ova (Simmons and Russell, 1965). Since fertilized mouse ova normally acquire maternal serum proteins during oviductal passage (Glass, 1963) they are presumably equally as vulnerable as spermatozoa to humoral immunity.

Implantation

At the time of attachment of the rabbit blastocyst to the uterus, the trophoblast is syncytial and the uterine epithelium is transformed into a symplasma. Fusion results in the co-existence of fetal and maternal nuclei in a common cytoplasm (Larsen, 1961). In contrast, fusion between fetal and maternal cells is not apparent in the implanting armadillo blastocyst (Enders, 1963). Cell hybridization can be artificially induced in tissue culture (Ephrussi and Sorieul, 1962; Harris *et al.,* 1966). Control of the metabolic regulatory mechanisms of the resulting *heterokaryon* is essentially unilateral, with dominance of the active cell (Harris *et al.,* 1966). Cell hybridization at implantation, especially if it were to occur generally among mammals, may have important implications for the function and antigenicity of the developing syncytiotrophoblast.

The phenomenon of "syngeneic preference" (the terms allogeneic inhibition, F_1 hybrid effect, and hybrid resistance are roughly synonymous), detected as the better growth of transplanted cells in isogeneic animals than in various semi-isologous F_1 hybrids (Hellström and Möller, 1965; Möller and Möller, 1965, 1966), might play a role in implantation. The suppression by cortisone of allogeneic inhibition (Hellström *et al.,* 1965) might also be pertinent. Antigenic dissimilarity between mother and fetus influences placental size: the placenta is larger in heterozygous than in homozygous pregnancies (Billington, 1964, 1965). At least part of the increased size of the allogeneic placenta is due to a specific maternal immune response, as shown by a reduction in placental size when the mother is specifically tolerant of the foreign fetal antigens (James, 1965). The increased placental size on crossing might also reflect the inherently greater growth potential of F_1 tissue since the hybrid fetuses are larger than homozygous fetuses (McCarthy, 1965).

Role of Placenta

The usual absence of maternal allograft reactivity directed against the fetus is generally attributed to a barrier function of the placenta (Medawar, 1953; Woodruff, 1958; Billingham, 1964). Lack of expression of histocompatibility antigens by the fetus, and maternal immunological inertia may play contributory roles (Medawar, 1953).

Although the placenta exhibits a greater diversity of structure and function than any other mammalian organ, it possesses one function common to all species: the separation of the maternal circulation from the fetal circulation. By acting as a mechanical barrier to the exchange of formed elements of the blood between mother and fetus, the placenta contributes to the prevention of maternal immunization by fetal antigens, and it also prevents maternal immunologically competent cells from entering and damaging the fetus.

Mouse trophoblast is relatively deficient in transplantation antigens. Trophoblastic growths from transplanted eggs are only weakly if at all antigenic to allogeneic (Kirby, 1960, 1963a, 1963b; Simmons and Russell, 1962, 1963) or even xenogeneic recipients (Kirby, 1962) and totally lack isoantigenic activity necessary for absorption of isohemagglutinins (Schlesinger, 1964). Transplantation studies of placenta or cell suspensions prepared from whole placenta have shown that paternal isoantigens are represented in the F_1 hybrid placenta (Simmons and Russell, 1962; Tyan and Cole, 1962; Uhr and Anderson, 1962; Hašková, 1963; van der Werf, 1963; Schlesinger, 1964) except in some strain combinations (Hašková, 1961, 1962, 1963). Human placental villi are deficient in blood group substances (Oettingen and Witebsky, 1928; Witebsky and Reich, 1932; Montemagno and di Stefano, 1964; Thiede *et al.*, 1965), in contrast to endothelia, epithelia and many organs of fetuses which are rich in A, B and H blood group antigens (Szulman, 1964). On the other hand, the human placenta is rich in leucocyte isoantigens (Bruning *et al.*, 1964, 1965).

A structural feature common to the hemochorial placentas of several species is the continuous lining layer of syncytiotrophoblast forming the boundary between fetal and maternal tissues (Enders, 1965b; Wynn and Davies, 1965). The syncytiotrophoblast displays no intrinsic proliferative activity. In man and the rhesus monkey it is derived from the underlying germinal layer, the cytotrophoblast (Richart, 1961; Galton, 1962; Midgley *et al.*, 1963; Enders, 1965a; Tao and Hertig, 1965).

The limiting layer of trophoblast is cellular in the hemochorial placenta of the rabbit (Larsen, 1962), rat (Jollie, 1964) and mouse (Kirby and Bradbury, 1965), and the cells are approximated by tight junctions (Enders, 1965b). Attention has been focused on the syncytiotrophoblast largely because of its unique structure and unique location. It will be interesting to ascertain whether syncytial transformation confers an immunological advantage on trophoblast. However, the existence of hemochorial placentas with cellular limiting trophoblast would tend to discount the immunological value of this type of adaptation. Transplantation antigens determined at the H-2 locus in the mouse are closely associated with the cellular surface membranes and with the membranes of the endoplasmic reticulum (Herberman and Stetson, 1965). Should the syncytiotrophoblast be shown to lack isoantigenic activity it will be interesting to learn how this is accomplished in view of its well developed endoplasmic reticulum (Enders, 1965b; Wynn and Davies, 1965).

It has been suggested that the syncytial nature of the invading trophoblast constitutes the fundamental difference between implantation and invasion of tissues by cancer, since the latter is cellular (Böving,

1959). The cellular nature of invasive mouse trophoblast was recently established by electronmicroscopy (Kirby and Malhotra, 1964), and it is consequently difficult to assess the general significance of syncytial transformation, either during implantation or in the definitive placenta. It is also uncertain whether decidua possesses special immunological or antigenic properties, a possibility raised by the finding of decidual resistance to invasion by certain transplanted tumors (Wilson, 1963).

The presence of fibrinoid in the placenta of many species has prompted the suggestion that this material might behave as an inert immunological barrier between fetus and mother (Bardawil and Toy, 1959; Kirby *et al.*, 1964; Bradbury *et al.*, 1965). In the mouse, the fibrinoid layer is conspicuously thicker in the placenta of the interstrain hybrid fetus and this increase is due not to heterozygosity but to the genetic dissimilarity of mother and fetus (Kirby *et al.*, 1964). Kirby *et al.* (1964) postulate that removal of the investing fibrinoid from the mouse placenta might reveal trophoblast capable of sensitizing allogeneic hosts. Billingham and Silvers (1962a) attribute the immunologically privileged position of the hamster cheek pouch as a site for transplants to the connective tissue underlying the pouch mucosa which prevents the passage of antigen to the host, and they suggest that a similar mechanism might be responsible for the immunological barrier properties of the placenta. It is therefore of interest that the intercellular matrix of the hamster cheek pouch skin is histochemically similar to placental fibrinoid (Kirby *et al.*, 1964; Bradbury *et al.*, 1965).

The placenta of the rat and mouse survives isogeneic transplantation for only a strictly limited time which suggests that it possesses an intrinsic life span little longer than the normal gestation period (Lichton, 1965; Simmons and Weintraub, 1965). Although the transplanted chorio-allantoic placenta is incapable of continued growth, the grafted yolk-sac membrane frequently undergoes diverse differentiation (Payne and Payne, 1961). Allogeneic grafts of placental fragments in man are invariably rejected, even by recipients with advanced cancer, unless transplantation is followed by intensive estrogenic stimulation in which case there is often considerable trophoblastic proliferation (Lajos *et al.*, 1964). These authors summarize previous studies of placental grafts. Simmons and Russell (1964b) present a useful bibliography of the literature on transplantation of trophoblast and definitive placenta.

The diaplacental passage of both cells and transplantation antigens is probably not an uncommon occurrence in normal pregnancy and hence perhaps the single most important immunological function of the placenta, apart from its barrier activity, is to withstand immunological attack should the mother become sensitized. Pregnancy in the mouse, rat or rabbit is not affected by specific maternal presensitization to the foreign transplantation antigens of the fetus (Mitchison, 1953; Woodruff,

1958; Lanman *et al.*, 1962, 1964; Lanman and Herod, 1965). Furthermore, even repeated heterospecific pregnancies fail to weaken a prevailing state of specific transplantation immunity, thereby indicating that the placenta cannot absorb specific immune agents from the maternal circulation (Heslop *et al.*, 1954; Medawar and Sparrow, 1956; Woodruff, 1958).

Deportation of trophoblast occurs in both normal and abnormal human pregnancy (Schmorl, 1893, 1905; Park, 1958; Bardawil and Toy, 1959; Douglas *et al.*, 1959; Thomas *et al.*, 1959; Attwood and Park, 1961; Iklé, 1961, 1964; Toy and Tedeschi, 1961; Wagner *et al.*, 1964; Jäämeri *et al.*, 1965; Hamilton and Boyd, 1966) but has not yet been reported in other species. It has been suggested that the invasion of maternal blood by trophoblast serves to induce a state of specific immunological unresponsiveness by constantly exposing the mother to an excess of fetal antigen (Douglas *et al.*, 1959; Thomas *et al.*, 1959). Salvaggio *et al.* (1960) detected syncytiotrophoblast and decidua-like cells in cord blood, but Iklé (1961) was unable to find any trophoblastic material in cord blood. Hamilton and Boyd (1966) describe "stromal trophoblastic buds," ectopic trophoblastic material in the villous connective-tissue cores of the human placenta, and tentatively suggest such buds might be capable of entering fetal vessels. The immunological significance of trophoblast deportation in either direction awaits clarification.

ANTIGENICITY OF FETUS

The degree of maturation of fetal antigenicity depends to some extent on the duration of gestation. Medawar (1953) has drawn attention to the possible advantage of fetal antigenic immaturity in species with constantly recurring pregnancies. Nevertheless, even the late development of antigens need confer no exemption from harmful maternal isoimmunity since "the immunological sins of one foetus may be expiated by the next" (Medawar, 1953).

Blood group antigens are present in the tissues of human embryos at 6-weeks ovulation age (Szulman, 1964). The development of immunological competence by both man and sheep relatively early in fetal life (Silverstein, 1964; van Furth *et al.*, 1965; Miller, 1966) indicates that all transplantation and other generally accessible antigens must already be represented; otherwise autoimmune reactions might follow the appearance of each new antigen (Triplett, 1962). Even in species with short gestation periods there is often considerable development of fetal antigenicity. In the mouse, transplantation antigens have been demonstrated from early stages of development by various methods (Woodruff, 1958; Hašková, 1959, 1961; Hašek *et al.*, 1962; Simmons and Russell, 1962, 1965; Tyan and Cole, 1962; Doria, 1963; Möller, 1963; Edidin, 1964a, b; Schlesinger, 1964, 1965), although other methods have sometimes failed to detect the presence of isoantigens in fetal tissues (Mitchison, 1953; Möller, 1961; Pizarro *et al.*, 1961). Potentially immunologically reactive cells have been demonstrated in fetal mouse thymus, spleen, liver and placenta (Dancis *et al.*, 1962; Douglas *et al.*, 1962; Tyan and Cole, 1963, 1964; Dancis *et al.*, 1966; but see Simmons and Russell, 1964a), and it is therefore not surprising that under suitable conditions it is possible to induce transplantation immunity in newborn mice (Howard and Michie, 1962; Howard *et al.*, 1962; Billingham and Silvers, 1962b; Brent and Gowland, 1963). The fetal rat can become sensitized to foreign transplantation antigens (Billingham *et al.*, 1965a) and therefore it presumably already possesses the full adult complement of transplantation antigens. The mouse fetal thymus exhibits the same level of H-2 isoantigenic activity as the adult organ, whereas other fetal tissues including the liver, spleen, heart, lung and kidney display only a fraction of the activity of the same organs in the adult (Schlesinger, 1964). This uneven distribution of isoantigenic activity might explain the apparent discrepancy between the possession of immunological competence by the fetus and the ability of various fetal tissues to withstand allogeneic and even xenogeneic transplantation. The relative success of such fetal or *brephoplastic* grafts has excited considerable attention (Toolan, 1958; Baxter *et al.*, 1958; Payne and Payne, 1961; Billingham, 1964; Billingham and Silvers, 1964; Willis and Hou, 1964). Billingham and Silvers (1964) showed that perinatal skin grafts were immunogenetically inferior to grafts of adult skin although they were equally susceptible to independently elicited states of sensitivity. Since tolerance may be induced in adult hosts by chronic intravenous exposure to transplantation antigens (Billingham and Silvers, 1963, 1964), vascularized solid tissue homografts in immunologically privileged sites occupy a particularly favorable position to induce tolerance. The success of brephoplastic grafts might therefore be due both to their reduced antigenicity and to the induction of actively acquired tolerance in the host.

Apart from fragments of placenta, the fetal cells most likely to enter the mother under ordinary circumstances are the formed elements of the blood, and these cells carry a variety of antigens: blood group substances on erythrocytes, leucocyte isoantigens, and transplantation antigens on leucocytes, particularly lymphocytes. Hence, although the fetus, especially in species with short gestation periods, may possess some relatively antigenically immature tissues, the fetal cells most likely to be encountered by the mother are those most likely to be antigenic.

Another point to be considered is the effectiveness of the route of immunization in transplantation immunity. For instance, in the rabbit, both leucocytes and epidermal cells fail to elicit transplantation immunity when injected intravenously although spleen cells are effective by this route. All routes of administration are equally effective in the mouse and guinea pig (Billingham *et al.*, 1957). In the ordinary course of pregnancy, peripatetic fetal and placental cells are most likely to gain entry into the maternal blood stream. In instances of fetal regression, however, resorption is partially mediated by phagocytic leucocytes which congregate in lymphoid tissue situated, in the rabbit, antimesometrially within the serous coat of the uterus. This would seem to constitute an effective route for maternal sensitization to foreign antigens of the fetus (Henderson, 1954).

MATERNAL IMMUNOLOGICAL REACTIVITY

Pregnancy lowers natural resistance to certain infections in man, possibly at least partially due to a decrease in the serum properdin concentration (Homer and McNall, 1961). Loss of maternal antibody to the fetus (Gitlin *et al.*, 1964) might be a contributory factor (*cf.* Campbell, 1965). Adrenocorticosteroid hormones are notorious for their suppression of antibody formation (Kass and Finland, 1953). In pregnancy, however, although the plasma concentration of these hormones is increased, the binding capacity of the alpha-globulin responsible for their transport is also raised, with the net result that blood levels of active hormone are supposedly not dissimilar from non-pregnant states (Steinbeck, 1963; Graves and Agersborg, 1964). Estrogen activity, on the other hand, rises steadily throughout pregnancy (Brown, 1963), and this would be expected to increase the level of plasma unbound cortisol, without, however, changing the proportion of unbound to total cortisol in the plasma or inducing hyperadrenocorticism (Plager *et al.*, 1964). The administration of cortisone prolongs allograft survival in mice but fails to weaken pre-existing immunity (Medawar and Sparrow, 1956). On the other hand, immunological responsiveness is increased by estrogens, primarily as the result of enhanced phagocytic activity (White, 1963; Nicol *et al.*, 1964).

Temporary but pronounced thymic atrophy occurs during pregnancy in the mouse, and lymph nodes undergo transient and less marked weight loss after parturition (Pepper, 1961). The pelvic lymph nodes in pregnant women lack discrete germinal centers but the significance of this change is not clear (Nelson and Hall, 1964, 1965). A mild degree of immunological inertia to allografts during pregnancy occurs in man (Andresen and Monroe, 1962; Bardawil *et al.*, 1962), rabbit (Heslop *et al.*, 1954), and rat (Anderson, 1965), but the delayed graft vascularization peculiar to pregnancy (Medawar and Sparrow, 1956) may be at least partially responsible.

Unlike the anterior chamber of the eye, the brain, and the hamster cheek pouch, the rodent uterus is not a privileged site for allogeneic grafts. Transplants to the uterus sensitize the host and are fully susceptible to a state of immunity, even during continuing heterospecific pregnancy when the fetuses and grafts share antigens foreign to the mother (Schlesinger, 1962). The ability of the fetus to develop in extrauterine sites such as the peritoneal cavity also indicates that the uterus is not essential as an immunological shield for the maintenance of pregnancy.

The maternal production of sufficient humoral antibody is necessary both to maintain her own immunological defenses and to equip the effectively unprotected fetus. This dual function may dictate qualitative as well as merely quantitative changes of immunological reactivity. It is becoming increasingly apparent that the heterogeneity of immune response serves a useful purpose (Uhr, 1964). The fetal demand for antibody and the special limitations imposed by the transmission parameters of the placenta may be responsible for appropriate adaptations in the quality of antibody production during pregnancy. It may also be surmised that maternal antibody production is different from that in the non-pregnant state on the basis of the endocrinological changes and in view of the alterations of other humoral proteins, particularly those concerned with blood coagulation (Phillips, 1963) and with hormone transport (Dowling *et al.*, 1956; Slaunwhite and Sandberg, 1959).

Since the phenomenon of "Leihimmunität" was first studied by Ehrlich in 1892, the paramount importance of "borrowed immunity" for the newborn infant has become increasingly recognized. Attention has been directed primarily to the various pathways of transmission of maternal antibodies to the fetus (see Pfaundler, 1907–1908; Brambell, 1958; Hemmings and Brambell, 1961; Freda, 1962; de Muralt, 1962; Berger and Novick, 1964; Dancis, 1965). The route is exclusively transplacental in man although antibodies in the colostrum, while not absorbed by the fetal intestine, are involved in combating infections initiated within the intestinal tract. At the opposite end of the spectrum, antibody transfer in ungulates is accomplished entirely postnatally by the gastrointestinal absorption of antibodies in colostrum. Intermediate patterns include antibody transmission across the yolk-sac splanchnopleure and via the amniotic fluid and fetal gut. In general, the extent of placental antibody transfer is correlated with the degree of attenuation of the placental membrane interposed between the maternal and fetal circulations. However, in rodents, which possess hemochorial placentas, postnatal transmission is important. Both in rabbit (Brambell *et al.*, 1960) and man (Gitlin *et al.*, 1964), the fetus concentrates F_c fragment (Fragment III) of antibody to a greater extent than other portions of the antibody molecule when these are administered separately to the mother. This suggests that despite the different placental structures involved (yolk-sac splanchnopleure in the rabbit and chorio-allantoic membrane in man) a common mechanism of active transport exists in both species based on the F_c fragment, or nonantibody but antigenic portion of the IgG (7 S γ_2-

globulin) antibody molecule, which has characteristics unique to this class of antibody. Proteins with lower molecular weights than IgG (156,000), such as transferrin (90,000), albumin (65,000), and orosomucoid (44,000), cross the human placenta much less readily than IgG (Gitlin *et al.*, 1964). Macroglobulin antibody (IgM or 19 S γ_1-globulin) is virtually incapable of traversing the human placenta (Gitlin *et al.*, 1964) but it is transmitted readily from mother to fetus in the rabbit (Hemmings and Jones, 1962).

The species diversity of antibody transmission characteristics of the placenta influences perinatal immunological behavior. For instance, in equine pregnancy, particularly after cross breeding horses and donkeys, the mare may become sensitized to foreign red cell antigens of the fetus. Harm to the offspring, however, only occurs postnatally and follows the gastrointestinal absorption of antibody in the colostrum, and the same holds for the pig (see Roberts, 1957). Erythroblastosis fetalis is producible experimentally in the rabbit by immunizing the doe to blood group antigens of the fetus (see Reif *et al.*, 1964). Since both maternal IgG and IgM are transmitted to the fetus, the responsible antibody presumably belongs to either class. In man, only those maternal isohemagglutinins which belong to the IgG antibody class are potentially harmful to the fetus (Kochwa *et al.*, 1961; Freda and Carter, 1962; Freda, 1962; Fong *et al.*, 1966). The only other primate so far reported possibly to develop erythroblastosis fetalis, a tamarin, displayed continuing postnatal hemolysis suggestive of antibody transmission in colostrum in addition to placental transfer (Gengozian *et al.*, 1966).

In man, although placental impermeability to macroglobulin antibody is beneficial to the fetus in that it excludes some potentially harmful isohemagglutinins, leucoagglutinins, autoantibodies, and antibodies of allergy, it also prevents the passage of other maternal antibodies which would be valuable to the fetus, particularly agglutinins to the gram-negative enteric bacilli, antibody to type 3 poliomyelitis and complement-fixing antibodies involved in immunity to tuberculosis and syphilis (see Berger and Novick, 1964). This explains the susceptibility, for instance, of the newborn infant to infection with such bacteria as *E. coli* (Gitlin *et al.*, 1963).

Pregnancy exerts a profound effect on some autoimmune diseases. The high frequency of remission of rheumatoid arthritis in pregnancy (Hench, 1938) led to the discovery of the therapeutic value of cortisone (Hench *et al.*, 1949), although in view of the basically unaltered level of circulating active hormone in pregnancy (Steinbeck, 1963) the involvement of immunological mechanisms cannot be entirely discounted (Nelson, 1965). The beneficial effect of pregnancy on other autoimmune diseases is well known but in most cases the cause of the improvement is a subject of speculation (Scott, 1966). An altered quality of maternal immunological reactivity due either to a basic change in the mechanism of immune response or to selective removal of maternal antibody by the fetus might conceivably be implicated.

In maternal cases of idiopathic thrombocytopenic purpura, myasthenia gravis, and thyrotoxicosis, associated with IgG autoantibodies or other agents such as long-acting thyroid stimulator in thyrotoxicosis (Adams, 1965), the newborn infant is liable to display a transient form of the disease as the result of passage of the autoantibody to the fetus (Scott, 1966), and in the case of maternal idiopathic thrombocytopenic purpura fetal intracranial hemorrhage is not uncommon (Heys, 1966). Transient neonatal diabetes might be caused in a similar fashion (Scott, 1966). The distinctive serological factors associated with other diseases of immunological aberration may cross the placenta; however, the effects on the fetus are usually minimal and transient and hence there is no indication that such autoantibodies play a pathogenetic role in the development of these diseases (Beck and Rowell, 1963).

Specific Immunological Interactions Between Mother and Fetus

Modifications of the maternal immune response by fetal antigens may, on the one hand, require diligent search for their disclosure while, on the other hand, they may lead to dire consequences for both fetus and mother. Maternal antigens are capable of altering the immunological status of the fetus but the manifestations of such interaction are unlikely to be severe.

Rhesus isoimmunization results in hemolytic disease in 1 in 200 of newborn infants. The fetal bleeds into the maternal circulation which are presumably responsible for most cases of sensitization occur during delivery (McConnell, 1966), and this explains the rarity of the disease among first born. The affected child thus pays the penalty for the loss of integrity of an older sibling's placenta. Levine (1943, 1958) observed that maternal sensitization to the Rh antigen D is unlikely to occur when there is ABO incompatibility, because the fetal erythrocytes are removed from the maternal circulation before they can provoke an immune response. Groups in Liverpool and New York have developed highly effective preventive measures consisting of the administration within 36 hours of delivery of anti-D gamma globulin which similarly removes fetal red cells from the maternal circulation (McConnell, 1966; Gorman et al., 1966). There is no evidence to support the hypothesis that Rh negative children of Rh positive mothers develop permanent tolerance of the Rh antigen D, as was once thought (Booth et al., 1953; Owen et al., 1954; Solomon and Yokoyama, 1961).

Leucoagglutinins are present in about 20 per cent of multiparous women (Payne and Rolfs, 1958; van Rood et al., 1958, 1959; Jensen, 1962; Payne, 1962, 1964). Fetal leucocytes, which are present in the peripheral blood from very early in gestation (Playfair et al., 1963), may enter the maternal circulation, and the placenta is another source of leucocyte isoantigens (Bruning et al., 1964, 1965). Payne (1964) was unable to demonstrate the existence of either an immunoneutropenia in 39, or depressed levels of circulating platelets in 10, newborn infants whose mothers possessed leucoagglutinins, even though transplacental passage of leucoagglutinins had occurred in 15 of the newborns. Halvorsen (1965) reports two cases of neonatal leucopenia due to fetomaternal leucocyte incompatibility. Several previously reported cases are discussed by Payne (1964) who questions whether they are strictly comparable to her own series. In view of the profound immunosuppressive effect of heterologous anti-lymphocyte serum in mice (Monaco et al., 1966), maternal isoimmunity might be anticipated occasionally to depress fetal immunological reactivity in a similar fashion.

With the development of serological techniques for the detection of antibodies to platelet antigens, a few cases of neonatal purpura due to maternal isoantibody to fetal platelet antigens were recognized (Shulman et al., 1962). Nevertheless, neonatal thrombocytopenia due to maternal isoantibody, like neonatal immunoneutropenia (Payne, 1964), appears to be a relatively uncommon condition in man, although, as mentioned above, newborn infants of mothers with idiopathic thrombocytopenic purpura may develop a transient form of the disease (Scott, 1966). In some instances of this autoimmune type of neonatal thrombocytopenic purpura the possibility exists of an additional isoimmune component (Scott, 1966). Thrombocytopenia due to maternal isoimmunization has been described in piglets. Purpura and bleeding begin five days after birth since, presumably, isoantibody reaches the young only in the colostrum (Stormorken et al., 1963; Nordstoga, 1965).

The infant relies on its "borrowed immunity" for several weeks after birth. A combination of factors, including an element of immunological immaturity and the presence of circulating antibody, prevents the young infant from producing significant quantities of antibody, and it is not until from one to three months that the rate of antibody synthesis by the infant begins to equal the rate of degradation of maternally derived antibody, which has a half life of approximately 28 days (Gitlin, 1966). During this transition between the decay of passively acquired antibody and the onset of active immunity the total gamma-globulin concentration falls to between 300 and 600 mg/100 ml plasma. In a small number of infants, the onset of active immunity is delayed until the age of four to six months and in exceptional cases until one to three years (Fudenberg and Hirschhorn, 1965; Gitlin, 1966). An immunological etiology has been proposed (Fudenberg and Hirschhorn, 1965). The fetus is capable of synthesizing a very small amount of IgG antibody (Fudenberg and Fudenberg, 1964; Mårtensson and Fudenberg, 1965). Since the immunoglobulins carry genetically determined antigens (see Fudenberg, 1965), there is a possibility that the mother might be sensitized either by foreign fetal Gm factors on the heavy polypeptide chains specific to IgG or by Inv factors on the light chains common to all immunoglobulin classes. Although fetal IgG may cross the placenta, albeit not as readily as IgG passes in the opposite direction (Gitlin *et al.*, 1964), the resulting maternal antibody usually belongs to the IgM class and is hence incapable of crossing the placenta. Occasionally, however, the maternal isoantibody is present in the IgG component and therefore might gain entry to the fetus (Fudenberg *et al.*, 1964; Fudenberg and Fudenberg, 1964; see Fudenberg and Hirschhorn, 1965). Unfortunately, evidence is presently lacking to explain the mechanism of damage by maternal isoantibody to fetal cells responsible for IgG synthesis (Gitlin, 1966). In rabbits, maternal isoantibodies to the paternal IgG immunoglobulin allotype severely impair the ability of the young heterozygote to develop immunoglobulins of the paternal allotype. However, a compensatory increase of the IgG allotype controlled by the allelic gene leads to apparently normal total IgG levels (Mage and Dray, 1965).

Allogeneic recipients of immunologically competent cells may develop a severe and often fatal graft-versus-host reaction, especially when the host is for one reason or another incapable of eliminating the attacking cells. In the neonatal mouse, the condition is known as runt disease and it is characterized by weight loss, dermatitis, eventual lymphoid atrophy, diarrhea, cachexia and death (Billingham and Brent, 1959). Billingham and Brent (1959) predicted similar consequences might follow the incorporation of maternal lymphocytes into the fetus. It has been shown that the placenta may allow maternal leucocytes to enter the fetus both in man (Desai and Creger, 1963) and in rabbit (Oehme *et al.*, 1966). In contrast, the mouse placenta is highly resistant to maternal-fetal passage even of erythryocytes (Finegold and Michie, 1961), although maternal erythrocytes may enter the human fetus (Zarou *et al.*, 1964; but see Donovan and Lund, 1966). There have been several reports of either proven or tentative instances of maternal-fetal lymphoid cell chimerism in human infants (Bain and Scott, 1965—see Scott, 1966; Kadowaki *et al.*, 1965; Lee *et al.*, 1965; Taylor and Polani, 1965; Turner *et al.*, 1966). The apparent rarity of the condition might reflect the capacity of the fetus to reject grafts of maternal tissue, and it is therefore interesting that maternal lymphoid cell chimerism has been reported in a child with thymic alymphoplasia, who as a fetus was presumably unable to reject the invading maternal cells (Kadowaki *et al.*, 1965).

Maternal lymphocytes have been tentatively implicated in the etiology of Hodgkin's disease (Green *et al.*, 1960) and systemic lupus erythematosus

(Dameshek, 1960) but as yet there is no direct evidence in support of these ingenious hypotheses.

Genetically determined protein polymorphism is a potential source of immunological interactions between mother and fetus. Dürwald *et al.* (1965) document the existence of precipitating antibodies with specificity directed against the Ag(a+) antigenic determinants of beta-lipoprotein in two of 157 multiparous women. Other well-defined allotypic systems, such as transferrin polymorphism (Seppälä, 1965), might be less likely to be involved in immunological interactions if, as is the case with transferrin in man, the placenta is relatively impermeable to the antigen (Gitlin *et al.*, 1964).

When BALB/c strain female mice, sensitized to C57BL/6 IgG immunoglobulin, were mated with C57BL/6 males there was a high incidence of both fetal and maternal deaths. In the reciprocal situation, no live progeny resulted but no maternal deaths occurred (Lieberman and Dray, 1964). Maternal death may have followed fetal death, but the possibility that maternal pathology occurred first cannot be excluded (Lieberman and Dray, 1964), especially in view of the known harmful effects of antigen-antibody reactions (Dixon, 1962–1963; Lee, 1963). Certainly in man, maternal damage may follow maternal-fetal immunological interaction. The association of pre-eclamptic toxemia with Rh isoimmunization provides a good example (Scott, 1966). An immunological etiology of other types of pre-eclamptic toxemia and eclampsia has been proposed many times (*e.g.*, Obata, 1919; Kalmus, 1946; Penrose, 1946; Seegal and Loeb, 1946; Lin, 1947; Hulka and Brinton, 1963; Mühe and Bünte, 1965); but although the causes of these conditions are presently only poorly understood, there is no convincing evidence of immunological components (*e.g.*, Prall and Kantor, 1966). Hulka *et al.* (1961, 1963) showed that the sera of post-partum women contain a globulin with an affinity for syncytiotrophoblast, but the significance of this finding awaits clarification.

The existence of antigenic relationships between placenta and other organs, particularly kidney, has excited considerable attention. Antisera to placenta prepared in heterologous species may produce abortion, nephritis and pulmonary lesions, and are known to be cytotoxic (Seegal *et al.*, 1955; Pressman and Korngold, 1957; Olivelli, 1958; Oliveira and Laus-Filho, 1962; Steblay, 1962; Boss, 1965). It has not been possible to demonstrate circulating antibodies against placental basement membrane or cytoplasmic antigens in the sera of normal gravidas or toxemic patients (Boss, 1965) although the occurrence of antibody to placental polysaccharide in the sera of toxemic women has been reported (Kaku, 1953).

Fetal and placental substances foreign to the mother have been identified in maternal sera by immunological techniques (Olivelli and Ruggieri,

1958; MacLaren *et al.,* 1959). Several fetal and placental materials which could be potentially antigenic to the mother have been characterized in maternal sera, among them placental lactogen (chorionic growth hormone-prolactin) (Josimovich and Atwood, 1964; Kaplan and Grumbach, 1965; Tallberg *et al.,* 1965) and alkaline phosphatase (Beckman *et al.,* 1966a; Birkett *et al.,* 1966). However, not all pregnancy enzymes in maternal sera are derived from the fetus (Beckman *et al.,* 1966b).

A tendency seems to exist for choriocarcinoma to arise in compatible matings, as shown by the high incidence of ABO compatibility (Scott, 1962; but see Schmidt and Hertz, 1961) and the increased likelihood of the disease among inbreeding communities (Azar, 1962). Although some women with metastatic choriocarcinoma may possess agglutinins against leucocytes and platelets from the husband (Mathé *et al.,* 1964) not all patients contain leucoagglutinins in their sera (Robinson *et al.,* 1963). Specific tolerance of skin grafts from the husband has been described (Robinson *et al.,* 1963). Immunotherapy, either by direct sensitization of the patient with tissues from the husband, or with heterologous antisera, has yielded some promising results (Doniach *et al.,* 1958; Cinader *et al.,* 1961; Hackett and Beech, 1961; Robinson *et al.,* 1963; Mathé *et al.,* 1964). Doniach *et al.* (1958), and Cinader *et al.* (1961), suggest that the cause for the better results of chemotherapy of choriocarcinoma in females, compared with male patients, may be the addition of an immune response to tumor antigens liberated by the treatment.

Good and Zak (1956) describe the case of an agammaglobulinemic woman who was able to produce a small amount of antibody during pregnancy. Plasma cells and antibody were present in the placenta but not in the fetus. Recently, it has been shown that pregnancy restores immunological competence of neonatally thymectomized mice, presumably by passage into the mother of the humoral factor elaborated by the fetal thymus (Osoba, 1965). The possibility cannot be ignored that, rather than resulting from exceptional circumstances, these fetal contributions to maternal immunological economy might be representative of the normal situation.

Infection is relatively uncommon in the fetus (Blanc, 1961), and metastatic tumors from the mother are rare (McGowan, 1964), presumably due to the effectiveness of the placental barrier, to the existence of maternal immunity, and to the potential immunological reactivity of the fetus.

The offspring of guinea pigs fed simple chemical allergens during pregnancy had a slightly reduced capacity to become sensitized to the homologous allergen (Baer *et al.,* 1958). Similarly, tolerance to a protein was induced in the young of rabbits injected with the protein during pregnancy (Trench *et al.,* 1964). The mechanism of transfer of immunological tolerance from parents to offspring, which has been reported in only a few instances, may be due to a direct effect of antigen on the egg (Skowron-Cendrzak, 1961; Hort, 1962; Guttmann and Aust, 1963). As is the case with Rh antigen, the ability of infants to make anti-A agglutinins is uninfluenced by the blood group of the mother (Hraba *et al.,* 1962). In contrast, fetal haptoglobin synthesis might be influenced by the maternal genotype (Siniscalco *et al.,* 1963).

Congenital malformations may be produced experimentally by various tissue antisera and this subject was recently reviewed by Brent (1966). The possibility that an immunological reaction might trigger parturition is often raised. If this were the case, prolonged gestation would be the rule in all highly inbred, homozygous, animals; tolerance does not affect the duration of gestation (Husain and Ketchel, 1965).

The status of allograft sensitivity of either mother or fetus may be altered by encounter with the other's histocompatibility antigens during pregnancy. In mice, multiparity may lead to the appearance of isohemagglutinins against fetal H-2 antigens (Herzenberg and Gonzales, 1962; Goodlin and Herzenberg, 1964; Kaliss and Dagg, 1964), specific allograft sensitivity and enhancement (Kaliss and Dagg, 1964). On the other hand, repeated heterospecific pregnancies involving smaller histocompatibility differences may cause maternal tolerance (Breyere and Barrett, 1960a, b, 1961; Prehn, 1960; Breyere and Burhoe, 1963; Lengerová and Vojtíšková, 1962, 1963) due to complete and permanent tolerance of some but not necessarily all of the paternal transplantation antigens (Breyere and Burhoe, 1963). Tolerance of the Y transplantation antigen in multiparous mice is not associated with demonstrable cellular chimerism and therefore subcellular antigen chimerism is presumably involved (Billingham *et al.,* 1965b). The conceptus is the primary source of antigen, and neither semen nor parturition contribute significantly to the induction of tolerance by parity (Porter and Breyere, 1964). In man, too, both maternal tolerance and sensitivity have been reported (Peer *et al.,* 1958, 1963; Bardawil *et al.,* 1962; Marchant *et al.,* 1964). Humoral antibodies involved in allograft reactivity may cross into the fetus or suckling animal without apparent harmful effects (Halasz and Orloff, 1963; Kaliss *et al.,* 1963; Lanman and Herod, 1965). The transfer of actual allograft sensitivity to the fetus in rats has been attributed to a process of active immunization resulting from passage of antigen across the placenta (Stastny, 1965). Billingham *et al.* (1965a) have shown that the rat may become sensitized *in utero* to maternal transplantation antigens. Newborn rabbits may be tolerant of maternal skin grafts (Iványi and Démant, 1965) or they may display enhanced reactivity to allografts (Najarian and Dixon, 1962). A number of attempts have been made to modify allograft reactivity by damaging the placenta with such agents as hyaluronidase, histamine and X-irradiation. The results of these studies are compared with the findings after undisturbed pregnancy in Table 1.

Table 1

MOTHER–OFFSPRING ALLOGRAFT RELATIONSHIP

Species	Altered Immunological Status of: *		References
	Mother	Offspring	
Man	Tolerance	Tolerance	Peer et al., 1958; 1963.
	—	? Tolerance	Murray et al., 1965.
	Immunity (to husband)	—	Bardawil et al., 1962; Marchant et al., 1964.
	Leucoagglutinins, etc.	—	See text.
Mouse	Tolerance (multiparous)	—	Breyere and Barrett, 1960a, b, 1961; Prehn, 1960; Breyere and Burhoe, 1963; Lengerová and Vojtíšková, 1962, 1963; Porter and Breyere, 1964.
	0	—	Medawar and Sparrow, 1956.
	—	0	Billingham et al., 1954; Lustgraaf and Eichwald, 1959; Billingham and Silvers, 1960; Jones and Krohn, 1962; Moulton et al., 1960 (irrad.).
	Hemagglutinins	—	See text; Goodlin et al., 1964 (ces. section).
Rat	Tolerance	Tolerance	Rogers et al., 1960.
	—	Tolerance	Lengerová, 1957 (irrad.); Kečkeš and Allegretti, 1963 (irrad.).
	—	Immunity	Billingham et al., 1965a.
Rabbit	—	0	Billingham et al., 1956.
	0	0	Lanman et al., 1963.
	—	Tolerance	Nathan et al., 1960 (hyaluronidase); Iványi and Démant, 1965.
	Tolerance	Tolerance	Najarian and Dixon, 1963 (hyaluronidase or histamine).
Guinea Pig	—	Tolerance	Billingham et al., 1956.
	—	0	Billingham and Silvers, 1965.
Cattle	Immunity	0	Billingham et al., 1952; Billingham and Lampkin, 1957.
Sheep	0	0	Galton, 1965.

* 0: no specific change in immunological status; —: not studied.

Conclusions

Faced with the genetic diversity of species and the existence in mammals of a sophisticated immunological mechanism, pregnancy finds itself in a singular biological position and has therefore resorted to unique strategy. By interposing a physical barrier between mother and fetus, the placenta undertakes to allow free exchange of nutrients and metabolites while at the same time acting in a highly restrictive manner to substances which might be construed by either party as "immunological." The trophoblast has probably renounced the antigenic markers which are considered the heritage of most nucleated tissues. Since loss of antigenicity might mean escape from the type of surveillance envisaged by Burnet (1957) to be involved in the homeostasis of proliferating cells necessary for the prevention of neoplasia, the placenta may have forfeited the capacity to outlive the allotted period of intrauterine fetal life.

Maternal sensitivity may be first expressed in the offspring at any stage of development, from before fertilization to after birth. The affected fetus is not necessarily to blame for the state of maternal sensitivity since this may be due to the isohemagglutinins normally present in the blood or it may have arisen in the course of disease affecting the mother but quite unconnected with pregnancy. In many other cases, maternal immunization is by proxy and the affected fetus is paying the penalty for the loss of placental integrity during delivery of an older sibling. Maternal sensitivity to foreign fetal allotypes inherited from the sire forms the basis of most immunological interactions. Allograft or tissue transplantation immunity, however, is apparently unlikely to have serious consequences, although it is presently unknown whether it is involved in such conditions as abortion, toxemia, and eclampsia.

Although the placenta in all species functions primarily to provide optimum conditions for the development of the fetus, it nevertheless exhibits a remarkable diversity of characteristics. The yolk sac and the allantois make variable contributions to the definitive placenta, although placentation is often entirely chorio-allantoic in higher mammals. Even the chorio-allantoic placenta displays a more striking variation in morphology than any other mammalian organ, and functionally it is equally as varied as shown for example by the many patterns of antibody transmission. The two most consistent attributes of the placenta are its barrier function in separating the maternal and fetal circulations and its reduced isoantigenicity.

The ability of the conceptus to thrive in a potentially hostile immunological environment has excited speculation concerning the possibility of achieving similar results with therapeutic tissue transplants in man. Artificial membranes permit the survival of allografts by mimicking the barrier function of the placenta but fail to ensure adequate nutrition. Until recently it was believed to be difficult or impossible to modify the expression of transplantation antigens by mature tissues, but the finding of reduced isoantigenicity of skin of patients with advanced neoplasia (Amos *et al.,* 1965) questions this premise.

Improved knowledge of mechanisms of implantation, placentation and placental function will not only serve to increase our understanding of the relationships between mother and fetus but may also indicate possible approaches to the problem of procuring permanent allograft survival in man.

Glossary

Tissue Transplantation Terminology (from Snell, 1964)

ALLOGENEIC (adj.): of disparate or foreign origin or genotype, though not from outside the species.

ALLOGRAFT (n.): a graft between genetically disparate individuals of the same species, or between inbred strains (Syn. *homograft, allogeneic graft*).

ALLOTYPE (n.): a Mendelizing or unit factor variant demonstrable because of its immunological properties.

ALLOTYPIC (adj.): pertaining to an allotype.

HISTOCOMPATIBILITY (adj.): relevant to the growth or failure to grow of tissue or tumor transplants.

HISTOCOMPATIBLE (adj.): a relation between the genotypes of donor and host such that a graft cannot be resisted.

ISOGENEIC (adj.): of same origin; used in contrast to *allogeneic* (Syn. *syngeneic*).

ISOGENIC (adj.): genetically identical.

ISOGRAFT (n.): a graft between genetically identical individuals, or within inbred strains.

XENOGENEIC (adj.): originating in a foreign species.

XENOGRAFT (n.): a graft between species.

References

Adams, D. D.: Pathogenesis of the hyperthyroidism of Grave's disease. Brit. Med. J. *i:*1015, 1965.

Amos, D. B., B. G. Hattler, and W. W. Shingleton: Prolonged survival of skin-grafts from cancer patients on normal recipients. Lancet *i:*414, 1965.

Anderson, J. M.: Immunological inertia in pregnancy. Nature *206:*786, 1965.

Andresen, R. H., and C. W. Monroe: Experimental study of the behavior of adult human skin homografts during pregnancy. Am. J. Obstet. Gynec. *84:*1096, 1962.

Ashton, G. C., and G. R. Fallon: β-globulin type, fertility and embryonic mortality in cattle. J. Reprod. Fertil. *3*:93, 1962.

Attwood, H. D., and W. W. Park: Embolism to the lungs by trophoblast. J. Obstet. Gynaec. Brit. Cwlth. *68*:611, 1961.

Azar, H. A.: Cancer in Lebanon and the Near East. Cancer *15*:66, 1962.

Baer, R. L., S. A. Rosenthal, and B. Hagel: The effect of feeding simple chemical allergens to pregnant guinea pigs upon sensitizability of their offspring. J. Immunol. *80*:429, 1958.

Bain, A. D., and J. S. Scott: Mixed gonadal dysgenesis with XX/XY mosaicism. The evidence for the occurrence of fertilisation by two spermatozoa in man. Lancet *i*:1035, 1965.

Bardawil, W. A., G. W. Mitchell, Jr., R. P. McKeogh, and D. J. Marchant: Behavior of skin homografts in human pregnancy. I. Habitual abortion. Am. J. Obstet. Gynec. *84*:1283, 1962.

Bardawil, W. A., and B. L. Toy: The natural history of choriocarcinoma: problems of immunity and spontaneous regression. Ann. N.Y. Acad. Sci. *80*:197, 1959.

Barth, R. F., and P. S. Russell: The antigenic specificity of spermatozoa. I. An immunofluorescent study of the histocompatibility antigens of mouse sperm. J. Immunol. *93*:13, 1964.

Baxter, H., M. A. Goldstein, and G. C. McMillan: The fate of human fetal homo- and heterografts. Plastic Reconstr. Surg. *22*:516, 1958.

Beck, J. S., and N. R. Rowell: Transplacental passage of antinuclear antibody. Lancet *i*:134, 1963.

Beckman, L., G. Björling, and C. Christodoulou: Pregnancy enzymes and placental polymorphism. I. Alkaline phosphatase. Acta Genet. *16*:59, 1966a.

——, ——, ——: Pregnancy enzymes and placental polymorphism: II. Leucine aminopeptidase. Acta Genet. *16*:122, 1966b.

Berger, M., and O. Novick: Antibody transfer from mother to fetus. Fortschr. Geburtsh. Gynäk. *17*:30, 1964.

Billingham, R. E.: Transplantation immunity and the maternal-fetal relation. New Engl. J. Med. *270*:667, 720, 1964.

Billingham, R. E., and L. Brent: Quantitative studies on tissue transplantation immunity. IV. Induction of tolerance in newborn mice and studies on the phenomenon of runt disease. Phil. Trans. Roy. Soc. B *242*:439, 1959.

Billingham, R. E., L. Brent, and P. B. Medawar: Quantitative studies on tissue transplantation immunity. III. Actively acquired tolerance. Phil. Trans. Roy. Soc. B *239*:357, 1956.

Billingham, R. E., L. Brent, P. B. Medawar, and E. M. Sparrow: Quantitative studies on tissue transplantation immunity. I. The survival times of skin homografts exchanged between members of different inbred strains of mice. Proc. Roy. Soc. B *143*:43, 1954.

Billingham, R. E., L. Brent, and N. A. Mitchison: The route of immunization in transplantation immunity. Brit. J. Exp. Path. *38*:467, 1957.

Billingham, R. E., and G. H. Lampkin: Further studies in tissue homotransplantation in cattle. J. Embryol. Exp. Morph. *5*:351, 1957.

Billingham, R. E., G. H. Lampkin, P. B. Medawar, and H. L. Williams:

Tolerance to homografts, twin diagnosis, and the freemartin condition in cattle. Heredity *6*:201, 1952.

Billingham, R. E., J. Palm, and W. K. Silvers: Transplantation immunity of gestational origin in infant rats. Science *147*:514, 1965a.

Billingham, R. E., and W. K. Silvers: Studies on tolerance of the Y chromosome antigen in mice. J. Immunol. *85*:14, 1960.

———, ———: Studies on cheek pouch skin homografts in the Syrian hamster. In: Ciba Found. Symp. on Transplantation, 90, Little, Brown, Boston, 1962a.

———, ———: Some factors that determine the ability of cellular inocula to induce tolerance of tissue homografts. J. Cell. Comp. Physiol. *60*, Suppl. 1: 183, 1962b.

———, ———: Sensitivity to homografts of normal tissues and cells. Ann. Rev. Microbiol. *17*:531, 1963.

———, ———: Studies on homografts of foetal and infant skin and further observations on the anomalous properties of pouch skin grafts in hamsters. Proc. Roy. Soc. B *161*:168, 1964.

———, ———: Re-investigation of the possible occurrence of maternally induced tolerance in guinea pigs. J. Exp. Zool. *160*:221, 1965.

Billingham, R. E., W. K. Silvers, and D. B. Wilson: A second study on the H-Y transplantation antigen in mice. Proc. Roy. Soc. B *163*:61, 1965b.

Billington, W. D.: Influence of immunological dissimilarity of mother and foetus on size of placenta in mice. Nature *202*:317, 1964.

———: The invasiveness of transplanted mouse trophoblast and the influence of immunological factors. J. Reprod. Fertil. *10*:343, 1965.

Birkett, D. J., J. Done, F. C. Neale, and S. Posen: Serum alkaline phosphatase in pregnancy: an immunological study. Brit. Med. J. *i*:1210, 1966.

Blanc, W. A.: Pathways of fetal and early neonatal infection. Viral placentitis, bacterial and fungal chorioamnionitis. J. Pediat. *59*:473, 1961.

Booth, P. B., I. Dunsford, J. Grant, and S. Murray: Haemolytic disease in first-born infants. Brit. Med. J. *ii*:41, 1953.

Boss, J. H.: Antigenic relationships between placenta and kidney in humans. Am. J. Obstet. Gynec. *93*:574, 1965.

Böving, B. G.: The biology of trophoblast. Ann. N.Y. Acad. Sci. *80*:21, 1959.

Boyd, J. D.: Some aspects of the relationship between mother and child. Ulster Med. J. *28*:35, 1959.

Bradbury, S., W. D. Billington, and D. R. S. Kirby: A histochemical and electron microscopical study of the fibrinoid of the mouse placenta. J. Roy. Microscopical Soc. *84*:199, 1965.

Brambell, F. W. R.: The passive immunity of the young mammal. Biol. Rev. *33*:488, 1958.

Brambell, F. W. R., W. A. Hemmings, C. L. Oakley, and R. R. Porter: The relative transmission of the fractions of papain hydrolyzed homologous γ-globulin from the uterine cavity to the foetal circulation in the rabbit. Proc. Roy. Soc. B *151*:478, 1960.

Brent, L., and G. Gowland: Immunological competence of newborn mice. Transplantation *1*:372, 1963.

Brent, R. L.: Immunologic aspects of developmental biology. Adv. Teratol. *1*:81, 1966.

Breyere, E. J.: Nature's homografts. Medical Times *93*:16, 1965.

Breyere, E. J., and M. K. Barrett: "Tolerance" in postpartum female mice induced by strain-specific matings. J. Nat. Cancer Inst. *24*:699, 1960a.

———, ———: Prolonged survival of skin homografts in parous female mice. J. Nat. Cancer Inst. *25*:1405, 1960b.

———, ———: Tolerance induced by parity in mice incompatible at the H-2 locus. J. Nat. Cancer Inst. *27*:409, 1961.

Breyere, E. J., and S. O. Burhoe: The nature of the "partial" tolerance induced by parity. J. Nat. Cancer Inst. *31*:179, 1963.

Brown, J. B.: Oestrogens in the human female. In: Modern Trends in Human Reproductive Physiology, ed. Carey, H. M., Washington: Butterworths, *1*:49, 1963.

Bruning, J. W., A. van Leeuwen, and J. J. van Rood: Purification of leukocyte group substances from human placental tissue. Transplantation *2*:649, 1964.

———, ———, ———: On the nature of leucocyte iso-antigens from human placental tissue. Proc. 10th Congr. Int. Soc. Blood Transf., Stockholm 1964, 177, 1965.

Burnet, Sir M.: Cancer—a biological approach. Brit. Med. J. *i*:779, 841, 1957.

Campbell, C. H.: Anaphylaxis in pregnant and mother mice following active and passive sensitization. Cornell Veterinarian *55*:545, 1965.

Cinader, B., M. A. Hayley, W. D. Rider, and O. H. Warwick: Immunotherapy of a patient with choriocarcinoma. Canad. Med. Assn. J. *84*:306, 1961.

Dameshek, W.: What is systemic lupus? Some comments on its pathogenesis and course. Arch. Int. Med. *106*:162, 1960.

Dancis, J.: The role of the placenta in fetal survival. Pediat. Clin. N. Am. *12*:477, 1965.

Dancis, J., G. W. Douglas, and J. Fierer: Immunologic competence of mouse placental cells in irradiated hosts. Am. J. Obstet. Gynec. *94*:50, 1966.

Dancis, J., B. D. Samuels, and G. W. Douglas: Immunological competence of placenta. Science *136*:382, 1962.

Desai, R. G., and W. P. Creger: Maternofetal passage of leukocytes and platelets in man. Blood *21*:665, 1963.

Dixon, F. J: The role of antigen-antibody complexes in disease. In: The Harvey Lectures *58*:21, 1962–3.

Doniach, I., J. H. Crookston, and T. I. Cope: Attempted treatment of a patient with chorioncarcinoma by immunization with her husband's cells. J. Obstet. Gynaec. Brit. Emp. *65*:553, 1958.

Donovan, J. C., and C. J. Lund: Transplacental passage of maternal erythrocytes. Am. J. Obstet. Gynec. *95*:834, 1966.

Doria, G.: Development of homotransplantation antigens in mouse hemopoietic tissues. Transplantation *1*:311, 1963.

Douglas, G. W.: The immunologic role of the placenta. Obstet. Gynec. Survey *20*:442, 1965.

Douglas, G. W., B. D. Samuels, and J. Dancis: Immunologic competence of mouse placental cells. Am. J. Obstet. Gynec. *84*:1126, 1962.

Douglas, G. W., L. Thomas, M. Carr, N. M. Cullen, and R. Morris: Trophoblast in the circulating blood during pregnancy. Am. J. Obstet. Gynec. *78*:960, 1959.

Dowling, J. T., N. Freinkel, and S. H. Ingbar: Thyroxine-binding by sera of pregnant women, newborn infants, and women with spontaneous abortion. J. Clin. Inv. *35*:1263, 1956.

Dürwald, W., D. Leopold, and K.-H. Krämer: The formation of precipitating antibodies after multiple pregnancies. Vox Sang. *10*:94, 1965.

Edidin, M.: Transplantation antigens in the mouse embryo. The fate of early embryo tissues transplanted to adult hosts. J. Embryol. Exp. Morph. *12*:309, 1964a.

———: Transplantation antigen levels in the early mouse embryo. Transplantation *2*:627, 1964b.

Edwards, A. W. F.: Human sex ratio and maternal immunity to male antigen. Nature *198*:1106, 1963.

Edwards, R. G., L. C. Ferguson, and R. R. A. Coombs: Blood group antigens on human spermatozoa. J. Reprod. Fertil. *7*:153, 1964.

Ehrlich, P.: Ueber Immunität durch Vererbung und Säugung. Z. Hyg. Infect.-Kr. *12*:183, 1892.

Eichwald, E. J., B. Wetzel, E. C. Lustgraaf, and C. McCabe: The effect of male-specific sensitization on the sex ratio of subsequent litters. Transplantation *2*:657, 1964.

Enders, A. C.: Fine structural studies of implantation in the armadillo. In: Delayed Implantation, ed. Enders, A. C., 281, University of Chicago Press, 1963.

———: Formation of syncytium from cytotrophoblast in the human placenta. Obstet. Gynec. *25*:378, 1965a.

———: A comparative study of the fine structure of the trophoblast in several hemochorial placentas. Am. J. Anat. *116*:29, 1965b.

Ephrussi, B., and S. Sorieul: Mating of somatic cells *in vitro*. Univ. Mich. Med. Bull. *28*:347, 1962.

Finegold, M., and D. Michie: Experiments on the maternal-foetal barrier in the mouse. I. A test for the transmission of maternal erythrocytes across the mouse placenta following X-irradiation. J. Embryol. Exp. Morph. *9*:618, 1961.

Fong, S. W., A. Nuckton, and H. H. Fudenberg: Characterization of maternal isoagglutinins in ABO hemolytic disease of the newborn. Blood *27*:17, 1966.

Freda, V. J.: Placental transfer of antibodies in man. Am. J. Obstet. Gynec. *84*:1756, 1962.

Freda, V. J., and B.-A. Carter: Placental permeability in the human for anti-A and anti-B isoantibodies. Am. J. Obstet. Gynec. *84*:1351, 1962.

Fudenberg, H. H.: The immune globulins. Ann. Rev. Microbiol. *19*:301, 1965.

Fudenberg, H. H., and B. R. Fudenberg: Antibody to hereditary human gamma-globulin (Gm) factor resulting from maternal-fetal incompatibility. Science *145*:170, 1964.

Fudenberg, H. H., and K. Hirschhorn: Agammaglobulinemia: some current concepts. Med. Clin. N. Amer. *49*:1533, 1965.

Fudenberg, H. H., E. R. Stiehm, E. C. Franklin, M. Meltzer, and B. Frangione: Antigenicity of hereditary human gamma globulin (Gm) factors—biological and biochemical aspects. Cold Spring Harbor Symp. Quant. Biol. *29*:463, 1964.

van Furth, R., H. R. E. Schuit, and W. Hijmans: The immunological development of the human fetus. J. Exp. Med. *122*:1173, 1965.

Galton, M.: DNA content of placental nuclei. J. Cell Biol. *13*:183, 1962.

————: Parent-offspring homograft relationship in sheep. Transplantation *3*:39, 1965.

Gengozian, N., C. C. Lushbaugh, G. L. Humason, and R. M. Kniseley: 'Erythroblastosis foetalis' in the primate, *Tamarinus nigricollis*. Nature *209*:731, 1966.

Gitlin, D.: Current aspects of the structure, function, and genetics of the immunoglobulins. Ann. Rev. Med. *17*:1, 1966.

Gitlin, D., J. Kumate, J. Urrusti, and C. Morales: The selectivity of the human placenta in the transfer of plasma proteins from mother to fetus. J. Clin. Inv. *43*:1938, 1964.

Gitlin, D., F. S. Rosen, and J. G. Michael: Transient 19S gamma$_1$-globulin deficiency in the newborn infant, and its significance. Pediatrics *31*:197, 1963.

Glass, L. E.: Transfer of native and foreign serum antigens to oviductal mouse eggs. Am. Zool. *3*:135, 1963.

Good, R. A., and S. J. Zak: Disturbances in gamma globulin synthesis as "experiments of nature." Pediatrics *18*:109, 1956.

Goodlin, R. C., and L. A. Herzenberg: Pregnancy induced hemagglutinins to paternal H-2 antigens in multiparous mice. Transplantation *2*:357, 1964.

Goodlin, R. C., L. Herzenberg, and S. De'ath: Isoimmunization associated with cesarean section in the mouse. Am. J. Obstet. Gynec. *90*:776, 1964.

Gorman, J. G., V. J. Freda, W. J. Pollack, and J. G. Robertson: Protection from immunization in Rh-incompatible pregnancies: a progress report. Bull. N.Y. Acad. Med. *42*:458, 1966.

Graves, L. R., Jr., and H. P. K. Agersborg, Jr.: Adrenal cortical activity in pregnancy and its relation to toxemia. A review. Obstet. Gynec. Survey *19*:399, 1964.

Green, I., M. Inkelas, and L. B. Allen: Hodgkin's disease: a maternal-to-foetal lymphocyte chimaera? Lancet *i*:30, 1960.

Guttmann, R. D., and J. B. Aust: A germplasm-transmitted alteration of histocompatibility in the progeny of homograft tolerant mice. Nature *197*:1220, 1963.

Hackett, E., and M. Beech: Immunological treatment of a case of choriocarcinoma. Brit. Med. J. *ii*:1123, 1961.

Halasz, N. A., and M. J. Orloff: Transplacental transmission of homotransplantation antibodies. J. Exp. Med. *118*:353, 1963.

Halvorsen, K.: Neonatal leucopenia due to fetomaternal leucocyte incompatibility. Acta Paediat. Scand. *54*:86, 1965.

Hamilton, W. J., and J. D. Boyd: Specializations of the syncytium of the human chorion. Brit. Med. J. *i*:1501, 1966.

Harris, H., J. F. Watkins, C. E. Ford, and G. I. Schoefl: Artificial heterokaryons of animal cells from different species. J. Cell Sci. *1*:1, 1966.

Hašek, M., V. Hašková, A. Lengerová, and M. Vojtíšková: Mother-foetus immunological relationship as an exceptional homograft model. In: Ciba Found. Symp. on Transplantation, 118, Little, Brown, Boston, 1962.

Hašková, V.: Transplantation immunity and immunological tolerance and the study of antigenicity of tissues and their derivatives. International Colloquium on the Biological Problems of Grafting, University of Liège, 95, 1959.

———: The relationship between the tissues of mother and foetus and tissue incompatibility. Folia Biol. 7:322, 1961.

———: Transplantation non-antigenicity of the foetal placenta. Nature *193*:278, 1962.

———: Differences in the antigenic effectiveness of the foetal part of mouse placenta depending on the strain combination employed. Folia Biol. *9*:99, 1963.

Hellström, K. E., I. Hellström, and G. Haughton: Abrogation of allogeneic inhibition by cortisone. Science *149*:82, 1965.

Hellström, K. E., and G. Möller: Immunological and immunogenetic aspects of tumor transplantation. Progr. Allergy *9*:158, 1965.

Hemmings, W. A., and F. W. R. Brambell: Protein transfer across the foetal membranes. Brit. Med. Bull. *17*:96, 1961.

Hemmings, W. A., and R. E. Jones: The occurrence of macroglobulin antibodies in maternal and foetal sera of rabbits as determined by gradient centrifugation. Proc. Roy. Soc. B *157*:27, 1962.

Hench, P. S.: Ameliorating effect of pregnancy on chronic atrophic (infectious rheumatoid) arthritis, fibrositis, and intermittent hydrarthrosis. Proc. Mayo Clinic *13*:161, 1938.

Hench, P. S., E. C. Kendall, C. H. Slocumb, and H. F. Polley: Effect of hormone of adrenal cortex (17-hydroxy-11-dehydrocorticosterone; compound E) and of pituitary adrenocorticotropic hormone on rheumatoid arthritis; preliminary report. Proc. Mayo Clinic *24*:181, 1949.

Henderson, M.: Foetal regression in rabbits; experimental studies of histolysis and phagocytosis. Proc. Roy. Soc. B *142*:88, 1954.

Herberman, R., and C. A. Stetson, Jr.: The expression of histocompatibility antigens on cellular and subcellular membranes. J. Exp. Med. *121*:533, 1965.

Herzenberg, L. A., and B. Gonzales: Appearance of H-2 agglutinins in outcrossed female mice. Proc. Nat. Acad. Sci. *48*:570, 1962.

Heslop, R. W., P. L. Krohn, and E. M. Sparrow: The effect of pregnancy on the survival of skin homografts in rabbits. J. Endocrin. *10*:325, 1954.

Heys, R. F.: Child bearing and idiopathic thrombocytopenic purpura. J. Obstet. Gynaec. Brit. Cwlth. *73*:205, 1966.

Homer, R. S., and E. G. McNall: Natural resistance to infectious diseases during pregnancy: possible relationship to serum properdin concentration. Am. J. Obstet. Gynec. *81*:29, 1961.

Hort, J.: Contribution to the question of the transfer of immunological tolerance from parents to offspring. Folia Biol. *8*:267, 1962.

Howard, J. G., and D. Michie: Induction of transplantation immunity in the newborn mouse. Transpl. Bull. *29*:1, 1962.

Howard, J. G., D. Michie, and M. F. A. Woodruff: Transplantation tolerance and immunity in relation to age. In Ciba Found. Symp. on Transplantation, 138, Little, Brown, Boston, 1962.

Hraba, T., A. Májský, Z. Vítová, and V. Matoušek: Influence of the mother's blood group on the formation of natural isoagglutinins by the child. Folia Biol. *8:*60, 1962.

Hulka, J. F., and V. Brinton: Antibody to trophoblast during early postpartum period in toxemic pregnancies. Am. J. Obstet. Gynec. *86:*130, 1963.

Hulka, J. F., V. Brinton, J. Schaaf, and C. Baney: Appearance of antibodies to trophoblast during the postpartum period in normal human pregnancies. Nature *198:*501, 1963.

Hulka, J. F., K. C. Hsu, and S. M. Beiser: Antibodies to trophoblasts during the post-partum period. Nature *191:*510, 1961.

Husain, R., and M. M. Ketchel: Normal pregnancy and parturition in rats with acquired immunological tolerance to their mates. Nature *206:*522, 1965.

Iklé, A.: Trophoblastzellen im strömenden Blut. Schw. Med. Wschr. *91:*943, 1961.

Iklé, F. A.: Dissemination von Syncytiotrophoblastzellen im mütterlichen Blut während der Gravidität. Bull. Schw. Akad. Med. Wissenschaften *20:*62, 1964.

Iványi, P., and P. Démant: Prolonged survival of maternal skin grafts in newborn rabbits. Folia Biol. *11:*321, 1965.

Jäämeri, K. E. U., A. P. Koivuniemi, and E. O. Carpén: Occurrence of trophoblasts in the blood of toxaemic patients. Gynaecologia *160:*315, 1965.

James, D. A.: Effects of antigenic dissimilarity between mother and foetus on placental size in mice. Nature *205:*613, 1965.

Jensen, K. G.: Leucocyte antibodies in serums of pregnant women. Serology and clinic. Vox Sang. *7:*454, 1962.

Jollie, W. P.: Fine structural changes in placental labyrinth of the rat with increasing gestational age. J. Ultrastruct. Res. *10:*27, 1964.

Jones, E. C., and P. L. Krohn: Effect of the maternal environment on strain-specific differences in the ovaries of new-born mice. Nature *195:*1064, 1962.

Josimovich, J. B., and B. L. Atwood: Human placental lactogen (HPL), a trophoblastic hormone synergizing with chorionic gonadotropin and potentiating the anabolic effects of pituitary growth hormone. Am. J. Obstet. Gynec. *88:*867, 1964.

Kadowaki, J.-I., R. I. Thompson, W. W. Zuelzer, P. V. Woolley, Jr., A. J. Brough, and D. Gruber: XX/XY lymphoid chimaerism in congenital immunological deficiency syndrome with thymic alymphoplasia. Lancet *ii:* 1152, 1965.

Kaku, M.: Placental polysaccharide and the aetiology of the toxaemia of pregnancy. J. Obstet. Gynaec. Brit. Emp. *60:*148, 1953.

Kaliss, N., and M. K. Dagg: Immune response engendered in mice by multiparity. Transplantation *2:*416, 1964.

Kaliss, N., M. K. Dagg, and J. H. Stimpfling: Maternal transfer of isoantibody in mice. Transplantation *1:*535, 1963.

Kalmus, H.: Genetical antigenic incompatibility as a possible cause of the toxaemias occurring late in pregnancy. Ann. Eugen. *13:*146, 1946.

Kaplan, S. L., and M. M. Grumbach: Serum chorionic "growth hormone-prolactin" and serum pituitary growth hormone in mother and fetus at term. J. Clin. Endocrin. Metab. *25:*1370, 1965.

Kass, E. H., and M. Finland: Adrenocortical hormones in infection and immunity. Ann. Rev. Microbiol. *7:*361, 1953.

Kečkeš, S., and N. Allegretti: Induction of tolerance to maternal skin-grafts in rats irradiated during the foetal life. Int. J. Rad. Biol. *7:*561, 1963.

Kirby, D. R. S.: Development of mouse eggs beneath the kidney capsule. Nature *187:*707, 1960.

————: Reciprocal transplantation of blastocysts between rats and mice. Nature *194:*785, 1962.

————: Development of the mouse blastocyst transplanted to the spleen. J. Reprod. Fertil. *5:*1, 1963a.

————: The development of mouse blastocysts transplanted to the scrotal and cryptorchid testis. J. Anat. *97:*119, 1963b.

Kirby, D. R. S., W. D. Billington, S. Bradbury, and D. J. Goldstein: Antigen barrier of the mouse placenta. Nature *204:*548, 1964.

Kirby, D. R. S., and S. Bradbury: The hemo-chorial mouse placenta. Anat. Rec. *152:*279, 1965.

Kirby, D. R. S., and S. K. Malhotra: Cellular nature of the invasive mouse trophoblast. Nature *201:*520, 1964.

Kochwa, S., R. E. Rosenfield, L. Tallal, and L. R. Wasserman: Isoagglutinins associated with ABO erythroblastosis. J. Clin. Inv. *40:*874, 1961.

Lajos, L., J. Görcs, J. Székely, I. Csaba, and S. Domány: The immunologic and endocrinologic basis of successful transplantation of human trophoblast. Am. J. Obstet. Gynec. *89:*595, 1964.

Lanman, J. T.: Transplantation immunity in mammalian pregnancy; mechanisms of fetal protection against immunologic rejection. J. Pediat. *66:*525, 1965.

Lanman, J. T., J. Dinerstein, and S. Fikrig: Homograft immunity in pregnancy: lack of harm to the fetus from sensitization of the mother. Ann. N.Y. Acad. Sci. *99:*706, 1962.

————, ————, ————: The survival time of skin homografts exchanged between mother and offspring in rabbits. Transplantation *1:*509, 1963.

Lanman, J. T., and L. Herod: Homograft immunity in pregnancy. The placental transfer of cytotoxic antibody in rabbits. J. Exp. Med. *122:*579, 1965.

Lanman, J. T., L. Herod, and S. Fikrig: Homograft immunity in pregnancy. Survival rates in rabbits born of ova transplanted into sensitized mothers. J. Exp. Med. *119:*781, 1964.

Larsen, J. F.: Electron microscopy of the implantation site in the rabbit. Am. J. Anat. *109:*319, 1961.

————: Electron microscopy of the chorioallantoic placenta of the rabbit. I. The placental labyrinth and the multinucleated giant cells of the intermediate zone. J. Ultrastruct. Res. *7:*535, 1962.

Lee, L.: Antigen-antibody reaction in the pathogenesis of bilateral renal cortical necrosis. J. Exp. Med. *117:*365, 1963.

Lee, S. L., F. Rosner, I. Rivero, F. Feldman, and A. Hurwitz: Refractory anemia with abnormal iron metabolism. Its remission after resection of hyperplastic mediastinal lymph nodes. New Engl. J. Med. *272*:761, 1965.

Lengerová, A.: Effect of irradiation during embryogenesis on relationship between maternal organism and offspring from aspect of tissue compatibility. Folia Biol. *3*:333, 1957.

Lengerová, A., and M. Vojtíšková: Postpartum reactivity of female mice to male-specific antigens. Folia Biol. *8*:21, 1962.

———, ———: Prolonged survival of syngeneic male skin grafts in parous C57BL mice. Folia Biol. *9*:72, 1963.

Levine, P.: Serological factors as possible causes in spontaneous abortions. J. Hered. *34*:71, 1943.

———: The influence of the ABO system on Rh hemolytic disease. Human Biol. *30*:14, 1958.

Lichton, I. J.: Survival and endocrine function of rat placenta implanted to the spleen. Endocrinology *76*:1068, 1965.

Lieberman, R., and S. Dray: Maternal-fetal mortality in mice with isoantibodies to paternal γ-globulin allotypes. Proc. Soc. Exp. Biol. Med. *116*:1069, 1964.

Lin, H. A. C.: Is toxemia of pregnancy an allergic reaction? Am. J. Obstet. Gynec. *54*:97, 1947.

Lloyd, C. W., and J. Weisz: Some aspects of reproductive physiology. Ann. Rev. Physiol. *28*:267, 1966.

Lustgraaf, E. C., and E. J. Eichwald: Maternal influence in skin grafting. Transpl. Bull. *24*:437, 1959.

MacLaren, J. A., R. D. Thornes, C. C. Roby, and D. E. Reid: An immunologic characteristic of the serum of normal pregnancy. Am. J. Obstet. Gynec. *78*:939, 1959.

Mage, R., and S. Dray: Persistent altered phenotypic expression of allelic γG-immunoglobulin allotypes in heterozygous rabbits exposed to isoantibodies in fetal and neonatal life. J. Immunol. *95*:525, 1965.

Marchant, D. J., W. A. Bardawil, G. W. Mitchell, Jr., and E. Carey: Observations on the behavior of skin homografts in human pregnancy. Fertil. Steril. *15*:272, 1964.

Mårtensson, L., and H. H. Fudenberg: Gm genes and γ_G-globulin synthesis in the human fetus. J. Immunol. *94*:514, 1965.

Mathé, G., J. Dausset, E. Hervet, J. L. Amiel, J. Colombani, and G. Brule: Immunological studies in patients with placental choriocarcinoma. J. Nat. Cancer Inst. *33*:193, 1964.

McCarthy, J. C.: Genetic and environmental control of foetal and placental growth in the mouse. Animal Production *7*:347, 1965.

McConnell, R. B.: The prevention of Rh haemolytic disease. Ann. Rev. Med. *17*:291, 1966.

McGowan, L.: Cancer and pregnancy. Obstet. Gynec. Survey *19*:285, 1964.

McLaren, A.: Does maternal immunity to male antigen affect the sex ratio of the young? Nature *195*:1323, 1962.

Medawar, P. B.: Some immunological and endocrinological problems raised by the evolution of viviparity in vertebrates. Symp. Soc. Exp. Biol. 7 (Evolution): 320, 1953.

Medawar, P. B., and E. M. Sparrow: The effects of adrenocortical hormones, adrenocorticotrophic hormone and pregnancy on skin transplantation immunity in mice. J. Endocrin. *14*:240, 1956.

Midgley, A. R., Jr., G. B. Pierce, Jr., G. A. Deneau, and J. R. G. Gosling: Morphogenesis of syncytiotrophoblast *in vivo:* an autoradiographic demonstration. Science *141*:349, 1963.

Miller, J. F. A. P.: Immunity in the foetus and the new-born. Brit. Med. Bull. *22*:21, 1966.

Mitchison, N. A.: The effect on the offspring of maternal immunization in mice. J. Genet. *51*:406, 1953.

Möller, E., and G. Möller: Contact-induced cytotoxicity and its relation to cellular immunity. Vox Sang. *11*:299, 1966.

Möller, G.: Studies on the development of the isoantigens of the H-2 system in newborn mice. J. Immunol. *86*:56, 1961.

————: Phenotypic expression of isoantigens of the H-2 system in embryonic and newborn mice. J. Immunol. *90*:271, 1963.

Möller, G., and E. Möller: Plaque-formation by non-immune and X-irradiated lymphoid cells on monolayers of mouse embryo cells. Nature *208*:260, 1965.

Monaco, A. P., M. L. Wood, J. G. Gray, and P. S. Russell: Studies on heterologous anti-lymphocyte serum in mice. II. Effect on the immune response. J. Immunol. *96*:229, 1966.

Montemagno, U., and M. di Stefano: Localizzazione delle sostanze emogruppali A e B nella placenta umana mediante tecnica immunofluorescente. Arch. Ostet. Ginec. *69*:319, 1964.

Morton, J. R., D. G. Gilmour, E. M. McDermid, and A. L. Ogden: Association of blood-group and protein polymorphisms with embryonic mortality in the chicken. Genetics *51*:97, 1965.

Moulton, M. A., J. Stimpfling, and J. B. Storer: Attempts to induce tolerance to maternal tissue by irradiation of fetal mice. Transpl. Bull. *26*:454, 1960.

Mühe, E., and H. Bünte: Untersuchungen der Placenta—dem Modell eines Homotransplantates: Nephritis nach Injektion von Antikörpern gegen die homologe Placenta. Zschr. Ges. Exp. Med. *139*:770, 1965.

de Muralt, G.: La maturation de l'immunité humorale chez l'homme. Helvetica Medica Acta *29* (Suppl. 42):1, 1962.

Murray, J. E., R. Gleason, and A. Bartholomay: Fourth report of the Human Kidney Transplant Registry: 16 September 1964 to 15 March 1965. Transplantation *3*:684, 1965.

Najarian, J. S., and F. J. Dixon: Homotransplantation immunity of neonatal rabbits. Proc. Soc. Exp. Biol. Med. *109*:592, 1962.

————, ————: Induction of tolerance to skin homografts in rabbits by alterations of placental permeability. Proc. Soc. Exp. Biol. Med. *112*:136, 1963.

Nathan, P., Gonzalez, E., and B. F. Miller: Tolerance to maternal skin grafts in rabbits induced by hyaluronidase. Nature *188*:77, 1960.

Nelson, J. H., Jr.: Alterations in immune mechanisms in pregnancy. Clin. Obstet. Gynec. *8*:263, 1965.

Nelson, J. H., Jr., and J. E. Hall: Studies on the thymolymphatic system in humans. I. Morphologic changes in lymph nodes in pregnancy at term. Am. J. Obstet. Gynec. *90*:482, 1964.

————, ————: Studies on the thymolymphatic system in humans. II. Morphologic changes in lymph nodes in early pregnancy and during the puer-perium. Am. J. Obstet. Gynec. *93:*1133, 1965.

Nicol, T., D. L. J. Bilbey, L. M. Charles, J. L. Cordingley, and B. Vernon-Roberts: Oestrogen: the natural stimulant of body defence. J. Endocrin. *30:*277, 1964.

Nordstoga, K.: Thrombocytopenic purpura in baby pigs caused by maternal isoimmunization. Path. Vet. *2:*601, 1965.

Obata, I.: On the nature of eclampsia. J. Immunol. *4:*111, 1919.

Oehme, J., H. Hundeshagen, and C. Eschenbach: Über die Passage markierter Leukocyten vom Muttertier zum Feten—zugleich ein Beitrag zur Runt-Disease. Klin. Wschr. *44:*430, 1966.

von Oettingen, K., and E. Witebsky: Plazenta und Blutgruppe. Münch. Med. Wschr. *75:*385, 1928.

de Oliveira, H. L., and J. A. Laus-Filho: Pulmonary lesions produced by nephrotoxic anti-placenta serum. Int. Arch. Allergy *20:*298, 1962.

Olivelli, F.: Sull'azione dei sieri antiplacentari nella cavia gravida. Min. Ginec. *10:*131, 1958.

Olivelli, F., and P. Ruggieri: Sulla presenza e sul significato degli antigeni placentari nel siero di donna gravida. Min. Ginec. *10:*953, 1958.

Osoba, D.: Immune reactivity in mice thymectomized soon after birth: normal response after pregnancy. Science *147:*298, 1965.

Owen, R. D., H. R. Wood, A. G. Foord, P. Sturgeon, and L. G. Baldwin: Evidence for actively acquired tolerance to Rh antigens. Proc. Nat. Acad. Sci. *40:*420, 1954.

Papiernik-Berkhauer, E.: Les relations immunitaires entre la mère et le foetus. Rev. Franç. Études Clin. Biol. *11:*239, 1966.

Park, W. W.: Experimental trophoblastic embolism of the lungs. J. Path. Bact. *75:*257, 1958.

Payne, J. M., and S. Payne: Placental grafts in rats. J. Embryol. Exp. Morph. *9:*106, 1961.

Payne, R.: The development and persistence of leukoagglutinins in parous women. Blood *19:*411, 1962.

———: Neonatal neutropenia and leukoagglutinins. Pediatrics *33:*194, 1964.

Payne, R., and M. R. Rolfs: Fetomaternal leukocyte incompatibility. J. Clin. Inv. *37:*1756, 1958.

Peer, L. A., W. Bernhard, and J. C. Walker, Jr.: Full-thickness skin exchanges between parents and their children. Am. J. Surg. *95:*239, 1958.

Peer, L. A., I. S. Walia, H. W. Gordon, and W. G. Bernhard: Different tolerance of a mother to skin grafts from her identical twin daughters. Plastic Reconstr. Surg. *31:*478, 1963.

Penrose, L. S.: On the familial appearances of maternal and foetal incompatibility. Ann. Eugen. *13:*141, 1946.

Pepper, F. J.: The effect of age, pregnancy and lactation on the thymus gland and lymph nodes of the mouse. J. Endocrin. *22:*335, 1961.

Pfaundler, M.: Die Antikörperübertragung von Mutter auf Kind. Arch. Kinderhielkunde *47:*260, 1907–1908.

Phillips, L. L.: Modifications of the coagulation mechanism during pregnancy.

In: Modern Trends in Human Reproductive Physiology, ed. Carey, H. M., Washington: Butterworths, *1*:190, 1963.

Pizarro, O., G. Hoecker, P. Rubinstein, and A. Ramos: The distribution in the tissues and the development of H-2 antigens of the mouse. Proc. Nat. Acad. Sci. *47*:1900, 1961.

Plager, J. E., K. G. Schmidt, and W. J. Staubitz: Increased unbound cortisol in the plasma of estrogen-treated subjects. J. Clin. Inv. *43*:1066, 1964.

Playfair, J. H. L., M. R. Wolfendale, and H. E. M. Kay: The leucocytes of peripheral blood in the human foetus. Brit. J. Haematol. *9*:336, 1963.

Porter, J. B., and E. J. Breyere: Studies on the source of antigenic stimulation in the induction of tolerance by parity. Transplantation 2:246, 1964.

Prall, R. H., and F. S. Kantor: Serum complement in eclamptogenic toxemia. Am. J. Obstet. Gynec. *95*:530, 1966.

Prehn, R. T.: Specific homograft tolerance induced by successive matings and implications concerning choriocarcinoma. J. Nat. Cancer Inst. *25*:883, 1960.

Pressman, D., and L. Korngold: Localizing properties of anti-placenta serum. J. Immunol. *78*:75, 1957.

Reif, A. E., H. J. Norris, N. Mahoney, E. R. Klein, and L. M. McVety: Experimental erythroblastosis foetalis in rabbits due to A incompatibility. Brit. J. Exp. Path. *45*:226, 1964.

Renkonen, K. O.: Decreasing sex-ratio by birth order. Lancet *i*:60, 1963.

Renkonen, K. O., O. Mäkelä, and R. Lehtovaara: Factors affecting the human sex ratio. Nature *194*:308, 1962.

Richart, R.: Studies of placental morphogenesis. I. Radioautographic studies of human placenta utilizing tritiated thymidine. Proc. Soc. Exp. Biol. Med. *106*:829, 1961.

Roberts, G. F.: Comparative Aspects of Haemolytic Disease of the Newborn. London: Heinemann, 1957.

Robinson, E., J. Shulman, N. Ben-Hur, H. Zuckerman, and Z. Neuman: Immunological studies and behaviour of husband and foreign homografts in patients with chorionepithelioma. Lancet *i*:300, 1963.

Rogers, B. O., A. P. Raisbeck, D. L. Ballantyne, Jr., and J. M. Converse: The genetics of skin homografting in rats between brothers, sisters, parents and grandparents. Trans. Int. Soc. Plastic Surgeons 421, 1960.

van Rood, J. J., J. G. Eernisse, and A. van Leeuwen: Leucocyte antibodies in sera from pregnant women. Nature *181*:1735, 1958.

van Rood, J. J., A. van Leeuwen, and J. G. Eernisse: Leucocyte antibodies in sera of pregnant women. Vox Sang. *4*:427, 1959.

Salvaggio, A. T., G. Nigogosyan, and H. C. Mack: Detection of trophoblast in cord blood and fetal circulation. Am. J. Obstet. Gynec. *80*:1013, 1960.

Schlesinger, M.: Uterus of rodents as site for manifestation of transplantation immunity against transplantable tumors. J. Nat. Cancer Inst. *28*:927, 1962.

——: Serologic studies of embryonic and trophoblastic tissues of the mouse. J. Immunol. *93*:255, 1964.

——: Immune lysis of thymus and spleen cells of embryonic and neonatal mice. J. Immunol. *94*:358, 1965.

Schmidt, P. J., and R. Hertz: Blood group factors in women with choriocarcinoma as compared with those of their husbands. Am. J. Obstet. Gynec. *82*:651, 1961.

Schmorl, G.: Pathologisch-anatomische Untersuchungen über Puerperal-Eklampsie. Leipzig: Vogel, 1893.

————: Ueber das Schicksal embolisch verschleppter Placentarzellen. Zbl. Gynäk. *29*:129, 1905.

Scott, J. S.: Choriocarcinoma. Observations on the etiology. Am. J. Obstet. Gynec. *83*:185, 1962.

————: Immunological diseases and pregnancy. Brit. Med. J. *i*:1559, 1966.

Seegal, B. C., M. W. Hasson, E. C. Gaynor, and M. S. Rothenberg: Glomerulonephritis produced in dogs by specific antisera. I. The course of the disease resulting from injection of rabbit antidog-placenta serum or rabbit antidog-kidney serum. J. Exp. Med. *102*:789, 1955.

Seegal, B. C., and E. N. Loeb: The production of chronic glomerulonephritis in rats by the injection of rabbit anti-rat-placenta serum. J. Exp. Med. *84*:211, 1946.

Seppälä, M.: Distribution of serum transferrin groups in Finland and their inheritance. Ann. Med. Exp. Biol. Fenn. *43* (Suppl. 4):1, 1965.

Shulman, N. R., R. H. Aster, H. A. Pearson, and M. C. Hiller: Immunoreactions involving platelets. VI. Reactions of maternal isoantibodies responsible for neonatal purpura. Differentiation of a second platelet antigen system. J. Clin. Inv. *41*:1059, 1962.

Silverstein, A. M.: Ontogeny of the immune response. Science *144*:1423, 1964.

Simmons, R. L., and P. S. Russell: The antigenicity of mouse trophoblast. Ann. N.Y. Acad. Sci. *99*:717, 1962.

————, ————: The immunologic problem of pregnancy. Am. J. Obstet. Gynec. *85*:583, 1963.

————, ————: Failure to demonstrate immunological competence in term mouse placental cells. Transplantation *2*:431, 1964a.

————, ————: Bibliography of transplantation of chorioallantoic placenta and trophoblast. Transplantation *2*:551, 1964b.

————, ————: Histocompatibility antigens in transplanted mouse eggs. Nature *208*:698, 1965.

Simmons, R. L., and J. Weintraub: Transplantation experiments on placental ageing. Nature *208*:82, 1965.

Siniscalco, M., L. Bernini, G. la Torretta, C. del Bianco, and S. Marsico: Preliminary data suggesting a possible influence of the mother's genotype on foetal haptoglobin synthesis. Acta Genet. *13*:235, 1963.

Skowron-Cendrzak, A.: Transfer of tolerance to skin grafts in F_1 C57BL inbred mice. Nature *189*:595, 1961.

Slaunwhite, W. R., Jr., and A. A. Sandberg: Transcortin: a corticosteroid-binding protein of plasma. J. Clin. Inv. *38*:384, 1959.

Snell, G. D.: The terminology of tissue transplantation. Transplantation *2*:655, 1964.

Solomon, J. M., and M. Yokoyama: Actively acquired tolerance in fetal-maternal combinations: a review. Transfusion *1*:383, 1961.

Stastny, P.: Accelerated graft rejection in the offspring of immunized mothers. J. Immunol. *95*:929, 1965.

Steblay, R. W.: Localization in human kidney of antibodies formed in sheep against human placenta. J. Immunol. *88*:434, 1962.

Steinbeck, A. W.: The adrenal cortex and reproductive functions. In: Modern Trends in Human Reproductive Physiology, ed. Carey, H. M., Washington: Butterworths, *1*:122, 1963.

Stormorken, H., R. Svenkerud, P. Slagsvold, H. Lie, and J. Lundevall: Thrombocytopenic bleedings in young pigs due to maternal isoimmunization. Nature *198*:1116, 1963.

Szulman, A. E.: The histological distribution of the blood group substances in man as disclosed by immunofluorescence. III. The A, B, and H antigens in embryos and fetuses from 18 mm in length. J. Exp. Med. *119*:503, 1964.

Tallberg, T., E. Ruoslahti, and C. Ehnholm: Immunological studies in human placental proteins and the purification of human placental lactogen. Ann. Med. Exp. Fenn. *43*:67, 1965.

Tao, T.-W., and A. T. Hertig: Viability and differentiation of human trophoblast in organ culture. Am. J. Anat. *116*:315, 1965.

Taylor, A. I., and P. E. Polani: XX/XY mosaicism in man. Lancet *i*:1226, 1965.

Thiede, H. A., J. W. Choate, H. H. Gardner, and H. Santay: Immunofluorescent examination of the human chorionic villus for blood group A and B substance. J. Exp. Med. *121*:1039, 1965.

Thomas, L., G. W. Douglas, and M. C. Carr: The continual migration of syncytial trophoblasts from the fetal placenta into the maternal circulation. Trans. Assoc. Am. Physicians *72*:140, 1959.

Toolan, H. W.: Studies of adult and embryonic skin homografts on conditioned or normal rabbits, with emphasis on the possible role of the ground substance. Ann. N.Y. Acad. Sci. *73*:546, 1958.

Toy, B. L., and L. G. Tedeschi: Embolic trophoblast in peripheral circulation during pregnancy. Proc. Soc. Exp. Biol. Med. *106*:865, 1961.

Trench, C. A. H., P. S. Gardner, and C. A. Green: Induction of immunological tolerance to human gamma-globulin in rabbits using the maternal route of inoculation. Immunology *7*:567, 1964.

Triplett, E. L.: On the mechanism of immunologic self recognition. J. Immunol. *89*:505, 1962.

Turner, J. H., N. Wald, and W. L. G. Quinlivan: Cytogenetic evidence concerning possible transplacental transfer of leukocytes in pregnant women. Am. J. Obstet. Gynec. *95*:831, 1966.

Tyan, M. L., and L. J. Cole: Development of transplantation isoantigens in the mouse embryo plus trophoblast. Transpl. Bull. *30*:136, 1962.

———, ———: Mouse fetal liver and thymus: potential sources of immunologically active cells. Transplantation *1*:347, 1963.

———, ———: Sources of potential immunologically reactive cells in certain fetal and adult tissues. Transplantation *2*:241, 1964.

Tyler, A.: Approaches to the control of fertility based on immunological phenomena. J. Reprod. Fertil. *2*:473, 1961.

Uhr, J. W.: The heterogeneity of the immune response. Science *145*:457, 1964.

Uhr, J. W., and S. G. Anderson: The placenta as a homotransplant. Nature *194*:1292, 1962.

Wagner, D., R. Schunck, and H. Isebarth: Der Nachweis von Trophoblastzellen im strömenden Blut der Frau bei normaler und gestörter Gravidität. Gynaecologia *158*:175, 1964.

van der Werf, B. A. M.: Pregnancy as a homograft. Acta Physiol. Pharmacol. Neerlandica *12*:182, 1963.

White, A.: Hormonal influences on immune mechanisms. Ann. Allergy *21*:417, 1963.

Willis, R. A., and L. T. Hou: The structure of long-surviving homografts of embryonic lung tissue in rats. J. Path. Bact. *87*:71, 1964.

Wilson, I. B.: A tumour tissue analogue of the implanting mouse embryo. Proc. Zool. Soc. Lond. *141*:137, 1963.

Witebsky, E., and H. Reich: Zur gruppenspezifischen Differenzierung der Placentarorgane. Klin. Wschr. *11*:1960, 1932.

Woodruff, M. F. A.: Transplantation immunity and the immunological problem of pregnancy. Proc. Roy. Soc. B *148*:68, 1958.

Wynn, R. M., and J. Davies: Comparative electron microscopy of the hemochorial placenta. Am. J. Obstet. Gynec. *91*:533, 1965.

Zarou, D. M., H. C. Lichtman, and L. M. Hellman: The transmission of chromium-51 tagged maternal erythrocytes from mother to fetus. Am. J. Obstet. Gynec. *88*:565, 1964.

REPRODUCTION AT HIGH ALTITUDES

J. Metcalfe,[1] M. J. Novy,[2] and E. N. Peterson [3]

*Heart Research Laboratory, University of Oregon Medical School,
Portland, Oregon*

Our concern with reproduction at high altitudes has arisen from an interest in placental oxygen transfer. The mammalian fetus, encased in a thick cyst of myometrium, establishes a chain of oxygen supply for its needs and persuades its mother to supply her end of that chain with quantities of well-oxygenated blood adequate for fetal development and survival. In contrast to other nutrients (amino acids, glucose, fats) which can be provided to the fetus in adequate amounts even if maternal blood flow is greatly reduced, oxygen transfer appears to be much more nearly flow-limited even at sea level (Guyton, 1963). One can hypothecate that the nutrient most critical for fetal (and species) survival in viviparous warm-blooded animals is oxygen because it is needed nearly continuously and in substantial quantities. How is the mammalian fetus provided with its needs for this element when its mother resides at high altitude where the primary problem, even for the air-breathing adult, is to obtain an adequate supply of oxygen in an environment with lowered atmospheric oxygen tension? Oxygen supply mechanisms to the fetus may be put under stress in such circumstances, and may fail. We propose to discuss the normal chain of oxygen supply to the mammalian fetus, modifications in the chain and its component links during maternal residence at altitudes from 10,000 to 15,000 feet above sea level, and evidence relevant to the subject of reproductive failure at high altitude.

Following the schema of Hurtado (1964), Figure 1 shows the chain of oxygen supply from ambient air to fetal tissue. At each link there is a decline in the activity of oxygen when expressed in terms of oxygen tension. Let us first examine the links as they operate at sea level.

[1] Professor of Medicine, Oregon Heart Association Chair of Cardiovascular Research.

[2] Research Associate, National Institute Child Health and Human Development.

[3] Special Research Fellow in Medicine, supported by the National Institute Child Health and Human Development.

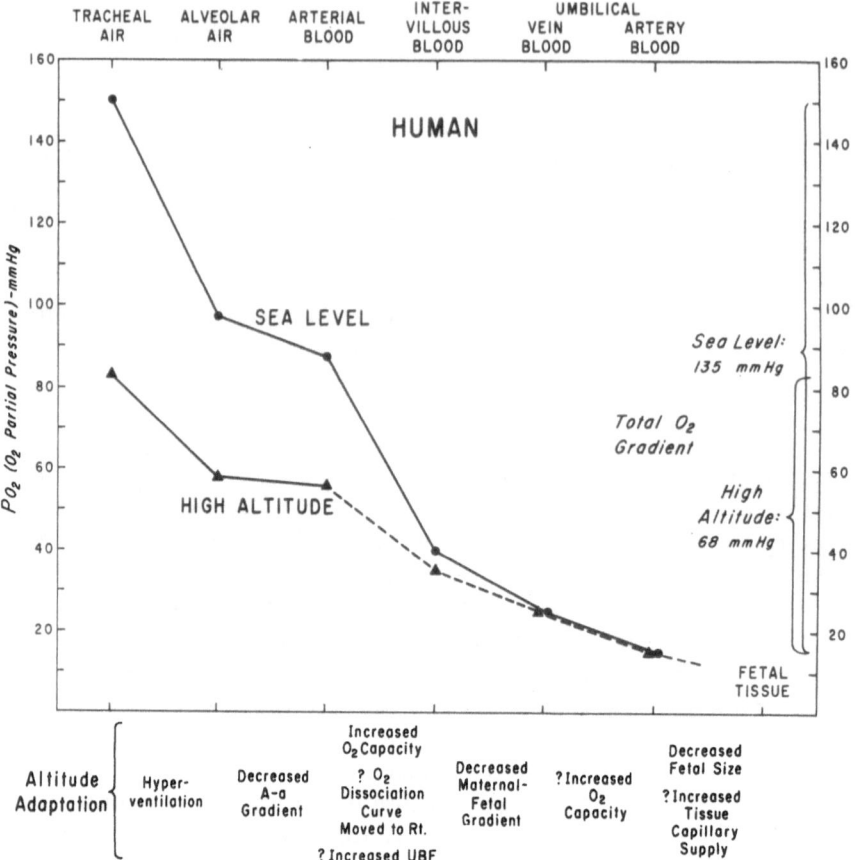

Fig. 1. Representative values for the oxygen tensions in tracheal air, alveolar air, and arterial blood of humans at sea level and high altitude. Values for blood in the intervillous spaces and the umbilical vessels are based on less reliable data (no data exist for these values at high altitude), but plausible values have been chosen for discussion. The slope of each line connecting two successive points expresses the gradient of oxygen tension between them, and the values for the total gradient are given at the right for sea level and for high altitude. The individual gradients and their sum are reduced at high altitude because of the adaptive mechanisms listed at the bottom of the figure.

1. First, ventilation of the lungs transfers oxygen from the external air to the maternal alveoli by convection. The fall in oxygen tension at this link is 60 mm Hg due to the displacement of some of the inhaled oxygen by water vapor and carbon dioxide. The decline is minimized by maternal hyperventilation which occurs during normal human pregnancy at sea level and at high altitude.

2. At the next link oxygen crosses the alveolar-capillary membrane

of the lungs by diffusion, a process which under resting circumstances requires a gradient of oxygen tension of approximately 10 mm Hg.

3. The next link is the transfer of arterial blood to the pregnant uterus. At this link convection again is called into play, powered by the maternal heart which drives the oxygen-laden blood in generous quantities to the pregnant uterus. Within the intervillous spaces oxygen tension is again diminished by admixture of the influent arterial blood with maternal blood which has already lost some oxygen by the process of gas exchange. Estimates of the oxygen tension in human intervillous space blood at term average 40 mm Hg (Bartels *et al.*, 1962).

4. Oxygen diffuses across the placental membrane from the maternal blood in the intervillous spaces into fetal blood in the villar capillaries. In the human the placental membrane is relatively thin (at least from a microscopic view) but it imposes a resistance to gas transfer which requires a further tension gradient of approximately 15 mm Hg. Thus, fetal blood in the umbilical vein has an oxygen tension of 25 mm Hg. This is the most oxygen-rich blood available to the fetus at sea level and has led to suggestions that the fetus in utero has some similarity to an adult at high altitude (Barcroft, 1947).

5. Umbilical vein blood is carried to the fetal tissues by a conveyor system (the fetal circulation) which is established remarkably early in pregnancy. The fetal circulation contains mechanisms for the preferential distribution of oxygen-rich blood, not only from the standpoint of volume flow but also by a streaming of the viscous blood, so that despite incomplete atrial and ventricular septa the fetal brain can enjoy blood richer in oxygen than that sent to the lower portions of the fetal body. In conditions of threatened hypoxia, the degree of this discrimination can be magnified (Dawes, 1962).

6. The final link in the chain of oxygen supply to the sites of electron transfer near the mitochondria is again diffusion, this time from capillary blood to the cells of the fetus. The magnitude of this diffusion gradient is variable and undefined but, according to our present concepts, its range must be limited by losses of oxygen tension further up in the chain.

When one considers the length of this oxygen conveyor system and the complexity of the mechanisms necessary to its proper functioning, one is tempted to sigh with relief at the fact of one's own individual survival, even to birth. Couldn't it have been done in a simpler way? Well, it has been done in simpler ways but not so successfully: the egg-layers and pouch-bearers seem doomed to extinction or servitude.

There are safety features which under some circumstances relieve the oxygen conveyor system. One of these is the ability of fetal tissues to engage in anaerobic metabolism, allowing the placenta to remove the end products either by aerobic metabolism in placental tissues or by transfer to maternal blood (Huckabee *et al.*, 1962a) (Huckabee *et al.*, 1962b).

The rate of fetal oxygen consumption in sheep decreases progressively with decreased rates of umbilical blood flow and oxygen saturation (Dawes, 1962). Another, and related, safety mechanism is the ability of fetal tissues at least in some species to survive total oxygen deprivation for remarkably long periods with subsequent recovery (Mott, 1961). There is no objective evidence, however, that these mechanisms operate in normal pregnancy at high altitude (Huckabee *et al.*, 1959).

Having described the chain of oxygen supply to the fetus and the component links of that chain at sea level, what can we say about the problem at altitude? First, that if the drop in oxygen pressure from ambient air to umbilical venous blood were inexorably fixed at 125 mm Hg the oxygen tension in umbilical venous blood would reach zero with maternal sojourn at altitudes above 7000 feet. Second, that the chain has been shown to be modified at almost every link. The adaptations minimizing drops in oxygen tension are listed below the appropriate link in the chain of oxygen supply in Figure 1.

1. First, hyperventilation occurs in all high-altitude residents, so that the loss in oxygen tension between ambient air and alveolar air is decreased. Human pregnancy is itself accompanied by hyperventilation: at high altitude the maternal respiratory center is doubly driven by pregnancy and by altitude. Hyperventilation at sea level seems to give little benefit to fetal oxygen supply because of the shape of the blood oxyhemoglobin dissociation curve, but at altitude maternal hyperventilation significantly increases the concentration of oxygen in maternal arterial blood.

2. The next resistance to oxygen flow is at the alveolar-capillary membrane. The hyperventilation seen at high altitude is accompanied in long-term residents by a decrease in the alveolar-capillary gradient of oxygen tension (Hurtado, 1964). The basis of this lowered resistance to oxygen transfer from air to blood is not known.

Now we come to a part of the series of resistances about which human data are lacking, and we have indicated this deficiency in our knowledge by dotting the lines in Figure 1.

3. Our work with sheep at Morococha (altitude 15,000 feet) showed that the rate of maternal uterine blood flow is increased by about one-third at high altitude. If the same increase in uterine blood flow occurs in the human, it would elevate the mean oxygen tension within the intervillous spaces. Two hematologic adjustments occur in the human resident at high altitude which would also act to sustain the oxygen tension in intervillous space blood. These adjustments are an increase in the hemoglobin concentration of maternal blood and a shift of the maternal blood oxyhemoglobin dissociation curve to the right, both changes documented by Hurtado (1964) as occurring in human high-altitude residents. Figure 2 attempts to explain the operation of these two com-

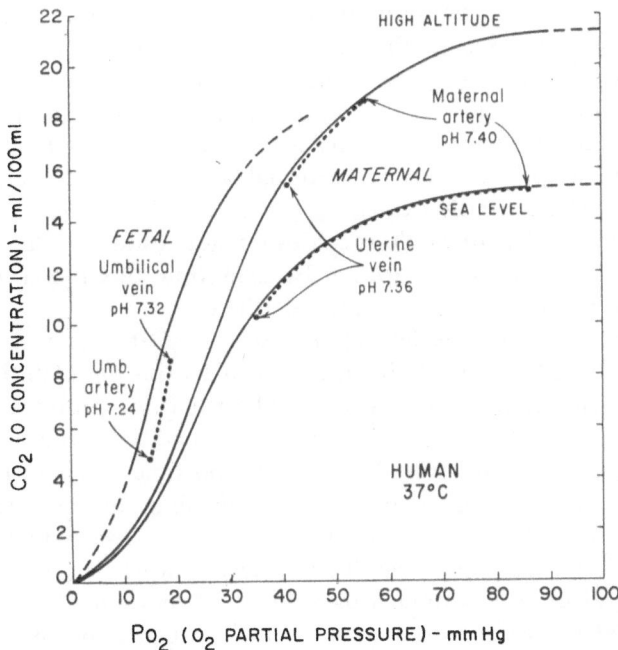

Fig. 2. Oxyhemoglobin dissociation curves for fetal and maternal blood of humans. The ordinate expresses oxygen concentration (ml O_2/100 ml blood) in blood rather than per cent saturation. The higher of the two curves for maternal blood represents a theoretical case at high altitude with an increased blood hemoglobin concentration and the curve shifted to the right. Values for oxygen concentrations in blood are located for the fetus and the sea-level mother according to data in the literature (Metcalfe *et al.*, 1966). The maternal values for high altitude are located assuming that the rate of uterine blood flow is increased by 50 per cent compared to sea level.

pensatory mechanisms. To draw it, we have used our data (Metcalfe *et al.*, 1955) to place the arterial and venous points on the maternal curves. The ordinate expresses the oxygen concentration of blood (rather than per cent saturation which, for our purposes, would be misleading). At high altitude the oxygen tension in maternal arterial blood is reduced, but its oxygen concentration is even higher than at sea level, thanks to the increased hemoglobin concentration and to maternal hyperventilation which increases alveolar oxygen tension by about 5 mm Hg. Also, because the curve relating oxygen tension to saturation is shifted to the right, more of the blood's oxygen is held at higher levels of oxygen tension. As a consequence, the mean oxygen tension in intervillous space blood is only slightly diminished at high altitude. Using the Barcroft equation (Hurtado, 1964), the mean oxygen tension in the intervillous space blood at sea level is 52 mm Hg. Even if the sea-level rate of uterine blood flow were maintained at high altitude, the mean intervillous oxygen tension would be only 10 mm Hg less at 15,000 feet, compared with the sea-level value. If the rate of uterine blood flow increases, the intervillous oxygen tension is even better maintained: a 50 per cent increase in uterine blood flow would maintain the intervillous blood oxygen tension at 47 mm Hg, only 5 mm Hg less than that calculated for sea level despite a drop in maternal arterial oxygen tension of at least 30 mm Hg. Parenthetically, such calculations of mean oxygen tension in blood in the intervillous spaces do not result in values identical (or even similar) to those found by sampling the blood directly, a discrepancy which need not be discussed here.

4. The most significant finding in our sheep studies was concerned with the next link in the chain: the oxygen tension in umbilical venous blood was maintained at levels identical with those found at sea level. Therefore, the gradient between maternal blood and umbilical venous blood was decreased at high altitude. The possible mechanisms for this accomplishment include an increase in diffusing surface area between maternal and fetal blood in the placenta and a decrease in the average thickness of the placental membrane. Recent work by Tominaga and Page (1966) gives an anatomical basis for the hypothesis that thinning of the placental membrane does occur at low oxygen tensions.

As a result of this series of adaptations the fall in oxygen tension is diminished at each link of the chain from ambient air to fetal blood, so that in sheep the mean oxygen tension in umbilical venous blood is not detectably different in fetuses at 15,000 feet from that seen in the same vessel in their sea-level counterparts. The same is true of umbilical arterial blood. There is, in fact, no evidence (from well-acclimatized sheep) that fetal tissue oxygen tensions are reduced at high altitude, thanks to the adaptive mechanisms which we have discussed. On the other hand, evidence from several species indicates fetal adaptations. First, in high-altitude sheep fetal blood contains a higher hemoglobin concentration than that seen at the same stage of pregnancy in sea-level fetuses. Second, Becker *et al.* (1955) have presented evidence from puppies born at high altitude of an increase in capillary vascularization in brain, heart and peripheral muscle. Third, a study of birth statistics from Leadville, Colorado (Lichty *et al.*, 1957), showed that human babies weigh less at birth at high altitude. No reduction in fetal weight was seen in our sheep, but we found not one instance of twin pregnancy in about 50 high-altitude ewes. It seems likely that different species employ different fetal mechanisms of adaptation to high altitude, a suggestion that is supported by other data from Leadville which showed no evidence of an increased hemoglobin concentration in the blood of the small high-altitude infants. Reynafarje (1959), after a thorough study of bone marrow from newborn infants, could find no evidence of increased erythropoiesis at high altitude. As Hurtado (1964) so aptly puts it, the human infant born at high altitude behaves in this respect as a newcomer. Indeed, modifications in the mother may be sufficient in some species to avoid the need for fetal adaptation. With reference to diminished fetal weight at high altitude, Naeye (1965) has shown that retardation of fetal growth secondary to deficient materno-fetal exchange is characterized by a reduction in cell size (that is, a selective diminution in cytoplasmic mass) rather than by reduction in cell number; in such a case oxygen consumption per kilogram may be considerably higher than in normal infants (Silverman and Sinclair, 1966). If this is true, a smaller inter-capillary diffusion distance would be of advantage from the standpoint of oxygen supply to fetal cells.

Finally, several adaptive mechanisms have been demonstrated to operate in the tissues of adult animals at high altitude and may possibly occur in the fetus. These include an increased tissue concentration of myoglobin, an increased resistance of tissues to lactic acid, and an increased cytochrome oxidase activity (Hurtado, 1964).

Our essential assignment is to discuss whether reproductive performance is diminished among residents of high altitude. From a theoretical standpoint, one would expect that this would be the case. If the chain of oxygen supply is strained in every link at high altitude—by increases in maternal ventilation, in maternal blood hemoglobin concentration, in an increased rate of uterine circulation—one would hypothesize a greater danger of the chain breaking. For instance, those mothers who are handicapped by iron deficiency or heart disease, or limitation to pulmonary ventilation, would seem less likely to succeed in reproduction at high altitude. Let us look at the evidence for and against this hypothesis.

In his scholarly book called *Acclimatization in the Andes,* Carlos Monge (1948) dwells at length upon the problems of reproduction at high altitude. Testicular degeneration was noted in cats, rabbits and white rats who were taken to Morococha, Peru. Azoospermia was noted in sheep at 3000 meters, and six pairs of geese taken to the Jungfraujoch at 4000 meters laid their last eggs on the day of arrival and during six months of observation did not lay again. Monge summarizes the recorded history of the conquistadors with reference to reproduction in the Andes. He quotes Father Cobo who said, "We even notice that where the Indians are healthiest and where they multiply the most prolifically is in these same cold tempers which is quite the reverse of what happens to children of the Spaniards, most of whom when born in such regions do not survive . . . But where it is most noticeable is in those who have half, a quarter or any admixture of Indian blood; for these are all raised with the same loving care as the pure Spanish children and yet the more Indian blood they have the better they survive and grow; so that it is now a common saying based on everyday experience that babes having some Indian in them run less risk in the cold regions than those not having this admixture." Monge points out that one of the motives for the transfer of the Spanish capitol from Jauja (at 3300 meters altitude) to Lima (near sea level) was the difficulty of maintaining herds of livestock at higher altitudes. He quotes from the "Annals of the Imperial City of Potosi," written by Antonio de la Calancha in 1639. "Potosi, when it was first founded, contained 100,000 natives and 20,000 Spaniards taken with the fever for riches which made the city so famous. While the former (the Indians) went on reproducing with customary Indian fertility, the latter either did not succeed in having children or they did not survive. The birth of the first Spaniard did not take place until 53 years after the founding of the city and the birth was attributed to a miracle of St. Nicholas of Tolentino." Monge believes that the Spanish conquerors crossed their Spanish race with the native one, "assuring thus the element of Indian ancestry which permits life on the upland without limitation," and he makes a very strong case for the occurrence of relative sterility at high altitude in newly arrived animals and man. He points out that reproductive success is recovered by prolonged residence in the Andes, stressing intermarriage with the Andean Indians as a factor in human adaptation. Incidentally, the sheep whose high-altitude reproductive data we have already discussed were the offspring of generations of high-altitude sheep. It should not be inferred that sea-level sheep could make the same adjustments if taken up to similar levels. Perhaps this is the place to point out that reproductive success is an essential criterion of adaptation to a new environment. No matter how well the individual adapts in other ways, sterility is intolerable for survival of the species. However, after reading Monge's charming account, one is left wondering in all fairness whether such secondary factors as cold, undernutrition and economic deprivation, all of which complicate the high-altitude environment of man, were not equally important with low atmospheric pressure in handicapping human reproduction. Unfortunately, the same question may be raised about most subsequent studies of Andean man: because of the economic and academic instability of the South American countries, scientific data on high-altitude reproduction have not been available from those places where they should be most plentiful. Mazess (1965), from an analysis of Peruvian census reports, found that neonatal mortality in the Peruvian highlands was about double that reported for lowland Peru. These data seem to confirm the results of Grahn and Kratchman (1963) which showed a high degree of association between neonatal mortality and altitude at birth in the United States population. Racial and social, and economic differences between the high- and low-altitude groups of Peruvians were recognized, but even when attempts are made to discount extraneous factors, the association between altitude and a higher newborn death rate persists. This increased neonatal mortality is associated with a decreased birth weight and raises the question as to whether these babies, like others of low birth weight, show a high incidence of physical and mental handicaps even if they survive (Lubchenco *et al.*, 1963).

From the standpoint of species survival, two points need making. First, death of an individual at anytime before adolescence is reproductive failure. Mazess (1965) quotes Tschopik's figures from Chucuito (altitude 3870 meters); of 150 live births, 14 per cent died before six months, 25 per cent before three years, and 34 per cent before thirteen years. From the standpoint of continuing the species, all 34 per cent were wasted. The second, and related, point is that if more fetuses, babies and children die at altitude because of malnutrition, they still represent failures of reproduction at high altitude: death in utero due to maternal hypoxia is no more fatal than death soon after birth related to the complex socio-economic mechanisms which characterize high-altitude environments; judging by persistence, social and economic factors are as difficult to correct as those in the milieu interieur. Their separation and identification are, of course, practically important.

The question of congenital defects in association with high altitude is less clear than the fact of increased infant and childhood mortality. There is no doubt that congenital anomalies can be produced by exposure of mammalian mothers to reduced barometric pressure during pregnancy (Ingalls, 1952), but such teratogenic exposures are usually acute and severe and may not be analogous with residence at high altitude. A preliminary report from Peru (Alzamora *et al.*, 1953) suggested that interatrial septal defect and patent ductus arteriosus were more frequent in children born at altitudes over 3000 meters, but this suggestion has never been confirmed to our knowledge. As Warkany (1960) has emphasized, local effects other than reduced atmospheric pressure may influence the incidence of persistency of the ductus arteriosus. Indeed, lack of oxygen may exert a protective effect against radiation-induced anomalies.

Because of the multiplicity of factors acting to influence human reproduction at high altitudes and because data on the subject have not given clear answers to the important questions concerning the health of offspring, analogies have been drawn with other situations which can be more clearly defined. Pregnant women with cyanotic congenital heart disease have a poor reproductive record: although some of this handicap is on a genetic basis (Neill and Swanson, 1961), children of cyanotic mothers have a substantially poorer prognosis than children with cyanotic fathers. In such circumstances, the risk of pregnancy failure is increased all along the line; there are more abortions, more premature births, more neonatal deaths (Metcalfe and Ueland, 1965). Another analogy has been drawn between maternal residence at high altitude and maternal anemia which in humans has been shown to be associated with an increased incidence of fetal deformity (Worcester *et al.*, 1950).

To summarize, evidence presently available to us suggests that all links in the complex chain which supplies oxygen to the mammalian fetus are under increased strain at high altitude. Adaptations which permit successful reproduction include maternal hyperventilation, increased maternal cardiac output, maternal polycythemia and increased permeability of the placental membrane. Since some, and perhaps all, of these require additional supplies of energy or nutrients one would predict a higher incidence of reproductive failure in the high places of the earth. This seems to be borne out by evidence for individuals newly arrived at altitude, but acclimatization does occur and for most organisms permits adequate reproductive success for survival of the species.

References

Alzamora, V., A. Rotta, G. Battilana, R. Abugattas, C. Rubio, J. Bouroncle, C. Zapata, E. Santa-Maria, T. Binder, R. Subiria, D. Paredes, B. Pando, and G. Graham: On the possible influence of great altitudes on the determination of certain cardiovascular anomalies: preliminary report. Ped. *12*:259, 1953.

Barcroft, J.: Researches on Pre-natal Life. Springfield, Ill.: Charles C. Thomas, 1947 (page 259).

Bartels, H., W. Moll, and J. Metcalfe: Physiology of gas exchange in the human placenta. Am. J. Obstet. Gynec. *84*:1714, 1962.

Becker, E. L., R. G. Cooper, and G. D. Hataway: Capillary vascularization in puppies born at a simulated altitude of 20,000 feet. J. Appl. Physiol. *8*:166, 1955.

Dawes, G. S.: The umbilical circulation. Am. J. Obstet. Gynec. *84*:1634, 1962.

Grahn, D., and J. Kratchman: Variation in neonatal death rate and birth weight in the United States and possible relations to environmental radiation, geology and altitude. Am. J. Human Genet. *15*:329, 1963.

Guyton, A. C.: Circulatory Physiology: Cardiac Output and Its Regulation. Philadelphia: W. B. Saunders Co., 1963 (page 314).

Huckabee, W. E., J. Metcalfe, H. Prystowsky, A. Hellegers, G. Meschia, and D. H. Barron: Uterine blood flow and metabolism in pregnant sheep at high altitude. Fed. Proc. *18*:72, 1959.

Huckabee, W. E., J. Metcalfe, H. Prystowsky, and D. H. Barron: Movements of lactate and pyruvate in pregnant uterus. Am. J. Physiol. *202*:193, 1962a.

——, ——, ——, ——: Insufficiency of O_2 supply to pregnant uterus. Am. J. Physiol. *202*:198, 1962b.

Hurtado, A.: Animals in high altitudes: resident man. In: Dill, D. B. (ed.): Handbook of Physiology, Section 4, Adaptation to the Environment. Washington, D.C.: American Physiological Society, 1964 (page 842).

Ingalls, T. H., F. J. Curley, and R. A. Prindle: Experimental production of congenital anomalies. New Eng. J. Med. *247*:758, 1952.

Lichty, J. A., R. Y. Ting, P. D. Bruns, and E. Dyar: Studies of babies born at high altitude. Am. Med. Assoc. J. Dis. Child. *93*:666, 1957.

Lubchenco, L. O., F. A. Horner, L. H. Reed, I. E. Hix, D. Metcalf, R. Cohig, H. C. Elliott, and M. Bourg: Sequelae of premature birth. Am. J. Dis. Child. *106*:101, 1963.

Mazess, R. B.: Neonatal mortality and altitude in Peru. Am. J. Phys. Anthrop. *23*:209, 1965.

Metcalfe, J., H. Bartels, and W. Moll: Gas exchange in the pregnant uterus. Physiol. Rev. *(In press)*

Metcalfe, J., S. L. Romney, L. H. Ramsey, D. E. Reid, and C. S. Burwell: Estimation of uterine blood flow in normal human pregnancy at term. J. Clin. Invest. *34*:1632, 1955.

Metcalfe, J., and K. Ueland: The heart and pregnancy. In: Hurst, J. W., and B. Logue: The Heart. New York: McGraw-Hill, 1965 (page 1094).

Monge, C.: Acclimatization in the Andes. Baltimore: The Johns Hopkins Press, 1948.

Mott, J. C.: The ability of young mammals to withstand total oxygen lack. Brit. Med. Bull. *17*:144, 1961.

Naeye, R. L.: Malnutrition: Probable cause of fetal growth retardation. Arch. Path. *79*:284, 1965.

Neill, C. A., and S. Swanson: Outcome of pregnancy in congenital heart disease. Circ. *24*:1003, 1961.

Reynafarje, C.: Bone marrow studies in the newborn infant at high altitudes. J. Ped. *54*:152, 1959.

Silverman, W. A., and J. C. Sinclair: Infants of low birth weight. New Eng. J. Med. *274*:448, 1966.

Tominaga, T., and E. W. Page: Accommodation of the human placenta to hypoxia. Am. J. Obstet. Gynec. *94*:679, 1966.

Warkany, J.: in Discussion of: Ingalls, T. H.: Environmental factors in causation of congenital anomalies. In: Wolstenholme, G. E. W. (ed.) Ciba Foundation Symposium on Congenital Malformations. Boston: Little, Brown and Co., 1960 (page 71).

Worcester, J., S. S. Stevenson, and G. Rice: 677 congenitally malformed infants and associated gestational characteristics. II. Parental factors. Ped. *6*:208, 1950.

FERTILITY AND REPRODUCTIVE
PERFORMANCE OF GROUPED MALE MICE

R. L. SNYDER *

*Penrose Research Laboratory, Zoological Society of Philadelphia, and
the Department of Pathology, University of Pennsylvania,
Philadelphia, Pennsylvania*

Reproductive failures are reported to be commonplace among mammals in nature. Agents responsible for these failures can be divided conveniently into two classes depending on their relation to population density. Thus, in Great Britain workers (Brambell, 1942, 1944; Brambell and Mills, 1944, 1947a, b; and Allen *et al.*, 1947) studying reproduction in wild rabbits found that approximately 60 per cent of all "litters" in a series collected were completely lost *in utero*. These workers, in general, favored a genetical explanation. Conaway and coworkers (1960) suggested that a high incidence of "total litter resorption" in swamp rabbits (*Sylvilagus aquaticus*) which followed sudden flooding of the habitat and resultant overcrowding was induced by an "adrenal stress syndrome." The same process of intrauterine resorption of embryos and fetuses has been found in varying degree among other species of wild mammals (Perry, 1945; Snyder and Christian, 1960; Snyder, 1962; Newson, 1966).

When laboratory populations of mice are supplied with surplus food and nesting materials, reproduction and mortality are density-dependent (Crew and Mirskaia, 1931; Brown, 1953; Southwick, 1955; Christian, 1956). One explanation for this phenomenon is that social conflicts, hence psychological stimuli, intensify as numbers of animals in contact increase. Endocrine responses to psychological stimuli mediated through the hypothalamus are believed to increase secretion of adrenocorticotrophin and decrease secretion of gonadotrophin (Christian *et al.*, 1965). These populations are self-regulated because reproductive rate is inversely related and mortality rate directly related to density. The role of endo-

* Supported in part by Grant HD-00543 from the USPHS, and a grant from the Smith Kline and French Foundation.

crines in the self-regulation of natural mammalian populations is uncertain at present.

Female mice in the laboratory are extremely sensitive to population pressures. Crowding increases intrauterine mortality of embryos and fetuses (Christian and LeMunyan, 1958; Helmreich, 1960) and induces long periods of pseudopregnancy (Mody and Christian, 1962). Exposure of recently impregnated females to the odor of urine from a strange male frequently blocks pregnancy (Bruce, 1960; Eleftheriou *et al.*, 1962). Crowding produces atrophy of the testes and the accessory sexual glands of male mice (Christian, 1955), but until now no one had studied fertility and reproductive performance of males in dense populations.

The present report describes fertility and reproductive performance of crowded male house mice. Breeding trials were conducted after 2, 6, 12, 20, and 50 weeks of crowding. Reproductive performance was based on percentage pregnancy, number of ovulations, number of visible implantations, survival of embryos and fetuses, litter size, and sex ratio of offspring. Controls were males of comparable age caged alone.

Semen was collected from both crowded and isolated mice by electrical stimulation. Volume of liquid ejaculate, sperm concentration and total number of spermatozoa per ejaculate were determined for each sample. Males were killed after 21 weeks of grouping and weights of the adrenals and the reproductive organs were compared with weights of these organs from males caged alone. The data were then analyzed to determine if changes in weights of the reproductive organs were reflected in volume of semen collected or in sperm concentration.

The model described in this paper appears to be satisfactory for the study of population density and male reproductive function in the laboratory. Reproductive function is evaluated at several levels, first, on the weights of the reproductive organs, second on the quantity of spermatozoa produced in electrically stimulated ejaculates, and third, on the actual reproductive performance measured by breeding trials.

MATERIALS AND METHODS

CROWDING EXPERIMENT I: Effects on Reproductive Performance

Brown house mice derived from wild stock were weaned at 3 weeks of age and caged alone in plastic cages (11" x 7") until needed for experiments. Crowding in this experiment was achieved by placing 20 males together in a stainless steel cage (21" x 17"). Food pellets were scattered on the cage floor and a water bottle was attached to each corner. This arrangement prevented dominant animals in a group from blocking subordinates from food and water. Between 8 and 9 A.M. each day, one animal in each group was selected at random and transferred to another group, so that every day each group lost one animal and gained a new one in its place. This procedure kept each group in a constant state of excitement and prevented the establishment of a stable social hierarchy.

A table of random numbers was used to assign 120 males to 6 groups and 110 males as controls. Controls were each caged alone in a plastic cage (11" x 7"). These males ranged in age from 23 to 173 days. The average age of the grouped mice was 114 days and that of the controls 118 days. Toes were clipped from both experimental and control mice for individual identification.

Reproductive performance was tested five times over a period of 60 weeks after 2, 6, 12, 20, and 50 weeks of grouping. The testing procedure consisted of placing each male alone with a female mouse for two weeks in one of the small plastic cages. A test female was virgin, sexually mature (at least 8 weeks old), and had been caged alone since weaning. Males were always introduced into the home cages of the females. Assignments were made from a table of random numbers. After 20 days cages were checked each morning for litters. Newborn mice were counted and weighed then killed with ether. Sex was determined by internal examination.

Two to three days after littering females were killed with ether. After fixation in formalin 6 μ sections of the ovaries taken 75 μ apart were stained with H and E. The number of current corpora lutea (corpora lutea of pregnancy) determined by microscopic study indicated the number of ovulations per pregnancy (Deanesly, 1930; Brambell, 1956; Snell, 1956) and the placental scars compared with the number of live young in the litter indicated the fate of each ovum. The site of placental attachment to the uterus is indicated by a large hemorrhagic spot after parturition. Blood is slowly resorbed after parturition and the spot becomes progressively smaller until it finally disappears. It is well known that three to four generations of placental scars are sometimes visible at one time in multiparous mice. Placental scars left by resorbed embryos and fetuses, owing to the fact that regression had commenced earlier, are smaller and darker in color than uterine scars of the same pregnancy left by full term viable fetuses delivered at parturition. Since the females in this experiment were primiparous, evidence of intrauterine mortality was provided by distinguishing the two kinds of scars. Experience has dictated a wait of two to three days after parturition before examination of the uterus to allow the swollen uterus time to shrink and to allow the diffuse hyperemia associated with parturition to subside. Smaller scars left by resorbed embryos and fetuses are often obscured by the swelling and hyperemia.

Females that had not littered within 35 days were killed and treated as above. Females that died were autopsied immediately. In this instance evidence of intrauterine mortality was provided by the presence of atrophied embryos and fetuses in various stages of resorption. Atrophied fetuses complete with amnionic sac were also sometimes delivered with live young.

CROWDING EXPERIMENT II: Effects on Adrenals, Reproductive Organs and Electroejaculation Test

Male brown house mice between 84 and 252 days of age were assigned randomly to five cages (21" x 17"), 20 mice per cage. One hundred mice were selected at the same time as controls and caged alone in plastic boxes (11" x 7"). Grouped mice were killed after 21 weeks. A suitable number of controls were selected randomly and killed for comparison of organ weights. Adrenals, testes, prostatic glands, and seminal vesicles were weighed on a Torsion balance after fixation in formalin.

Electroejaculation Test

The electroejaculation test followed a technique described recently for obtaining uncoagulated semen from intact mice (Snyder, 1966). Volume of liquid semen, sperm concentration and total number of spermatozoa per ejaculate were determined from representative samples of grouped and isolated mice during the 21 week interval.

Statistical Analysis of Data

Proportions and percentages were tested for significance by the method of Chi Square. Numerical data were analyzed by an Analysis of Variance designed for subclasses with unequal sample sizes (Snedecor, 1956).

RESULTS

CROWDING EXPERIMENT I: Effects on Reproduction

Mortality and Fertility

The mice in this experiment were grouped for 50 weeks and an additional 10 weeks were allotted to the five breeding trials. Thus, in addition to effects of grouping on reproductive function there was the added factor of chronological aging to be considered. Figure 1 shows the numbers of mice alive at the beginning of each test period and the percentage that were proven fertile. Ten of the grouped mice died during the first two weeks, but losses were relatively few until 26 weeks had elapsed. Fifty grouped mice died between weeks 26 and 58. During the 60 week period only 9 of the controls died. Fertility was maintained in the controls until after week 26, but showed a significant drop in the grouped mice by this time.

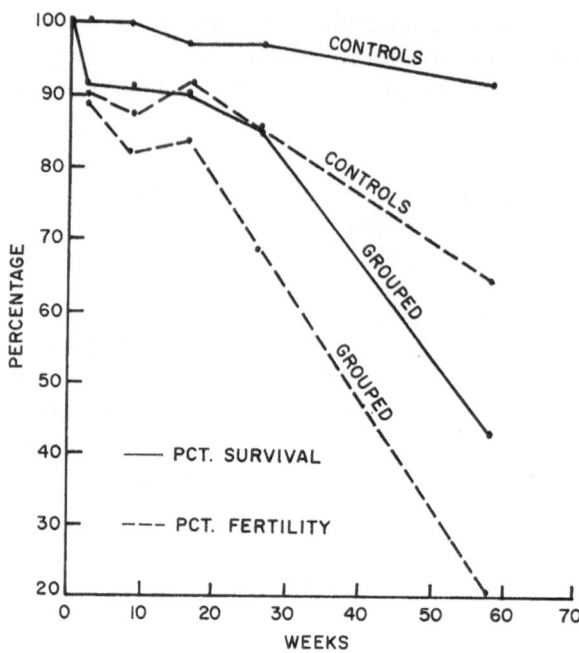

Fig. 1. Survival and fertility of grouped males compared with controls.

Ovulations Per Pregnancy

Mean numbers of corpora lutea per pregnancy are given in Table 1 for the first three breeding trials on the assumption that each corpus luteum represented one ovulation. An analysis of variance of these data revealed a significantly higher mean number of corpora lutea for the controls and a significant decrease in numbers of corpora lutea with decreasing age of the test females. The unbiased estimate of the mean treatment difference in corpora lutea per pregnancy was 0.588 (Snedecor, 1956). Weighted averages in Table 1 apply when mean values are each given equal weight in sample size.

Implantations Per Pregnancy

The next measure of reproductive performance was number of implantations per pregnancy indicated by the placental scars in the uteri of the females two to three days after parturition (Table 2). The analysis of variance indicated significantly lower mean values for females mated to grouped males and a significant change in mean implantations associated with time. The unbiased estimate of mean treatment difference was 0.35 implantations per pregnancy.

Table 1

MEAN CORPORA LUTEA, MEAN AGE, AND MEAN WEIGHT OF PREGNANT
TEST FEMALES (Number of pregnant females examined
are in parentheses)

Weeks Grouped	Mean Corpora Lutea		Mean Age of Females in Days		Mean Weight of Females in Grams	
	Controls	Grouped	Controls	Grouped	Controls	Grouped
2	12.70 (92)	11.83 (94)	133	129	26	26
6	12.04 (92)	11.57 (84)	110	119	25	25
12	11.46 (92)	11.06 (78)	68	65	23	22
Weighted Means:	12.07	11.49	104	104	25	24

Analysis of Variance—Corpora Lutea Per Pregnancy

Source	Degrees Freedom	Mean Squares	F-Value
Treatment	1	45.84	4.79 ($P < 0.05$)
Time	2	45.42	4.75 ($P < 0.05$)
Interaction	2	2.90	Not Significant
Individuals	526	9.57	—

Litter Size and Intrauterine Mortality

Females mated to grouped males produced smaller litters in all five breeding trials. The production of small litters seems to have been the result of a combination of factors. Females mated with controls had more ovulations and with the exception of the first and last breeding trials also had more implantations per pregnancy. The unbiased estimate of the mean treatment difference in resorptions per pregnancy was 0.212 (grouped > controls), but this difference was not quite significant at the 5 per cent confidence level. However, the proportion of developing embryos and fetuses that were resorbed was a better measure of intrauterine mortality during pregnancy because females bred to grouped males had fewer implantations to begin with. Overall, a significantly higher proportion of the embryos and fetuses of females bred to grouped males was lost (Table 3). Resorptions were proportionately higher among females mated with grouped males in every test except the second. Thus, smaller litter sizes for females mated to grouped males was attributed to fewer implantations and greater intrauterine mortality.

Loss of Ova Between Ovulation and Implantation

Table 4 shows the mean number of ova lost between ovulation and implantation. Loss of ova can be attributed to failure of fertilization

Table 2

MEAN IMPLANTATIONS, MEAN LITTER SIZE, AND MEAN NUMBER OF
RESORBED EMBRYOS AND FETUSES PER PREGNANT TEST FEMALE
(C = Controls,   G = Grouped)

Weeks Grouped	Implantations		Litter Size		Resorptions	
	C	G	C	G	C	G
2	7.99	8.28	7.06	6.82	0.93	1.46
6	8.22	7.70	6.94	6.56	1.28	1.14
12	8.09	7.35	7.02	6.22	1.07	1.13
20	8.26	7.63	6.86	5.87	1.40	1.76
50	7.28	7.55	6.34	6.00	0.94	1.55
Weighted Means:	7.97	7.70	6.85	6.29	1.12	1.41

Analysis of Variance—Implantations Per Pregnancy

Source	Degrees Freedom	Mean Squares	F-Value
Treatment	1	24.27	5.24 $(P < 0.01)$
Time	4	13.82	2.98 $(P < 0.01)$
Interaction	4	8.78	Not Significant
Individuals	799	4.63	—

Litter Size

Treatment	1	52.47	6.94 $(P < 0.01)$
Time	4	10.62	Not Significant
Interaction	4	4.54	Not Significant
Individuals	799	7.56	—

Table 3

PROPORTION OF IMPLANTED EMBRYOS ULTIMATELY RESORBED (Total
number of implantations given in parentheses)

Weeks Grouped	Controls	Grouped	Statistical P Values
2	.1164 (791)	.1763 (811)	$P < 0.005$
6	.1557 (789)	.1481 (685)	N.S.*
12	.1323 (793)	.1537 (669)	N.S.
20	.1695 (760)	.2307 (534)	$P < 0.01$
50	.1291 (473)	.2053 (83)	N.S.
Average Overall:	.1414 (3606)	.1751 (2782)	$P < 0.005$

Overall $X^2 = 43.49$, degrees of freedom = 9, $P < 0.005$.
* N.S. = Not Significant.

Table 4

LOSSES BETWEEN OVULATION AND IMPLANTATION (Number of corpora lutea are indicated in parentheses)

Weeks Grouped	Mean Number Ova Lost Per Pregnancy		Proportion of Ova That Failed to Implant	
	Controls	Grouped	Controls	Grouped
2	4.71	3.55	.3709 (1168)	.3001 (1112)
6	3.82	3.87	.3173 (1108)	.3345 (972)
12	3.37	3.71	.2941 (1054)	.3354 (863)
Weighted Means:	3.96	3.71	.3274	.3233

or to the death of the implanted embryo before the conceptus or the placenta was visible macroscopically. The proportion of ova lost is listed for females used in the first three tests. Losses fluctuated between 29 and 37 per cent of the ovulations and were apparently not influenced by treatment of the males or age of the females.

Impregnated females that lost all of their fertilized ova without leaving placental scars would be listed as nulliparous. Also "partial litter losses," which were not represented by visible placental scars, would not be counted as intrauterine mortality. In either case, as pointed out by Brambell and Mills (1944), this would lead to estimates of intrauterine mortality that were lower than were actually the case.

Sex Ratio of Offspring

Sex ratios of the offspring in the various categories are listed in Table 5. Proportionately more female offspring were produced by grouped

Table 5

SEX RATIOS OF OFFSPRING EXPRESSED AS PROPORTION MALES (Number of offspring examined in parentheses)

Weeks Grouped	Controls	Grouped	Statistical P Values
2	.5215 (650)	.5257 (622)	N.S.*
6	.5097 (616)	.4352 (540)	$P < 0.005$
12	.5285 (613)	.5000 (464)	N.S.
20	.5615 (618)	.4543 (372)	$P < 0.005$
50	.5095 (369)	.5200 (50)	N.S.
Totals:	.5276 (2866) †	.4829 (2048) †	$P < 0.005$

Overall $X^2 = 24.99$, degrees of freedom = 9.

* N.S. = Not Significant.

† 477 (8.8%) newborn mice were killed and partially eaten by mothers before they could be collected for determination of sex.

males. The difference amounted to 4.47 per cent between experimental and control litters. Sex ratios differed significantly after 6 and 20 weeks of grouping.

CROWDING EXPERIMENT II: Effects on Adrenals, Reproductive Organs, and Electroejaculation Test

Organ Weights

Mean weights of the adrenals and reproductive organs are listed in Table 6 for the survivors of Crowding Experiment II. Mortality among the 5 groups of 20 mice was attributed to two factors, fighting and the electroejaculation test. Losses were 9 per cent from fighting and 23 per cent from electric shock. Ejaculates were obtained from all mice listed

Table 6

MEAN WEIGHTS OF BODY, ADRENALS, TESTES, SEMINAL VESICLES, AND PROSTATE AND RESULTS OF ELECTROEJACULATION TESTS OF GROUPED AND CONTROL MALES COMPARED (Body weights are in grams and organ weights are in milligrams)

Group (N)	Body	Adre-nals	Testes	Semi-nal Vesi-cles	Pros-tate	Vol-ume Ejacu-late (mm³)	Sperm Densi-ties (1000/ mm³)	Total Sperm (Thou-sands)
			CONTROLS					
A (15)	31.8	4.25	161	189	20.4	2.94	509	1480
B (13)	30.5	4.04	172	188	21.1	3.32	778	2178
C (11)	30.9	3.50	142	178	21.1	3.33	611	1960
D (11)	30.0	3.71	153	180	20.0	3.27	442	1334
E (13)	28.5	3.85	156	176	19.2	3.81	650	2441
Overall Average:	30.4	3.90	157	183	20.3	3.34	589	1886
			GROUPED					
A (14)	30.7	5.57	126	134	15.6	2.99	382	836
B (11)	31.3	6.07	122	125	16.5	2.63	421	1066
C (15)	32.6	5.35	125	124	13.9	3.15	419	1000
D (11)	30.2	5.93	120	114	14.6	3.19	168	525
E (16)	30.6	5.73	141	102	13.6	3.14	790	2075
Overall Average:	31.1	5.70	128	119	14.7	3.07	447	1182

in Table 6; thus, the differences in organ weights must reflect the effects of grouping not the electrical stimulation. This statement does not imply that electro-shock would not affect organ weights, however. Electro-ejaculation tests were commenced 30 days after the mice were grouped. Controls for each group were provided at the same time by testing alternately a grouped mouse then a control. Group A was tested first, followed by group B, and so on, until group E was completed. Tests were conducted as follows: Group A—June 21–July 27; B—Aug. 4–27; C—Aug. 31–Sept. 13; D—Sept. 13–16; and E—Sept. 27–Oct. 1.

Inspection of the data in Table 6 shows the usual effects of grouping on the adrenals and reproductive organs. There was a 46 per cent increase in adrenal weight, an 18 per cent decrease in testicular weight, a 35 per cent decrease in the weight of the seminal vesicles, and a 27 per cent decrease in prostatic weight when grouped males were compared to control males.

Results of Electroejaculation Tests

Results of the electroejaculation tests listed in Table 6 show an 8 per cent decrease in volume of liquid ejaculate, a 24 per cent decrease in sperm density, and a 37 per cent decrease in number of spermatozoa per ejaculate when grouped males are compared to controls. The unbiased estimate of the mean difference in sperm concentration between controls and grouped mice was 146 thousand. This difference was not statistically significant. The unbiased estimate of the mean treatment difference in number of spermatozoa per ejaculate was 754 thousand ($P < 0.01$). The analysis of variance (Table 7) also indicated significant differences associated with the time during which the electroejaculation tests were conducted.

The data were further analyzed to see if the test gave consistent results over the period of time covered by this experiment. Sperm density measurements for the controls were analyzed separately. The differences from

Table 7

ANALYSIS OF VARIANCE (Data on total number of spermatozoa per ejaculate)

Source	Degrees Freedom	Mean Square	F-Value
Treatment	1	19,467,684,000	8.85 ($P < 0.01$)
Times	4	8,569,872,000	3.90 ($P < 0.01$)
Interaction	4	630,925,000	0.29
Individuals	127	2,200,185,000	

one period to the next were not significant. When sperm concentrations of the grouped mice were similarly analyzed, a significant difference between groups was indicated. The results of the electroejaculation tests from group E were unlike those of the other groups. Sperm concentrations were highest in group E and mean volume of ejaculate and numbers of spermatozoa per ejaculate were similar to values obtained from control mice. Table 6 gives no definite clues for this inconsistency although the mean testicular weight was highest in this group of mice. Crowding effects are known to vary considerably in groups of similar composition (Christian *et al.*, 1965). Mortality was lowest in this group, also it was tested last. Thus, crowding effects possibly were not so pronounced in this group; also there might have been some recovery in testicular function with time.

Relation of Testicular Weights to Semen Charactertistics

Correlation coefficients were calculated for the relationship between testicular weight and each of three semen measurements. Testicular weight was directly related to sperm concentration ($r = 0.20$, $P < 0.05$) and to number of spermatozoa per ejaculate ($r = 0.21$, $P < 0.05$) but showed no relation to semen volume ($r = 0.06$). An unexpected relationship between semen volume and sperm concentration was indicated by a negative correlation coefficient of 0.18. Thus, the largest ejaculates collected by electrical stimulation more often than not contained the lowest numbers of spermatozoa.

DISCUSSION AND CONCLUSIONS

Both chronological aging and crowding were found to affect fertility and reproductive performance of male house mice. The effects were manifest either because of primary alterations of spermatogenesis and spermiogenesis or because of alterations of female reproductive physiology in response to male behavior. As a basis for discussion, the two possibilities can be outlined as two hypotheses. The one would attribute the observed changes primarily to alterations of male reproductive physiology, while the second would consider alterations of female reproductive physiology as at least a contributing factor.

The first hypothesis can be outlined as follows: Social conflicts and actual internecine strife act as psychological stimuli to decrease secretion of gonadotrophins which result in testicular atrophy and reduction of the size of the accessory sex glands. Spermatogenesis and spermiogenesis and in turn, sperm quality and sperm concentration are reduced. Infertility ensues when quality and quantity of spermatozoa drop below certain levels. Semen quality could influence conception, embryo survival, litter size, and sex ratio of the offspring if vitality of the fertilizing spermatozoa were a crucial factor in determining vitality of the conceptus. Sex ratio would be affected if proportionately more female determining spermatozoa occurred in the semen or if mortality were differentially higher among male embryos and fetuses.

Evidently, there are conflicting results concerning the relation of semen quality to fertility in humans (Hartman, 1965). Freund (1962) concluded in his review that there was no definite evidence that any of the commonly measured semen characteristics were directly related to fertility in a cause and effect relationship. On the other hand, extensive studies by MacLeod and Gold (summarized in Hartman's paper) have definitely shown that the better the "quality" of the semen, the higher the fertility. These workers further report that, in general, poor semen predisposes to abnormalities or accidents of pregnancy. A considerable body of evidence from studies of lower animals indicates a definite relationship between reduced fertility and such characteristics as deficient sperm numbers, abnormal sperm morphology, and reduced sperm vitality (Hammond, 1952).

Support for the first hypothesis is provided by the results of the experiments with brown house mice. A significant increase in the adrenal weights of grouped mice is presumptive evidence for a pituitary-adrenal response to stressful stimuli. A concurrent reduction in the level of secretion of gonadotrophins is indicated by the reduction in testicular weights. In general, reduction in testicular weight was associated with lower sperm concentration and fewer spermatozoa in ejaculates collected by electrical stimulation. The low correlation coefficients for this association were likely due to the wide range in sperm concentration in ejaculates collected by the electro-shock technique (Snyder, 1966). The large variance (standard deviation) of measurements of semen characteristics is probably one reason for the paucity of unequivocal evidence for a relationship between semen quality and fertility. Because of this and because of the many factors that affect fertility, large samples would be required to show a significant association between fertility and any one characteristic of the semen.

The second hypothesis is based on the possibility that female mice were adversely affected by the presence of males that had been grouped. Grouped males presumably would behave differently than those that had been caged alone. Since the majority of the mice in a group of 20 males assume a subordinate position in the social hierarchy, it is not difficult to imagine that a discernible behavior pattern could develop. Psychological stimuli could then alter the secretory pattern of gonadotrophins and affect ovarian and uterine function.

Support for the second hypothesis comes from studies which show extreme sensitivity of the female mouse to environmental stimuli (Bruce, 1960; Eleftheriou *et al.*, 1962; Chipman *et al.*, 1966).

Certain results of the experiments with brown house mice also can be interpreted to support the second hypothesis. Trial females bred to males that had been caged alone had significantly more ovulations than those bred to grouped males. Ovulation rate is related to age, but in these trials females were matched for age. Thus, it would seem that ovulation rate was influenced by the type of male present. Endocrine responses of this nature could also affect the uterine environment, thus increasing intrauterine mortality or favoring the development and survival of one sex over the other. The last possibility is suggested in two recent papers. The rate of reduction of triphenyltetrazolium to formazan was used as an index of estrogenic effects, and examined in relation to reproductive performance (Schultze, 1965). Female rats with high uterine reaction rates produced litters with a higher percentage of female pups than females with a low level of reaction. In this study, litter size was not related to uterine metabolic activity. Modification of the sex ratio in rats was accomplished by Geiringer (1961) by the administration of ACTH to mothers during early pregnancy. The number of females in the litters was increased by about 10 per cent over that in control litters.

At present, data are insufficient to support any definite conclusions regarding the functional basis for the reduced fertility and altered reproductive performance of grouped male house mice. However, the two hypotheses outlined in this paper seem to offer satisfactory models for further study of the psychological and physiological mechanisms involved. Perhaps the most important result of these experiments was the realization that psychological responses on the part of the female to male behavior must be considered in studies of this nature.

The problems of sampling and control of experiments are infinitely more difficult when dealing with natural populations. Nevertheless, there are several studies to indicate that reproductive failures among feral mammals are often related to population pressures (see Christian *et al.*, 1965). Many scientists are inclined to attribute such losses to diet, to genetics, to unfavorable climate, or to other density-independent factors. At present, it would be difficult to reconcile these opposing views because of lack of conclusive data on causation and the paucity of critical measurements of food supply and physiology. Still, it seems fair to point out that one should be aware that reproductive failures encountered in nature might be due to either density-dependent or density-independent agents or even to a combination of the two classes of agents. In any case, efforts should be made to obtain data of a correlative nature in order to substantiate conclusions.

COMPARATIVE ASPECTS OF REPRODUCTIVE FAILURE

ERRATA PAGE 343

Scrapie and visna are not the same; "visna" should be deleted from the paragraph heading (and the table of contents). The two sentences comprising lines 19-22 should be deleted. In their place should be:

"Scrapie can be serially induced in sheep and mice, but the nature of the agent is undefined. The agent of visna, a clinically similar ovine disease, is a virus and its pathogenesis may resemble lymphocytic choriomeningitis virus infection in mice (Gudnádottir and Pálsson, 1965)."

The reference to NINDB monograph #2 should be: "Slow, Latent, and Temperate Virus Infections" edited by Gajdusek, Gibbs, and Alpers, published 1965, by the U. S. P. H. S.

References

Allen, P., F. W. R. Brambell, and I. H. Mills: Studies on sterility and prenatal mortality in wild rabbits. 1. The reliability of estimates of prenatal mortality based on counts of corpora lutea, implantation sites and embryos. J. Exp. Biol. *23*:312, 1947.

Brambell, F. W. R.: Intra-uterine mortality in the wild rabbit (*Oryctolagus cuniculus*). Proc. Roy. Soc. B *130*:462, 1942.

——: The reproduction of the wild rabbit (*Oryctolagus cuniculus*). Proc. Zool. Soc. Lond. *114*:1, 1944.

——: Ovarian Changes. In: Marshall's Physiology of Reproduction (ed.) A. S. Parkes. London, New York, and Toronto: Longmans, Green and Co., Vol. I: Part 1, 3rd Ed., 1956.

Brambell, F. W. R., and I. H. Mills: Prenatal mortality. Nature *153*:558, 1944.

——, ——: Studies on sterility and prenatal mortality in wild rabbits. II. The occurrence of fibrin in the yolk-sac contents of embryos during and immediately after implantation. J. Exp. Biol. *23*:332, 1947a.

——, ——: Studies on sterility and prenatal mortality in wild rabbits. III. The loss of ova before implantation. J. Exp. Biol. *24*:192, 1947b.

Brown, R. Z.: Social behavior, reproduction and population changes in the house mouse. Ecol. Monogr. *23*:217, 1953.

Bruce, H. M.: A block to pregnancy in the mouse caused by proximity of strange males. J. Reprod. Fertil. *1*:96, 1960.

Chipman, R. K., J. A. Holt, and K. A. Fox: Pregnancy failure in laboratory mice after multiple short-term exposure to strange males. Nature *210*:653, 1966.

Christian, J. J.: Effect of population size on the adrenal glands and reproductive organs of male mice in populations of fixed size. Am. J. Physiol. *182*:292, 1955.

——: Adrenal and reproductive responses to population size in freely growing populations. Ecology *37*:258, 1956.

Christian, J. J., and C. D. LeMunyan: Adverse effects of crowding on reproduction and lactation of mice and two generations of their progeny. Endocrin. *63*:517, 1958.

Christian, J. J., J. A. Lloyd, and D. E. Davis: The role of endocrines in the self-regulation of mammalian populations. Recent Progr. in Hormone Res. *21*:501, 1965.

Conaway, C. H., T. S. Baskett, and J. E. Toll: Embryo resorption in the swamp rabbit. J. Wildl. Mgt. *24*:197, 1960.

Crew, F. A., and L. Mirskaia: Effect of density on adult mouse populations. Biol. General *7*:239, 1931.

Deanesly, R.: The corpora lutea of the mouse, with special reference to fat accumulation during the oestrous cycle. Proc. Roy. Soc. B *106*:578, 1930.

Eleftheriou, B. E., F. H. Bronson, and M. X. Zarrow: Interaction of olfactory and other environmental stimuli on implantation in the deer mouse. Science *137*:764, 1962.

Freund, M.: Interrelation among the characteristics of human semen and factors affecting semen-specimen quality. J. Reprod. Fertil. *4:*143, 1962.

Geiringer, E.: Effect of ACTH on sex ratio of the albino rat. Proc. Soc. Exp. Biol. Med. *106:*752, 1961.

Hammond, J.: Fertility. In: Marshall's Physiology of Reproduction, A. S. Parkes (ed.). London, New York and Toronto: Longmans, Green and Company, Vol. II, 3rd ed., 1952.

Hartman, C. G.: Correlations among criteria of semen quality. Fertil. and Steril. *16:*632, 1965.

Helmreich, R. L.: Regulation of reproductive rate by intrauterine mortality in the deer mouse. Science *132:*417, 1960.

Mody, J. K., and J. J. Christian: Adrenals and reproductive organs of female mice kept singly, grouped, or grouped with a vasectomized male. J. Endocrin. *24:*1, 1962.

Newson, R. M.: Reproduction in the feral coypu (*Myocastor coypus*). In: Comparative Biology of Reproduction in Mammals. Symposia of the Zoological Society of London, No. 15. Edited by I. W. Rowlands, New York and London: Academic Press Inc., 1966.

Perry, J. S.: The reproduction of the wild brown rat (*Rattus norvegicus* Erxleben). Proc. Zool. Soc. Lond. *115:*19, 1945.

Schultze, A. B.: Litter size and proportion of females in the offspring of multiparous rats with varying uterine metabolic levels. J. Reprod. Fertil. *10:*145, 1965.

Snedecor, G. W.: Statistical Methods. Ames, Iowa: Iowa State College Press, 5th Ed., 1956.

Snell, G. D.: Biology of the Laboratory Mouse. New York: Dover Publications, Inc., 1956.

Snyder, R. L.: Reproductive performance of a population of woodchucks after a change in sex ratio. Ecology *43:*506, 1962.

————: Collection of mouse semen by electroejaculation. Anat. Rec., *155:*11, 1966.

Snyder, R. L., and J. J. Christian: Reproductive cycle and litter size of the woodchuck. Ecology *41:*647, 1960.

Southwick, C. H.: The population dynamics of confined house mice supplied with unlimited food. Ecology *36:*212, 1955.

CLOSING REMARKS

T. C. JONES

Department of Pathology, Harvard Medical School,
Veterinary Pathologist, Angell Memorial Hospital,
Boston, Massachusetts

We have now completed our fifth day of this conference which soon shall come to an end. Our group consists of people of many lands, diverse backgrounds, and widely dispersed interests, but unified by an interest in some aspect of reproduction.

We have shared our work and ideas during this week and each of us must be taking away something more than we brought. I hope that all of us have found it interesting; many have found it stimulating and at least some have found it inspiring. I think that most of us are grateful for the opportunity to renew old friendships and to make new ones.

It is not possible for me to make an adequate review of this week's program and I shall not try to do so but I would like to describe a few impressions of my own. At the very outset, we could see some of the fascinating information that can be obtained by careful observation, by a prepared and inquisitive mind, even of the species most difficult to study—man. We have also been shown some of the exquisite detail that may be disclosed by the skilled application of precision instruments. Brilliantly planned and executed experiments were described to us and their significant additions to knowledge duly recorded. The most precise and significant experiments, we must admit, come not from the mind of man but are performed by nature. Our problem is to understand them.

The advantages of studying reproduction and its failures in many species have become obvious during this week, as we have looked at some of nature's experiments. The unity of biologic process becomes finally apparent as we uncover their diverse variety. Many of the great gaps in knowledge of the details of reproductive failure have become obvious to us. It is also apparent that in order to understand reproductive pathology we must master anatomic and physiologic details and

473

before we fully understand the normal processes of reproduction, we must also know all about their failures.

I believe I express the feeling of all members when I thank the speakers for their generous contributions of time and effort; Dartmouth College for the excellent facilities and atmosphere which we have enjoyed here; Jane Juergens and Maxine Reifer for operation of the registration desk and many helpful courtesies; our projection staff behind the scenes, Dave Hartwell and Ken Aldrich; and the staff of the department of pathology for attending to many details.

Our gratitude must be expressed to those who have underwritten the financial support of this conference; Charles River Breeding Laboratories; Eli Lilly Research Laboratories; Geigy Pharmaceuticals; Lakeview Hamster Colony; Lederle Laboratories; The National Institute of Child Health & Human Development; The Population Council; Schering Corporation; Smith, Kline & French Foundation; Syntex Company, and Upjohn Company.

We are indebted to Kurt Benirschke, who conceived the idea for this conference, inspired all of the rest of us to help, and worked endlessly to assure the success of this week. Most of all, our thanks to Marion, Ingrid, Rolf and Stephen Benirschke for contributing so much to the delightfully warm and friendly atmosphere in which we have lived the past week.

INDEX